ACS SYMPOSIUM SERIES **614**

Microelectronics Technology

Polymers for Advanced Imaging and Packaging

Elsa Reichmanis, EDITOR
AT&T Bell Laboratories

Christopher K. Ober, EDITOR
Cornell University

Scott A. MacDonald, EDITOR
IBM Almaden Research Center

Takao Iwayanagi, EDITOR
Hitachi Research Laboratory

Tadatomi Nishikubo, EDITOR
Kanagawa University

Developed from a symposium sponsored by the ACS Division of Polymeric Materials: Science and Engineering, Inc., and the Polymers for Microelectronics Division of the Society of Polymer Science, Japan, at the 209th National Meeting of the American Chemical Society, Anaheim, California, April 2–6, 1995

American Chemical Society, Washington, DC 1995

Library of Congress Cataloging-in-Publication Data

Microelectronics technology: polymers for advanced imaging and
packaging: developed from a symposium sponsored by the ACS Division
of Polymeric Materials: Science and Engineering, Inc., and the Polymers
for Microelectronics Division of the Society of Polymer Science, Japan,
at the 209th National Meeting of the American Chemical Society,
Anaheim, California, April 2–6, 1995 / Elsa Reichmanis... [et al.], editor.

p.   cm.—(ACS symposium series; 614)

Includes bibliographical references and index.

ISBN 0–8412–3332–2

1. Microelectronic packaging—Materials—Congresses.
2. Polymers—Congresses. 3. Photoresists—Congresses.

I. Reichmanis, Elsa, 1953–   . II. American Chemical Society.
Division of Polymeric Materials: Science and Engineering, Inc.
III. Kobunshi Gakkai (Japan). Polymers for Microelectronics Division.
IV. American Chemical Society. Meeting (209th: 1995: Anaheim,
Calif.) V. Series.

TK7874.M475   1995
621.381′046—dc20   95–44669
                                                              CIP

This book is printed on acid-free, recycled paper.

Foreword

THE ACS SYMPOSIUM SERIES was first published in 1974 to provide a mechanism for publishing symposia quickly in book form. The purpose of this series is to publish comprehensive books developed from symposia, which are usually "snapshots in time" of the current research being done on a topic, plus some review material on the topic. For this reason, it is necessary that the papers be published as quickly as possible.

Before a symposium-based book is put under contract, the proposed table of contents is reviewed for appropriateness to the topic and for comprehensiveness of the collection. Some papers are excluded at this point, and others are added to round out the scope of the volume. In addition, a draft of each paper is peer-reviewed prior to final acceptance or rejection. This anonymous review process is supervised by the organizer(s) of the symposium, who become the editor(s) of the book. The authors then revise their papers according to the recommendations of both the reviewers and the editors, prepare camera-ready copy, and submit the final papers to the editors, who check that all necessary revisions have been made.

As a rule, only original research papers and original review papers are included in the volumes. Verbatim reproductions of previously published papers are not accepted.

Contents

Preface.. xi

CHEMICALLY AMPLIFIED RESIST
MATERIALS AND PROCESSES

1. **An Analysis of Process Issues with Chemically Amplified Positive Resists** 4
   O. Nalamasu, A. G. Timko, Elsa Reichmanis, F. M. Houlihan, Anthony E. Novembre, R. Tarascon, N. Münzel, and S. G. Slater

2. **The Annealing Concept for Environmental Stabilization of Chemical Amplification Resists**... 21
   Hiroshi Ito, Greg Breyta, Donald C. Hofer, and R. Sooriyakumaran

3. **Structure–Property Relationship of Acetal- and Ketal-Blocked Polyvinyl Phenols as Polymeric Binder in Two-Component Positive Deep-UV Photoresists** 35
   C. Mertesdorf, N. Münzel, P. Falcigno, H. J. Kirner, B. Nathal, H. T. Schacht, R. Schulz, S. G. Slater, and A. Zettler

4. **Lithographic Effects of Acid Diffusion in Chemically Amplified Resists**... 56
   C. A. Mack

5. **Acid Diffusion in Chemically Amplified Resists: The Effect of Prebaking and Post-Exposure Baking Temperature**..................... 69
   Jiro Nakamura, Hiroshi Ban, and Akinobu Tanaka

6. **Correlation of the Strength of Photogenerated Acid with the Post-Exposure Delay Effect in Positive-Tone Chemically Amplified Deep-UV Resists**... 84
   F. M. Houlihan, E. Chin, O. Nalamasu, J. M. Kometani, and R. Harley

7. Following the Acid: Effect of Acid Surface Depletion on Phenolic Polymers 110

James W. Thackeray, Mark D. Denison,
Theodore H. Fedynyshyn, Doris Kang, and Roger Sinta

8. Water-Soluble Onium Salts: New Class of Acid Generators for Chemical Amplification Positive Resists 124

Toshio Sakamizu, Hiroshi Shiraishi, and Takumi Ueno

9. Photoacid and Photobase Generators: Arylmethyl Sulfones and Benzhydrylammonium Salts 137

J. E. Hanson, K. H. Jensen, N. Gargiulo, D. Motta,
D. A. Pingor, Anthony E. Novembre, David A. Mixon,
J. M. Kometani, and C. Knurek

10. Functional Imaging with Chemically Amplified Resists 149

Alexander M. Vekselman, Chunhao Zhang,
and Graham D. Darling

11. Hydrogen Bonding in Sulfone- and N-Methylmaleimide-Containing Resist Polymers with Hydroxystyrene and Acetoxystyrene: Two-Dimensional NMR Studies 166

Sharon A. Heffner, Mary E. Galvin, Elsa Reichmanis,
Linda Gerena, and Peter A. Mirau

12. NMR Investigation of Miscibility in Novolac—Poly(2-methyl-1-pentene sulfone) Resists 180

Sharon A. Heffner, David A. Mixon, Anthony E. Novembre,
and Peter A. Mirau

13. Styrylmethylsulfonamides: Versatile Base-Solubilizing Components of Photoresist Resins 194

Thomas X. Neenan, E. A. Chandross, J. M. Kometani,
and O. Nalamasu

14. 4-Methanesulfonyloxystyrene: A Means of Improving the Properties of tert-Butoxycarbonyloxystyrene-Based Polymers for Chemically Amplified Deep-UV Resists 207

J. M. Kometani, F. M Houlihan, Sharon A. Heffner, E. Chin,
and O. Nalamasu

15. Dienone—Phenol Rearrangement Reaction: Design Pathway for Chemically Amplified Photoresists 228

Ying Jiang, John Maher, and David Bassett

NOVEL CHEMISTRIES AND APPROACHES
FOR SUB-0.25-μm IMAGING

16. **Single-Layer Resist for ArF Excimer Laser Exposure Containing Aromatic Compounds** **239**
    Tohru Ushirogouchi, Takuya Naito, Koji Asakawa,
    Naomi Shida, Makoto Nakase, and Tsukasa Tada

17. **Design Considerations for 193-nm Positive Resists** **255**
    Robert D. Allen, I- Y. Wan, Gregory M. Wallraff,
    Richard A. DiPietro, Donald C. Hofer, and Roderick R. Kunz

18. **Top-Surface Imaged Resists for 193-nm Lithography** **271**
    Roderick R. Kunz, Susan C. Palmateer, Mark W. Horn,
    Anthony R. Forte, and Mordechai Rothschild

19. **Silicon-Containing Block Copolymer Resist Materials** **281**
    Allen H. Gabor and Christopher K. Ober

20. **A Top-Surface Imaging Approach Based on the Light-Induced Formation of Dry-Etch Barriers** **299**
    U. Schaedeli, M. Hofmann, E. Tinguely, and N. Münzel

21. **Plasma-Developable Photoresist System Based on Polysiloxane Formation at the Irradiated Surface: A Liquid-Phase Deposition Method** **318**
    Masamitsu Shirai, Norihiko Nogi, Masahiro Tsunooka,
    and Takahiro Matsuo

22. **New Polysiloxanes for Chemically Amplified Resist Applications** **333**
    J. C. van de Grampel, R. Puyenbroek, A. Meetsma,
    B. A. C. Rousseeuw, E. W. J. M. van der Drift

23. **Environmentally Friendly Polysilane Photoresists** **355**
    James V. Beach, Douglas A. Loy, Yu-Ling Hsiao,
    and Robert M. Waymouth

POLYMER DIELECTRICS FOR MICROELECTRONIC APPLICATIONS

24. **Fluoropolymers with Low Dielectric Constants: Triallyl Ether–Hydrosiloxane Resins** **369**
    Henry S.-W. Hu, James R. Griffith, Leonard J. Buckley,
    and Arthur W. Snow

25. **Photophysics, Photochemistry, and Intramolecular Charge Transfer of Polyimides** 379
   Masatoshi Hasegawa, Yoichi Shindo, and Tokuko Sugimura

26. **Structure, Properties, and Intermolecular Charge Transfer of Polyimides**.. 395
   Masatoshi Hasegawa, Junichi Ishii, Takahumi Matano,
   Yoichi Shindo, Tokuko Sugimura, Takao Miwa, Mina Ishida,
   Yoshiaki Okabe, and Akio Takahashi

27. **Application of Polyisoimide as a Polyimide Precursor to Polymer Adhesives and Photosensitive Polymers** 413
   Amane Mochizuki and Mitsuru Ueda

28. **Polyimide Nanofoams Prepared from Styrenic Block Copolymers** 425
   J. L. Hedrick, T. P. Russell, C. Hawker, M. Sanchez,
   K. Carter, Richard A. DiPietro, and R. Jerome

29. **Internal Acetylene Unit as a Cross-Link Site for Polyimides**.......... 439
   Tsutomu Takeichi and Masaaki Tanikawa

30. **Vapor-Depositable Polymers with Low Dielectric Constants** 449
   J. A. Moore, Chi-I Lang, T.-M. Lu, and G.-R. Yang

31. **Plasma Polymerization in Direct Current Glow: Characterization of Plasma-Polymerized Films of Benzene and Fluorinated Derivatives** 471
   Toshihiro Suwa, Mitsutoshi Jikei, Masa-aki Kakimoto,
   and Yoshio Imai

32. **Syntheses and Properties of Allylated Poly(2,6-dimethyl-1,4-phenylene ether)** 485
   Yoshiyuki Ishii, Hiroji Oda, Takeshi Arai,
   and Teruo Katayose

33. **Synthesis and Photochemistry of a 2,6-Dialkoxyanthracene-Containing, Side-Chain-Substituted Liquid-Crystalline Polymer**.. 504
   David Creed, Charles E. Hoyle, Anselm C. Griffin,
   Ying Liu, and Surapol Pankasem

34. **Hybrid Polyimide–Polyphenylenes by the Diels–Alder Polymerization Between Biscyclopentadienones and Ethynyl-Terminated Imides** 518

Uday Kumar and Thomas X. Neenan

35. **Polysiloxane Thermoplastic Polyurethane Modified Epoxy Resins for Electronic Application** 527

Tsung-Han Ho and Chun-Shan Wang

INDEXES

Author Index 545

Affiliation Index 546

Subject Index 547

Preface

POLYMERIC MATERIALS HAVE FOUND WIDESPREAD USE in the electronics industry in both the manufacturing processes used to generate today's integrated circuits and as component structures in the completed devices. The broad applicability of polymers arises from the ability to design and synthesize such materials with the precise functionalities and properties required for a given application. Notably, polymeric materials have been used as lithographic imaging materials (resists) and dielectric, passivation, and insulating materials.

Radiation sensitivity is the key property required of materials used for imaging the individual elements of an integrated circuit. Known as the microlithographic process, it is the linchpin technology used to fabricate electronic devices and is critically dependent on the polymer–organic materials chemistry used to generate the radiation-sensitive imaging material known as the resist. As the lithographic technologies evolve to allow fabrication of smaller and more compact circuit elements, new resist chemistries and processes will be needed.

Low dielectric polymeric materials are used as adhesives, encapsulants, substrates, and thin-film dielectric layers. The ease of processing polymeric materials is stimulating the design and development of materials with improved performance. Polymers with lower dielectric constants, higher thermal stability, photo-imageability, low dielectric loss, and low thermal expansion coefficient are especially sought after.

We hope that this book reveals the diversity of the chemical challenges facing the microelectronics industry as well as the rapid progress and evolution of research in advanced polymeric materials for microelectronics.

This volume is based upon the symposium co-sponsored by the American Chemical Society (ACS) Division of Polymeric Materials: Science and Engineering, Inc. (PMSE), and the Society of Polymer Science, Japan, held in Anaheim in 1995. We thank the PMSE Division and the Society of Polymer Science for their sponsorship of the meeting, the authors for their efforts and contributions to both the symposium and this volume, and the referees for their assistance in reviewing each manuscript. Additionally, financial support for the symposium from OCG Microelectronic Materials, Inc., Hoechst-Celanese, Inc., the Shipley Company, AT&T Bell Laboratories, the PMSE Division, and the Petroleum Research Fund administered by the ACS, is gratefully acknowledged.

Special thanks to Sheryl Tarantino for her help in preparing and organizing this book. We also thank Michelle Althius, Anne Wilson, and the production staff of the ACS Books Department for their efforts in assembling the volume.

ELSA REICHMANIS
AT&T Bell Laboratories
600 Mountain Avenue
Murray Hill, NJ 07974

CHRISTOPHER K. OBER
Department of Materials Science and Engineering
Cornell University
Ithaca, NY 14853–1501

SCOTT A. MACDONALD
IBM Almaden Research Center
650 Harry Road
San Jose, CA 95120

TAKAO IWAYANAGI
The First Material Department
Hitachi Research Laboratory
Hitachi, Ibaraki 319–12, Japan

TADATOMI NISHIKUBO
Department of Applied Chemistry
Kanagawa University
Yokohama, Kanagawa 221, Japan

August 24, 1995

CHEMICALLY AMPLIFIED RESIST MATERIALS AND PROCESSES

Elsa Reichmanis

The intense drive towards designing and fabricating integrated circuits having individual elements which are less than 0.3 µm can be viewed as truly a global effort. The microelectronics business is driven by the need to build devices which contain an increasing number of individual circuit elements on a semiconductor material (1,2). The increased density enables the device to perform more functions in a shorter period of time than previously possible while maintaining a constant real estate area. The ability to shrink the feature size is critically dependent upon the technologies involved in the delineation of the circuit pattern. These technologies are part of the overall microlithographic fabrication process (3-5).

Parallel to the design of resist chemistries and associated processes is the development of highly sophisticated equipment used to print the circuit patterns. The predominant technology today is "conventional photolithography" (i and g-line photolithography employing 350-450 nm light). Incremental improvements in tool design and performance have allowed the continued use of 350-450 nm light to produce ever smaller features (6). Additionally, the same basic positive photoresist consisting of a diazonaphthoquinone photoactive compound and a novolac resin has continued to be the resist of choice (7,8). The alternative lithographic technologies, will, however, require new resist chemistries. The leading technological alternatives to conventional photolithography are short wavelength (190-250 nm) photolithography, direct-write or projection electron-beam, proximity x-ray or scanning ion-beam lithography (1,5,9).

0097–6156/95/0614–0001$12.00/0

The focus of the following chapters concerns the design of polymer/organic materials and chemistries that may prove useful in radiation sensitive resist films. Such resists must be carefully designed to meet the specific requirements of each lithographic technology. Although these requirements vary according to the radiation source and device processing sequence, the following resist properties are common to all lithographic technologies: sensitivity, contrast, resolution, optical density, etching resistance, and purity (10,11). These properties can be achieved by careful manipulation of materials' structure, molecular properties and synthetic methods.

One approach to providing resists with high sensitivity and contrast involves the concept of chemical amplification which employs the photogeneration of an acidic species that catalyzes many subsequent chemical event such as deblocking of a protective group or crosslinking of a matrix resin. The overall quantum efficiency of such reactions is, thus, effectively much higher than that for initial acid generation. A chemically amplified resist is generally composed of three or more elements; i) a matrix polymer, ii) a photoacid generator, and iii) a moiety capable of effecting differential solubility between the exposed and unexposed regions of the film either through a crosslinking reaction or other molecular transformation. These elements may be either discrete molecular entities that are formulated into a multicomponent resist system or elements of a single polymer.

The high sensitivity associated with chemically amplified resists is the result of the catalytic action of the acid during a post-exposure reaction. Such catalytic action, while improving the photospeed also enhance the resist's sensitivity to airborne basic contaminants, basic moieties on substrate surface and hotplate non-uniformity. Such deamplification of the photoacid by airborne or substrate related contaminants results in "T-tops" and "foot" formation, respectively. High catalytic chain length also dictates the need for extremely tight control of hotplate uniformity to achieve good linewidth control. In addition, high protection of the base soluble phenolic moieties with bulky groups such as t-boc (t-butoxycarbonyl) results in excessive shrinkage at the exposed areas during the post-exposure bake (PEB), poor adhesion and moderate etch selectivity.

Both the acid catalyzed crosslinking and deprotection reactions are critically dependent on post-exposure bake temperature, time, and method. In addition, it is essential that the acid concentration remain unchanged and uniformly distributed in the exposed areas between exposure and post exposure bake (called the post-exposure delay PED). The control of these parameters represent the significant difference between conventional resists and chemically amplified resists.

In order to process chemically amplified resists with good control it is necessary to provide a clean, amine-free environment. This is most easily accomplished with a wafer track system that is interfaced to the exposure tool, enclosed in an environmental chamber that has an activated charcoal filtration system. This type of cluster lithography system has many other advantages in controlling the overall process yields and are additionally becoming common for conventional resist systems. Chemical approaches to achieving improved PED latitude include using more readily removable protective groups and alternative acid generator chemistries.

In addition to needing an understanding of the chemical reactions and kinetics of each step, it is important to appreciate that *diffusion* of the acid is also important. Diffusion controlled reactions are well known and can be difficult to control. Not only is the reaction rate of the acid-induced deprotection controlled by temperature, but so is the diffusion distance (catalytic chain length) and rate of some undesirable side reactions that prematurely consume the acid. An appreciation of the chemistry and chemical kinetics leads one to predict that several process parameters associated with the post-exposure bake will need to be optimized if these materials are to be used in a submicron lithographic process.

Since the conception of chemical amplification mechanisms for microlithographic applications over a decade ago, increasing attention has been given to such processes in that they provide advantages in terms of sensitivity and contrast with minimal increase in process complexity. Additionally, a given chemistry may find application to more than on lithography technology. The original work in chemically amplified resists has spawned many research efforts to define chemistries appropriate for matrix materials and photogenerators of catalysts, primarily strong acids. These efforts have led to the development and commercialization of several positive and negative-tone chemically amplified resists. Significant challenges exist today that relate to both the fundamental and applied materials chemistry, and to understanding the processes associated with the use of chemically amplified materials. Understanding these issues is required in order to effectively introduce chemically amplified resists into a device manufacturing environment.

LITERATURE CITED

1. "Electronic and Photonic Applications of Polymers", *ACS Advances in Chemistry Series*, **218**, Bowden, M. J., Turner, R. R., Eds., ACS, Washington, D.C., 1988.
2. Powell, M. W., *Solid State Technology*, 1989, **332(3)**, 66.
3. Sze, S. M., "VLSI Technology", McGraw-Hill, New York, NY, 1983.
4. Wolf, S., Tauber, R. N., "Silicon Processing for the VLSI Era", Lattice Press, Sunset Beach, CA, 1986.
5. Thompson, L. F., Willson, C. G., Bowden, M. J., "Introduction to Microlithography", *ACS Symposium Series 219*, ACS, Washington, D.C. 1983, pp. 2-85.
6. McCoy, J. H., Lee, W., Varnell, G. L., *Solid State Technology*, 1989, **32(3)**, 87.
7. Reichmanis, E., Thompson, L. F., *Chemical Reviews*, **89**, 1273.
8. Wilson, C. G., In "Introduction to Microlithography", *ACS Symposium Series*, **219**, Thompson, L. F., Willson, C. G., Bowden, M. J., Eds., ACS, Washington, D.C., 1983, pp. 88-159.
9. Takigawa, T., J. *Photopolymer Sci. and Technol.*, 1992, **5(1)**, 1.
10. Thompson, L. F., Bowden, M. J., In "Introduction to Microlithography", *ACS Symposium Series*, **219**, ACS, Washington, D.C., 1983, pp. 162-214.
11. Moreau, W. M., "Semiconductor Lithography, Principles, Practices and Materials", Plenum, NY, 1988.

RECEIVED August 14, 1995

Chapter 1

An Analysis of Process Issues with Chemically Amplified Positive Resists

O. Nalamasu[1], A. G. Timko[1], Elsa Reichmanis[1], F. M. Houlihan[1],
Anthony E. Novembre[1], R. Tarascon[1], N. Münzel[2], and S. G. Slater[3]

[1]AT&T Bell Laboratories, 600 Mountain Avenue, Murray Hill, NJ 07974
[2]OCG Microelectric Materials AG, Klybeckstrasse 141, Basel,
CH–4002, Switzerland
[3]OCG Microelectronic Materials Inc., 200 Massasoit Avenue,
East Providence, RI 02914

Chemically amplified (CA) resist materials exhibit excellent resolution, sensitivity and process latitude and hence are the materials of choice for advanced lithographic technologies. The performance of CA positive resists, however, was critically dependent on controlling the time elapsed between the exposure and post-exposure bake (post-exposure delay, PED) process steps. The high catalytic chain lengths characteristic to CA positive resists, also enhance the resist's sensitivity to airborne basic contaminants, basic moieties on the substrate surface and hotplate's uniformity. Such deamplification of the photoacid by airborne or substrate bound basic contaminants results in "T-tops" and "foot" formation, respectively. High protection of the base soluble phenolic moieties with bulky groups such as t-boc (t-butoxycarbonyl) resulted in excessive shrinkage at the exposed areas during the post-exposure bake (PEB), poor adhesion and inadequate etch selectivity. This paper will review and analyze several of these problems and comment on how the issues were alleviated by the design of the CA positive resists that have lower protection ratios, weaker acid generating PAGs and lower catalytic chain lengths at PEB. This paper will also present our analysis of the PED problem by a variety of spectroscopic techniques and lithography in controlled ambients and detail the effect of several chemical and process parameters on the PED stability. The substrate contamination effects for the CA resists are illustrated and some possible paths of deactivation are proposed for titanium nitride and silicon nitride substrates. Additionally, results will be presented for CAMP6 resist illustrating the issues with CA resists and results for ARCH (Adavnced Resist CHemically amplified) resist demonstrating the resolution or minimization of PED stability, substrate contamination, CD sensitivity and shrinkage issues.

0097–6156/95/0614–0004$12.00/0
© 1995 American Chemical Society

New resist materials and processes are necessary to pattern ≤ 0.25 μm design rule circuits with advanced deep-UV, X-ray and e-beam lithographic technologies. Chemically amplified positive resist systems were introduced to meet high sensitivity and high resolution requirements of these technologies (*1,2*). The inherent sensitivity typical to most chemically amplified (CA) positive systems emanates from the catalytic action of the photo- or radiolytically generated acid during the post-exposure bake (PEB) process step. Catalytic chain lengths of >1000 have been reported in the literature for positive CA resists (3-5). The high catalytic chain lengths (3-5) attained during the PEB, while improving the photospeed also amplify the resist's sensitivity to airborne basic contaminants, basic moieties on substrate surface and hotplate's uniformity. Any depletion or deactivation of photoacid during the PED or by the substrate surface manifests in the form of "T-tops" and "foot" formation, respectively (*6,7*). High catalytic chain length also dictates the need for an extremely tight control of hotplate uniformity to achieve good linewidth control. In addition, high protection of the base soluble phenolic moieties with bulky groups such as t-boc (t-butoxycarbonyloxy) resulted in excessive shrinkage at the exposed areas during the post-exposure bake (PEB), poor adhesion resulted due to stress in the film, and inadequate etch selectivity was observed due to the rapid removal of the protecting groups during the pattern transfer (*7-9*).

This paper will detail the results of an analytical study of PED phenomenon using (*10*) the PTBSS (poly(t-butoxycarbonyloxystyrene sulfone) polymer with triphenylsulfonium hexafluoroarsenate and nitrobenzyl ester PAGs. We will also describe and review the effect of different process and chemical parameters on the PED stability and illustrate methods of alleviating the PED effects. The paper also proposes possible mechanistic pathways to account for the "foot" formation observed for CA positive resists with titanium nitride and silicon nitride substrates.

Experimental

Synthesis. PTBSS copolymer (composition TBS/SO_2: 3.00/1; Mw: 105 K, D:1.6) and PAGs (triphenylsulfonium hexafluoroarsenate, nitrobenzyl ester sulfonate) syntheses have been described elsewhere (*11*). CAMP is the commercial name for the resists formulated with PTBSS copolymer. CAMP6 is the commercial name for the resists formulated with PASTBSS (poly(4-acetoxystyrene-4-t-butoxycarbonyloxystyrene sulfone) polymer.

Lithography. CAMP and ARCH photoresists were jointly developed by AT&T and OCG Microelectronic Materials Inc. and are available from OCG Microelectronic Materials Inc.. CAMP and CAMP6 are two component (PTBSS or PASTBSS co- and terpolymer formulated with a nitrobenzyl ester PAG) whereas ARCH is a three component system containing a base soluble resin, a PAG and a dissolution inhibitor with protecting groups. PTBSS copolymer was also formulated with triphenyl sulfonium hexafluoroarsenate (hereafter referred to as CAMP with onium salt) PAG. The resist solutions were applied on HMDS (hexamethyl disilazane) primed substrates. The wafers were softbaked to remove the residual solvent. Film thickness measurements were done on a nanospec thickness gauge (Nanometrics, Inc.) or a Dektak model IIA profilometer. Deep-UV (248 nm) exposures were performed using either a GCA Laserstep® prototype system with an 0.35 NA lens, 5X reduction optics or a GCA XLS deep-uv exposure tool (0.53 NA, 4x reduction optics, Cymer laser) interfaced with an MTI track. The e-beam exposures were done with a Jeol-JBX5DII operating at 50 keV and 10 nm spot size and X-ray exposures were carried out on a proximity X-ray stepper with an exposure spectrum of 0.9-1.8 nm centered at 1.4 nm. Some PED experiments were performed in a controlled environmental chamber using a homogenized beam from a Lambda Physik 103 G KrF laser. The exposed wafers were then post-exposure baked with or without any PED. In some experiments, an additional t-butyl methacrylate-methacrylic acid copolymer protective overcoat (600 A°) was applied on the resist film prior to exposure.

The quantum yields were determined in films of poly(methyl methacrylate) containing different PAG's as previously described (5, 12).

The dissolution rates were measured by a Perkin-Elmer development rate monitor (DRM) and were analyzed on a personal computer.

Spectroscopic Characterization. IR (Infra Red) spectra were recorded either on a Digilab FTS60 FT-IR, a Nicolet Model 8020 Galaxy FT-IR spectrometer. RBS spectra were recorded using a 2.120 MeV He^{+2} ion beam at a backscattering angle of 162°. The spectra were accumulated for a total ion dose of 40 μC using a 10 nA beam current. The number of As atoms/cm^2 in the sample film was calculated by comparison of the accumulated data for TBSS/PAG films to spectra obtained for a standard Si wafer implanted with a known dose of Sb. X-ray photoelectron spectroscopic (XPS) data were acquired on a Phi 5400 ESCA system with Mg K_α x-rays (1253.6 ev). Data acquisition was limited to less than one hour for all samples to minimize the effect of radiation induced changes in the film. All data reported here were taken at an electron angle of 45°. Laser Ablation Microprobe Mass Spectrometer (LAMMS) data were acquired on a Laser Ionization Mass Analyzer (LIMA) manufactured by Cambridge Mass Spectrometry-Kratos Analytical. This system is equipped with a frequency quadrupled Nd:YAG laser (266 nm) to vaporize and ionize a given sample for time-of-flight (tof) analysis . An entire tof mass spectrum is obtained from each laser shot with a sensitivity of a few ppm for most elements in the periodic table. UV spectra were taken with a Hewlett Packard 8452A diode array spectrophotometer. Absolute thickness measurements of resists were obtained with a Dektak model IIA profilometer.

HMDS sampling and detection method. Air samples were typically collected over 12-14 hours on silica gel tubes. The typical amount of air collected was between 90 and 120 liters. Desorption of HMDS was attempted with carbon disulfide and toluene solvents. Best results were obtained with toluene. Since

HMDS hydrolyzes very quickly to hexamethyl silanol which subsequently forms hexamethyldisiloxane, GC-FID method was developed to detect hexamethyldisiloxane. Hexamethyldisiloxane had a very good FID response. The detection limit of the method is 1 µg/sample. By this method, 15 ppb of HMDS could reliably detected. With the 12-14 hrs. sampling times, the detection limit of the method for HMDS concentration is about 5 ppb.

Results and discussion

Post-exposure Delay (PED). PED phenomenon was investigated by a combination of bulk and surface sensitive spectroscopic methods and by lithography experiments in controlled ambient. The effect of photoacid strength, protecting group chemistry, exposure tool environment, dissolution rate selectivity and resist thickness on the swing curve (E_{max} or E_{min}) on the CA positive resists' PED stability is examined below.

Spectroscopic analysis and Lithography in controlled ambient. Film thickness measurements for CAMP indicated that thickness loss in the exposed areas due to deprotection did not vary with PED time upto 1 hour for a given exposure dose. SEM analysis, however, clearly showed surface residue and "T-tops" formation in as little as 30 sec. PED time (Figure 1). For a given PED, the surface residue formation was found to be more severe for smaller features. With CAMP onium salt formulation, 0.5 µm trenches were completely "bridged" at the surface within a 3 min. PED.

RBS analysis was used to determine arsenic (As), and sulfur (S) concentration and distribution profiles in the CAMP onium salt resist films at every process step. The As peak served as a marker for both the PAG and the photoacid whereas S provided information regarding the PAG and polymer. Depth profiles clearly ruled out any redistribution or loss of PAG during the process sequence as the As concentration in the unexposed areas was identical within experimental error after each process step. The RBS spectra of the wafers with PED = 1 and 30 mins after PEB were superimposable indicating that there is no loss or detectable redistribution of PAG or photoacid during the PED. However, the wafer with 1 min. PED was aqueous base soluble, whereas the wafer with 30 min. PED was aqueous base insoluble.

In IR experiments, the peak at 1756 cm^{-1} corresponding to the t-boc carbonyl stretch served as a good marker to monitor the t-boc deprotection reaction as a function of process events as well as PED time variation. A plot of exposure dose as a function of t-boc concentration and normalized film thickness (Figure 2) after development revealed that >95% t-boc deprotection was necessary to induce aqueous base solubility. The results also revealed that small increase in t-boc concentration (1-2%) can turn the resist from aqueous base-soluble to aqueous base-insoluble. IR Intensities of t-boc group measured in thin films showed substantial increase with PED time indicating incomplete deprotection with increased PED time (Figure 3).

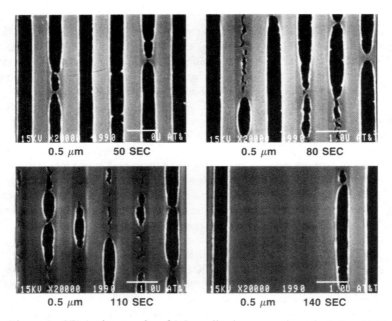

Figure 1: SEM micrographs of 0.5 µm line/space gratings depicting PED induced surface residue in CAMP resist.

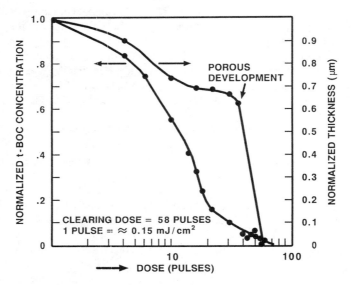

Figure 2: A plot showing the effect of exposure dose on the normalized t-boc concentration and normalized resist thickness for CAMP resist.

While the IR analysis revealed that t-boc deprotection may be the cause of surface insoluble residue formation with PED, XPS and LAMMS techniques conclusively proved it to occur at the resist surface. In the case of XPS, sampling depth was ~40 A°, thus allowing for a clear differentiation between surface vs. bulk reactions by comparing the results to those from IR spectroscopy. The most clearly separated peak in the XPS spectrum of PTBSS was the C_{1s} line associated with the carbonyl carbon of the t-boc group (Figure 4). The binding energy for this is 5.9 eV higher than those of the other carbons. Some overlap is apparent with satellite peak due to energy losses resulting from electron coupling with π-$\pi*$ transition of the aromatic ring. Curve fitting, however, allowed separation and determination of intensity changes in the carbonyl group that resulted from the PED. In CAMP with onium salt, after exposure and PEB, the intensity of the carbonyl C_{1s} peak was reduced by about 65% in the exposed resist. This is in contrast to the IR results which indicate that complete t-boc group removal occurred after PEB. With 30 min. PED, the carbonyl peak retains most of its original intensity with no further change in its intensity with a 3 day PED. In CAMP (PTBSS formulation with a nitrobenzyl ester), most of the t-boc groups are removed after exposure and PEB. A PED of 30 min. showed no significant increase in t-boc intensity. With a 3 day PED, the carbonyl peak regained most of its original intensity.

LAMMS tof (time-of-flight) positive ion mass spectral data of bridged structures, that result from PED, indicate to the presence of significant amounts of t-boc groups. This corroborates the IR results that indicate the aqueous base insolubility to be a result of the incomplete t-boc deprotection. The LAMMS analysis of the exposed and immediately post-exposure baked regions prior to development, however, showed complete removal of t-boc groups.

To summarize PED analytical analysis, SEM pictures showing "bridges" and "T-top" formation with PED clearly revealed it to be a surface phenomenon. RBS analysis showed no loss or redistribution of PAG or its photoproducts during the processing thus eliminating the loss of photoacid at the surface to be a contributor to surface residue formation. IR spectroscopy elucidated the strong relationship between resist's aqueous base solubility and t-boc removal. It clearly showed that >95% t-boc removal is necessary to induce aqueous base solubility and that small increases in t-boc concentration can turn the resist from aqueous base soluble into aqueous base insoluble (Figure 3). IR results showed that t-boc concentration increases with PED and XPS established it to occur first at the resist surface. LAMMS corroborated the IR and XPS results showing the presence of t-boc groups in the "bridges" that formed during the PED.

Lithography experiments in controlled ambient were performed to determine the cause of PED. The experiments excluded variations in relative humidity or Oxygen to be the cause of PED effects. However, when experiments were conducted in basic environment with as little as 15 ppb of HMDS, resist performance deteriorated dramatically. Other airborne basic contaminants that have a deleterious effect on the resist sensitivity, PED stability and overall image quality include NMP (N-methylpyrrolidone), a typical solvent used in resist strippers; DMF (N,N-dimethyl formamide), a solvent commonly found in "weld-on" chemicals for

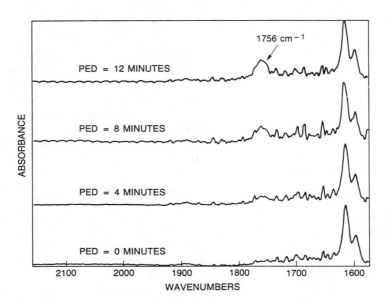

Figure 3: IR spectra depicting the increase in carbonyl stretch (t-boc concentration) at 1756 cm^{-1} with increased PED time intervals between exposure and PEB for CAMP resist.

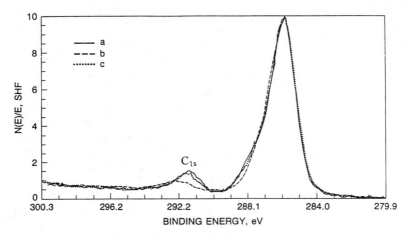

Figure 4: XPS spectra of thin films of PTBSS/triphenylsulfonium hexafluoroarsenate (CAMP with onium salt) depicting variation in carbonyl C_{1s} intensity a) before exposure, b) after exposure and immediate   PEB and c) after exposure and PEB with a 3 day PED interval.

PVC water piping, ammonia, hydrolysis product of HMDS, and N-morpholine, cyclohexylamine that are used as anti-oxidants in air handling systems. Even fresh latex paints contain sufficient quantities of amine to have pronounced negative effect on the resist PED stability.

Since ppb levels of airborne basic contaminants are proved to affect the resist performance dramatically, several approaches (*13*) have been developed to improve the resist PED stability. Two most successful approaches to dramatically alleviate the airborne contamination effects and improve the PED stability constitute: performing resist processing in a tightly controlled environment that is filtered off airborne basic contaminants, and isolation of the resist surface from the immediate processing environment by means of a thin, weakly acidic protective overcoat on the top of the resist (Figure 5). Additives in resist formulation and additives in developers were also implied to have improved the PED stability.

Photoacid Strength vs. PED stability. Previously, we have shown (*5*) that there is a reciprocal relationship between the clearing dose (D_p) and the product of absorbance/μm (ABS/μm), quantum yield (Φ), and the catalytic chain length for the removal of *t*-boc groups (Chain Length), (Equation 1).

$$D_p \ \alpha \ 1/(ABS/\mu m*\Phi*Chain\ Length) \quad (1)$$

This relationship is further refined by using the fraction of light absorbed per micron (%ABS/μm) to yield equation 2, which estimates more accurately the absorption of the photons by the resist film .

$$D_p \ \alpha \ 1/(\%ABS/\mu m*\Phi*Chain\ Length) \quad (2)$$

Relative chain lengths can then be determined using the equation 3 for a given family of PAGs in a polymer system.

$$Chain\ Length \ \alpha \ 1/(\%ABS/\mu m*\Phi*D_p) \quad (3)$$

The PED stability was investigated as a function of the acid strength for PTBSS with 2-trifluoromethyl-6-nitrobenzyl arylsulfonates. Since the nucleophilicity of the sulfonate anion should decrease with increasing σ and acid strength, the chain length should increase with increasing σ. The logarithm of the chain length should then be proportional to Hammett sigma values (Equation 4).

$$Log(Chain\ Length) \ \alpha \ \rho\sigma \quad (4)$$

Since the chain length is a measure of the differing rates of t-boc removal, a strong correlation was expected between the σ (acid strength) and catalytic chain length. The PED experiment was performed under four different conditions. The two conditions that represent minimum and maximum interaction of the photoacid with airborne basic contaminants were: processing with overcoat and no PED, and processing without an overcoat with 30 min. PED, respectively. The catalytic chain length determined with the overcoat application was found to be a linear function

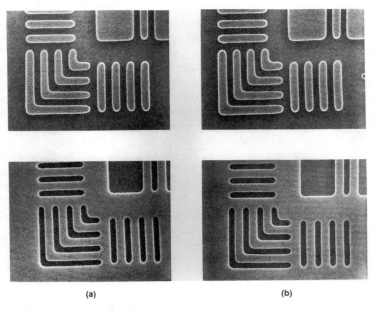

(a)   (b)

Figure 5: SEM photomicrographs of CAMP processed with a thin overcoat [(a) no PED, b) 15 min. PED] that isolates resist surface from the immediate processing environment and airborne contaminants thus improving the resist's PED stability.

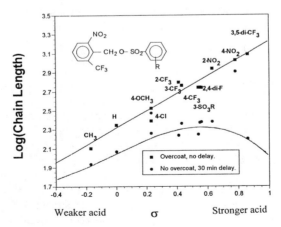

Figure 6: Plots of Log(Chain Length) versus the Hammett sigma values for substitution at the benzenesulfonyl moiety of 2-trifluoromethyl-6-nitrobenzyl benzenesulfonate in PTBSS resist matrix. The data is presented for resists processed with a) overcoat and no PED and b) without PED and 30 min. PED.

of the acid strength, as expected (Figure 6). This confirmed that without significant interaction with basic contaminants resist sensitivity is governed primarily by the photoacid strength (nucleophilicity of the acid anion). The catalytic chain length increased linearly with increased σ.

The results from the resist processed without an overcoat and exposed to airborne contaminants for 30 minute (PED = 30 minutes) showed a leveling of the expected increase in the relative chain length with the photoacid strength (Figure 6). This is most probably a result of the deactivation of the stronger acids by the airborne basic contaminants as well as other side reactions from the stronger nucleophile.

A few of the PAGs, however, effected PED stability in a more complex way. For example, 4-Cl, 4-OCH$_3$ and 4-NO$_2$ substituted PAGs showed less susceptibility to environmental effects than what is expected from their Hammett sigma values and acid strength. It appears that polar 4-substituents on our PAG system helped improve the PED stability. This may be due to the interaction of the airborne contaminants with PAG itself (*14*). Since undissociated PAG is the second most abundant component in the resist formulation, the ability of the PAG to interact with airborne basic contaminants can be an important pathway to improve the PED stability. This will both decrease the diffusion of airborne contaminants into the resist film as well as decrease the probability of basic airborne contaminants encountering the sulfonic acid and thus contributing to improved PED stability. All three of the PAG's which are resistant to airborne contaminants have polar 4-substituents susceptible to dipolar interactions. Conversely, most of the materials showing poor resistance to the PED effect either have groups that are weakly polarizable (F, CF$_3$) or ones with poor dipoles (H, CH$_3$) and would consequently form poor dipolar interactions.

Protecting Group. While several acid-labile protecting groups have been shown to be effective for masking the base-solubilizing functionalities of the polymer, the majority of the reported work for CA positive resist applications, was centered on the use of t-boc protecting groups or carbonate groups. The t-boc groups are thermally stable, and have been shown to be very effective dissolution inhibitors in concentrations as small as 5-10 % on the polymer, as discussed above. The activation energy for t-boc removal, however is quite high and requires stronger acids and higher PEB temperatures for good photospeed. Substitution of t-boc group with THP (tetrahydropyranyl) group or other acetal and ketals led to linewidth slimming with PED (*15-17*). This linewidth slimming can be dramatically alleviated by reducing the catalytic chain length by a combination of using PAGs that generate weaker photoacids, reducing the protecting groups on the polymer, employing the additives that are capable of interacting with the airborne contaminants. PED stability can also be substantially improved by using PAG materials that are designed to generate intermediate compounds rather than photoacids upon exposure. These intermediate compounds generate acid during the PEB.

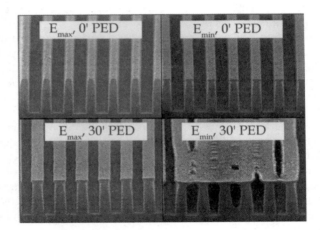

Figure 7: PED latitude for an ARCH formulation at E_{max} and E_{min} thicknesses on the swing curve. The SEMs were shown for 0.25 μm l/s pairs at PED=0 and 30 minutes.

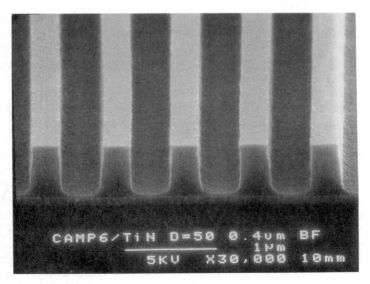

Figure 8: SEM pictures (0.5 μm l/s) showing a very very strong interaction for CAMP6 with titanium nitride as evidenced by the "foot" at the substrate.

Exposure Tool Environment. Since PED stability is dependent on the base concentration in the processing environment, intuitively, PED stability is also expected to vary with the exposure tool environment. We have examined the PED stability of ARCH in deep-uv, proximity X-ray and e-beam exposure tools *(18)*. ARCH showed no change in its linewidth or image quality over a 15 hr. PED in the vacuum environment of the e-beam tool. In contrast, ARCH exhibited less than 1 hr. PED stability in both the Helium environment of proximity X-ray tool and activated charcoal filtered air of the deep-uv exposure tool. The e-beam results indicate that resist components or their photoproducts are non-volatile even at the high vacuum of e-beam tool and that high vacuum probably resulted in very low concentration of airborne basic contaminants. Less than 1 hr. PED stability in the proximity X-ray tool may allude to the presence of a finite amount of airborne basic contaminants in that environment.

Dissolution Rate Selectivity. While higher dissolution rate selectivities (R_e/R_o where R_e the dissolution rate of the fully exposed region and R_o represents the rate of the unexposed regions) result in high resolution and improved contrast for CA positive resists, very high dissolution contrast can also adversely affect PED latitude *(19)* . An optimum dissolution ratio of $1e^{+3}$ to $1e^{+4}$ was suggested for CA positive systems based on the results obtained for t-boc protected poly(vinylphenol) system.

We have measured dissolution rate selectivities of approximately 10^4 and 10^3 for CAMP6 *(20)* and ARCH resists, respectively. The PED latitudes for CAMP6 and ARCH were <5 mins. and 30 mins., respectively indicating that a lower dissolution ratio may be beneficial for improving the PED latitude. PED latitude, however, is a very complex function of polymer molecular weight, distribution, protection chemistry and PAG chemistry. With a given polymer system and PAG chemistry, there is strong evidence for improved PED stability with lower dissolution selectivities. The lower dissolution selectivity, however can result in thickness erosion and needs to be balanced with the other desired properties in the resist.

Swing Curve Effects. We and others have shown that PED stability is also effected by the resist thickness. We have observed the PED stability to be better at the E_{max} than at the E_{min} of the swing curve for ARCH (Figure 7). The difference in PED stability between E_{max} and E_{min} may be a product of the higher catalytic chain length (lower resolution dose) and the presence of a node at the resist surface at E_{min} of the swing curve.

Substrate Contamination. CA resists such as CAMP, like the majority of the current commercial chemically amplified positive resists show a strong "foot" on titanium nitride and silicon nitride substrates (Figure 8). It is known that there is considerable amount of hydrogen is present in silicon nitride films and that the hydrogen content varied with deposition method and conditions. IR (Infra Red) studies indicated that a significant portion of this hydrogen exists in the form of N-

H bonds (*21,22*). These basic moieties at the substrate surface presumably neutralize the photoacid to result in "foot" formation.

In the case of interaction with titanium nitride, the cause of "foot' formation is less well understood. It has been previously shown that a thin silicon oxide layer between titanium nitride and CA resist eliminates this problem indicating the "foot" formation to be caused by the titanium nitride surface. One likely possible path for acid deactivation may result from the reduction of protons by the electrons in the conduction band at the titanium nitride film surface.(23) Previous studies on titanium nitride films have indicated that the surface of these films are titanium oxidelike due to oxidation. (*24*)

The optimized ARCH material, with its lower catalytic chain length and lower dissolution selectivity between the exposed and unexposed areas, however shows no noticeable "foot" on titanium nitride or silicon nitride substrates (Figure 9). We are currently studying the correlation between the effect of photoacid strength and interaction of resist with the film surface as evidenced by the magnitude of "foot".

Shrinkage and Etch Resistance.   Shrinkage at PEB in the exposed areas is not only dependent on the size and amount of the protecting group but also on the byproducts of the deprotection reaction (*2,7*). In the t-boc resist systems, the deprotection results in volatile carbon dioxide and isobutene molecules. In resist systems with protecting groups such as trimethylsilyl ethers, the silanols released

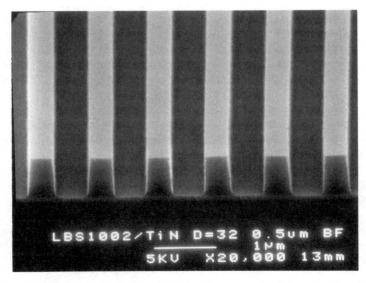

Figure 9: ARCH resist (0.5 μm l/s) showing no noticeable interaction (no foot) with titanium nitride substrate.

after deprotection, undergo condensation resulting in no significant loss of volatiles or shrinkage. With the protecting groups that do release volatiles at PEB, the shrinkage can be minimized and etch resistance can be improved by decreasing the size of the protecting group and/or decreasing the protection ratio in the base resin.

CD sensitivity to PEB. Since the photoacid diffusion distance and the deprotection turnover rate (catalytic chain length) are PEB temperature dependent, variations in PEB temperature affect the CD linewidth. Most of the commercially available CA positive resists show ≥ 10 nm/°C during the PEB for 0.35 µm line/space features (*18*).

We have previously investigated PTBSS polymer with nitrobenzyl ester and onium salt PAGs. The sensitivity with nitrobenzyl ester PAG was a strong function of the PEB temperature. PTBSS with triphenyl sulfonium hexafluorarsenate registered no change in its sensitivity over the 90-130 °C PEB range (*25*). Resist materials with low or no sensitivity to PEB temperature have been described in literature. For example, ARCH resist material shows no change in linewidth of 0.30 µm features over 12°C PEB range (*18*).

Summary

Issues with CA positive resist technology have been described. The issues have been studied at a fundamental level by several research groups. The newer CA resists exhibit good adhesion, photospeed, etch selectivity and vastly improved PED latitude (Figure 10). Additionally, we have demonstrated that resist systems that exhibit little or no "footing" on semiconductor substrates and that show very little change in the CD linewidth with PEB temperature variations (Δlw/°C).

We have studied the post-exposure delay (PED) for poly(4-t-butoxycarbonyloxystyrene-sulfone) (PTBSS) formulated with various photoacid generators (PAG) by a variety of analytical techniques along with lithography in controlled environments. The analysis revealed that PED is a surface phenomenon and results from the incomplete deprotection of the protecting groups on the polymer due to the photoacid deactivation at the resist surface by airborne basic contaminants in concentrations as low as 15 ppb. The effect of photoacid strength, protecting group chemistry and process parameters was examined on the PED latitude. Patterns of CA resists also show a "foot" on titanium nitride and silicon nitride surfaces presumably due to either neutralization or inactivation of the photoacid at the substrate surface.

The analytical studies we have done on PED phenomenon and resist-substrate surface interactions enabled us to design resist systems that have significantly superior PED latitude and that show little interaction with the semiconductor substrates.

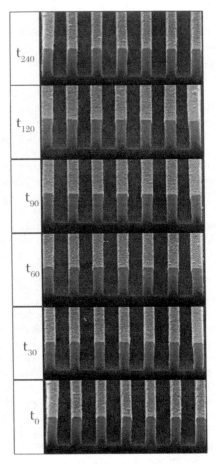

Figure 10: An optimized resist formulation incorporating several PED stability concepts showing evidence of no linewidth variation or "t-tops" formation even after a 4 hr. PED.

Acknowledgments.

We like to thank J.M. Kometani, A. Timko, M. Cheng, D. R. Stone, S. Vaidya, L. Heimbrook, F. Baiocchi, T.X. Neenan, R.S. Kanga, T. Ohfuji and others for their technical contributions and helpful discussions.

References.
1. Reichmanis, R.; Houlihan, F.M.; Nalamasu, O.; Neenan, T.X.; Chem. Mater., 1992, 3, 394,
2. Lamola, A., Szmanda, C.R., Thackeray, J. W. Solid State Technol., 1991, 34(8), 53.
3. McKean, D.R.; Schaedeli, U.; MacDonald, S.A.; In "Polymers in Microlithography", ACS Symposium Series 412, Reichmanis, E., MacDonald, S.A.; Iwayanagi, T. Eds., ACS, Washington, D.C., 1981, pp. 27-38.
4. Seligson, D., Das, S. Gaw, H., J.Vac.Sci.Technol., 1988, B6(6), 2303.
5. Houlihan, F.M.; Neenan, T.X.; Reichmanis, R.; Kometani, J.M.; Chin, E.; Chem.Mater., 1991, 3, 462..
6. Nalamasu, O.; Reichmanis, E.; Timko, A.G.; Novembre, A.E.; Tarascon, R.; Houlihan, F.M.; Munzel, N.; Holzwarth, H.; Falcigno, P.; Slater, S.G.; Frey, M.D.; Proceedings of the 10th International Conference on Photopolymers, Society of Plastic Engineers, Mid-Hudson section, October 30 - November 2, 1994, Ellenville, N.Y.; pp. 380-385.
7. MacDonald, S. A., Clecak, N.J., Wendt,, H. R., Willson, C. G., Snyder, C. D., Knors, C.J, Peyoe, N., Maltabes, J. G., Morrow, J., McGuire, A.E., and Holmes, S. J., Proc. SPIE, 1991, 1466, 2.
8. Nalamasu, O.; Cheng, M.; Kometani, J.M.; Vaidya, S.; Reichmanis, E.; Thompson, L.F.; Proc. SPIE, 1990, 1262, 32.
9. Nalamasu, O., Reichmanis, E., Cheng, M., Pol, V., Kometani, J. M., Houlihan, F. M., Neenan, T.X., Bohrer, M.P., Mixon, D.A., Thompson, L. F., Proc. SPIE, 1991, 1466, 13.
10. Nalamasu, O., Vaidya, S., Kometani, J.M., Reichmanis, E., Thompson, L.F., Proc. Reg. Tech. Conf. on   Photopolymers, Ellenville, NY Oct. 28-30, 1991, 225.
11.Tarascon, R. G. Reichmanis, E., Houlihan, F. M., Shugard, A., Thompson, L.F. Polym. Eng. Sci., 1989, 29(3), 850.
12.Neenan, T.X., Houlihan, F.M., Reichmanis, E., Kometani, J. M., Bachman, B.J., Thompson, L. F., Macromolecules, 1990, 23, 145.
13. Reichmanis, E.; Thompson, L.F.; Nalamasu, O.; Blakeney, A.; Slate, S. Microlithography World, 1991, 1, 2 .
14. Houlihan, F.M.; Chin, E.; Nalamasu, O.; Kometani, J.M.; Harley, R.; Proceedings of the 10th International Conference on Photopolymers, Society of Plastics Engineers Inc., Mid-Hudson Section, Oct. 31 - Nov. 2, 1994, Ellenville, NY, pp. 48-57.
15. Hattori, T.; Schlegel, L.; Imai, A.; Hayashi, N.; Ueno, T.; J. Photopolym. Sci. and Technol., 1993, 6, 497.
16. Murata, M.; Kobayashi, E.; Yumoto, Y.; Miura, T.; Yamaoka, T.; J.Photopolym. Sci.and Technol., 1991, 4, 497.
17. Huang, W-S.; Kwong, R.; Katnani, A.; Khojasteh, M.; Proc. SPIE, 1994, 2195, 37.

18. Nalamasu, O., Reichmanis, E.; Timko, A.G.; Tarascon, R.; Novembre, A.E.; Slater, S.; Holzwarth, H.; Falcigno, P.; Munzel, N.; Microelectronic Engg., 1995, 27, 367.
19. Itani, T.; Iwasaki, H.; Fujimoto, M.; Kasama, K.; Proc. SPIE, 1994, 2195, 126.
20. Ohfuji, T.; Nalamasu, O.; Stone, D.R.; NEC Research & Development, 1994, 35, 7.
21. Wolf, S.; Tauber, R.N.; Silicon Processing for the VLSI Era, Vol. 1, pp.191-195, Lattice Press, Sunset Beach, CA, 1986.
22. Claasen, W.A.P.; Valkenburg, W.G.J.N.; Willemsen, M.F.C.; Wilgert, M.W.V.D.; J. Electrochem. Soc., 1985, 132, 893.
23. Tompkins, H.G., J. Appl. Phys., 1991, 70, 3876.
24. Serpone, N.; Pelizetti, E.; Photocatalysis, Fundamentals and Applications, John Wiley & Sons, New York, 1989.
25. Nalamasu, O.; Cheng, M.; Timko, A.G.; Pol, V.; Reichmanis, R.; Thompson, L.F.; J. Photopolym. Sci. Tech., 1991, 4, 299.

RECEIVED September 11, 1995

Chapter 2

The Annealing Concept for Environmental Stabilization of Chemical Amplification Resists

Hiroshi Ito, Greg Breyta, Donald C. Hofer, and R. Sooriyakumaran

IBM Almaden Research Center, 650 Harry Road,
San Jose, CA 95120–6099

The postexposure bake delay problem of chemical amplification resists, which manifests itself as a line width change and/or T-top/skin formation, is primarily caused by surface contamination by airborne basic substances. This paper describes our annealing concept to solve this serious problem, which is based on reduction of the resist film free volume by carrying out postapply bake above its glass transition temperature. As the diffusivity of small molecules is an exponential function of the free volume, this approach dramtically reduces contaminant absorption and provides excellent environmental stabilization. Use of *meta*-isomer resins to lower glass trnsition temperatures for better annealing is described first to prove the validity of our annealing concept. In addition, a new positive resist named ESCAP is proposed, which is processed at unconventionally high temperatures to achieve good annealing and contamination resistance.

The chemical amplification concept proposed in 1982 (*1,2*) has become the predominant foundation for the design of advanced resist systems for lithographic imaging as the minimum feature size of semiconductor devices shrinks to <0.5 μm. The amplification effect is achieved by employing an acid (generated by irradiation) as a catalyst to carry out a cascade of chemical transformations in the resist films. Among varieties of acid-catalyzed imaging mechanisms reported (*3-5*), deprotection of polymer pendant groups to induce a polarity change from a nonpolar to polar state (solubility change), as embodied in IBM's tBOC resist (Scheme I) (*1,6*), has attracted the most attention as a basis for designing aqueous base developable positive deep UV (248 nm) resists for replacement of the classical diazonaphthoquinone/novolac resist.

In spite of the extensive world-wide research efforts, however, full implementation of chemical amplification resists in manufacturing has been severely hampered by the so-called delay problem (or the latent image instability), which is manifested as a "skin" or "T-top" in positive-working systems. The formation of the insoluble surface layer has been ascribed to contamination by airborne basic substances such as amines and *N*-methylpyrrolidone (NMP) using a ^{14}C labeling technique (*7*). Due to the catalytic nature of the imaging mechanisms, a trace amount, on the order of 10 ppb, of airborne basic substances absorbed in the resist film interferes with the desired acid-catalyzed reactions.

Some engineering solutions to the contamination problem have been proposed, which include

- purification of the enclosing atmosphere using activated carbon filtration (*7*),
- application of a protective overcoat to seal out airborne contaminats (*8-10*),
- incorporation of a stabilizing additive into resist formulations (*11-13*).

In addition, reduction of the activation energy of the deprotection reaction has been reported to provide the image stability because elimination of postexposure bake (PEB) results in elimination of the delay problem (*14*).

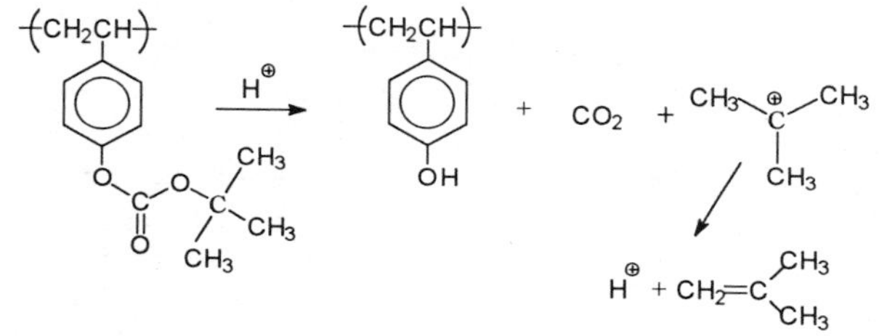

Scheme I Acid-catalyzed deprotection for polarity change

Our approach has been to first understand the propensity of thin polymer films to absorb airborne NMP (*15*). Based on these studies, we have concluded that it is primarily glass transition temperatures (T_g) of the polymers that determine the contaminant uptake and that reduction of the free volume of the resist film by annealing enhances the contamination resistance (*15-18*). This paper describes our annealing concept for environmental stabilization of chemical amplification resists (Scheme II).

**Annealing → Reduced Free Volume → Reduced Contaminant Uptake
→ Environmental Stabilization**

Scheme II Annealing concept for environmental stabilization

Experimental

m-t-Butoxycarbonyloxystyrene (BOCST) was synthesized in a fashion similar to its *p*-isomer by reacting 3-hydroxybenzaldehyde with di-*t*-butyl dicarbonate followed by the Wittig olefination and then subjected to radical polymerization with benzoyl peroxide (BPO) in toluene at 60 °C (Scheme III) (*16,17*). Copolymers of BOCST with hydroxystyrene (HOST) were prepared by partial protection, with di-*t*-butyl dicarbonate in the presence of a base in acetone, of polyhydroxystyrene (PHOST), which in turn had been obtained by heating the PBOCST powder at 180 °C on a Kugelrohr apparatus (Scheme III) (*17*). Copolymers of 4-HOST with *t*-butyl acrylate (TBA) were synthesized by radical copolymerization of 4-acetoxystyrene (ACOST) with TBA using BPO as the initiator in toluene and subsequent selective hydrolysis of the acetoxy group with ammonium hydroxide in methanol (Scheme IV) (*18*).

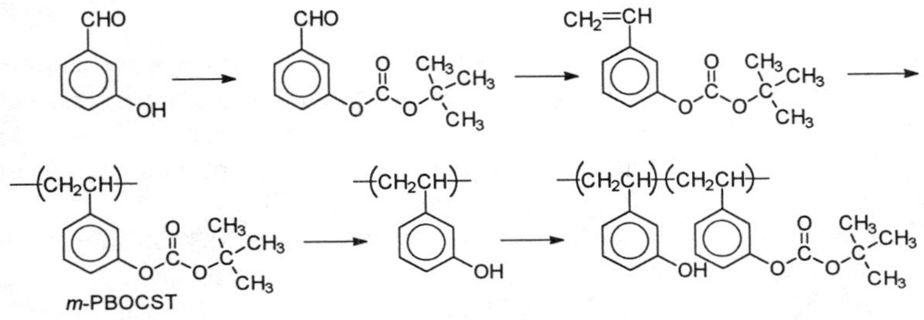

Scheme III   Synthesis of *m*-PBOCST and *m*-poly(BOCST-co-HOST)

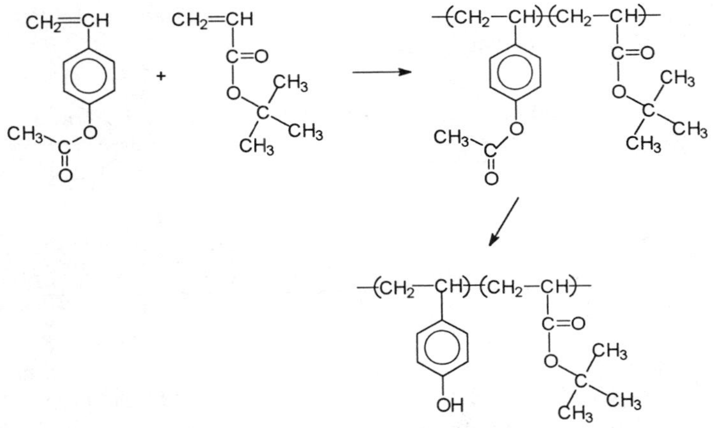

Scheme IV Synthesis of poly(TBA-co-HOST)

The casting solvent employed for the resist formulations was propylene glycol methyl ether acetate (PMA).

Results and Discussion

Our approach to environmental stabilization has been to incorporate a stabilization mechanism in the resist itself, which then could eliminate the cumbersome air filtration and/or protective top coat. As illustrated in Scheme II, our annealing concept for environmental stabilization involves heating resist films above their T_g to reduce the free volume. It should be noted that a small reduction of the free volume can result in a profound decrease in the diffusivity of small molecules (such as airborne contaminants) in polymer films as the diffusivity is an exponential function of the free volume.

Annealing of resist films can be achieved either by lowering T_g of resist polymers or by increasing the postapply bake (PAB) temperature. We employed the first approach to demonstrate the validity of our annealing concept and the second approach for further proof and to design a viable positive resist system.

Reduction of Glass Transition Temperature. Although T_g of polymer films can be reduced by lowering molecular weights and/or adding plasticizers, we have decided to utilize *meta*-isomers for comparison with more commonly employed *para*-isomers (*16,17*), which was expected to introduce a minimum number of undesirable variables in the comparison in contrast to the other methods to decrease T_g.

While th *p*- and *m*-PBOCSTs exhibit identical thermal deprotection behaviors and very similar UV absorption spectra, their glass transition temperatires are vastly different; 130 (M_n=56,600 and M_w=92,700) and 85 °C (M_n=34,400 and M_w=57,200), respectively. Thus, the commonly-employed PAB temperature of 90-100 °C is lower than the T_g of *p*-PBOCST and higher than that of the *meta*-isomer, which makes a drastic difference in densification (free volume reduction) and therefore in NMP uptake, as demonstrated in Table I.

Table I *p*- and *m*-PBOCST Films Baked at 100 °C for 5 min

isomer	T_g (°C)	residual PMA (wt%)	NMP uptake (ng/wafer)	thickness (nm)	$n_{||}$	$n_\perp$
para	130	0.60	931	1416	1.5074	1.5104
meta	85	0.55	99	1317	1.5170	1.5167

When baked at 100 °C for 5 min, these polymer films contain almost no residual casting solvent (0.6 and 0.55 %) but absorb NMP at vastly different rates (931 vs 99 ng/wafer for the *para*- and *meta*-isomers, respectively) (*15-17*). The *meta*-isomer film has a significantly higher refractive index (n) than the *para*-isomer film ($\Delta n_{||}$=~0.008, $\Delta n_\perp$=~0.005) when heated at 100 °C, indicating that the former film is more annealed and densified than the latter film. Glassy polymers at T_g have a minimum free volume of ca. 2.5 % (fraction of free volume f=0.025). The refractive index difference (~0.4 %) of the isomeric PBOCST films baked at 100 °C suggests that the *p*-isomer film has ~40 % more free volume than the *m*-isomer film. Mobility of small molecules in polymer films is proportional to exp(-B/f) and therefore the *m*-isomer allows contaminants to diffuse much more slowly than the *p*-isomer, by a factor of 10^3~10^6 depending on the size of the diffusant (B=0.5-1.0) (Figure 1). Thus, absorption of airborne contaminants in thin polymer films is controlled by the degree of annealing.

The greater resistance of the *m*-isomer toward airborne NMP contamination has been confirmed lithographically (*16,17*). A *m*BOC resist based on *m*-PBOCST (M_n=19,600 and M_w=53,400) and 4.75 wt% triphenylsulfonium hexafluoroantimonate was compared lithographically side-by-side with IBM's tBOC resist consisting of *p*-

PBOCST and 4.75 wt% of the same acid generator in terms of NMP contamination (Figure 2). The resist films (ca. 1 μm thick) were prebaked at 100 °C for 5 min, exposed on a Perkin Elmer Micralign 500 mirror projection scanner in the deep UV mode, treated with 50-100 ppm of NMP for 5 min in a bell jar, postbaked at 90 °C for 90 sec, and developed in a negative mode with anisole. As Figure 1 clearly indicates, the *m*BOC resist treated with NMP before postexposure bake printed very well uniformly all over the 5" wafer whereas the NMP-treated tBOC resist exhibited a massive film loss during the anisole development due to poor conversion in the acid-catalyzed deprotection reaction. Thus, it is clear that the *m*BOC resist is far more resistant to NMP contamination than the tBOC resist due to its lower T_g and better annealing during prebake.

Although NMP uptake is not affected by the residual PMA concentration (*19*), the comparison of the *m*- and *p*-isomers which contain only a trace and the same amount of the residual casting solvent has provided unequivocal evidence for the higher contamination resistance of the lower T_g polymer.

We next compared BOCST-HOST copolymers made from *p*- and *m*-isomers (Scheme III), as summarized in Table II. The BOCST concentrations in the copolymers were adjusted to give similar dissolution rates in aqueous base. *m*-PHOST is slightly less soluble in aqueous base than its *para*-isomer (*20*) and requires a smaller degree of partial protection. The T_g's of the copolymers were estimated from the values for the homopolymers because the direct T_g measurement of these copolymers is hampered by thermal deprotection below T_g (*21*). When heated at 100 °C for 5 min, the two isomer films contain about the same amount of PMA (9.2 and 11.3 wt%) (*17*). However, the NMP uptake is drastically different, with the *p*-isomer absorbing 4.6 times more NMP than the *m*-isomer film. The higher T_g copolymer absorbs more NMP. Due to ca. 10 % residual PMA, T_g of the *m*-isomer film could be lower than the bake temperature of 100 °C.

Table II Poly(BOCST-co-HOST) Baked at 100 °C for 5 min

isomer	BOCST (mol%)	est. T_g (°C)	residual PMA (wt%)	NMP uptake (ng/wafer)
para	25	168	9.2	1596
meta	20	135	11.3	350

Aqueous base developable positive resists were formulated by dissolving these HOST-BOCST copolymers and 8 wt% of a *N*-sulfonyloxyimide acid generator in PMA. After prebake at 90 °C for 1 min, the coated 5" wafers were stored for 15 min in activated-carbon-filtered air or in filtered airstream containing 10 ppb NMP, exposed on a Perkin Elmer 500 in the deep UV mode, postbaked at 90 °C for 90 sec, and devloped with a tetramethylammonium hydroxide aqueous solution for 1 min. As demonstrated in Figure 3, the *m*-copolymer resist treated with 10 ppb NMP before exposure performed as well as the control stored in the purified air whereas the *p*-copolymer resist exhibited massive skins and T-top profiles when stored in the NMP-containing air (*17*). The improved environmental stability of the *meta*-copolymer

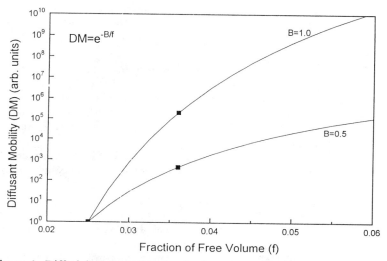

Figure 1   Diffusivity of small molecules in polymer films as a function of free volume.

Figure 2   NMP contamination of tBOC resists based on *para-* (top) and *meta-* PBOCST (bottom); the resist films exposed on a Perkin Elmer Micralign 500 scanner were treated with 50-100 ppm NMP, postbaked, and developed with anisole in a negative mode. (Reproduced with permission from ref. 16. Copyright 1993 Society of Photo-Optical Instrumentation Engineers.)

resist is not due to a reduced photospeed; the *meta*-isomer resist received ca. 11 % less UV dose than the *para*-siomer system in this contamination study.

Thus, these two examples utilizing a *meta*-isomer to lower T_g clearly indicate that lower T_g resist films absorb much less NMP (or other airborne contaminants) and are lithographically more robust. Novolac-based chemical amplification resists are more stable toward contamination than PHOST-based systems due to their lower T_g. The additive approach to environmental stabilization mentioned in Introduction may in fact work on the basis of better annealing due to plasticization (reduction of T_g) by the additives.

High Temperature Bake. Since low T_g resists suffer from thermal flow during high temperature fabrication processes, though more stable environmentally, we have decided to design a thermally stable positive resist system that can be baked at unconventionally high temperatures to achieve good annealing. This approach, however, demands a thermally and hydrolytically stable resin and acid generator, presenting a challenge.

A typical chemical amplification positive resist design is schematically illustrated in Scheme V. The resist system consists of a base-soluble functionality such as phenol or carboxylic acid, a protected dissolution inhibiting group, and an acid generator. The base-soluble and protected functionalities can be chemically connencted together (copolymer) or mechanically mixed (blend). The unexposed film is not soluble in

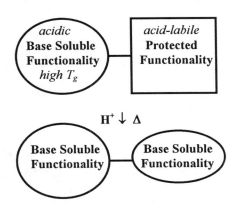

Scheme V Chemically amplified positive resist

aqueous base due to the dissolution inhibition effect of the protected functionality, which is converted to a base-soluble form by reaction with a photochemically-generated acid and subsequent bake, providing positive images upon aqueous base development. Such resist films cannot be heated above T_g usually as the acid-labile protecting group cannot survive such a high temperature bake in the presence of the acidic base-solubilizing group which increases T_g through a hydrogen-bonding interaction. Figure 4 illustrates the reduced thermal stability of a protecting group in the presence of a phenolic functionality, employing poly(BOCST-co-HOST) as an

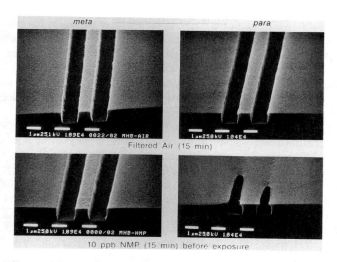

Figure 3 Effects of filtered air (top) and 10 ppb NMP (bottom) on positive images projection-printed in the *meta*- (left) and *para*-poly(HOST-co-BOCST) (right) resists. (Reproduced with permission from ref. 17. Copyright 1993 Technical Association of Photopolymers, Japan.)

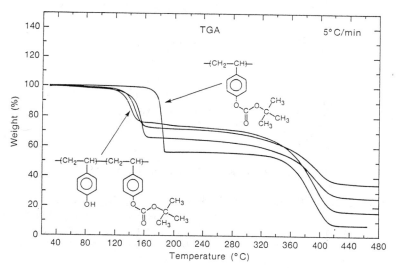

Figure 4 TGA curves of poly(HOST-co-BOCST) and PBOCST.

example (*21*). PHOST partially protected with the tBOC group undergoes thermal deprotection at ca. 130 °C while the fully protected polymer, PBOCST, is stable to ca. 190 °C and the T_g of the copolymers is higher than the deprotection temperature of 130 °C due to the high T_g of PHOST (160-180 °C) (*21*).

In contrast to the majority of the partially-protected PHOST's employed in the chemical amplification scheme, copolymers of TBA with HOST (Scheme IV) are stable to 180 °C as the TGA curve in Figure 5 indicates, allowing baking above their T_g of ~150 °C (*18*). The high thermal stability of the resin allows PAB at upto 180 °C.

Another important issue is the selection of a thermally and hydrolytically stable acid generator (*18*). Figure 6 shows TGA curves of two acid generators, triphenylsulfonium trifluoromethanesulfonate (triflate) and *N*-camphorsulfonyloxy-naphthalimide (CSN). The sulfonium salt derivatives are extremely stable thermally and were thus employed in our initial formulation. The naphthalimide-based acid generators are also attractive because of their excellent thermal and hydrolytic stabilities (*22*). Since this resist requires high temperature bake processes (140-160 °C), use of a bulky acid generator based on a bulky sulfonic acid is necessary to minimize diffusion and evaporation of an acid generator and/or a phtochemically generated acid during the heating processes (*22*). CSN generates a bulky and rather weak camphorsulfonic acid and is stable to 250 °C.

The HOST-TBA copolymer film absorbs NMP at a much slower rate (7.8 ng/min from airstream containing 10 ppb NMP) even when baked at 100 °C (below its T_g of ~150 °C) than a conventional positive chemical amplification resist (18 ng/min) (Table III), which is presumably due to the incorporation of the acrylate structure (solubility parameter consideration) (*15*). However, baking the copolymer film at 170 °C (above its T_g) results in a dramatic reduction of NMP uptake to 1.8 ng/min (Table III), which is similar to the rate for a bare silicon wafer (1.1 ng/min) (*15-18*). Thus, the high temperature bake minimizes NMP absorption in the TBA-HOST copolymer film due to reduced free volume. The spin-cast copolymer film is completely isotropic ($n_\parallel$=1.5388 and $n_\perp$=1.5395) and contains 7.9 wt% residual PMA when baked at 100 °C for 5 min. Heating the film to 170-180 °C reduces the PMA residue to 0.51 % and results in 9.4 % thickness reduction with an increase in the refractive index by ~0.5 % ($n_\parallel$=1.5481 and $n_\perp$=1.5485) (Table III) (*18*).

Table III Poly(TBA-co-HOST) Films Baked at 100 and 170-180 °C for 5 min

bake temp (°C)	residual PMA (wt%)	NMP uptake (ng/min)	thickness (nm)	$n_\parallel$	$n_\perp$
100	7.9	7.8	2084	1.5388	1.5395
170-180	0.51	1.8	1886	1.5481	1.5485

A similar observation was made on *p*-PHOST (*22*), which absorbs a large amount of NMP due to its high T_g (*15*). A film of *p*-PHOST (M_w=14,000 and T_g=168 °C) is anisotrpic ($n_\parallel$=1.5872 and $n_\perp$=1.5887) when baked at 100 °C for 5 min,

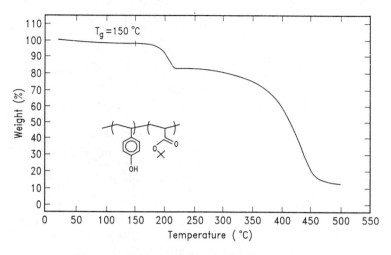

Figure 5 TGA curve of poly(HOST-co-TBA).

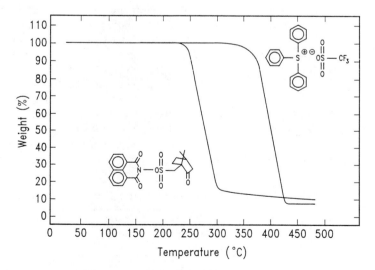

Figure 6 TGA curves of *N*-camphorsulfonyloxynaphthalimide and triphenylsulfonium trifluoromethanesulfonate.

indicating poor annealing, and contains 20.4 wt% residual PMA. Baking the film at 180 °C for 5 min removes the anisotropy ($n_{\parallel}$=1.5984 and $n_{\perp}$=1.5989), increases the refractive index by 0.6-0.7 %, and reduces the NMP uptake by ~40 %.

The poly(TBA-co-HOST) resist was lithographically compared side-by-side with the state-of-the-art positive chemical amplification resist APEX using a GCA KrF excimer laser stepper in an uncontrolled laboratory atmosphere (*16-18*). When postbaked 2 hr after exposure, the new resist produced images that are indistinguishable from the control images obtained with no delay while the comparison resist exhibited a massive skin layer when postbaked 1 hr after exposure. Furthermore, we have observed an overnight stability of the latent image as Figure 7 demonstrates. In this experiment, one exposed wafer was immediately postbaked and developed with a 0.255 N tetramethylammonium hydroxide aqueous solution (Figure 7 top) and the other wafer exposed under the same conditions (21 mJ/cm^2) was allowed to stand on a bench top for ca. 18 hr and then postbaked and developed (Figure 7 bottom). The 0.35 μm line/space images obtained with an overnight delay were identical to the control, indicating excellent contamination resistance of the new resist. The APEX resist exhibits a skin layer when PEB is carried out 30 min after exposure in this environment.

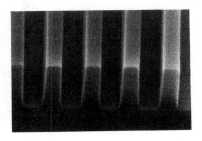

no delay

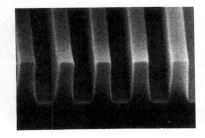

18 hr delay

Figure 7 Positive 0.35 μm line/space patterns printed in ESCAP at 21 mJ/cm^2 on Micrascan II with no PEB delay (top) and an overnight delay (bottom).

The ESCAP resist is environmentally stable due to good annealing achieved by high temperature bake. The reduction of the free volume can rsult in decreased diffusion of a photochemically generated acid in conjunction with the use of a bulky acid generator producing a bulky acid. The ESCAP resist provided a 1.2 μm depth-of focus and 20 % exposure latitude for 0.25 μm line/space patterns on an ASM-L 0.50 NA KrF excimer laser stepper using a 0.27 N tetramethylammonium hydroxide aqueous solution as the developer. Figure 8 shows 0.25 μm line/space patterns obtained at 28 mJ/cm^2 using a Micrascan II. Although the ESCAP resist employs a rather weak camphorsulfonic acid, its sensitivity is quite high, which is due to the high PEB temperature and conversion of ester to carboxylic acid (more soluble than phenol in aqueous base). Due to the high T_g and the high thermal deprotection temperature of the resin, the ESCAP image is devoid of thermal flow at 150 °C for at least 15 min (Figure 9), which is a very important advantage in device fabrication in comparison with other chemical amplification positive resists which suffer from image degradation at <130 °C due to thermal deprotection.

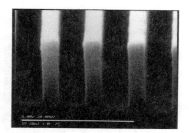

Figure 8   Positive 0.25 μm line/space paterns printed in ESCAP at 28 mJ/cm^2 on Micrascan II.

Figure 9   ESCAP image maintaining its integrity when heated at 150 °C for 15 min.

Acknowledgments

The authors thank H. Truong for her GPC measurements and thermal analyses, R. Johnson for his NMR support, W. Conley, W. Brunsvold, H. Levinson, and D. Taitano for their assistance in Micrascan exposures, and T. Fischer for his process optimization on a ASM-L KrF excimer laser stepper.

Literature Cited

1. Ito, H.; Willson, C. G.; Fréchet, J. M. J. *Digest of Technical Papers of 1982 Symposium on VLSI Technology* **1982**, 86.
2. Ito, H.; Willson, C. G. *Technical Papers of SPE Regional Technical Conference on Photopolymers* **1982**, 331.
3. Iwayanagi, T.; Ueno, T.; Nonogaki, S.; Ito, H.; Willson, C. G. in *Electronic and Photonic Applications of Polymers*; Bowden, M. J.; Turner, S. R., Eds.; Advances in Chemistry Series 218; American Chemical Society: Washington, D. C.,1988, p. 107.
4. Reichmanis, E.; Houlihan, F. M.; Nalamasu, O.; Neenan, T. X. *Chem. Mater.* **1991**, *3*, 394.
5. Ito, H. in *Radiation Curing in Polymer Science and Technology*; Fouassier, J. P.; Rabek, J. E., Eds.; Elsevier: London, 1993, Vol. IV, Chapter 11.
6. Ito, H.; Willson, C. G. in *Polymers in Electronics*; Davidson, T., Ed.; Symposium Series 242; American Chemical Society: Washington, D. C., 1984, p. 11.
7. MacDonald, S. A.; Clecak, N. J.; Wendt, H. R.; Willson, C. G.; Snyder, C. D.; Knors, C. J.; Deyoe, N. B.; Maltabes, J. G.; Morrow, J. R.; McGuire, A. E.; Holmes, S. J. *Proc. SPIE* **1991**, *1466*, 2.
8. Nalamasu, O.; Cheng, M.; Timko, A. G.; Pol, V.; Reichmanis, E.; Thompson, L. F. *J. Photopolym. Sci. Technol.* **1991**, *4*, 299.
9. Kumada, T.; Tanaka, Y.; Ueyama, A.; Kubota, S.; Koezuka, H.; Hanawa, T.; Morimoto, H. *Proc. SPIE* **1993**, *1925*, 31.
10. Oikawa, A.; Santoh, N.; Miyata, S.; Hatakenaka, Y.; Tanaka, H.; Nakagawa, K. *Proc. SPIE* **1993**, *1925*, 92.
11. Röschert, H.; Przybilla, K.-J.; Spiess, W.; Wengenroth, H.; Pawlowski, G. *Proc. SPIE* **1992**, *1672*, 33.
12. Funhoff, D. J. H.; Binder, H.; Schwalm, R. *Proc. SPIE* **1992**, *1672*, 46.
13. Przybilla, K. J.; Kinoshita, Y.; Kubo, T.; Masuda, S.; Okazaki, H.; Padmanaban, M.; Pawlowski, G.; Roeschert, H.; Spiess, W.; Suehiro, N. *Proc. SPIE* **1993**, *1925*, 76.
14. Huang, W.-S.; Kwong, R.; Katnani, A.; Khojasteh, M. *Proc. SPIE* **1994**, *2195*, 37.
15. Hinsberg, W. D.; MacDonald, S. A.; Clecak, N. J.; Snyder, C. D.; Ito, H. *Proc. SPIE* **1993**, *1925*, 43.
16. Ito, H.; England, W. P.; Clecak, N. J.; Breyta, G.; Lee, H.; Yoon, D. Y.; Sooriyakumaran, R.; Hinsberg, W. D. *Proc. SPIE* **1993**, *1925*, 65.

17. Ito, H.; England, W. P.; Sooriyakumaran, R.; Clecak, N. J.; Breyta, G.; Hinsberg, W. D.; Lee, H.; Yoon, D. Y. *J. Photopolym. Sci. Technol.* **1993**, *6*, 547.

18. Ito, H.; Breyta, G.; Hofer, D.; Sooriyakumaran, R.; Petrillo, K.; Seeger, D. *J. Photopolym. Sci. Technol.* **1994**, *7*, 433.

19. Hinsberg, W. D.; MacDonald, S. A.; Snyder, C. D.; Ito, H.; Allen, R. D. in *Polymers for Microelectronics*; Thompson, L. F.; Willson, C. G.; Tagawa, S., Eds.; Symposium Series 537; American Chemical Society: Washington, D. C., 1994, p. 101.

20. Sooriyakumaran, R.; Ito, H.; Mash, E. A. *Proc. SPIE* **1991**, *1466*, 419.

21. Ito, H. *J. Polym. Sci., Polym. Chem. Ed.* **1986**, *24*, 2971.

22. Ito, H.; Breyta, G.; Hofer, D.; Fischer, T.; Prime, B. *Proc. SPIE* **1995**, *2438*, in press.

23. Hinsberg, W. D.; Lee. H. personal communication.

RECEIVED August 8, 1995

Chapter 3

Structure–Property Relationship of Acetal- and Ketal-Blocked Polyvinyl Phenols as Polymeric Binder in Two-Component Positive Deep-UV Photoresists

C. Mertesdorf[1], N. Münzel[1], P. Falcigno[1], H. J. Kirner[1], B. Nathal[1], H. T. Schacht[1], R. Schulz[1], S. G. Slater[2], and A. Zettler[1]

[1]OCG Microelectronic Materials AG, Klybeckstrasse 141, Basel, CH–4002, Switzerland
[2]OCG Microelectronic Materials Inc., 200 Massasoit Avenue, East Providence, RI 02914

Acetals have been used as protecting groups in chemically amplified positve DUV-photoresists in order to render phenol derivatives aqueous base insoluble. In the present study acetals and ketals were investigated as pendant blocking groups in polyvinyl phenols. The impact of the protecting group structure on the protecting group decomposition and the glass transition was studied. Mechanistic considerations were used to explain the protecting group stabilities observed. Two-component resist formulations containing these polymers and a photo acid generator were prepared and exposed on a high NA GCA and a low NA Canon excimer laser stepper. Resolution, DOF, post exposure delay latitude, thermal flow stability and resist decomposition were determined. Secondary reactions of the protecting groups, released in the resist film, were briefly discussed. Some resists exhibited 0.23 μm line/space resolution and a good focus latitude. Approximately two hours post exposure delay latitude and a thermal flow stability of 140°C were obtained with an optimized system.

It is believed that IC device manufacturers will switch from conventional to deep-UV lithography at design rules below 0.3 μm. Most commercial positive deep-UV resists are based on partially protected polyvinyl phenols. Unlike conventional diazonaphthoquinone/novolak technology, here an acid is liberated in a primary photochemical event which catalyzes deblocking of the masked phenols in a subsequent reaction. The technical hurdles typically associated with such chemically amplified resists, such as delay time sensitivity, seemed to be insurmountable. Widely reported problems included dimensional changes of the latent resist image (*1,2*) and the formation of an insoluble surface layer (*3*). These profile deteriorations emerged

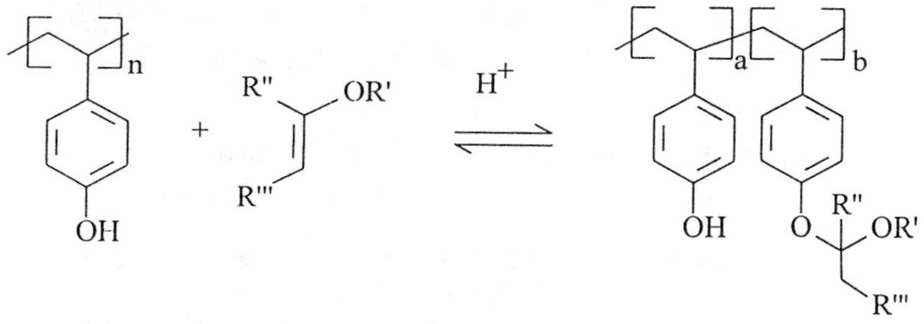

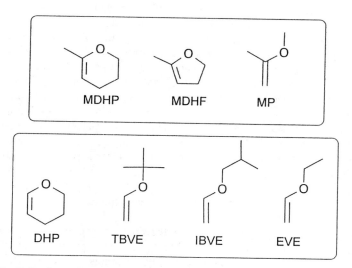

Figure 1. Synthesis of acetal (R''=H) and ketal protected polyvinyl phenols. The degree of protection in mol% is given by (b/a+b) x 100 with n=a+b. Reproduced with permission from reference 37.

Figure 2. Selection of the enol ethers used as protecting groups. Reproduced with permission from reference 37.

upon prolonged intervals between distinct process steps, such as exposure and post exposure bake. However, considerable progress has been made recently, based on a better understanding of the mechanisms causing these deficiencies. Basic material properties, as, for example, the free volume of the resist binder (*4, 5*) and the diffusion length of the acid within the polymer matrix (*6,7*) play a decisive role in terms of surface inhibition and dimensional change, respectively. The deleterious influence of these factors becomes critical as the dimensions shrink. Manufacturing of 256 MB DRAM, for example, requires accurate control of 0.25 μm features throughout the process.

A variety of different protecting groups known in organic synthesis (*8*) have been applied to positive deep-UV photoresists. Among these, carbonates (*9-14*), ethers (*15*), silylethers (*16*) and acetals (*2, 17-19*) are the most frequently used phenol blocking groups. Acid labile esters were attached onto the phenol side group via an ether linkage (*20-22*). Upon deprotection the carboxylic acid is released.

In this study protecting groups of moderate stability, such as acetals and ketals, were considered. These groups were attached as pendant protecting groups onto a polyvinyl phenol backbone. The chemical structure of the protecting groups was varied. Polymer properties and lithographic performance of two-component resist mixtures containing these polymers and a photo acid generator, were evaluated.

Experimental

Materials. Acetal and ketal protected resins were prepared according to a literature procedure (*23*) by an acid catalyzed addition reaction of enol ethers and polyvinyl phenol (PVP) as shown in Figure 1. Enol ethers were obtained commercially. 3,4-Dihydro-6-methyl-2H-pyran (MDHP) was prepared in house according to Perkin (*24*). Figure 2 gives a selection of the enol ethers applied. Different degrees of protection were obtained according to Figure 1 by selecting the appropriate ratio of the starting materials. Upon conversion of the enol ethers listed in the upper box of Figure 2, ketal protected PVP's were derived. Acetal blocked polymers were obtained by treatment of PVP with the vinyl ethers summarized in the lower box. The structural difference between the families of ketals and acetals is confined to the nature of the substituent R'' (Figure 1).

Disulfones (*25*) and sulfonium salts (*26*) were used as the photoacid generators (PAG's). Synthesis of the appropriate PAG's was performed according to literature procedures.

Thermal properties of the polymers were recorded on a Perkin Elmer 7 series thermal analyzer. Degrees of protection were obtained either from the weight loss observed in the thermogravimetric run (TGA), or from inverse gated decoupling ^{13}C-NMR (Bruker AM-400). Good agreement was found between the two methods. Molecular weights were determined by gel permeation chromatography (GPC) using a Hewlett-Packard-liquid chromatograph (HP 1090) equipped with three PSS-gel GPC columns (Polymer Standards Service: 10^{-4}Å, 10^{-3}Å, 10^{-2}Å, 5 μm). Calibration was based on polystyrene standards obtained from PSS. UV detection operating at 254 nm was applied.

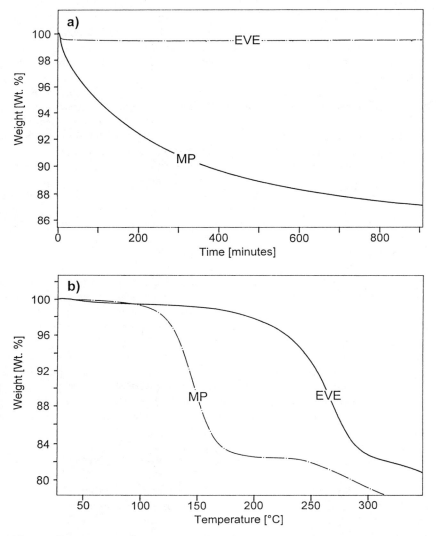

Figure 3. TGA traces of acetal (EVE) and ketal (MP) protected PVP (same degree of protection): a) isothermal at 90°C, b) heating rate 10°C/min. Reproduced with permission from reference 37.

Lithography. Resist formulations were prepared by dissolving the polymer along with the PAG in propylene glycol monomethyl ether acetate (PGMEA) or in methyl-3-methoxypropionate (MMP). All solutions were filtered through 0.2 µm teflon membrane filters. The substrates were vapor primed with HMDS either in a Y.E.S. oven at 125°C or by liquid dispense prior to resist coating.

Resist solutions were spin coated onto silicon substrates. Film thickness was measured interferometrically after soft bake (SB) using a refractive index of 1.59 (@633 nm). Different film thicknesses between 0.6 and 0.95 µm, corresponding to maximum and minimum light incoupling were evaluated.

Exposures were performed on a Canon FPA 4500 (NA=0.37, 5x reduction optics) or on a GCA XLS (NA=0.53, 4x reduction) DUV stepper. Both steppers were equipped with KrF excimer lasers, operating at 248.4 nm.

After a post exposure bake (PEB), the wafers were developed with OPD 4262 (0.262 N tetramethyl ammonium hydroxide) by a 60 s immersion, or by a spray/puddle process (3-5 s/20-50 s). A 20 s rinse with deionized water and a spin dry were applied after development.

Results and Discussion

Impact of the Protecting Group Structure on Thermal Decomposition. The chemical structure of the protecting groups, as outlined in Figures 1 and 2, is one of the factors which governs their rate of decomposition. Another factor, which will be addressed later, is the amount of residual unprotected phenols, or vice versa, the degree of protection (see Figure 1). If one thinks in terms of the residual free phenolic hydroxyl groups in the copolymer acting as a weakly acidic environment, the situation can be considered as low molecular weight acetal or ketal compounds dissolved in a weakly acidic solution. Data concerning the hydrolysis of acetals and ketals in an acidic aqueous media have been known for over 70 years (*27*). The authors found that acetone diethyl ketal hydrolyzed 2500 times faster than acetaldehyde diethyl acetal, which proceeds opposite to their rate of formation. Formation is driven by the electrophilic activity of the carbonyl carbon, which is known to be higher in the case of the aldehyde. Acidolysis of acetals versus ketals corresponds also to the general observation that the hydrolysis rates of ethers increase with an increasing number of carbons linked to the α-carbon adjacent to the ether oxygen (*28, 29*). This goes with the propensity of the alkyl group to be released as a cation from the ether oxygen via the intermediate oxonium ion state. Due to inductive effects, stability of tertiary cations is superior to secondary and primary ones. Steric effects play a further role. Strain is relieved by protonation of the oxygen atom and subsequent cleavage of the oxygen-carbon bond (*30, 31*).

The same stability sequence is observed for acetals (EVE) and ketals (MP), when attached to the PVP backbone, as measured by TGA (Figure 3). The ketal protected polymer loses the majority of its protecting groups, when stored for 15 hours at 90°C. In contrast to this result, the weight loss observed for the acetal resin is only 0.4% under the same conditions (Figure 3a). This weight loss can be attributed to loss of adsorbed water. In TGA runs recorded with a heating rate of 10°C/min, decomposition onset of the acetal is approximately 75°C higher than the onset recorded for the ketal analog (Figure 3b). Due to the impact of residual phenolic sites

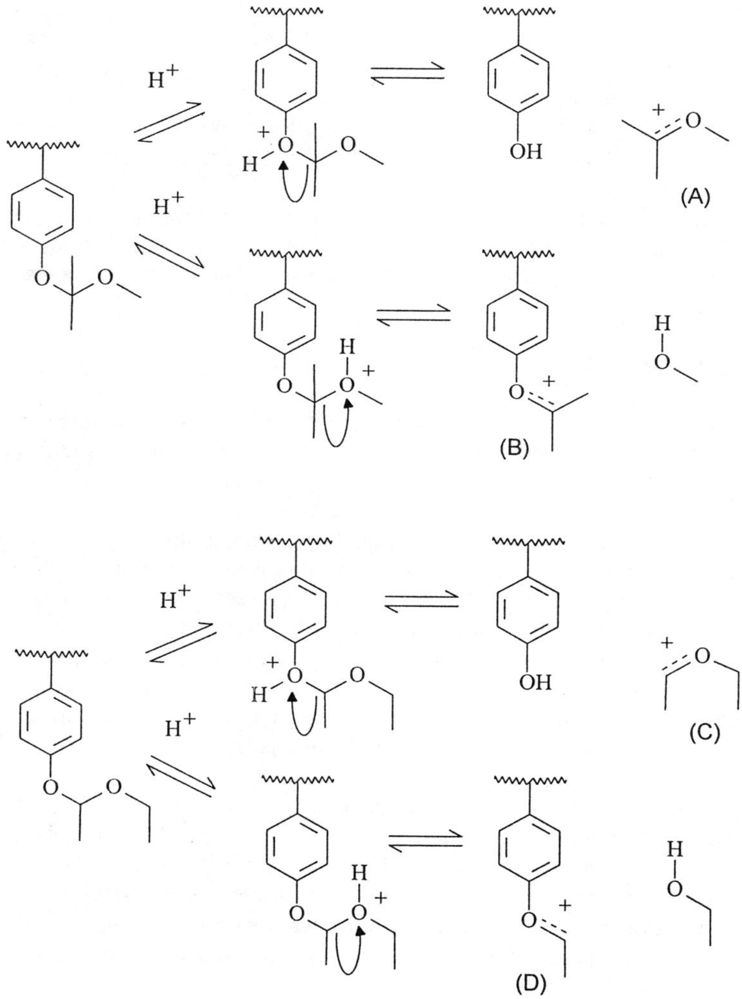

Figure 4. Mechanism of proton assisted decomposition of EVE and MP protected PVP.

on the polymer stability (*32, 33*), the same degrees of protection were considered. EVE and MP have a similar structure with respect to the nature of R' and R'''.

A mechanism of proton assisted decomposition of EVE and MP protected PVP, which accounts for these observations, is depicted in Figure 4. Protonation can either occur at the alkyl phenyl ether oxygen or at the aliphatic ether oxygen site, which depends on their relative base strengths. In any of the two cases, the intermediate oxocarbenium ion is considerably more stabilized in the case of the ketal (A, B) than in the case of the acetal (C, D). Moreover, it is reasonable to assume that ketal oxygens are more basic than acetal oxygens, according to a base strength study of some aliphatic ethers in aqueous sulfuric acid (*30*).

As mentioned above, protecting group decomposition is further affected by the amount of residual unprotected phenols in the copolymer. This is illustrated in Figure 5 for the MP-PVP system. The temperature at which deprotection occurs increases with increasing degree of protection. This dependency has already been described in detail for the t-BOC-PVP system (*32, 33*) and was also found for the acetal and ketal protected resins investigated in this study. This seems to be a general relationship which is independent from the nature of the protecting group. More likely, it is due to a common structural element within such polymers, as the weakly acidic phenols, which act as a deprotection catalyst (auto-catalytic effect (*32, 33*)). Phase separations, due to immiscible polymers formed during the protection reaction, were not observed in the present study. Irregularities of decomposition temperatures within a series of partially protected t-BOC-PVP's were caused by such phase separations (*32*).

The TGA weight loss in the temperature range given in Figure 5 corresponds well with the theoretically expected weight loss, if a total thermolysis of the protecting groups is considered. Although somewhat higher values were obtained by NMR, NMR and TGA correlate well. The small difference can be ascribed to some non volatile decomposition products formed upon heating. The values given were derived from TGA.

Isothermal recording at 90°C (Figure 5b) reveals that the rate of blocking group thermolysis is not linearly dependent on the degree of protection. MP-PVP's with 73, 67, 59 and 52 mol% protection exhibit a linear relationship between weight loss and thermolysis time, with increasing slope for decreasing blocking. In contrast to this, 34 mol% protected MP-PVP shows already a strongly nonlinear behavior with 40 % of its total protecting groups lost after 200 min. In the case of 67 mol% protection, only 7.5 % of the total blocking groups are released under the same conditions. Figure 6 shows a plot of the ratio of released and total blocking groups versus the degree of protection for respective MP-PVP's stored for 200 min at 90°C. The curve is fitted well by an exponential function. At degrees of protection below 50 mol% rapid decomposition occurs and the stability required is not provided. If this auto-acceleration is due to a hyperacidity effect, as known from high ortho novolaks (*34*), is very unlikely. In PVP's with an all-para structure, phenols are radially distributed from the backbone and intramolecular hydrogen bonding is less likely to occur. However, weaker consecutive hydrogen bonds between individual chains may develop upon deprotection, which may increase acidity and thus the deprotection rate in a non linear fashion.

Release of the ketal-carbon-oxygen or acetal-carbon-oxygen bonds, as depicted in Figure 4, explains the stability difference observed between ketals and acetals.

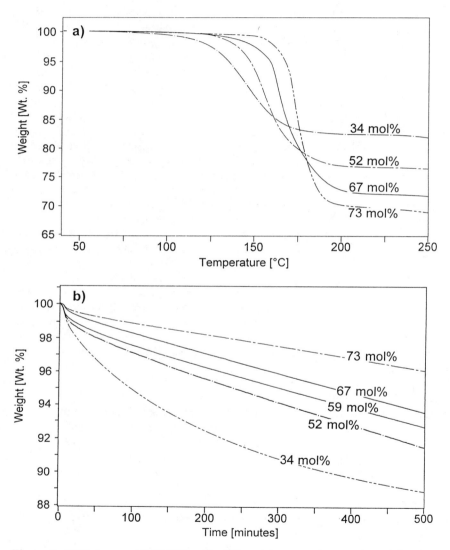

Figure 5. TGA traces of MP-PVP with different degrees of protection: a) heating rate 10°C/min, b) isothermal at 90°C. Reproduced with permission from reference 37.

Cleavage of the O-R' bond could be competitive when R' is a tertiary radical, in accordance with the high stability of the intermediate tertiary carbon cation of R' formed (*29*). In fact, the decomposition temperature drops, if a primary carbon radical R' (EVE) is substituted by a tertiary one (TBVE). In the case of TBVE-PVP, decomposition onset was shifted to lower temperatures by roughly 50°C with respect to EVE-PVP (Figure 7). A mechanistic consideration of this competitive path is given in Figure 8. Proton catalyzed decompostion yields the considerably stabilized tertiary butyl cation intermediate and the hemi acetal. The hemi acetal is known to hydrolyze very fast to acetaldehyde (*28*) releasing, in this particular case, the phenol moiety.

Six membered cyclic structures can adopt a strain free chair or boat configuration. Taking this into consideration, Tdec's of MDHP protected PVP's are expected to be in the range of non-cyclic ketals with primary alkoxy carbons R', as for example MP-PVP. However, lower decompositon onsets were found with the cyclic protecting group (MDHP), as shown in Figure 9. It is likely that the ring attached to the phenol via its 6-position is not able to be completely strain free, due to restrictions implied by the backbone. The same tendency was observed, when the respective cyclic and non-cyclic acetals, such as DHP-PVP and EVE-PVP, were compared, with EVE-PVP being the more stable system.

Five membered rings of the tetrahydrofuran type are rather planar and therefore strained, which makes them much more prone to proton catalyzed ring opening ether cleavage than the tetrahydropyran ring. Arnett at al. discussed the contribution of strain effects in cyclic ethers to their base strengths (*31*). An interpretation of their observation of tetrahydrofuran being considerably more basic than tetrahydropyran is given in terms of electron correlation forces between non-bonded electron pairs on the oxygen and the electrons engaged in bonding the hydrogens to the adjacent carbons. The repulsion of the oxygen lone pair for nearby bonding pairs is reduced upon coordination with a proton. This effect is supposed to be considerably larger in the case of tetrahydrofuran whose unshared electrons are constrained to orbitals that are eclipsed by the adjacent bonds to hydrogen. In the case of tetrahydropyran, the orbitals are staggered to the adjacent carbon-hydrogen bonds. Bonding of the cyclic ethers to the PVP-backbone changes their conformation and thus the base strengths of the endocyclic oxygens. Also introduction of the methyl group - when considering ketals with respect to acetals - affects the basicities of furan and pyran type protecting groups to a different extent. However, the relative basicities are maintained in the ketal series, as deduced from the protecting group stabilities observed. Thus, MDHF protected PVP is less stable than its pyran analog (MDHP-PVP). According to Hesp et al., who investigated the acetal analogs (DHF-PVP and DHP-PVP), a higher photospeed can be expected in the case of the more basic furan protecting group (*18, 19*). In the present study photospeeds of the cyclic ketals were not determined, due to fast decomposition of MDHF and MDHP protected PVP occurring in the TGA experiments which gives rise to a low stability of the corresponding resist formulations.

Crosslinking between individual polymer chains has to be considered in the protection reaction outlined in Figure 1. Crosslinking occurs, when a polymer immobilized hydroxy group adds, for example, to the oxocarbenium ion intermediate (see Figure 4: A,B,C and D) which forms from the protected phenol moiety. Further routes can be imagined, as the foregoing formation of a phenyl vinyl ether group via

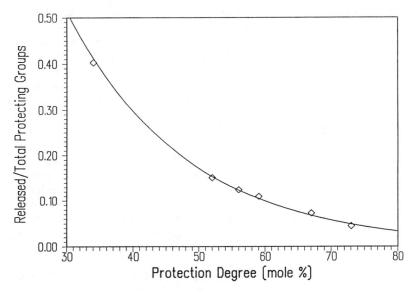

Figure 6. Thermolysis of MP-PVP dependent on the degree of protection; values derived from Figure 4b (200 min at 90°C). Reproduced with permission from reference 37.

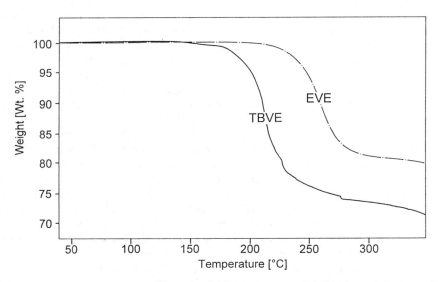

Figure 7. TGA traces of EVE and TBVE protected PVP (same degree of protection): heating rate 10°C/min. Reproduced with permission from reference 37.

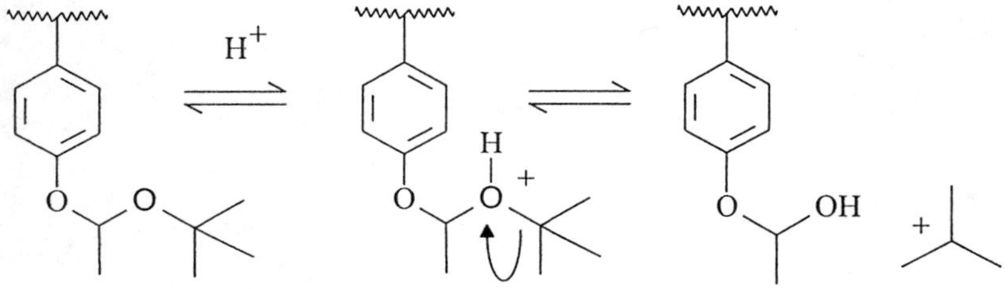

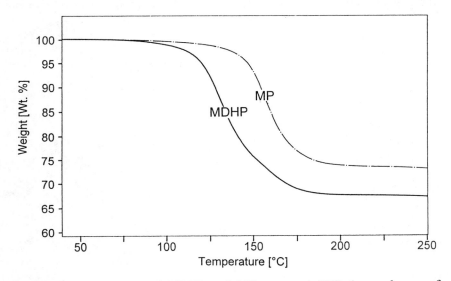

Figure 8. Mechanism of competitive proton assisted decomposition path of TBVE protected PVP.

Figure 9. TGA traces of MDHP and MP protected PVP (same degree of protection): heating rate 10°C/min. Reproduced with permission from reference 37.

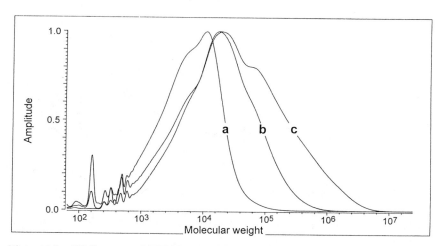

Figure 10. GPC traces of TBVE-PVP dependent on the amount of catalyst used in the synthesis (x 10^{-4} mol%): a) 31, b) 256, c) 784; values given are relative to PVP.

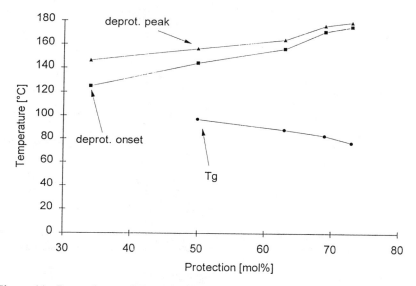

Figure 11. Dependence of Tg and Tdec (onset and peak) on the protection degree of MP-PVP; values derived from DSC and TGA: heating rate 10°C/min. Reproduced with permission from reference 37.

proton release from the oxocarbenium ion and its subsequent conversion with phenols from adjacent polymer chains. The extent of any route depends on the rate constants of the individual reaction step. According to our results, the acidolytic lability of the O-R' linkage is one of the factors which determines the rate of crosslinking. This would mean that acidolysis of the O-R' linkage with formation of the hemi-acetal is the rate determining step, as depicted, for example, in Figure 8. The GPC traces given in Figure 10 illustrate crosslinking as occurring in the synthesis of TBVE-PVP. An increase of the molecular weight is observed at long reaction time (60 h) and higher catalyst concentration. This cannot be explained by simply an addition of the vinyl ethers to some of the phenol side groups. Rough conditions (prolonged reaction time, high amount of catalyst) can cause gelation of the reaction mixture.

It is interesting to note that DHP does not show branching under the conditions applied. Presumably, this is due to the O-R' subunit being linked to the acetal carbon (R' and R''' together form a trimethylene bridge) which makes the back reaction (reformation of pyran-ring) likely to occur.

Impact of the Protecting Group Structure on the Glass Transition. The glass transition (Tg) governs the flow stability of the profiles generated in a resist film. A lack in flow stability causes failures during pattern transfer in the plasma etch. On the other hand, it is desirable to have a process window available between Tg and decomposition, to be able to effectively anneal the resist film at temperatures high enough above Tg, without decomposing the material. Annealing of the resist film reduces the free volume and thus the penetration of ambient contaminants (*4*). Further, the diffusion length of the acid should be reduced, if the free volume is decreased.

Protecting groups with large and flexible carbon radicals (R' and R''') are expected to affect packing of the polymer backbone more than small, compact radicals do. This results in a larger mobility of the chain segments for a given temperature, due to less interchain interactions, which translates to a lower Tg.

How Tg depends on the degree of protection is given in a plot in Figure 11 for MP-PVP together with the respective decomposition temperatures (see preceding section). Upon increasing the ratio of protected phenol units, the Tg decreases. This Tg dependency is known for partially protected PVP's and has already been published for t-BOC-PVP (*33*). In the case of 35 mol%, MP-PVP decomposes before the Tg is reached. In order to have a process window available between Tg and decomposition, degrees of protection of at least 50 mol% are required for MP-PVP (see also preceding section).

An example of the lithographic performance derived with a two-component resist based on MP-PVP (55 mol% protection) and 3% PAG is given in the next section (Figure 14). The resist shows high resolution and vertical wall profiles. However, as already pointed out in the previous section, fast decomposition of MP-PVP limits shelf life of the resist composition. Partial decomposition was also observed on the hotplate during processing. The degree of protection was decreased in the unexposed regions after the post exposure bake. It turned out that, despite the considerably high degree of protection of MP-PVP and its low Tg (see Figure 11), thermal flow stability of the resist was higher than 130°C. This can be ascribed to the fact that the resist decomposes on the hotplate and the Tg gradually increases upon volatilization of MP from the resist film (see next section).

A large window between Tg and decomposition was found, when protecting groups of higher stability, such as acetals, were attached to the PVP backbone. Figure 12 shows decomposition and Tg versus degree of protection-plot obtained for the EVE-PVP system. In the case of compact protecting groups, such as MP, Tg changes only slightly when the degree of protection is varied over a wide range. From Figure 11 it can be deduced that this range is roughly 20°C between 50 mol% and 73 mol% protection. In contrast to this, Tg of the copolymer is more degraded by the EVE protecting group than by an equal amount of MP. This tendency is continued when applying IBVE which distorts packing of the polymer backbone even more.

A considerably small Tg degradation was observed with the cyclic protecting groups, such as DHP. This is due to the number of possible conformations being limited in the case of cyclic hydrocarbons with respect non-cyclic systems composed of same number of carbons. The results is a smaller distortion of backbone packing with the conformation limited cyclic protecting groups.

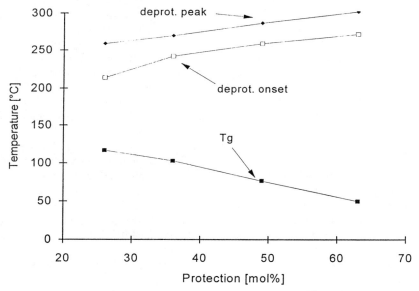

Figure 12. Dependence of Tg and Tdec (onset and peak) on the protection degree of EVE-PVP; values derived from DSC and TGA: heating rate 10°C/min. Reproduced with permission from reference 37.

Secondary Reactions of the Protecting Groups. Since acetal/ketal-formation/decomposition is an equilibrium, secondary reactions of the protecting groups liberated in the resist film have to be considered. Secondary reactions were only sparingly discussed in the literature and correlated with the lithographic performance (*35, 36*). They depend on the volatility of the protecting group fragment(s) and the concentration of further reactants affecting equilibrium (acid, PVP-phenol, water).

In the case of cyclic blocking groups (DHP, MDHP and MDHF) only one fragment is released upon acidolysis, since the alcohol component (O-R') is covalently linked to the acetal or ketal carbon atom. A study of how DHP-decomposition products affect resist behavior has been published by Sakamizu et al. for DHP-PVP/Novolak (*35*). Non-cyclic protecting groups (MP, TBVE, IBVE and EVE) can decompose into two fragments, e.g. acetone and methanol in the case of MP. The primary fragment is determined by the oxygen site, which is protonated first, and by the direction of cleavage. This, in turn, is dependent on factors, such as the basicity of the oxygens, the relief of strain and the stability of the intermediate oxonium ion or carbocation etc.. The situation is outlined in Figure 13 for MP-PVP, with MP being the primary fragment.

According to GC-MS measurements, MP (1) is released upon pyrolyzing MP-PVP (Mertesdorf, C.; Mareis, U.; Schaedeli, U. unpublished data.). MP (1) is the primary decomposition product if protonation and subsequent cleavage occurs at the phenyl alkyl ether oxygen site with elimination of a proton from the oxocarbenium ion intermediate (Figure 13a). Under the conditions present within the exposed parts of the resist film, MP (1) is in an equilibrium state with acetone (4). Acetone (4) can undergo self-condensation in an aldol-type reaction. Acetone (4) and its condensates (6), (7) were detected by GC-MS of an exposed resist film (Mertesdorf, C.; Mareis, U.; Schaedeli, U. unpublished data.). Of course, MP must not nessessarily be the primary decomposition product in the resist film, as it is the case when pyrolyzing the pure resin. However, secondary reactions of the protecting groups released have been outlined in Figure 13a, starting with MP for simplicity. MP (1) undergoes rapid polymerization in the presence of strong acids, e.g. sulfonic acids. In the absence of water, cationic polymerization (Figure 13b) has to be considered as a further possible reaction, due to the MP-acetone equilibrium (Figure 13a) being shifted to the MP side (1).

Lithography

The SEM micrograph of 0.25 µm line/space features, generated with a two component resist containing MP-protected polyvinyl phenol and 3% PAG, is shown in Figure 14. Strong standing waves are apparent and are found to increase with the degree of protection. Standing waves are characteristic for a diffusion limited system. The standing wave phenomenon could be caused by a portion of the MP molecules formed being consumed in an oligomerization process which is initiated by excess of the photo generated acid. As long as a growing chain (8) is not terminated by a protic reagent containing nucleophilic sites (e.g. water, alcohol), the proton consumed is not

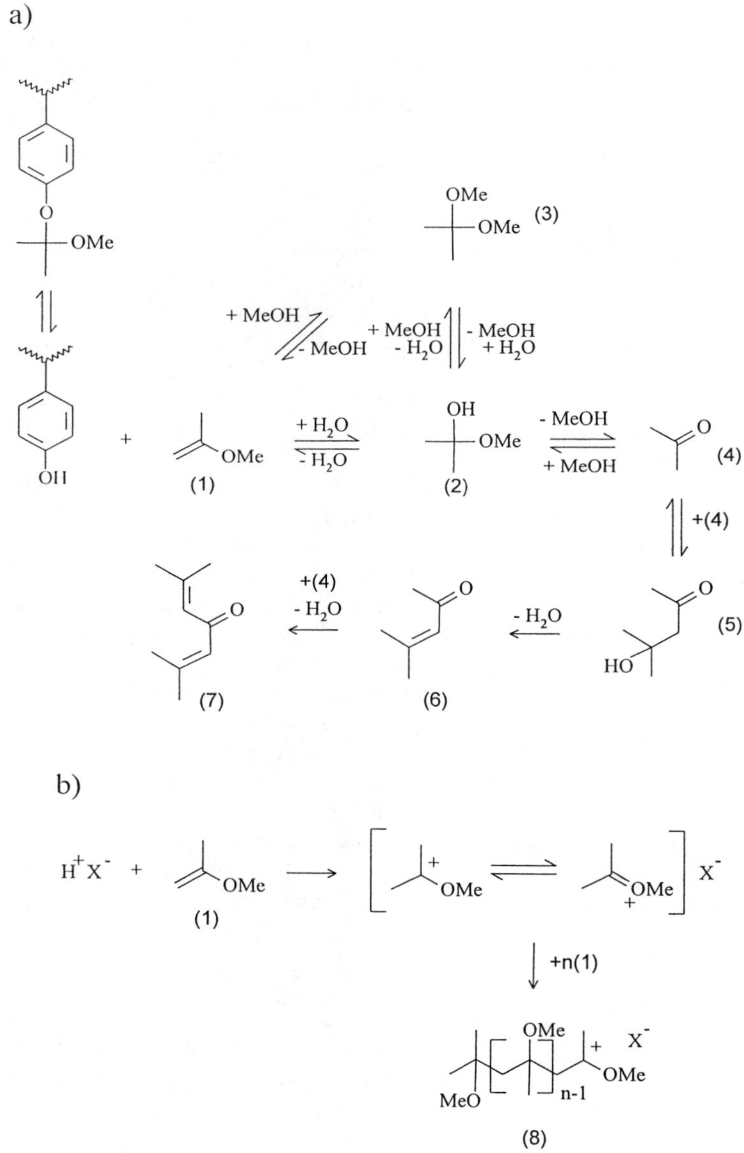

Figure 13.　Acid-catalyzed-secondary reactions of MP-PVP. Reproduced with permission from reference 37.

liberated. The diffusion of the growing chain decreases with increasing chain length, which could be the reason for the limited diffusion. It can be expected that the MP-oligomers suggested do not act as strong dissolution inhibitors in the exposed regions of the resist film.

Although high resolution (0.23 μm), appreciable DOF and delay time robustness was achieved with MP-PVP based resists, tradeoffs with respect to shelf life and decomposition during processing limited their applicability. These tradeoffs were surmounted by using acetal chemistry, which is much more tolerant with respect to thermal decomposition. Processing and lithographic performance of acetal based resists, designated *Acetal-1* and *Acetal-2* are given below. These resists are compatible with process conditions used for conventional photoresists. Table I summarizes the conditions applied for *Acetal-1*.

Table I. Process conditions used for *Acetal-1*

Process Step	Process Conditions
Softbake (SB)	60 s 130°C, vacuum hot plate
Film thickness	0.82 μm
Exposure	GCA XLS DUV stepper, NA = 0.53, ~18 mJ cm^{-2}
Post exposure bake (PEB)	60 s 110°C, vacuum hotplate
Development	OPD 4262 (0.262 N TMAH), 5 s spray/30 s puddle

Soft bake (SB) is performed above Tg and below the thermal decomposition of the resist. A bake window of approximately 30°C was used for SB optimization. Although the resists are characterized by a large post exposure bake (PEB) latitude, optimization of the PEB temperature is critical in order to minimize proximity effects and CD shrinkage during post exposure delay (PED). The resists are compatible with 2.38 % aqueous TMAH developers. Less than 2 nm of dark erosion were observed after a 60 s development. Track development was not yet optimized, however, good results were obtained with a 5 s spray/30s single puddle process with *Acetal-1*.

The coded mask line width of bright and dark field line space (l/s) features and the isolated lines and trenches were printed down to 0.23 μm on silicon with the NA 0.53 exposure tool. As shown in Figure 15 and 16, 0.23 μm l/s were linearly resolved within a ± 10% CD with *Acetal-1* and *Acetal-2*. Figure 17 gives the depth of focus (DOF) obtained with *Acetal-1* which is 0.8 μm at 0.25 μm l/s (bright field), with the non-optimized spray/puddle development process.

PED experiments were performed at different sites with different steppers. Although deblocking proceeds without PEB at room temperature, a PEB step is required as a reset in order to minimize CD change. Roughly 2 h PED latitude were obtained within a ± 10% CD with the two resists.

Acetal-1 provides a thermal flow stability of 140°C on a vacuum hotplate, as shown in Figure 18.

——— 0.5 μm

Figure 14. SEM micrograph of 0.25 μm line/space features (NA=0.53) obtained with a 2-component resist containing MP-PVP (55 mol% protection) and 3% PAG. Reproduced with permission from reference 37.

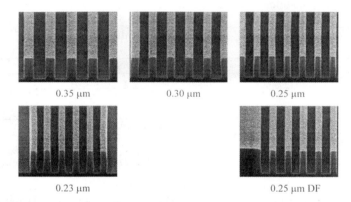

Figure 15. Resolution/linearity of *Acetal-1*. Reproduced with permission from reference 37.

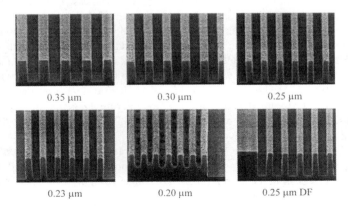

Figure 16. Resolution/linearity of *Acetal-2*: film thickness = 0.82 μm, NA = 0.53, Dev.: 0.262 TMAH im..

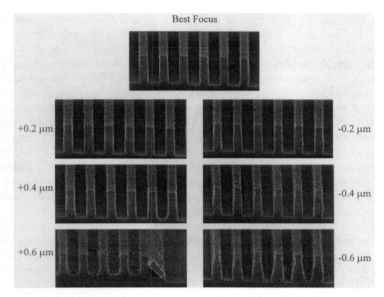

Figure 17. DOF of *Acetal-1*: 0.25 μm l/s (bright field). Reproduced with permission from reference 37.

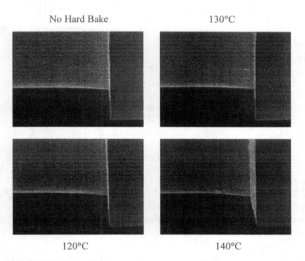

Figure 18. Thermal flow stability of *Acetal-1*: 120 s on a vacuum hotplate. Reproduced with permission from reference 37.

Conclusion

Different enol ethers yielding acetal and ketal protected PVP's were investigated as pendant blocking groups in positive DUV-resists. As expected, ketal polymers decompose much faster than their structural acetal analogs. Within a ketal or acetal family, decomposition rates vary according to their structure which can be ascribed to factors, such as the base strength of the oxygens, the relief of strain and the stability of cationic intermediates. Tg of PVP is degraded by indroduction of the pendant protecting group. Degradation is dependent on the degree of protection and on the conformational freedom of the respective ketal or acetal moiety. Secondary reactions of the protecting groups released in the resist film have to be considered in the structural design of acetal and ketal protected PVP's. Although ketals showed appreciable resolution, a considerable tradeoff with respect to shelf life and process stability was determined. Preoptimized resist formulations based on acetal chemistry exhibit 0.23 μm line/space resolution, 0.8 μm focus latitude at 0.25 μm resolution and approximately two hours post exposure delay latitude.

Acknowledgments. We would like to acknowledge C. De Leo, D. Frey and J.P. Unterreiner for lithographic work. Thank you to O. Nalamasu (AT&T Bell Labs) and A.G. Timko (AT&T Bell Labs) for exposures on the high NA stepper and U. Schaedeli (Ciba) and T.X. Neenan (AT&T Bell Labs) for helpful discussions.

References

1.   Padmanaban, M.; Kinoshita, Y.; Kudo, T.; Lynch, T.; Masuda, S.; Nozaki, Y.; Okazaki, H.; Pawlowski, G.; Przybilla, K.J.; Roeschert, H.; Spiess, W.; Suehiro, N. *J. Photopolym. Sci. Technol.* **1994**, *7*, 461.
2.   Hattori, T.; Schlegel, L.; Imai, A.; Hayashi, N.; Ueno, T. *J. Photopolym. Sci. Technol.* **1993**, *6*, 497.
3.   Hinsberg, W.D.; MacDonald, S.A.; Clecak N.J.; Snyder, C.D. *Proc. SPIE, Adv. Resist Technol. Process. IX* **1992**, *1672*, 24.
4.   Ito, H.; England, W.P.; Sooriyakumaran, R.; Clecak, N.J.; Breyta, G.; Hinsberg, W.D.; Lee, H.; Yoon, D.J. *J. Photopolym. Sci. Technol.* **1993**, *6*, 547.
5.   Ito, H.; Breyta, G.; Hofer, D.; Sooriyakumaran, R.; Petrillo, K.; Seeger, D. *J. Photopolym. Sci. Technol.* **1994**, *7*, 433.
6.   Yoshimura, T.; Nakayama, Y.; Okazaki, S. *J. Vac. Sci. Technol. B* **1992**, *10*, 2615.
7.   Asakawa, K. *J. Photopolym. Sci. Technol.* **1993**, *6*, 505.
8.   Greene, T.W. *Protecting Groups in Organic Synthesis;* John Wiley & Sons: New York, 1981.
9.   Eib, N.K.; Barouch, E.; Hollerbach, U.; Orszag, S.A. *Proc. SPIE, Adv. Resist Technol. Process. X* **1993**, *1925*, 186.
10.   Sturtevant, J.; Holmes, S.; Rabidoux, P. *Proc. SPIE, Adv. Resist Technol. Process. IX* **1992**, *1672*, 114.
11.   Ito H.; Willson, C.G. *Polym. Eng. Sci.* **1983**, *23*, 1012.
12.   Münzel, N.; Holzwarth, H.; Falcigno, P.; Schacht, H.T.; Schulz, R.; Nalamasu, O.; Timko, A.G.; Reichmanis, E.; Kometani, J.; Stone, D.R.; Neenan, T.X.; Chandross, E.A.; Slater, S.G.; Frey M.D.; Blakeney, A. *Proc. SPIE, Adv. Resist Technol. Process. XI* **1994**, *2195*, 47.

13. Nalamsu, O.; Timko, A.G.; Cheng, M.; Kometani, J.; Galvin, M.; Heffner, S.; Slater, S.G.; Blakeney, A.J.; Münzel, N.; Schulz, R.; Holzwarth, H.; Mertesdorf, C.; Schacht, T. *Proc. SPIE, Adv. Resist Technol. Process. X* **1993**, *1925*, 155.

14. Nalamasu, O.; Cheng, M.; Timko, A.G.; Pol, V.; Reichmanis, E.; Thompson, L.F. *J. Photopolym. Sci. Technol.* **1991**, *4*, 299.

15. Crivello, J.V.; Conlon, D.A.; Lee, J.L. *Polym. Mater. Sci. Eng.* **1989**, *61*, 422.

16. Murata, M.; Kobayashi, E.; Yamachika, M.; Kobayashi, Y.; Yumoto, Y.; Miura, T. *J. Photopolym. Sci. Technol.* **1992**, *5*, 79.

17. Hattori, T.; Schlegel, L.; Imai, A.; Hayashi, N.; T. Ueno, T. *Proc. SPIE, Adv. Resist Technol. Process. X* **1993**, *1925*, 146.

18. Hesp, S.A.M.; Hayashi, N.; Ueno, T. *J. Appl. Polym. Sci.* **1991**, *42*, 877.

19. Hayashi, N.; Hesp, S.M.A.; Ueno, T.; Toriumi, M.; Iwayanagi, T.; Nonogaki, S. *Polym. Mater. Sci. Eng.* **1989**, *61*, 417.

20. Onishi, Y.; Oyasato, N.; Niki, H.; Hayase, R.H.; Kobayashi, Y.; Sato, K.; Miyamura, M. *J. Photopolym. Sci. Technol.,* **1992**, *5*, 47.

21. Sinta, R.; Hemond, R.C.; Medeiros, D.R.; Rajaratnam, M.M.; Thackeray, J.W.; Canistro, D.; *United States Patent* **1993**, 5,258,257.

22. Thackeray, J.W.; Adams, T.; Fedynyshyn, T.H.; Georger, J.; Hemond, R.; Madeiros, D.; Mori, J.M.; Orsula, G.W.; Sinta, R.F.; Small, R.D. *J. Photopolym. Sci. Technol.* **1993**, *6*, 645.

23. Shiraishi, H.; Hayashi, N.; Ueno, T.; Sakamizu, T.; Murai, F. *J. Vac. Sci. Technol. B* **1991**, *9*, 3343.

24. Perkin, W.H. *J. Chem. Soc (London)* **1887**, *51*, 702.

25. Aoai, T.; Aotani, Y.; Umehara, A.; Kokubo, T. *J. Photopolym. Sci. Technol.* **1990**, *3*, 389.

26. DeVoe, R.J.; Olofson, P.M.; Sahyun, M.R.U. *Adv. Photochem.* **1992**, *17*, 313.

27. Skrabal, A.; Mirtl, K.H. *Z. Phys. Chem. (Leibzig)* **1924**, *111*, 98.

28. Skrabal, A.; Schiffrer, A. *Z. Phys. Chem. (Leibzig)* **1921**, *99*, 290.

29. Tronow, B.W.; Ladigina, L.W. *Ber. Dtsch. Chem. Ges.* **1929**, *62*, 2844.

30. Arnett, E.M.; Wu, C. Y. *J. Am. Chem. Soc.* **1962**, *84*, 1680.

31. Arnett, E.M.; Wu, C. Y. *J. Am. Chem. Soc.* **1962**, *84*, 1684.

32. Ito, H. *J. Polym. Sci., Part A* **1986**, *24*, 2971.

33. Paniez, P.J.; Rosilio, C.; Mouanda, B.; Vinet, F. *Proc. SPIE, Adv. Resist Technol. Process. XI* **1994**, *2195*, 14.

34. Dammel, R. In *Diazonaphthoquinone-based Resist*, O'Shea, D.C., Ed.; SPIE Tutorial Texts in Optical Engineering; SPIE Optical Engineering Press: Bellingham, Washington 1993, Vol. TT 11; 41-48.

35. Sakamizu, T.; Shiraishi, H.; Yamagushi, H.; Ueno, T.; Hayashi, N. *Jpn. J. Appl. Phys.* **1992**, *31*, 4288.

36. Funhoff, D.J.H.; Binder, H.; Dijkstra, H.J.; Goethals, M.; Krause, A.; Moritz, H.; Reuhman-Huisken, M.E.; Schwalm, R.; Van Driessche, V.; Vinet, F. *Proc. SPIE, Adv. Resist Technol. Process. X* **1993**, *1925*, 53.

37. Mertesdorf, C.; Münzel, N.; Holzwarth, H.; Falcigno, P.; Schacht, H.T.; Rohde, O.; Schulz, R.; Slater, G.S.; Frey, D.; Nalamasu, O.; Timko, A.G.; Neenan, T.X., "Structural Design of Ketal and Acetal Blocking Groups in 2-Component Chemically Amplified Positive DUV-Resists" in *Advances in Resist Technology and Processing XII*, Robert D. Allen, Editor, Proc. SPIE 2438, 84-98 **1995**.

RECEIVED August 9, 1995

Chapter 4

Lithographic Effects of Acid Diffusion in Chemically Amplified Resists

C. A. Mack

FINLE Technologies, P.O. Box 162712, Austin, TX 78716

Chemically amplified resists, based on generation of acid during exposure followed by an acid catalyzed reaction during a post-exposure bake which changes the solubility of the photoresist in developer, are discussed and their kinetics are modeled mathematically. Acid generation is assumed to follow standard first order kinetics. Amplification is modeled as a reaction of the acid with the polymer, first order in polymer reactive sites and of arbitrary order in acid. Acid loss is considered with four common pathways: atmospheric base contamination, evaporation of acid from the top of the film, neutralization of the acid at the substrate, and bulk acid quenching. All four are discussed and two are mathematically described. Diffusion of the acid is an integral part of the reaction process. In addition, diffusion is necessary to smooth out standing waves. However, excessive diffusion will cause a degradation of the latent image and a reduction in process latitude. The effects of acid diffusion are investigated by providing a full solution to the reaction-diffusion system of equations for the kinetics of chemically amplified resists. This reaction-diffusion model includes the possibility of a reaction dependent acid diffusivity.

Chemically amplified resists have emerged as the most likely class of resist chemistries for use in Deep-UV lithography. First proposed by Ito and Willson (*1*), these resists are based on the generation of acid during exposure to light, followed by an acid catalyzed reaction during a post-exposure bake which changes the solubility of the photoresist in developer. For such systems, one molecule of photogenerated acid can cause many (possibly hundreds) of subsequent reactions, thus the name "chemically amplified." An important aspect of the chemical mechanism of

0097–6156/95/0614–0056$12.00/0

amplification is the diffusion of the acid. This paper will explore the impact of diffusion on the lithographic properties of generic chemically amplified resists using mathematical modeling techniques. The acid generation will be modeled as a first order reaction, and the amplification will have an arbitrary order with respect to the acid concentration. The diffusion will be modeled with constant diffusivity as well as various forms of reaction dependent diffusivities. The complete system of equations will be solved as a reaction-diffusion system using finite difference techniques. Finally, the solution will be integrated with a complete lithographic modeling program to determine the impact of acid diffusion on lithographic performance.

Resist Kinetics

The kinetics of the exposure and catalyzed amplification of chemically amplified photoresists have been described elsewhere *(2,3)*, but will be reviewed here for a typical case. These resists are composed of a polymer resin (possibly "blocked" to inhibit dissolution), a photoacid generator (PAG), and possibly a crosslinking agent, dye or other additive. As the name implies, the photoacid generator forms a strong acid, H^+, when exposed to Deep-UV light. Ito and Willson first proposed the use of an aryl onium salt *(1)*, and triphenylsulfonium salts are now used extensively as PAGs. The reaction of a common PAG is shown below:

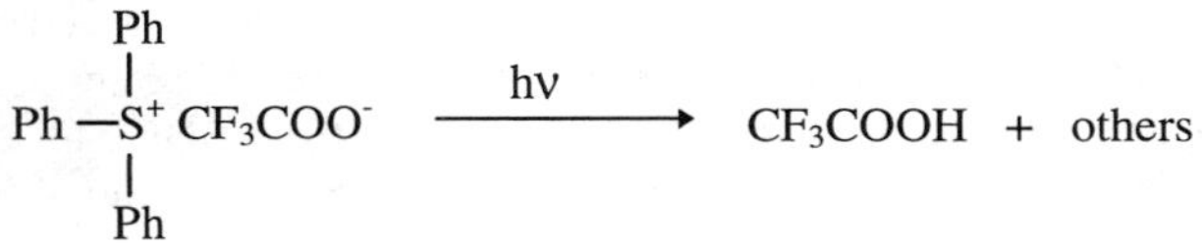

$$\begin{array}{c} Ph \\ | \\ Ph - S^+ \, CF_3COO^- \\ | \\ Ph \end{array} \xrightarrow{\;\; h\nu \;\;} CF_3COOH \; + \; others$$

The acid generated (trifluoroacetic acid) is a derivative of acetic acid where the electron-drawing properties of the fluorines are used to greatly increase the acidity of the molecule. The PAG is mixed with the polymer resin at a concentration of typically 5-15% by weight, with 10% as a typical formulation.

Sturtevant, et al. *(4)*, describe three possible mechanisms for the photoreaction of the PAG: direct absorption of a photon by the PAG, absorption of a photon by the polymer and subsequent electron transfer to the PAG, and photon absorption by the polymer resulting in fluorescence which then exposes the PAG. For direct photon absorption by the PAG, the kinetics of the reaction would be standard first order:

$$\frac{\partial G}{\partial t} = -CGI \tag{1}$$

where G is the concentration of PAG at time t (the initial PAG concentration is G_o), I is the exposure intensity, and C is the exposure rate constant. For constant intensity, the rate equation can be solved for G:

$$G = G_o \, e^{-CIt} \tag{2}$$

The acid concentration H is given by

$$H = G_o - G = G_o \left(1 - e^{-CIt}\right) \tag{3}$$

The other two non-direct photon absorption mechanisms could lead to different kinetics of exposure. However, for low exposure dose situations, equation (3) and other kinetics leads to near linear response with dose. It is highly likely that a combination of at least two of these mechanisms is occurring in most chemically amplified resists. However, equation (3) should adequately describe the overall dose dependence of acid generation using an effective exposure rate constant C.

Exposure of the resist with an aerial image $I(x)$ results in an acid latent image $H(x)$. A post-exposure bake (PEB) is then used to thermally induce a chemical reaction. This may be the activation of a crosslinking agent for a negative resist or the deblocking of the polymer resin for a positive resist. The reaction is catalyzed by the acid so that the acid is not consumed by the reaction and H remains constant. Ito and Willson first proposed the concept of deblocking a polymer to change its solubility *(1)*. A base polymer such as poly (*p*-hydroxystyrene), PHS, is used which is very soluble in an aqueous base developer. It is the hydroxyl groups which give the PHS its high solubility so by "blocking" these sites (by reacting the hydroxyl group with some non-ionizable groups) the solubility can be reduced. Ito and Willson employed a *t*-butoxycarbonyl group (*t*-BOC), resulting in a very slowly dissolving polymer. In the presence of acid and heat, the *t*-BOC blocked polymer will undergo acidolysis to generate the soluble hydroxyl group, as shown below.

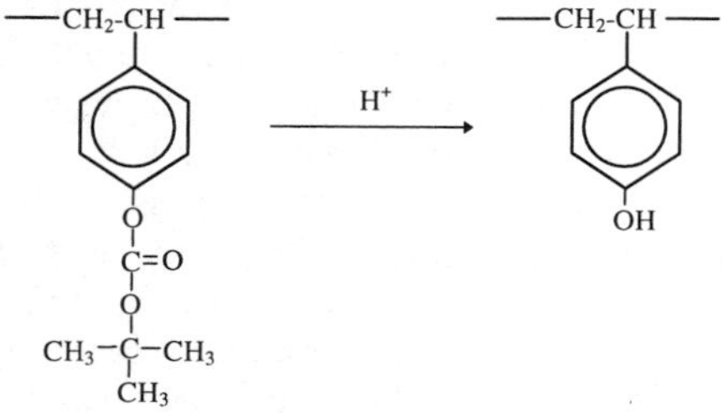

One drawback of this scheme is that the cleaved *t*-BOC is volatile and will evaporate, causing film shrinkage in the exposed areas. Larger molecular weight blocking groups can be used to reduce this film shrinkage to acceptable levels (below 10%). Also, the blocking group is such an effective inhibitor of dissolution, that nearly every blocked site on the polymer must be deblocked in order to obtain significant dissolution. Thus, the photoresist can be made more "sensitive" by only partially blocking the PHS. Typical photoresists use 10-30% of the hydroxyl groups blocked with 20% a typical value. Molecular weights for the PHS run in the range of 3000 to 5000 giving about 20 to 35 hydroxyl groups per molecule.

Using M as the concentration of some reactive site, these sites are consumed (i.e., are reacted) according to kinetics of some unknown order in H and first order in M:

$$\frac{\partial M}{\partial t'} = -K_{amp} M H^n \tag{4}$$

where K_{amp} is the rate constant of the crosslinking reaction and t' is the bake time. Simple theory would indicate that $n=1$ but the general form will be used here. Assuming H is constant, equation (4) can be solved for the concentration of reacted sites X:

$$X = M_o - M = M_o\left(1 - e^{-K_{amp}H^n t'}\right) \tag{5}$$

(Note: Although H^+ is not consumed by the reaction, the value of H is not locally constant. Diffusion during the PEB causes local changes in the acid concentration, thus requiring the use of a reaction-diffusion system of equations *(5)*. The approximation that H is constant is a useful one, however, which gives insight into the reaction. A more accurate reaction-diffusion approach will be presented in a following section.)

It is useful here to normalize the concentrations to some initial values. This results in a normalized acid concentration h and normalized reacted and unreacted sites x and m:

$$h = \frac{H}{G_o} \qquad x = \frac{X}{M_o} \qquad m = \frac{M}{M_o} \tag{6}$$

Equations (3) and (5) become

$$h = 1 - e^{-CIt}$$
$$m = 1 - x = e^{-\alpha h^n} \tag{7}$$

where α is a lumped "amplification" constant equal to $G_o^n K_{amp} t'$. The result of the PEB is an amplified latent image $m(x)$, corresponding to an exposed latent image $h(x)$, resulting from the aerial image $I(x)$.

Acid Diffusion

The above analysis of the kinetics of the amplification reaction assumed a locally constant concentration of acid H. Although this could be exactly true in some circumstances, it is typically only an approximation and is often a poor approximation. In reality, the acid diffuses during the bake. In one dimension, the standard diffusion equation takes the form

$$\frac{\partial H}{\partial t'} = \frac{\partial}{\partial x}\left(D_H \frac{\partial H}{\partial x}\right) \qquad (8)$$

where D_H is the diffusivity of acid in the photoresist. Solving this equation requires a number of things: two boundary conditions, one initial condition, and a knowledge of the diffusivity as a function of position and time.

The initial condition is the initial acid distribution within the film, $H(x,0)$, resulting from the exposure of the PAG. The two boundary conditions are at the top and bottom surface of the photoresist film. The boundary at the wafer surface is assumed to be impermeable, giving a boundary condition of no diffusion into the wafer. The boundary condition at the top of the wafer will depend on the diffusion of acid into the atmosphere above the wafer. Although such acid loss is a distinct possibility, it will not be treated here. Instead, the top surface of the resist will also be assumed to be impermeable.

The solution of equation (8) can now be performed if the diffusivity of the acid in the photoresist is known. Unfortunately, this solution is complicated by two very important factors: the diffusivity is a strong function of temperature and, most probably, the extent of amplification. Since the temperature is changing with time during the bake, the diffusivity will be time dependent. The concentration dependence of diffusivity results from an increase in free volume for typical positive resists: as the amplification reaction proceeds, the polymer blocking group evaporates resulting in a decrease in film thickness but also an increase in free volume. Since the acid concentration is time and position dependent, the diffusivity in equation (8) must be determined as a part of the solution of equation (8) by an iterative method. The resulting simultaneous solution of equations (4) and (8) is called a reaction-diffusion system.

The temperature dependence of the diffusivity can be expressed in a standard Arrhenius form:

$$D_o(T) = A_R \exp(-E_a / RT) \qquad (9)$$

where D_o is a general diffusivity, A_r is the Arrhenius coefficient and E_a is the activation energy. A full treatment of the amplification reaction would include a thermal model of the hotplate in order to determine the actual time-temperature history of the wafer *(6)*. To simplify the problem, an ideal temperature distribution will be assumed: the temperature of the resist is zero (low enough for no diffusion or reaction) until the start of the bake, at which time it immediately rises to the final bake temperature, stays constant for the duration of the bake, then instantly falls back to zero.

The concentration dependence of the diffusivity is less obvious. Several authors have proposed and verified the use of different models for the concentration dependence of diffusion within a polymer. Of course, the simplest form (besides a constant diffusivity) would be a linear model. Letting D_o be the diffusivity of acid in completely unreacted resist and D_f the diffusivity of acid in resist which has been completely reacted,

$$D_H = D_o + x\left(D_f - D_o\right) \tag{10}$$

Here, diffusivity is expressed as a function of the extent of the amplification reaction. Another common form is the Fujita-Doolittle equation *(7)* which can be predicted theoretically using free volume arguments. A form of that equation which is convenient for calculations is shown here:

$$D_H = D_o \exp\left(\frac{\alpha\, x}{1 + \beta\, x}\right) \tag{11}$$

where α and β are experimentally determined constants and are, in general, temperature dependent. Other concentration relations are also possible *(8)*, but the Fujita-Doolittle expression will be used in this work.

Acid Loss

Through a variety of mechanisms, acid formed by exposure of the resist film can be lost and thus not contribute to the catalyzed reaction to change the resist solubility. There are two basic types of acid loss: loss that occurs between exposure and post-exposure bake, and loss that occurs during the post-exposure bake.

The first type of loss leads to delay time effects -- the resulting lithography is affected by the delay time between exposure and post-exposure bake. Delay time effects can be very severe and, of course, are very detrimental to the use of such a resist in a manufacturing environment *(9,10)*. The typical mechanism for delay time acid loss is the diffusion of atmospheric base contaminates into the top surface of the resist. The result is a neutralization of the acid near the top of the resist and a corresponding reduced amplification. For a negative resist, the top portion of a line is not insolublized and resist is lost from the top of the line. For a positive resist, the effects are more devastating. Sufficient base contamination can make the top of the resist insoluble, blocking dissolution into the bulk of the resist. In extreme cases, no patterns can be observed after development. Another possible delay time acid loss mechanism is base contamination from the substrate, as has been observed on TiN substrates *(10)*.

The effects of acid loss due to atmospheric base contaminants can be accounted for in a straightforward manner *(11)*. The base diffuses slowly from the top surface of the resist into the bulk. Assuming that the concentration of base contaminate in contact with the top of the resist remains constant, the diffusion equation can be solved for the concentration of base, B, as a function of depth into the resist film:

$$B = B_o \exp\left(-(z/\sigma)^2\right) \tag{12}$$

where B_o is the base concentration at the top of the resist film, z is the depth into the resist ($z=0$ at the top of the film) and σ is the diffusion length of the base in resist. The standard assumption of constant diffusivity has been made here so that diffusion length goes as the square root of the delay time.

Since the acid generated by exposure for most resist systems of interest is fairly strong, it is a good approximation to assume that all of the base contaminant will react with acid if there is sufficient acid present. Thus, the acid concentration at the beginning of the PEB, H^*, is related to the acid concentration after exposure, H, by

$$H^* = H - B \qquad or \qquad h^* = h - b \tag{13}$$

where the lower case symbols again represent the concentration relative to G_o, the initial photoacid generator concentration.

Acid loss during the PEB could occur by other mechanisms. For example, as the acid diffuses through the polymer, it may encounter sights which "trap" the acid, rendering it unusable for further amplification. If these traps were in much greater abundance than the acid itself, the resulting acid loss rate would be first order.

$$\frac{\partial h}{\partial t'} = - K_{loss} h \tag{14}$$

where K_{loss} is the acid loss reaction rate constant. Of course, other more complicated acid loss mechanisms can be proposed, but in the absence of data supporting them, the simple first order loss mechanism will be used here.

Reaction-Diffusion System

The combination of a reacting system and a diffusing system where the diffusivity is dependent on the extent of reaction is called a reaction-diffusion system. The solution of such a system is the simultaneous solution of equations (4) and (8) using equation (3) as an initial condition and equation (10) or (11) to describe the reaction-dependent diffusivity. A convenient and straightforward method to solve such equations is the finite difference method (see, for example, reference *12*). The equations are solved by approximating the differential equations by difference equations. By marching through time and solving for all space at each time step, the final solution is the result after the final time step. A key part of an accurate solution is the choice of a sufficiently small time step. If the spatial dimension of interest is Δx, the time step should be chosen such that the diffusion length is less than Δx. For the calculations that follow, the time step was adjusted so that the maximum possible diffusion length during one time step was one third of the spatial increment Δx.

Modeling Results

The reaction-diffusion system described above was integrated into the comprehensive lithography simulation program PROLITH/2 *(13)*, including atmospheric base contamination and bulk acid loss. This software package predicts the light distribution within the resist during exposure by a projection optical system, solves the exposure rate equation (including the possibility of a film whose optical properties change during exposure), uses a 2-dimensional version of the reaction-diffusion solution described above, and then models the development as a function of the extent of amplification to produce a two-dimensional cross-section of a photoresist profile.

Using PROLITH/2 to predict lithographic results such as a resist linewidth or the dose-to-clear (E_o) provides a mechanism for exploring the effects of different diffusion properties. Figure 1 shows a simulation of the dose-to-clear as a function of the acid diffusivity assuming a constant diffusivity and a post-exposure bake time of 60 sec and using the modeling parameters given in Table I. The simulation used a silicon wafer substrate resulting in significant standing waves within the resist. As a result, a minimum amount of diffusion is required for the development process to punch through the standing waves and clear to the bottom of the resist. A diffusivity of about 10 nm^2/s (corresponding to a diffusion length of 35 nm) is needed to eliminate the effects of standing waves on E_o.

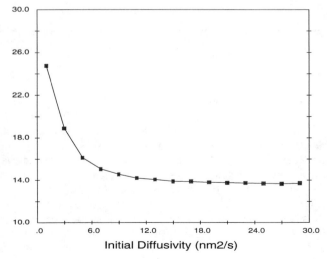

Figure 1. Assuming constant diffusivity for a 60 second post-exposure bake, a simulation of dose-to-clear versus acid diffusivity shows that a minimum amount of diffusion is required for the development process to punch through the standing waves *(14)*.

Table I. Typical PROLITH/2 input parameters for
simulation of chemically amplified resists

Imaging Tool:	CEL or Top ARC:
Wavelength = 248.0 nm	Not Used
Bandwidth = 0.0 nm	Intermediate Layers: none
Numerical Aperture = 0.6	Substrate: Silicon
Reduction Ratio = 4.0	
Image Flare = 0.00	Resist System: Positive
Aberrations: None	Thickness = 0.800 μm
Partial Coherence = 0.50	Absorption Parameter A = -0.10 1/μm
Linewidth = 0.25 μm	Absorption Parameter B = 0.30 1/μm
Pitch = 0.50 μm	Exposure Rate Const. C = 0.01 cm^2/mJ
Mask Bias = 0.0 μm	Refractive Index = 1.75
Focal Position = -0.20 μm	Development Model: Original Mack
	Max Develop Rate = 100.0 nm/s
Exposure Energy = 25.0 mJ/cm^2	Min Develop Rate = 0.05 nm/s
	Threshold M = -100.
Chem. Amp. PEB Parameters	Selectivity Parameter n = 5.00
PEB Bake Time = 60.0 sec	Relative Surface Rate = 0.10
Acid Reaction Order = 1	Inhibition Depth = 0.10 μm
Amplification Rate Const. = 0.10 s^{-1}	
Acid Loss Rate Constant = 0.010 s^{-1}	Development Time = 60.0 sec
Atmospheric Contamination: None	

Figure 2 shows how linewidth and sidewall angle of the photoresist profile are influenced by diffusivity. As both plots indicate, a minimum diffusivity is required to reduce the standing waves in order to obtain reasonable lithographic results. The resist profile simulations also show that excessive diffusion results in poor lithographic performance. As the diffusivity increases, the latent image in the resist begins to degrade, causing in this case an increase in the linewidth (as measured at the bottom of the profile) and a reduction of the sidewall angle.

From Figure 2 one could conclude that a diffusivity of about 15 nm^2/s to about 40 nm^2/s is required to give acceptable results. Confirmation of this range can be obtained by examining the effect of diffusion on process latitude. Figure 3 shows two focus-exposure matrices for diffusivities of 20 nm^2/s and 50 nm^2/s (corresponding to diffusion lengths of 49 and 69 nm, respectively). A loss of focus and exposure latitude is observed for the larger diffusion.

Resist Linewidth (microns)

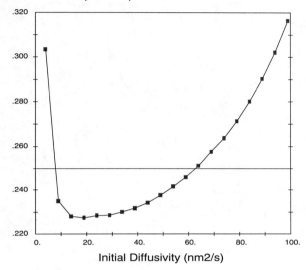

Sidewall Angle (degrees)

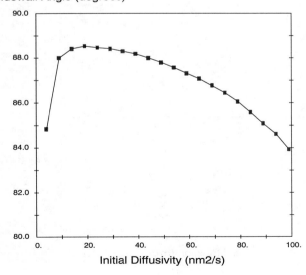

Figure 2. Effect of a constant diffusivity (60 second PEB) on resist linewidth and sidewall angle.

Resist Linewidth (microns)

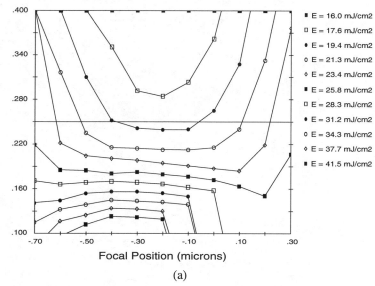

Focal Position (microns)

(a)

Resist Linewidth (microns)

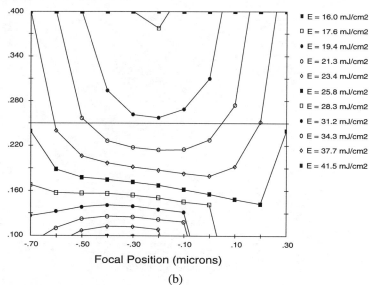

Focal Position (microns)

(b)

Figure 3. Focus-exposure matrices showing the effect of diffusivity on process latitude for (a) 20 nm^2/s and (b) 50 nm^2/s constant diffusivities and a 60 second PEB.

Conclusions

From the above simulation results, it is quite apparent that diffusion plays an important role in the lithographic behavior of chemically amplified resists. A minimum amount of diffusion is required to reduce the effects of standing waves. However, too much diffusion results in a degradation in the latent image and thus reduced lithographic performance. For the case of a constant diffusivity, diffusion lengths between 40 and 70 nm are required to meet the competing demands. If, however, the diffusivity is not constant, more latitude and improved performance can be obtained. What is required is an increase in the diffusivity as a function of the extent of reaction. The areas which are to be dissolved in the developer (the exposed areas of the positive resist) should have a high diffusivity to remove standing waves. The unexposed areas should have a low diffusivity to limit the degradation of the latent image. An ideal resist would have a diffusivity tailored to create a "diffusion well" in the exposed areas, which will not allow the diffusion of acid into the unexposed areas.

Acknowledgments

The author would like to thank Jeff Byers and John Sturtevant of SEMATECH, and John Petersen of Shipley Co. for many useful and enlightening discussions on chemically amplified resists.

References

1. H. Ito and C. G. Willson, "Applications of Photoinitiators to the Design of Resists for Semiconductor Manufacturing," in Polymers in Electronics, ACS Symposium Series 242 (1994) pp. 11-23.
2. R. A. Ferguson, J. M. Hutchinson, C. A. Spense, and A. R. Neureuther, *J. Vac. Sci. Technol.* B 8, 1423 (1990).
3. D. Ziger, C. A. Mack, and R. Distasio, Proc. SPIE **1466**, 270 (1991).
4. J. Sturtevant, S. Holmes, W. Conely, R. Bantu, W. Brunsvold, J. Byers, and S. Webber, "The Role of Photosensitization in DUV Resist Formulations," *SPE 10th International Conference on Photopolymers, Proc.*, (1994).
5. E. Barouch, U. Hollerbach, S. A. Orszag, M. T. Allen, and G. S. Calabrese, Proc. SPIE **1463**, 336 (1991).
6. C. A. Mack, D. P. DeWitt, B. K. Tsai, and G. Yetter, "Modeling of Solvent Evaporation Effects for Hot Plate Baking of Photoresist," *Advances in Resist Technology and Processing XI, Proc.*, SPIE Vol. 2195 (1994) pp. 584-595.
7. H. Fujita, A. Kishimoto, and K. Matsumoto, "Concentration and Temperature Dependence of Diffusion Coefficients for Systems Polymethyl Acrylate and n-Alkyl Acetates," *Transactions of the Faraday Society*, Vol. 56 (1960) pp. 424-437.

8. D. E. Bornside, C. W. Macosko and L. E. Scriven, "Spin Coating of a PMMA/Chlorobenzene Solution," *Journal of the Electrochemical Society*, Vol. 138, No. 1 (Jan., 1991) pp. 317-320.

9. S. A. MacDonald, et al., "Airborne Chemical Contamination of a Chemically Amplified Resist," *Advances in Resist Technology and Processing VIII, Proc.*, SPIE Vol. 1466 (1991) pp. 2-12.

10. K. R. Dean and R. A. Carpio, "Contamination of Positive Deep-UV Photoresists," *OCG Microlithography Seminar Interface '94, Proc.*, (1994) pp. 199-212.

11. T. Ohfuji, A. G. Timko, O. Nalamasu, and D. R. Stone, "Dissolution Rate Modeling of a Chemically Amplified Positive Resist," *Advances in Resist Technology and Processing X, Proc.*, SPIE Vol. 1925 (1993) pp. 213-226.

12. F. P. Incropera and D. P. DeWitt, <u>Fundamentals of Heat and Mass Transfer</u>, 3rd edition, John Wiley & Sons (New York: 1990).

13. FINLE Technologies, Austin, Texas

14. Jeff Byers, private communication.

RECEIVED September 11, 1995

Chapter 5

Acid Diffusion in Chemically Amplified Resists

The Effect of Prebaking and Post-Exposure Baking Temperature

Jiro Nakamura, Hiroshi Ban, and Akinobu Tanaka

NTT LSI Laboratories, 3–1 Morinosato Wakamiya, Atsugi-shi, Kanagawa 243–01, Japan

Evaluations of acid diffusion in chemically amplified resists using three different methods have revealed that higher prebaking and lower post-exposure baking (PEB) temperatures are effective in reducing the diffusion distance. Investigations into the influence of diffusion on resist characteristics show that there is a reciprocal relationship between sensitivity and resolution due to the effects of acid diffusion. This relationship is able to be formulated using the terms of activation energies for overall catalytic reactions and for acid diffusion.

Reactions in the formation of latent images in chemically amplified resists[1] take place in two consecutive steps: acid generation during exposure, and acid-catalyzed reactions during post-exposure baking (PEB). This mechanism makes the lithographic performance more sensitive to the process conditions than that of conventional resist systems. In particular, resist characteristics are greatly influenced by the acid diffusion during PEB[2-6]. Acid diffusion is essential to induce long chains of catalytic reactions. Appropriate diffusion also suppresses edge roughness caused by the nonuniform exposure that results from the effects of multi-interference in photolithography and from electron scattering in e-beam and X-ray lithographies. Excess acid diffusion, on the other hand, degrades latent image quality and thus, the replicated pattern profile. When developing high-performance resist materials and optimizing process conditions, it is therefore important to know how far catalytic acids diffuse in resist films, and to understand the relationship between acid diffusion and resist characteristics.

We have developed three ways to evaluate acid diffusion in resist films during PEB: the mask-contact method, the electrochemical method, and the ion conductivity method. This paper describes these method and presents the associated results. In addition, it discusses the influence of acid diffusion on lithographic performance based on experimental and calculated results.

0097–6156/95/0614–0069$12.00/0

Evaluation of Acid Diffusion

Experiment.
   a) Mask-Contact Replication Method.   A mask contact replication method was used to determine acid diffusion in a positive resist. The EXP (E-beam/X-ray Positive resist) *(7)* was spin-coated onto a 4-inch Si wafer, followed by sprinkling small pieces (about 100-1000 μm^2) of Ta, as an X-ray absorber mask, onto the wafer before drying. X-ray exposure, PEB, and development were carried out while keeping the absorber on the resist. After development, the undercut depth of the resists for various exposure doses from the absorber pattern edge was measured with scanning electron microscopy (SEM).

   b)   Electrochemical Method.   An electrochemical method was used for the negative resist. An interdigitated array (IDA) electrode was fabricated using Pt on a SiO_2 substrate by photolithography. The width of each electrode and of the gaps between the electrodes was 2 μm. Chemically amplified negative resist SAL 601 (Shipley. Co.) was spincoated onto the IDA electrode and prebaked on a hot plate at 100°C for 120 s. All areas of the resist film on the IDA electrode were irradiated with an e-beam and then post-exposure baked in an oven at 100°C for 30 min while direct current of 10 V was applied. After the resist film on the electrode was cut orthogonal to the electrodes to obtain a cross section, it was developed.

   c) Ion Conductivity Method.   The diffusion coefficient of protons in a novolac resin was evaluated by measuring ion conductivity. Novolac resin with 5 wt% p-toluenesulfonic acid was coated on an IC sensor and baked in an oven for 30 min at various temperatures and the dielectric loss factor of films was measured with an Eumetric System II Microdielectrometer (Micromet Instruments). The dielectric loss factor varies with the ion conductivity, the dipoles of polar groups in resists, and the electrode polarization. The effect of the electrode polarization can be corrected with appropriate software *(8)* in the microdielectrometer. The dielectric loss factor ε'' is given by

$$\varepsilon'' = \frac{\sigma}{\omega \varepsilon_0} + \frac{(\varepsilon_r - \varepsilon_u)\, \omega\, \tau_d}{1 + (\omega\, \tau_d)^2}$$

$$(1)$$

where σ is the ion conductivity, ε_u is the "unrelaxed" permittivity, ε_r is the "relaxed" permittivity, ε_0 is the permittivity in a vacuum, ω is the frequency of alternating current applied, and τ_d is the dipole relaxation time.*(8)* When the frequency is so low that $\omega\, \tau_d$ is much smaller than 1, the second term in Eq. (1) can be negligible. The diffusion coefficient D (cm^2/s) is given by

$$D = \frac{\sigma\, k\, T}{C\, q^2}$$

$$(2)$$

where C is the concentration of mobile ions, q is the valence of ions, k is the Boltzmann constant, and T is the absolute temperature.

Results and Discussion.
   a) Mask-Contact Method.   In an actual exposure, it is not easy to distinguish the contributions of other resolution-limiting factors from those of acid

diffusion. The mask contact replication method was therefore adopted to exclude the effects of other factors. Examples of resist patterns are shown in Fig. 1. No undercutting is seen in the pattern produced with the exposure dose of 150 mJ/cm^2, which is equal to the minimum dose necessary to clear all the resist of the exposed area, but the depth of the undercutting reaches 0.4 μm if the exposure dose is doubled. The undercutting must be caused by both electron-scattering and acid diffusion, because the other lithographic resolution limiting factors such as penumbral shadow and Fresnel diffraction have been eliminated in this method. The range of electrons scattering in materials has been reported to be about the same in materials having the same density[9]. The range of electrons scattering in EXP should be the same as that in PMMA because they have the same density of 1.2 g/cm^3. This means that the difference in undercutting between EXP and PMMA is caused by the diffusion of catalytic acids in the EXP, because PMMA is not a chemically amplified resist and is therefore unaffected by acid diffusion. Thus the range of acid diffusion was evaluated from the difference in undercutting between EXP and PMMA under various baking conditions. The results are plotted in Fig. 2, where the exposure dose is normalized by the sensitivity for each baking condition. The acid diffusion range increased with increasing PEB temperature (Fig. 2(a)). For example, the acid diffusion range at the normalized exposure dose of 2 increased from 0.1 to 0.4 μm by raising the PEB temperature from 55°C to 75°C. On the other hand, the diffusion range decreased with increasing prebaking temperature (Fig. 2(b)). These results for acid diffusion coincide with those of changes in replicated patterns in actual lithographies[5]. This confirms that acid diffusion plays a significant role in determining the width of replicated pattern.

b) Electrochemical Method.   Since the mask contact replication method can only be used for positive resists, we applied the electrochemical method to evaluate acid diffusion in a negative resist. The reaction on the electrodes and the schematic diagram for acid concentration in films are shown in Fig. 3. In this method, the electrochemical reduction of protons on the cathode is used to determine the diffusion range. The protons may be reduced to hydrogen molecules on the cathode surface when direct current is applied. If the proton reduction occurs at the cathode at a rate greater than it is supplied by diffusion, the diffusion layer of protons becomes longer with increasing direct current application time, decreasing the proton concentration near the cathode.

Examples of cross-sectionally developed resist patterns are shown in Fig. 4. A semicircular cavity with a radius of about 0.3 μm is seen near the edge of the cathode, while there are no such cavities near the anode or between the anode and cathode. It is clear from the results shown in Fig. 4 that the decrease in proton concentration near the edge of the cathode is greater than that at the center of the cathode.

In the SAL resist system, protons generated by exposure diffuse in the film and act as a catalyst for crosslinking reactions during PEB. Consequently, the semicircular cavity observed at the edge of the cathode seems to suggest that the extent of crosslinking reaction in this area is smaller than in other areas due to the lower concentration of catalytic protons near the cathode.

The growth of the diffusion layer due to the scarcity of protons near the cathode is given by

$$L = \sqrt{2\,D\,t} \qquad (3)$$

where L is the length of the diffusion layer and t is time[10]. Assuming that the cavity radius is equal to the thickness of the diffusion layer, the diffusion coefficient of a

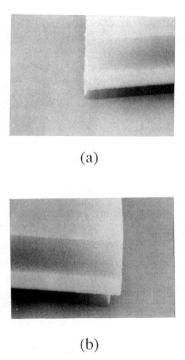

(a)

(b)

Fig. 1. Replicated resist pattern profile with the absorber pattern. Exposure doses are (a) 150 and (b) 300 mJ/cm^2.

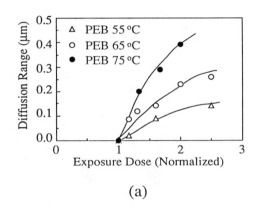

(a)

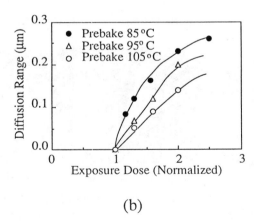

(b)

Fig. 2. Dependence of diffusion range on (a) PEB and (b) prebaking conditions.

electrochemical reaction
on cathodes

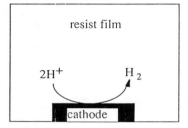

acid concentration in film

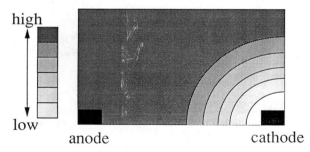

Fig. 3. Electrochemical reaction on the cathode surface and schematic diagram
of acid concentration.

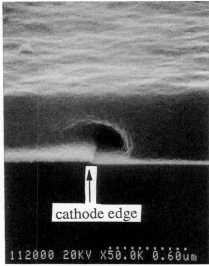

Fig. 4. (a) SEM photograph of cross-sectionally-developed pattern on the interdigitated array electrode. (b) Magnified view of cathode edge. All areas of the resist film were irradiated. The exposure dose was 3 µC/cm² and the applied voltage was 10 V.

proton is about 70 nm^2/s, which is consistent with the other result reported for SAL resist(*11*).

    c) Ion Conductivity Method. The dependence of the diffusion coefficient on the prebaking and PEB temperatures is shown in Fig. 5. The values plotted here are an average for protons and counter anions. The diffusion coefficient decreased as the prebaking temperature increased, probably due to a reduction in free volume caused by the evaporation of the residual solvent. The diffusion coefficient, however, increased with increasing PEB temperature. For example, when the prebaking temperature was 115°C, it increased from 1.5 to 20 nm^2/s by raising the PEB temperatures from 55°C to 75°C. This relationship between acid diffusion and baking conditions coincides with the results obtained with the mask-contact replication method. This method can be applied for both positive and negative resists and is a simpler and more direct way to determine acid diffusion in resists films.

Influence of Acid Diffusion on Resist Performance

The previous section describes how far catalytic acids diffuse in films. Here we discuss the relationship between this acid diffusion and resist characteristics. For calculation, it was assumed that a resist material is composed of a base-resin, a dissolution inhibitor, and an acid generator. The basic idea can be applied to other types of resist systems.

Model of Latent Image Formation. In chemically amplified resist systems, the reactions needed for the formation of a latent image, which here means the distribution of decomposed dissolution inhibitors, are not completed during exposure. The formation of latent image including the PEB process should thus be considered. In conventional resists, the lateral distribution of the decomposed dissolution inhibitor $C_I(x)$ is formulated by Eq. (4) as a function of the profile of light intensity on resist $I(x)$ by

$$C_I(x) = C_{I_0}\{1 - exp\,(-k_1\,I(x))\} \quad , \tag{4}$$

where C_{I0} and k_1 are constants. The concentration of catalytic acids $C_A(x)$ in arbitrary position x in the chemically amplified resists is given by

$$C_A(x) = C_{A_0}\{1 - exp\,(-k_2\,I(x))\} \quad , \tag{5}$$

where C_{A0} and k_2 are constants. Since chemically amplified resists are usually used under conditions where the decomposition of the acid generator is very small, the amount of generated acid is regarded to be roughly proportional to the exposure energy. Assuming that the rate of an acid-catalyzed reaction is proportional to one order of acid concentration, the rate equation is given by

$$\frac{\delta C_I}{\delta t} = k_3\,(1 - C_I)\,C_A \quad , \tag{6}$$

where k_3 is the rate constant. The acid concentration usually changes due to acid diffusion during PEB and thus should be regarded as a time-dependent factor. Integrating Eq. (6), the distribution of the decomposed dissolution inhibitor is given by

$$C_I(x) = C_{I_0}\{1 - exp(-k_3 M_A(x))\}$$

$$M_A(x) = \int_0^{t_{PEB}} C_A(x,t)\, dt$$

(7)

The change in the concentration of catalytic acid can be calculated on the basis of Fick's diffusion equation.

Latent Image Contrast.   By comparing Eq. (4) with (7), the integral value of catalytic acid concentration $M_A(x)$ in (7) is found to correspond to $I(x)$ in (4). The following discussion of the quality of the latent image is therefore made based on this integral value. In evaluating the quality of latent images of line-and-space patterns produced in conventional resists (not chemically amplified ones), the contrast of the latent image is usually defined by

$$Contrast = \frac{I_{max} - I_{min}}{I_{max} + I_{min}} ,$$

(8)

where I_{max} and I_{min} are the maximum and minimum values of $I(x)$. In chemically amplified resist systems, however, it is more reasonable to define contrast as

$$Contrast = \frac{M_{Amax} - M_{Amin}}{M_{Amax} + M_{Amin}} ,$$

(9)

where M_{Amax} and M_{Amin} are the maximum and minimum values of $M_A(x)$. The change in image contrast due to acid diffusion is shown in Fig. 6. It is assumed that the initial acid distribution used here is given by a sine function with a contrast of 0.7. Below a diffusion length of about 30 nm, the image contrast for a 0.35-μm lines-and-spaces pattern hardly degrades. Thus the quality of the latent image is dominated by the quality of the aerial image, which in turn is determined by the exposure system. The contrast values decreased rapidly, however, when the diffusion length exceeds 30 nm, suggesting that acid diffusion is the predominant factor determining image quality. The critical value shifts to a shorter diffusion length with decreasing pattern size. As will be shown later, a large diffusion length enhances resist sensitivity. There is thus an optimal diffusion length that maximizes sensitivity without decreasing resolution. Optimal diffusion length for a 0.35-μm lines-and-spaces pattern is about 30 nm. When the diffusion length is larger than this, there is a reciprocal relationship between sensitivity and resolution due to acid diffusion.

Another important role of acid diffusion is in reduction of edge roughness caused by the standing wave effect. The plot for a 37.5-nm line-and-space pattern in Fig. 6 corresponds to edge roughness with the case of an exposing light wavelength of 248 nm and a film refractive index of 1.65. To smooth this roughness, a diffusion length of over 30 nm is required.

Relationship between Resolution and Sensitivity.   Since the acid concentration generated by exposure and acid diffusion length during PEB have a proportional influence on the resolution and sensitivity of resists(12), they were respectively used as indicators of sensitivity and resolution. Although the acid loss caused by airborne contaminants and side-reactions affects the sensitivity and

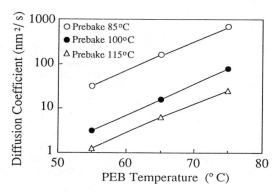

Fig. 5. Diffusion coefficient of acids in novolac resin as a function of PEB temperature for various prebaking temperatures.

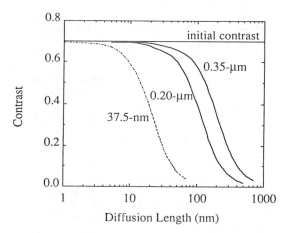

Fig. 6. Dependence of latent image contrast on the diffusion length of catalytic acids in line-and-space patterns.

resolution of most chemically amplified resists, we have not taken the influence of acid loss into consideration here.

First, the relationship between acid diffusion and the resist sensitivity is discussed when PEB time is changed for fixed materials and fixed PEB temperatures. A simple two-molecule reaction model was used to evaluate the effect of acid diffusion. A catalytic acid collides with dissolution inhibitors, resulting in the decomposition of inhibitors at a certain probability (σ). In the case of a fixed temperature, probability does not vary. The collision frequency (ω) of an acid with a dissolution inhibitor in the resist film is given by

$$\omega = 4 \pi (D_A + D_I) (r_A + r_I) C_A (1- C_I) \tag{10}$$

where D_A and D_I are the diffusion coefficients of the acid and inhibitor, r_A and r_I are their radii, and C_A and C_I are their concentrations(13). When dimension of pattern is much larger than that of acid diffusion length, the acid concentration is constant during PEB. Therefore, the reaction rate and C_I can be represented using σ by

$$\frac{d C_I}{d t} = 4 \pi \sigma (D_A + D_I) (r_A + r_I) C_A (1- C_I) \tag{11}$$

and

$$C_I = 1 - exp \{4 \pi \sigma (D_A + D_I) (r_A + r_I) C_A t \} \tag{12}$$

For given resist materials, the C_I needed for fixed development conditions is constant. Since the diffusion coefficient of the dissolution inhibitor is usually much smaller than that of the acid, C_A is inversely proportional to $(D_A t)$ or the square of the acid diffusion length. In other words, the sensitivity, which is defined by the exposure dose needed to clear the exposed film, can be expressed using the diffusion coefficient and PEB time, or diffusion length, as

$$Sensitivity = k_5 \frac{1}{D t} = k_5 \frac{2}{L^2}. \tag{13}$$

Here, k_5 is a constant and L is the acid diffusion length. This equation shows that the resist sensitivity, in an area larger than the diffusion area, becomes higher with increasing diffusion length. Although the concentration of acids in a small area is decreased due to their diffusion, the degree of change in acid concentration is usually much smaller than that of the change in diffusion coefficient(12). The relationship in Eq. (13) can therefore be used for most types of pattern.

In addition, when the PEB time is changed, a correlation between the resolution and sensitivity of produced resists can be formulated as

$$Resolution = \frac{k_6}{\sqrt{Sensitivity}}, \tag{14}$$

where k_6 is a constant. The experimentally obtained relationship between the resolution and the sensitivity of resist EXP is shown in Fig. 7. The sensitivity was changed by varying the PEB time from 30 to 300 s at a temperature of 75 °C. A reciprocal relationship between the resolution and the sensitivity of EXP is evident.

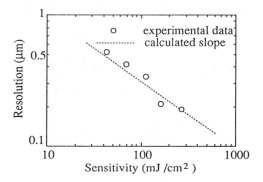

Fig. 7. Measured and calculated resolution of chemically amplified resist EXP as a function of sensitivity. The resolution is defined as the minimum size of pattern that can be resolved with exposure latitude of ±10%.

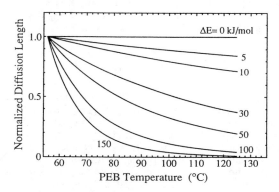

Fig. 8. Dependence of diffusion length on PEB temperature. ΔE is the difference in activation energies between overall acid-catalyzed reactions and diffusion of acids.

The slope obtained from Eq. (14) is shown by the dotted line as a comparison. Judging from the fairly good correlation of calculated results with experimental ones, the proposed model seems reasonable in explaining the role of acid diffusion in catalytic reactions.

Next, the relationship between resolution and sensitivity is discussed when PEB temperature is changed. The relationship between the two in Eq. (14) is applicable only when the reaction probability is constant for various temperatures. Constant reaction probability means that the activation energy for the overall acid-catalyzed reaction is equal to that for the acid diffusion, or in other words, acid diffusion controls the rate of acid-catalyzed reactions.

To evaluate this relationship more generally, the term for the activation energy is introduced into equations for reaction rate and diffusion length:

$$\frac{\delta C_I}{\delta t} = k_0\, e^{\frac{-E_{ar}}{RT}}\,(1 - C_I)\,C_A$$

$$(15)$$

$$L^2 = 2\,D_0\, e^{\frac{-E_{ad}}{RT}}\, t \quad,$$

$$(16)$$

where E_{ar} and E_{ad} are the activation energies for overall acid-catalyzed reaction and diffusion, k_0 and D_0 are the pre-exponential factors, and R and T are the gas constant and the absolute temperature. In the solutions, the activation energies for the diffusion are much smaller than for the overall reactions. On the other hand, the activation energies for the diffusion in films are usually much larger than those in solutions.

By eliminating the term for PEB time from Eqs. (15) and (16), the relationship between diffusion length and concentration of acids becomes

$$L^2 = 2\frac{D_0}{k_0}\, e^{\frac{E_{ar} - E_{ad}}{RT}}(-\ln\,(1 - C_I))\,\frac{1}{C_A}$$

$$(17)$$

In this case, the PEB time is changed to induce a fixed extent of acid-catalyzed reactions, depending on the PEB temperature applied. For a given resist material, the ratio of decomposed dissolution inhibitors needed for fixed development is constant. Equation (17) can therefore be rearranged into

$$L = K\, e^{\frac{E_{ar} - E_{ad}}{2\,RT}}\sqrt{\frac{1}{C_A}}$$

$$(K = \sqrt{\frac{2\,D_0}{k_0}(-\ln\,(1 - C_{I_D}))}\,)$$

$$(18)$$

where K is constant for a given material. Here, C_{I_D} is the decomposition ratio of dissolution inhibitors needed for development.

The diffusion length needed for fixed amounts of acids and catalytic reactions depends on PEB temperature, as shown in Fig. 8. This is for various values of difference ΔE between the activation energies of the overall acid-catalyzed reaction and acid diffusion. The diffusion length values plotted here are normalized by the value given at PEB of 55°C. This figure, therefore, shows what diffusion length is needed to induce a reaction of a given extent in the presence of acids of fixed concentration at various PEB temperatures. The diffusion length at higher temperatures for shorter PEB times is smaller than that at lower temperatures for longer periods when compared under the same catalytic chain length, while the diffusion length at higher temperatures is naturally larger that at lower temperatures when compared over the same PEB time. When the difference between both energies is 150 kJ/mol, diffusion length at 65°C is only about one-half that at 55°C. In this case, a higher PEB temperature will be preferable to enhance the resolution.

The relationship between acid concentration and diffusion length for various PEB temperatures is shown in Fig. 9. The graphs show diffusion length normalized by that for PEB at 55°C and for a normalized acid concentration of 10. When both activation energies are the same, the diffusion length is independent of PEB temperature and is inversely proportional to the square root of acid concentration. When activation energy for overall reaction is 30 kJ/mol larger than that for diffusion, required acid concentration increases at a lower temperature even at the same diffusion length.

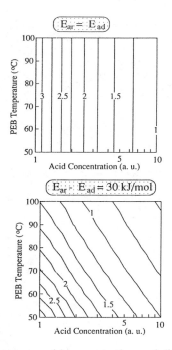

Fig. 9. Relationship between acid concentration and diffusion length needed to induce a given amount of catalytic reactions.

Conclusion

Acid diffusion in a resist film was evaluated using three different techniques: the mask contact, the electrochemical, and ion conductivity methods. Mask contact replication is useful for positive resists and the electrochemical method for negative resists. Ion conductivity method can be used to determine acid diffusion in both positive and negative resists and is a simpler and more direct way. It was found that higher prebaking temperatures and/or lower post-exposure baking temperatures are effective in reducing the acid diffusion distance that greatly influences replicated pattern width.

An evaluation of the influence of acid diffusion on lithographic performance revealed that there is an optimum diffusion length to produce the maximum sensitivity without deterioration of the resolution and that there is a reciprocal relationship between sensitivity and resolution that can be attributed to acid diffusion. By formulating a relationship between the activation energies of diffusion and overall acid-catalyzed reactions and the diffusion length needed to decompose a given degree of acid-catalyzed reactions, it was found that when these two activation energies are the same, the diffusion length is inversely proportional to the square root of the amount of generated acid. When the activation energy for the overall acid-catalyzed reaction is larger than that for the acid diffusion, a shorter diffusion length is required for higher PEB temperatures. Higher resolution should therefore be obtained by increasing the PEB temperature for fixed resist sensitivity.

Acknowledgments

The authors wish to thank Tetsushi Sakai, Yutaka Sakakibara and Tadahito Matsuda for their advice and encouragement.

Literature Cited

1) Ito, H.; Willson, C. G. *Polym. Eng. Sci.* **1983**, *23*, 1012.
2) McKean, D. R.; Schaedeli, U.; MacDonald, S. A. In *Polymers in Microlithography;* Reichmanis, E.; MacDonald, S. A.; Iwayanagi, T., Eds.; ACS Symp. Ser. 412; ACS: Washington, D. C. 1989, 27-38.
3) Schlegel, L.; Ueno, T.; Hayashi, N.; Iwayanagi, T. *J. Vac. Sci. & Technol.* **1991**, *B9*, 278.
4) Fedynyshyn, T.; Cronin, M.; Szmanda, C. *J. Vac. Sci. & Technol.* **1991**, *B9*, 3380.
5) Nakamura, J.; Ban, H.; Deguchi, K.; Tanaka, A. *Jpn. J. Appl. Phys.* **1991**, *30*, 2619.
6) Yoshimura, T.; Nakayama, Y.; Okazaki. S, *J. Vac. Sci. & Technol.* **1991**, *B10*, 2615.
7) Ban, H.; Nakamura, J.; Deguchi, K.; Tanaka, A. *J. Vac. Sci. & Technol.* **1991**, *B9*, 3387.
8) Day, D.; Lewis, J.; Lee, H.; Senturia, S. *J. Adhesion* **1985**, *18* , 73.
9) Everhart, T.; Hoff, P. *J. Appl. Phys.* **1971**, *42* , 5837.
10) Bard, A.; Faulkner, L. *Electrochemical Methods*; John Wiley & Sons : NY, 1980.
11) Yoshino, H.; Matsumoto, H. *Jpn. J. Appl. Phys.* **1992**, *31*, 4283.
12) Nakamura, J.; Ban, H.; Tanaka, A. *Jpn. J. Appl. Phys.* **1992**, *31*, 4294.
13) Tsuboi, S.: *Genzaikagaku* **1992**, No 3. 12.

RECEIVED July 17, 1995

Chapter 6

Correlation of the Strength of Photogenerated Acid with the Post-Exposure Delay Effect in Positive-Tone Chemically Amplified Deep-UV Resists

F. M. Houlihan, E. Chin, O. Nalamasu, J. M. Kometani, and R. Harley[1]

AT&T Bell Laboratories, 600 Mountain Avenue, Murray Hill, NJ 07974

The changes in clearing doses (D_p) for deep UV chemically amplified resists formulated with poly(4-t-butoxycarbonyloxystyrene-sulfone) were correlated with changing the photo-acid generator (PAG) component using a variety of 2-trifluoromethyl-6-nitrobenzyl benzenesulfonates. These changes in D_p's were also monitored under different environmental conditions, before post-exposure bake (PEB), to determine their sensitivities to airborne contaminants. It was found that if airborne contamination is eliminated through using a protective overcoat, then D_p decreased with increasing reactivity of the photo-released acid. Also, for most of the resist formulations, the sensitivity to airborne contamination increased with the strength of the photo-released acid. However, sensitivity to airborne contaminants was lessened in the presence of PAG's with substituents that can associate strongly with other molecules. Combining these effects, the order of resistance of resist formulations to airborne contaminants as a function of substituents on PAG on the aryl sulfonate moiety was 4-OCH$_3$ >4-NO$_2$ > 4-Cl > > 3-SO$_3$R > > 4-CH$_3$ ≥ 4-H > 2-CF$_3$ ≥ 3-CF$_3$ ≥ 4-CF$_3$ ≳ 2,4-diF > 2-NO$_2$> 3,5-diCF$_3$.

Scheme 1 depicts chemical amplification through the acidolysis reaction during post-exposure bake (PEB) initiated by photo-released acid in a typical *t*-butoxycarbonyl (*t*-BOC) based deep UV resist (*1*). One serious processing problem with such materials is the occurrence in exposed areas of surface depletion of photo-released acid by basic airborne contaminants (Scheme 2). This results in an ineffective removal of dissolution inhibiting *t*-BOC groups during PEB.

This depletion of acid manifests itself as a developer insoluble surface layer which leads to T-shaped lines (T-topping) resulting in loss of resolution of small features (*2,3*). Although airborne contamination may occur at any time during processing, its effect is most readily seen by increasing the time between exposure

---

[1]Current address: Stanford University, Stanford, CA 94305

0097–6156/95/0614–0084$13.50/0

and PEB; in this circumstance, it is called the post-exposure delay (PED) effect.

Methods have been described in the literature to reduce the PED effect by using protective overcoats (2), filtering the impurities from the air (3), and annealing away free volume from the polymer (4). We will describe here work that we have done to better understand the part the chemistry of the photo-acid generators (PAG's) plays in the PED effect. This information would be helpful in the rational design of PAG's less susceptible to the PED effect.

In our previous work on PAG's based on the 2-nitrobenzyl chromophore, we have shown that Hammett plots were useful in predicting the relative thermal stability of resist films in which PAG's of differing structures were added (5-8). This supported the reaction mechanism for thermal decomposition depicted in Scheme 3.

Hammett plots should also be useful in predicting lithographic properties such as lithographic sensitivity. However, to accomplish this one must first try to establish a relationship between lithography and chemistry. Previously, we have shown (7a) that there is a reciprocal relationship between the clearing dose (D_p) (as defined in (9)) and the product of absorbance/μm (ABS/μm), quantum yield (Φ), and the catalytic chain length for the removal of t-BOC groups (Chain Length). This relationship was improved by using the fraction of light absorbed per micron (%ABS/μm), (Equation 1) which is a better measure of the number of photons absorbed by the resist film (Figure 1), especially in more absorbent films.

$$D_p \propto 1/(\%ABS/\mu m * \Phi * \text{Chain Length}) \quad (1)$$

Figure 1 allows us to link empirically the relative values of Chain Length for t-BOC removal for a resist system to its D_p as shown in Equation 2.

$$\text{Chain Length} \propto 1/(\%ABS/\mu m * \Phi * D_p) \quad (2)$$

Assuming that the different photo-acids diffuse equally, then differing substituents on the arylsulfonate moiety of the PAG's may affect the Chain Length for t-BOC acidolysis through reactions which change directly the concentration of 'free' (10) H$^+$ or affect this concentration indirectly by changing the concentration of the intermediate t-butyl cation. Moreover, these reactions must be reversible to account for the observation that in our 2-nitrobenzylsulfonate based resists increasing PEB time also increases the Chain Length (7a). Reversible, interceptions of either 'free' H$^+$ or t-butyl cation would account for this.

Scheme 4 illustrates one possible scenario in which increased nucleophilicity and/or basicity of the sulfonate anion decreases 'free' H$^+$. The nucleophilicity (11) and the basicity (12,13) of arylsulfonates are known to decrease 2-3 of magnitude with increasing σ. Because arylsulfonic acids are know to be more poorly dissociated in non-polar environments (14), substituents effects could greatly affect the equilibrium in Scheme 4, changing the concentration of 'free' H$^+$ available in the relatively non-polar polymer matrix. Similarly, binding of nucleofugal anions (even triflate) in carbocationic-like processes is an accepted phenomenon which has been reviewed in the literature (15). This binding would be expected to proceeds well under S_N1 or S_N conditions especially in a non-polar environment such as the resist matrix. As an example, tosic acid cannot be used effectively in the cationic

Scheme 1

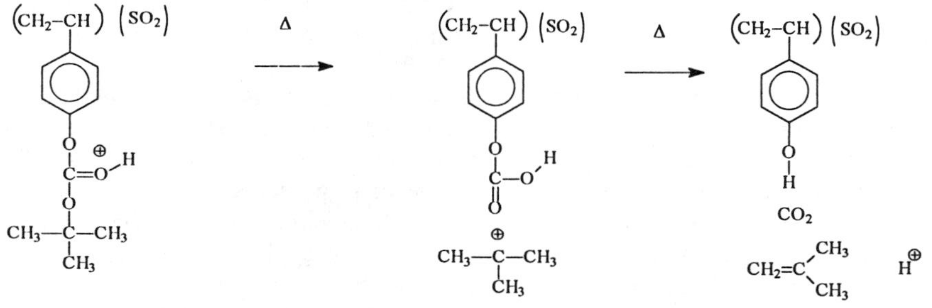

Scheme 2

Scheme 3

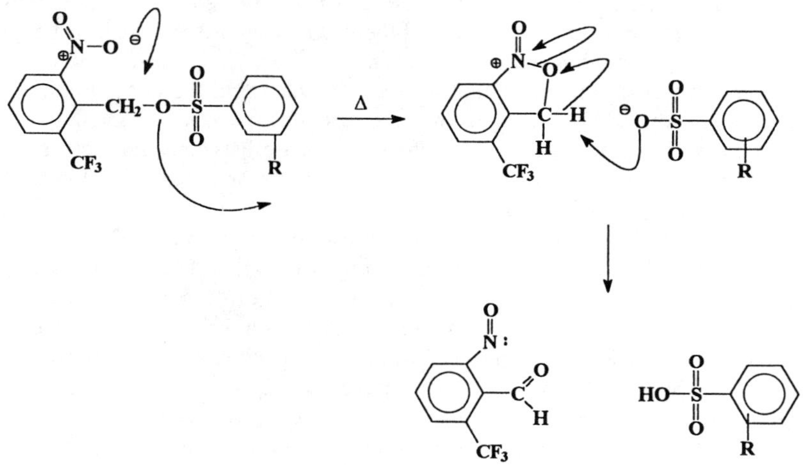

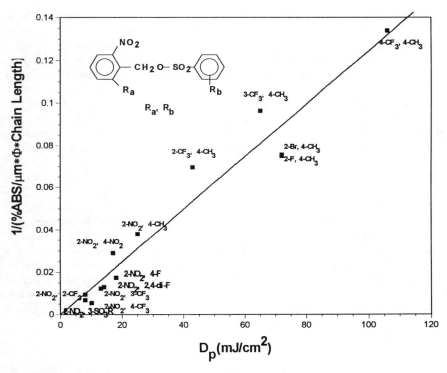

Figure 1

> *Plot of 1/(%ABS/μm*Φ*Chain Length) versus D_p for resists formulated with PTBSS and 2-nitrobenzyl benzenesulfonate PAG's.*

Scheme 4

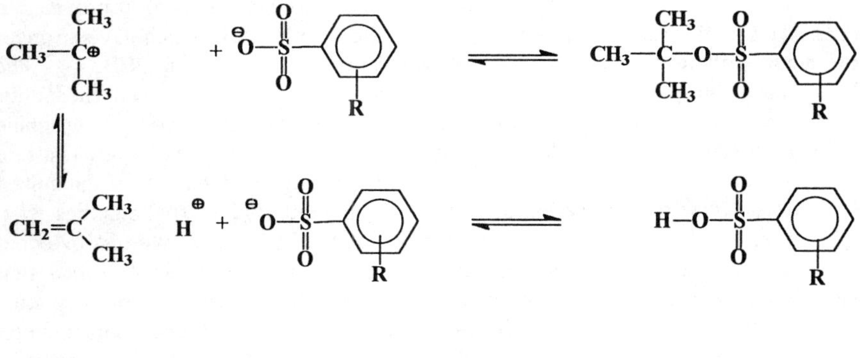

polymerization of expoxides because it is known to combine covalently with the propagating carbocation (*16*).

Another possibility is that the substituents change the basicity/nucleophilicity of the aryl moiety itself as shown in Scheme 5. Arguments similar to the one above may be made for the application of Equation 3 to this mechanism and to other possible mechanisms in which one postulates a   reversible process competitive to *t*-BOC acidolysis which consumes either protons or *t*-butyl cations, leveling the concentration of 'free' H^+.

Because the   Chain Length is a measure of the differing rates of   *t*-BOC removal, it is reasonable to expect that the Chain Length for acidolysis of *t*-BOC would increase with the Hammett sigma values in response to changing concentration of 'free' H^+. Given this, Log(Chain Length) should be proportional to σ (Equation 3).

Log(Chain Length) $\propto$ $\rho\sigma$ (3)

However, our previous measurement of Chain Lengths, by IR spectroscopy of resist systems formulated with 2,6-dinitrobenzyl arylsulfonates PAG's, showed that this equation predicts only the behavior of weaker photo-acids (*7a*). For stronger acids, a turnover of the effect was noted (Figure 2A). Equation 4 derived from 2 and 3 predicts that the same type of turnover effect should have been observed when plotting D_p lithographic data.   This behavior was indeed suggested as seen in Figure 2B. However, because PED conditions were not controlled in this previous work the data shows considerable scatter.

Log(1/(%ABS/μm*Φ*D_p)) $\propto$ $\rho\sigma$ (4)

These results indicated that the increased reactivity of the acid (decreased basicities and/or nucleophilicities of arylsulfonate such as depicted in Schemes 4 and 5) was being funneled into some other reaction pathway.   At the time, we proposed (*7a*) that one pathway might be the PED effect caused by airborne basic contaminants (*2,3*).   We report here further investigation to demonstrate conclusively the link between the PED effect and the reactivity of the acid employed in photo-imaging by implementing control of PED conditions. We have chosen for this study the 2-trifluoromethyl-6-nitrobenzyl chromophore, instead of the 2,6-dinitrobenzyl chromophore, because of the large enhancement in thermal stability (*7a*) it provides to PAG's. This allowed us to study resists formulated with PAG which gave stronger photo-acids without competitive thermal generation of acid during PEB.   The polymer matrix, poly(4-*t*-butoxycarbonyloxystyrene-sulfone) (PTBSS), was chosen for this study because it yields upon acidolytic cleavage the very hydrophilic poly(hydroxystyrene-sulfone). Poly(hydroxystyrene-sulfone) has an extremely fast rate of dissolution and cannot be significantly dissolution inhibited. This minimized contributions to the rate of dissolution from the PAG's differing hydrophobicities.

In this work we have found that by reducing the PED effect with an overcoat (*2*), the behavior expected from Equation 4 was observed.   This confirmed that without basic contaminants resist sensitivity is governed by the reactivity of the photo-acid due to 'free' H^+. Moreover, when resists are exposed to airborne contamination under controlled conditions, most of the PAG's show a PED effect which increases with the   reactivity of the photo-acid. A few of the PAG's which have polar,

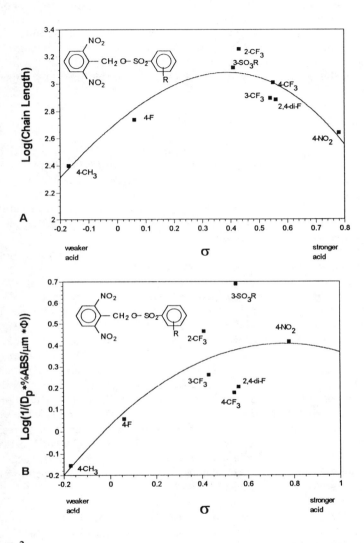

Figure 2
A) Plot of Log(Chain Length) versus the Hammett sigma values for
substitution at the benzenesulfonyl moiety of 2-6-dinitrobenzyl benzenesulfonate PAG's in a PTBSS resist matrix.
B) Plot of Log(1/(D_p *%ABS/μm *Φ) versus the Hammett sigma values
for substitution at the benzenesulfonyl moiety of 2-6-dinitrobenzyl
benzenesulfonate PAG's in a PTBSS resist matrix.

polarizable substituents, which have lone electron pairs, impart a greater resistance to airborne contaminants than expected.   This suggest that other interactions are interfering with the reaction of 'free' H^+ with airborne basic contaminants in the resist matrix.   Among the possible candidates for this are,   the polarizability   induced interaction of undissociated PAG with base, or that of the photo-acid with the resist matrix, or a basic interaction of the lone electron pairs on these substituents with 'free' H^+.

Experimental

Synthesis   Poly(4-*t*-butoxycarbonyloxystyrene-sulfone)   (composition   4-*t*-butoxycarbonyloxystyrene(TBS)/SO$_2$: 3.00/1;   Mw: 105 K, Mw/Mn:1.6) (PTBSS) was synthesized as previously described (*17*).   Tetramethylammonium   hydroxide (TMAH) was obtained from the Johnson Matthey   Company. The PAG's were made by using methods described previously (*7a,b*) and   their identities and purities were confirmed by ^{1}H NMR, IR spectroscopies, and LC/MS.   All other   chemicals were obtained from Aldrich Chemical Co.

Lithography   Scanning electron micrographs (SEM)   cross-sections were obtained on a JEOL   scanning electron microscope.   Exposures were done using a prototype deep-UV exposure tool (numerical aperture=0.35, 5X optics) operating at 248 nm.   All thickness measurements were done on a nanospec thickness gauge (Nanometrics, Inc.) or a Dektak model IIA profilometer.   Photoresist   for lithographic experiments consisted of 14 wt % of PTBSS dissolved in ethyl 3-ethoxypropionate, containing 6 mole % of PAG relative to the polymer's *t*-BOC groups.   The solution was filtered through a series of 1.0, 0.5, and 0.2-μm Teflon filters (Millipore, Inc.). Photoresists were spun (speed   2 K) onto hexamethyldisilazane-primed silicon substrates and prebaked at 105°C for 60 s.   The PEB was done at 115°C for 30 s.   Development was accomplished with 0.17 N aqueous TMAH.   Four PED procedures were employed. In the first two,   clearing doses were evaluated for   large features in a resist with and without a 30 minute delay before PEB.   Additionally, a *t*-butyl methacrylate-methacrylic acid copolymer protective overcoat (*2*) (600 A) was applied and   clearing dose measurements were also made with and without the time delay before PEB. The lithographic results are summarized in Table I.

Quantum Yield   The quantum yields were determined in films of poly(methyl methacrylate) containing different PAG's as previously described (*5,6a*). The results are summarized in Table II.

Absorbance Measurement UV spectra were taken with a Hewlett Packard 8452A diode array spectrophotometer.   Absolute thickness measurements of resists were obtained with a Dektak model IIA profilometer. ABS/μm was found by   dividing the absorbance at 248 nm   by the measured thickness of the resist. The %ABS/μm were calculated from ABS/μm. The results are summarized in Table II.

Thermal Analysis   Differential scanning calorimetry   (DSC) data for the solid samples were obtained using a Perkin-Elmer DSC-7 differential scanning calorimeter interfaced with a TAC 7 thermal analysis controller and a PE-7700 data station. All

Table I Lithographic Results of Post-Exposure Bake Study

PAG[a] R	$\Sigma\sigma$[b]	D_p no overcoat no PED mJ/cm^2	D_p no overcoat 30 min PED mJ/cm^2	D_p overcoat no PED mJ/cm^2	D_p overcoat, 30 min PED mJ/cm^2
4-OCH$_3$	-0.27(0.23)^c	50	54	48	50
4-CH$_3$	-0.17	145	159	108	115
H	0.00	97	113	60	63
4-Cl	0.23	62	69	51	58
2-CF$_3$	0.41^c	64	80	22	26
3-CF$_3$	0.43	51	58	23	25
4-CF$_3$	0.54	67	78	25	25
3-SO$_3$R	0.55	18	40	17	25
2,4-di-F	0.56^c	55	60	26	30
2-NO$_2$	0.63^c	22	36	10	18
4-NO$_2$	0.78	40	50	38	40
3,5-di-CF$_3$	0.86	26	46	6	8

a) For structural assignment please refer to Figure 3.
b) Values taken from reference 29
c) σ' from reference 30 used for situations in which resonance contribution is suppressed for steric (i.e. ortho) or electronic rea sons.

Table II Quantum Yields and Absorbance s of Resists

PAG[a] R	Quantum yield Φ	ABS/μm AU/μm	%ABS/μm %/μm
4-OCH$_3$	0.05	0.60	0.75
4-CH$_3$	0.11	0.22	0.40
H	0.11	0.23	0.41
4-Cl	0.11	0.25	0.44
2-CF$_3$	0.11	0.22	0.40
3-CF$_3$	0.11	0.23	0.41
4-CF$_3$	0.11	0.22	0.40
3-SO$_3$R	0.11	0.38	0.58
2,4-di-F	0.11	0.21	0.38
2-NO$_2$	0.08	0.89	0.87
4-NO$_2$	0.02	0.60	0.75
3,5-di-CF$_3$	0.10	0.76	0.83

a) For structural assignment please refer to Figure 3.

samples were heated from 30 to 450°C at a heating rate of 10°C/min. Samples ranged in mass from 1.20 to 2.00 mg and were encapsulated in aluminum pans. All measurements were obtained in ultra high purity (99.999%) N_2 with a gas flow rate of 20 cc/min. The results from thermal analysis of the PAG's are summarized in Table III.

Results and Discussion

Thermal Stability of PAG The decomposition behavior of the 2-trifluoromethyl-6-nitrobenzyl arylsulfonates was examined by DSC (Table III). Relative thermal stabilities of these materials were judged by measuring the DSC temperature for the maximum rate of the exotherm for decomposition (T_{min}). As shown previously, T_{min} is useful in predicting the relative order of resist stability to PEB (*6,7a,8*). Figure 3 shows that T_{min} decreases with increasing electron withdrawing ability (increasing σ or Σσ for multiple substitution) of substituents on the arylsulfonate moiety. However, even the PAG with the highest Σσ, 2-trifluoromethyl-6-nitrobenzyl 3,5-di-trifluoromethylbenzene sulfonate, still has a respectable thermal stability. As expected, unexposed resists formulated with these PAG's withstood PEB conditions without loss of *t*-BOC (i.e., no change in thickness or IR spectrum). This ensured that thermal decomposition differences between the PAG's were not affecting the measurements of relative Chain Lengths.

Lithographic Results The lithographic sensitivities (D_p's) of the resists were measured under four PED conditions: Using a protective overcoat on the resist with no PED time; using a protective overcoat with a 30 minute PED time; using no protective overcoat with no PED; and finally, using no protective overcoat with a 30 minute PED. These conditions are listed in order of increasing expected exposure to airborne basic contaminants. Table I summarizes these experiments. Several trends can be observed from this data even without correcting for differences in Φ or %ABS/μm between the resists. Under conditions of least exposure to airborne contamination (overcoat present, no PED), the PAG's liberating the most reactive photo-acids give the lowest D_p's (highest sensitivities). As the expected exposure to airborne contaminants worsens, all the resists show consistently increasing D_p's (decreasing sensitivity) symptomatic of the PED effect. The magnitude of this PED effect tends to be the greater for the more reactive photo-acids. Interestingly, the resists containing PAG's based on 4-chloro, 4-methoxy or 4-nitrobenzenesulfonates deviate from this general trend. These resist were far less susceptible to the PED effect.

Hammett Plots of the 'Relative' Chain Length To better understand these trends, $Log(1/(\%ABS/\mu m*\Phi*D_p))$ was plotted against σ (or Σσ for multiple substitution) for the four PED conditions. This was done because according to Equation 2, which was derived empirically from Figure 1, $1/(\%ABS/\mu m*\Phi*D_p)$ should be proportional to the Chain Length for acidolysis of *t*-BOC and is consequently a measure of 'relative' Chain Length. These plots show how this 'relative' Chain Length estimated from lithographic measurements varies with the expected reactivity of the acid, and levels of environmental contamination. Figures 4 A-D show these plots for the four PED conditions. In all of these plots, a resonance suppressed Hammett σ' substitution constant was employed for the 4-CH_3O substituent (Table I) because of the protonation of the 'basic' methoxy moiety

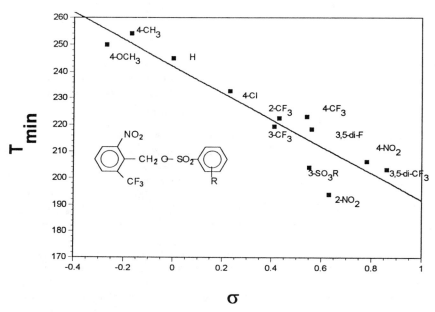

Figure 3

Plot of T_{min} versus Hammett sigma values for substitution at the benzene-sulfonyl moiety of 2-trifluoromethyl-6-nitrobenzyl benzenesulfonate PAG's

Scheme 5

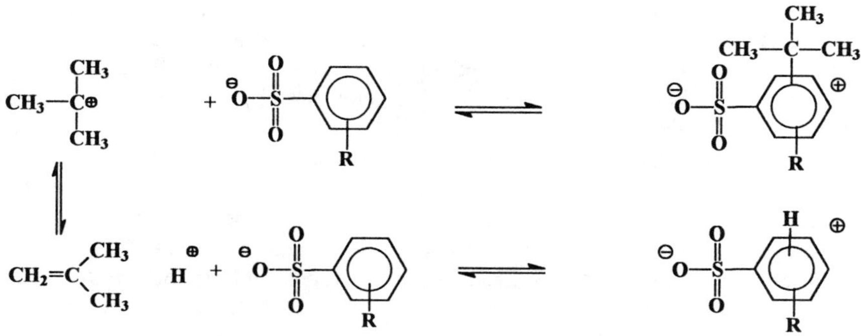

Table III Thermal Stabilities and Calculated Chain Lengths of PAG's

PAG[a] R	$\Sigma\sigma$[b]	T_{min} (°C)	Chain Length Overcoat, No PED	Chain Length No overcoat, 30 min PED
4-OCH$_3$	-0.27(0.23)[c]	250	340	302
4-CH$_3$	-0.17	254	129	88
H	0	245	225	119
4-Cl	0.23	233	249	184
2-CF$_3$	0.41[c]	220	634	174
3-CF$_3$	0.43	223	587	233
4-CF$_3$	0.54	223	558	179
3-SO$_3$R	0.55	204	560	238
2,4-di-F	0.56[c]	219	556	241
2-NO$_2$	0.63[c]	194	876	243
4-NO$_2$	0.78	206	1072	815
3,5-di-CF$_3$	0.86	203	1231	161

a) For structural assignment please see Figure 3
b) Taken from reference 29.
c) σ' from reference 30 used for situations in which resonance   contribution is
   surpressed for steric   (i.e. ortho) or electronic   reasons.

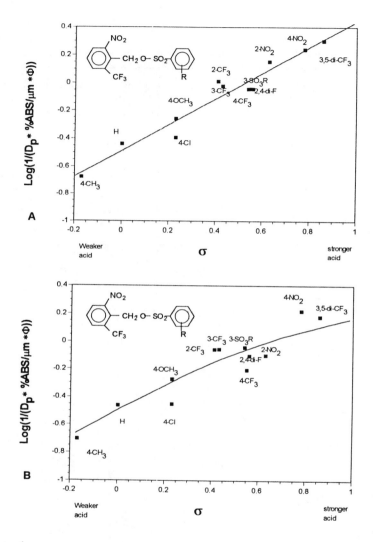

Figure 4

    A) Plot of $Log(1/(D_p *\%ABS/\mu m *\Phi)$ versus the Hammett sigma values for
    substitution at the benzenesulfonyl moiety of 2-trifluoromethyl-6-
    nitrobenzyl benzenesulfonate in a PTBSS resist matrix for
    processing done with an overcoat and without a PED.

    B) Same plot as A but for processing done with an overcoat and a 30
    min PED.

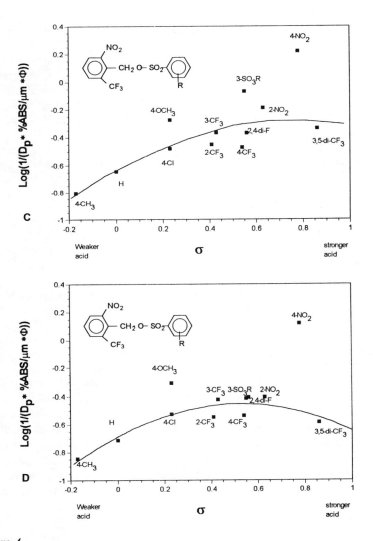

Figure 4

    C) Plot of Log(1/(D*p*%ABS/μm*Φ) versus the Hammett sigma values for substitution at the benzenesulfonyl moiety of 2-trifluoromethyl 6-nitrobenzyl benzenesulfonate in a PTBSS resist matrix for processing done without an overcoat and without a PED.

    D) Same plot as C but for processing done without an overcoat and with a 30 minute PED.

(Scheme 6) which effectively reduces its electron donating capability. This corrected for changes in the relative pK_a of 4-methoxybenzenesulfonic acid versus the other aryl sulfonic acids induced by the 'basic' methoxy.

In Figure 4A it is seen that when airborne contamination is minimized (overcoat, no PED), most of the PAG's give the expected linear increase of $Log(1/(\%ABS/\mu m*\Phi*D_p)$ with σ. This shows that, when environmental contamination is eliminated, the 'relative' Chain Length for t-BOC acidolysis decreases only when the expected reactivity of the acid is decreased (decreasing 'free' H^+) through side reactions induced by electron rich (smaller σ) substituents on the arylsulfonate anion (Schemes 4 and 5).

Figure 4D shows the same plot when the resist is exposed to maximum airborne contamination (no overcoat, 30 min PED). A 'turnover' of the expected increase in $Log(1/(\%ABS/\mu m*\Phi*D_p)$ with increasing reactivity of acid is observed. This is similar to what was observed ($7a$) for resist systems formulated with PAG's with a 2,6-dinitrobenzyl chromophore (Figure 2, A, B). However, there is much less 'scatter' in this data because PED conditions were controlled. This 'turnover' effect indicates that, as the reactivity of the acid increases, another process induced by the airborne contaminants is becoming competitive with the acidolysis of the t-BOC group. A leveling of the 'free' acid concentration by these contaminants is one possibility (Scheme 2). This will be discussed in more detail later.

Figure 4B shows that even with the protective overcoat, after 30 min PED, the turnover effect starts to manifest itself indicating that contamination is starting to leach through. Similarly, Figure 4C, in which PEB was done as quickly as possible but without a protective overcoat, shows that even with a very short PED some exposure to airborne contamination has occurred. The extent of the 'turnover' under these two conditions is intermediate to that observed in Figure 4 A and D.

Evaluation of the 'Turnover' Effect To evaluate how the 'turnover' changes with acid reactivity, the difference in 'relative' Chain Lengths (i.e. Δ $1/(\%ABS/\mu m*\Phi*D_p$)) was calculated for conditions of maximum and minimum exposure to airborne contaminants. A plot of the logarithm of Δ $1/(\%ABS/\mu m*\Phi*D_p$)) against σ was done to evaluate the effect of acid strength (Figure 5). For most of the substituents this Hammett plot shows a linear increase of $Log(\Delta$ $1/(\%ABS/\mu m*\Phi*D_p$)) with σ. This indicates that as the expected reactivity of the photo-acids increases (more free H^+) the effect of basic airborne contaminants becomes greater. There are several possible ways of explaining this:

Mechanistic Models for the 'Turnover' Effect. The 'turnover' effect may be kinetic, governed only by how quickly basic impurities encounter photo-acid in the viscous resist matrix during PED. Photo-acids with electron withdrawing substituents would give a higher concentrations of 'free' H^+ increasing the chance of an encounter during PED. Differences in concentration of 'free' H^+ of 2-3 orders of magnitude are expected between the strongest and weakest sulfonic acid that we have studied.

It may be an equilibrium effect in which basic impurities and photo-acid diffuse so quickly towards one another that an equilibrium is set up during PED. A leveling of the 'free' H^+ concentration would then occur. The extent of this leveling would depend on the difference in pK_a's between the photo-acid and the protonated basic impurities. In this way, the greater the pK_a of the photo-acid, the greater the extent of leveling would become.

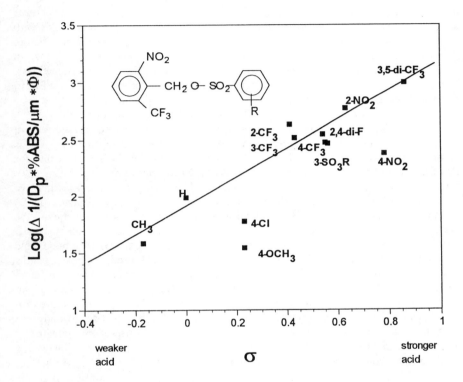

Figure 5

Plot of Log(ΔI/(Dp*%ABS/μm*Φ) versus the Hammett sigma values
for substitution at the benzenesulfonyl moiety of 2-trifluoromethyl-
6-nitrobenzyl benzenesulfonate in a PTBSS resist matrix.

Scheme 6

Alternatively, in a non-polar medium, such as the exposed resist before PEB, the sulfonic acids may be so poorly dissociated that no reaction is possible between the basic contaminants and photo-acid. It is known, for instance, that tosic acid has a pK_a of only 6.4 in propylene carbonate. The basic contaminants would then not 'see' any free H^+ during the PED time. During PEB the increase in temperature could decrease the pK_a of the stronger photo-acids (largest σ)--or that of their conjugate acids with the resist polymer-- sufficiently for them to react with the basic contaminants before they can evaporate from the surface of the resist (for an example of this see (18)). In contrast, the weaker sulfonic acids --or their conjugate acids-- would be consumed less, being unable to dissociate sufficiently rapidly during the initial stages of the PEB to compete with base volatilization. Therefore, the weaker sulfonic acid would still be available for the acidolytic cleavage of t-BOC during the later stages of the PEB.

Calculation of 'Absolute' Chain Length So far, we have discussed relative values of Chain Length as measured by the ratio $1/(D_p*\text{Chain Length}*\Phi)$. It is possible to obtain absolute values for the Chain Length since it is known that for PTBSS about 93% deprotection of t-BOC is needed to achieve base solubility (19). This calculation allows us to make an evaluation of two major assumptions that we have made in this study. The first is that equation 2, although it depends on D_p measurements done at high t-BOC conversion, still gives a good estimate of the average rate of t-BOC acidolysis during the PEB. The second is that all the resist formations achieve solubility at 93% deprotection of t-BOC regardless of any possible dissolution inhibition from the PAG. Table III gives a list of Chain Lengths calculated, from D_p's for conditions of minimum and maximum environmental contamination. If contamination is minimized, the Chain Length for both strong (~1000) and weak acids (~100) compare reasonably well to those in the literature (7a, 20). This confirms that our assumptions were reasonably valid. Not surprisingly, as with the relative Chain Length, the absolute Chain Length decreases dramatically for most of the resist formulations when they are exposed to increasing environmental contamination by lengthening the PED time (Table III). As before, the effect is generally greater as the strength of the acid released from the PAG is increased. Using the same argument as above, Hammett plots can be made illustrating the effect of changing substituents on Log(Chain Length) under conditions of changing environmental contamination (Figure 6), or of Log(Δ Chain Length) (Figure 7) to illustrate the changes in the PED effect with substitution pattern. These plots show exactly the same trends as we have previously discussed in the plots based on $1/(\%\text{ABS}/\mu m*\Phi*D_p)$, the 'relative' Chain Length.

Mechanistic Models for the Anomalous Substituent Effect. A few of the PAG's impart a PED behavior to the resists that is more complex. These materials, 4-methoxybenzenesulfonate, 4-chlorobenzenesulfonate, and 4-nitrobenzenesulfonate, are much less susceptible to the PED effect than would be predicted by a simple competition between the reactions depicted in Schemes 1, 2 and 4 (or 5). Three possible mechanisms will be postulated.

The first possible mechanism is a competitive absorption of basic airborne contaminants at the resist surface. Scheme 7 shows such a·scenario in which basic airborne contaminants can interact at the exposed resist surface with either polymer (21), PAG or liberated acid. Since the polymer component is not changed in the different resists we have evaluated, only the changes in undissociated PAG's or photo-released acids can be responsible for the changing behavior of the resists.

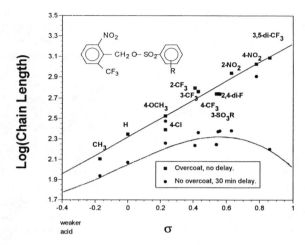

Figure 6

Plots of Log(Chain Length) versus the Hammett sigma values for
substitution at the benzenesulfonyl moiety of 2-rifluoromethyl-6-nitrobenzyl
benzenesulfonate in a PTBSS resist matrix. Resists processed either
with an overcoat and no PED or without an overcoat and with a 30 minute

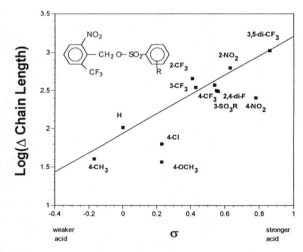

Figure 7

Plot of Log(ΔChain Length) versus the Hammett sigma values for
s ubstitution at the benzenesulfonyl moiety of 2-trifluoromethyl-6-nitrobenzyl
benzenesulfonate in a PTBSS resist matrix.

After the polymer, the PAG is the second most important component (6 % mole) of the resist.   Even in exposed regions 90-96% of the PAG remains unaltered. Thus, basic airborne contaminants will likely   meet with undissociated PAG before free sulfonic acid.   Therefore,   if the PAG associates strongly with the airborne basic contaminants,   this will   slow   the rate of the   reaction with acid by slowing diffusion of base into the film.   The importance of the diffusion rate of airborne basic impurities into the resist film has been established by others (22). In our case, all three of the PAG's which are resistant to airborne contaminants have polar 4-substituents susceptible to dipolar interactions.   Conversely,   most of the materials showing poor resistance to the PED effect either have groups that are weakly polarizable (F, CF_3) or ones with poor dipoles (H, CH_3) and would consequently form poor dipolar interactions.   One exception, the PAG with a $2-NO_2$ substituent--unlike the $4-NO_2$ substituted material-- suffers from a vicinal interaction with the SO_3R moiety which would make interactions with contaminants difficult.   The only remaining exception, the $3-SO_3R$ substituted PAG, does show an enhanced resistance to PED effect if the delay time is shortened (Figure 4C).   According to Scheme 7, this entails   that   although   the $3-SO_3R$ substituent can   interact with   airborne contaminants it does so less effectively than the other polar substituents.   A more detailed discussion of polarizability and its possible significance in explaining the PED effect will be done at the end of this section.

A second possible mechanism is that   photo-acids bearing polar polarizable group   could also undergo interaction with the resist matrix limiting its diffusion towards absorbed basic impurities (Scheme 8).   Thus, using arguments similar to those described above, this model also explains the differences in the magnitude of the PED effect observed due to the differing structure of the PAG resist component.

A third possibility is that the 4-MeO, 4-Cl and $4-NO_2$ substituents, which all have lone pairs of electrons, are acting   as basic moieties and are competing with airborne contaminants for the 'free' H^+ thereby slowing their rate of reaction with these and suppressing the turnover (Scheme 9). Although this may account for the behavior of the PAG with a 4-MeO (Scheme 6) substituent (we attempted to correct for this by employing σ') the other substituents bearing lone pairs, 4-Cl and $4-NO_2$ are unlikely to be protonated. For example, the pK_a of protonated nitrobenzene is only -12 far less than a protonated carbonyl group ~-7 (23). Moreover, there should not be a very big pK_a difference between 2-nitrobenzenesulfonate and 4-nitrobenzenesulfonate protonated PAG's (or the corresponding acids) but nonetheless resists formulated with these two materials show very different PED behavior as described above. However, hydrogen-bond formation decreasing the accessibility of 'free' H^+ cannot be discounted as a possibility since NO_2, MeO, Cl are hydrogen bond acceptors (24) and sulfonic acids are potent hydrogen bond donors (25). The intermolecular hydrogen bonding of 2-nitrobenzenesulfonic acid, would be sterically hindered by the vicinal SO_3H group.

Correlation Between Molecular Polarizability and the 'Anomalous' PED Substituent Effect. Molecular polarizability governs many physical properties of materials. For instance, Debye interactions (dipole-induced-dipole interactions) explain the mutual solubility of many polar and polarizable molecules (26). Debye interactions also explain why   polarizable aromatic molecules are easily extract by N-methypyrilidone from oil (26).   Similarly, N-methylpyrilidone from the air would be expected to have a greater affinity for PAG's with polar polarizable substituents (i.e. 4-MeO, $4-NO_2$ and 4-Cl). Increasing the polarizability of alkyl groups on amines

Scheme 7

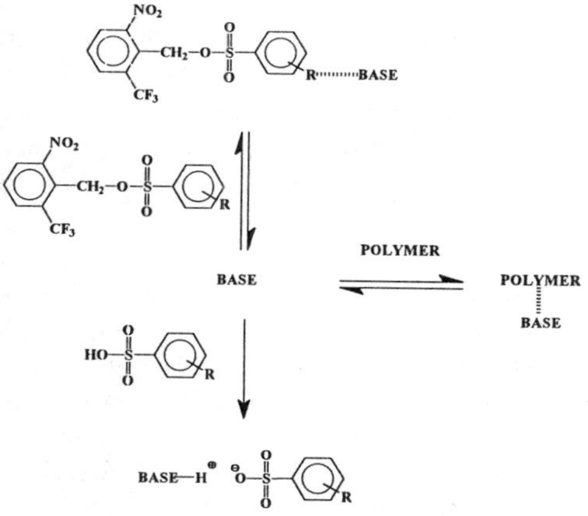

Scheme 8

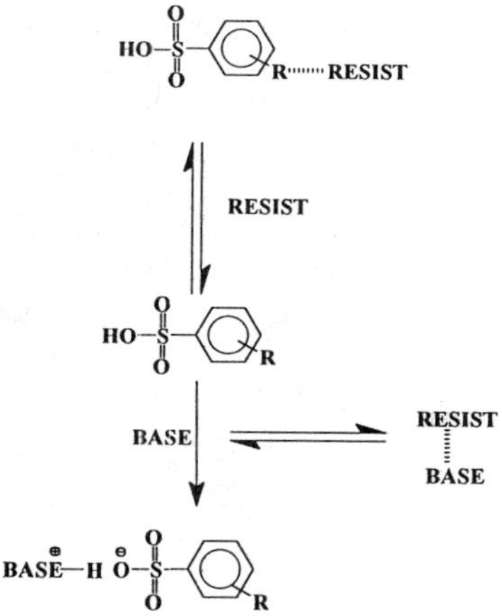

Scheme 9

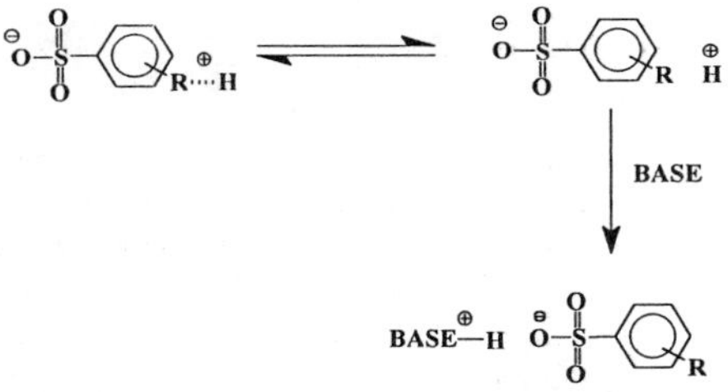

generally increases their gas phase basicity (27) (t-butylNH$_2$ > i-ProNH$_2$ > MeNH$_2$ > NH$_3$). Thus the PAG's with polar polarizable groups would also be expected to interact better with strong polarizable airborne bases such as (Me)$_3$N and (Me)$_2$NH.

Measurements of both the 'anomalous' PED substituent effect and the relative polarizabilities of the PAG's are needed to see if there is indeed any correlation between these two properties. The PAG's with polar substituents gave lower Log(Δ Chain Length) values than expected from the regression line found for the other PAG's in the plot of Log(Δ Chain Length) versus σ (Figure 7). The difference, ΔLog(ΔChain Length), between the observed value of Log(Δ Chain Length) and the value expected from the regression line established by the 'non-anomalous' substituents can be used as a measure of the effect of the polar polarizable groups on the PED effect. Physical properties such are the refractive index are often used a measure the polarizability of molecules. Unfortunately, refractive indexes of suitable model compounds could not be found. However, another example of a physical property governed by molecular polarizability is the nematic to isotropic (N-I) transition temperature in nematic liquid crystals (28). The N-I transition in liquid crystals is know to increase in the order 4-MeO >4-NO$_2$ >4-Cl > 4-CH$_3$ which is thought to indicate the increasing order of molecular polarizability (28). This substitution effect parallels the in the difference in PED effect we have observed (Δ log(Δ Chain Length) in the PAG's with polar polarizable 4-subsituents. This is demonstrated by a plot of this difference against a measure of the relative polarizability induced by the 4-substituents (the N-I transition temperature of a typical 4-substituted aryl nematic liquid crystal (28)) (Figure 8). This indicates that PAG's which have substitution patterns that induce better polarizability tend to give less of a PED effect.

This correlation between 'polarizability' and the unusual PED behavior tends to point towards either the first or second mechanism (Scheme 7 and 8) where the polarizability reduces the tendency for photo-acid to encounter base by either 'tying-up' base, photo-acid or both and preventing them from diffusing towards each other. However, a competition between the hydrogen bonding of the photo-acids and their reaction with airborne base contaminants is still a possibility.

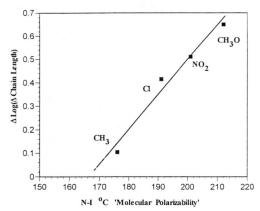

Figure 8

Plot of ΔLog(Δ Chain Length) versus T-N transition temperature 'Polarizability of substituents.'

The 'Anomalous' PED Effect and Lithographic Imaging of Small Features. Further evidence for the PAG induced improvement in resistance to the PED effect was found by looking at SEM cross-sections. This was done by comparing the resist formulation with 2-trifluoromethyl-6-nitrobenzyl 4-methoxybenzenesulfonate to that with a strong acid PAG ($\sigma > 0.54$) such as 2-trifluoromethyl-6-nitrobenzyl 2,4-difluorobenzenesulfonate. As can be seen in Figure 9 A, B, the resist with the strong acid PAG, even with a protective overcoat, gives considerable T-topping of 0.5 μm lines and spaces with a 30 minute PED. In contrast, the resist with the 4-methoxybenzenesulfonate PAG showed no T-topping (0.35 μm lines and spaces) after 30 minute PED (Figure 10 A, B). Furthermore, T-topping is not observed even when a protective overcoat is omitted, although some rounding of features is noticeable. This is probably caused by volatilization of acid. Although the other resistant PAG's also offer some protection against T-topping, as expected, the effect decreases in the order 4-CH$_3$O > 4-NO$_2$ > 4-Cl.

Conclusion

We have established the following decreasing order of resistance to the PED effect: 4-OCH$_3$ >4-NO$_2$ > 4-Cl > >3-SO$_3$R > > 4-CH$_3$ ≥ 4-H >2-CF$_3$ ≥ 3-CF$_3$ ≥ 4-CF$_3$ > 2,4-diF> 2-NO$_2$> 3,5-diCF$_3$. These trends have been rationalized as follows:

It has been shown that when airborne contamination is minimized, the resist sensitivity increases with the reactivity of the acid. Several mechanisms have been postulated to account for this increase in reactivity (changing acidity, nucleophilicity).

For PAG's that have substituents that do not interact well with other functionalities, the sensitivity of the resist to the PED effect increases with the reactivity of the acid. Several mechanisms have been postulated to account for this: It may be a kinetic effect, governed only by how quickly basic impurities come across photo-acid in the viscous resist matrix during PED. It may be a leveling of the 'free' H$^+$ concentration, which would increase with the difference in pK$_a$'s between the photo-acid and the protonated basic impurities. Finally, the weaker sulfonic could be reacting less with the impurities because they are unable to dissociate sufficiently rapidly during the initial stages of the PEB to compete with base volatilization.

The magnitude of the PED effect appears to be lessened in the presence of PAG's which have polar-polarizable substituents that can interact well with other moieties. Three possible mechanisms have been outlined to explain this effect: The undissociated PAG interacting with airborne bases, photo-acids interacting with the resist matrix and "free" H$^+$'s or hydrogen bonded photo-acids interacting with the substituents on the PAG. All three mechanisms could account for the effect, however, the correlation between the expected order of polarizability and the lessening of the PED effect tends to favor the first two possibilities. The poor basicity expected to be induced by the lone pairs of the 4-Cl and 4-NO$_2$ substituents is not a strong argument in favor of the third mechanism. However, one cannot discount the possibility a hydrogen bond interaction of the photo-acid (or some other H$^+$ complex).

A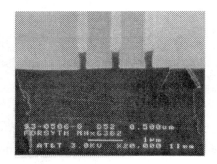

B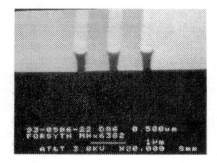

Figure 9:
> *SEM's of 0.50 mm lines and spaces obtained with a resist formulated*
> *with a strong acid PAG (s > 0.54) such as*
> *2 -trifluoromethyl-6-nitrobenzyl 2,4-difluorobenzenesulfonate.*
> *A) Resist with overcoat, no PED, Dp= 52 mJ/cm2.*
> *B) Resist with overcoat, 30 minute PED, Dp = 86 mJ/cm2*

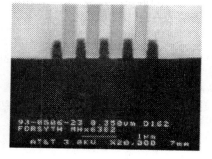

Figure 10

SEM's of 0.35 mm lines and spaces obtained with a resist formulated
with 2-trifluoromethyl-6-nitrobenzyl 4-methoxybenzenesulfonate.
A) Resist with overcoat, no PED, Dp= 162 mJ/cm2.
B) Resist with overcoat, 30 minute PED, Dp = 162 mJ/cm2.

To minimize the PED effect, an ideal PAG would both generate a less reactive photo-acid (small σ) and have a polarizable substituents that can interact well with other molecules. However, since a weaker acid would reduce sensitivity (unless PEB temperature or time is increased) then a better compromise might involve a PAG that generates a reactive photo-acid but is substituted with a polarizable functionalities that can hinder the PED effect.

References

1. Willson, C. G.; Frechet J. M. J.; Tessier, T; Houlihan, F.M., *J. Electrochem. Soc.* **1986**, 133(1), 181: Ito, H.; *Proc. KTI Microelectronics Seminar,* **1988**, 81: Reichmanis, E.; Houlihan, F. M.; Nalamasu, O.; Neenan, T.X., *Chem. of Mat.* **1991**, 3, 397: Lamola, A. A.; Szmanda, C.R.; Thackeray, J. W; *Solid State Technol.* 1991, 34(8), 53.

2. Nalamasu, O.; Reichmanis, E.; Cheng, M.; Pol, V.; Kometani, J. M.; Houlihan, F. M.; Neenan, T.X.; Bohrer, M.P.; Mixon, D.A.; Thompson, L. F.; *Proc. Soc. Photo-Opt Instr. Eng.* **1991**, 1466, 13.

3. MacDonald, S. A.; Clecak, N.J.; Wendt,, H. R.; Willson, C. G.; Snyder, C. D.; Knors, C.J; Peyoe, N.; Maltabes, J. G.; Morrow, J.; Mc Guire, A.E.; and Holmes, S. J., *Proc. Soc. Photo-Opt. Instr. Eng.* **1991**, 1466, 2.

4. Ito H.; England, W.P.; Clecak, N.J.; Breyta, G.; Lee, H.; Yoon, D.Y.; Sooriyakumaran, R.; and Hinsberg, W.D., *Proc. Soc. Photo-Opt. Instr. Eng.,* *1925, Proc. Soc. Photo-Opt. Instr. Eng.* **1993**, 1925, 65.

5. Houlihan, F.M.; Shugard,, A.; Gooden,, R.; Reichmanis, E.; *Macromolecules*, **1988**, 21, 2001.

6. Neenan, T.X.; Houlihan, F.M.; Reichmanis, E.; Kometani, J. M.; Bachman, B.J.; Thompson, L. F.; *Macromolecules*, **1990**, 23, 145.

7. a) Houlihan, F.M.; Neenan, T.X.; Reichmanis, E.; Kometani, J. M.; Chin, T., *Chem. Mater.*, **1991**, 3, 462: b) Houlihan, F. M.; Neenan, T. X.; Reichmanis, E.; **1993**, Us Patent 5,200,544.

8. Houlihan, F.M.; Chin, E.; Nalamasu, O.; Kometani,, J.M.;, Neenan, T. X.; Pangborn, A., *J. Photopolym. Sci. Techn.*, **1993**, 6(4), 515.

9. Thompson, L.F.; Bowden, M.J.; In *Introduction to Microlithography*; Editors, Thompson, L.F., Willson, C.G., Bowden, M.J.; ACS Symposium Series 219; Washington DC; 1983; Chapter 4; pp 170.

10. The authors do not imply that actual 'free' H^+ exists in the resist matrix but employ this euphemism to refer to acid which is dissociated from the sulfonate moiety and loosely associated with some other components of the resist matrix.

11. Furukawa, N., Fujihara, H.; in *The Chemistry of Sulphonic Acids, Esters and their Derivatives*; Editors, Patai, S., Rappoport, Z.; The Chemistry of Functional Groups; John Wiley and Son; New York, NY; 1991; Chapter 7; pp 266.

12. King J.F., In *The Chemistry of Sulphonic Acids, Esters and their Derivatives*; Editors, Patai, S., Rappoport, Z.; The Chemistry of Functional Groups; John Wiley and Son; New York, NY; 1991; Chapter 6; pp 250.

13. Perrin, D.D., Dempsey, B.; In *pK_a Prediction for Organic Acids and Bases*; Chapman and Hall; New York, NY; 1981; pp 128.

14. Izutsu, K.; In *Acid-Base Dissociation Constants in Dipolar Aprotic Solvents*; Blackwell Scientific Publications; Boston, Mass; 1990; pp 131.

15. Zefirov, N.S.; Koz'min, A.S., *Acc. Chem. Res*; **1985**; 18; 154.

16. Alm. R.; *Modern Paints and Coatings*; **1980**; V 70; No 10; 88.

17. Tarascon, R. G.; Reichmanis, E.; Houlihan, F. M.; Shugard, A.; Thompson, L.F., *Polym. Eng.   Sci.* **1989**, 29(3), 850.

18. This would occur if the effective pK_a of the arylsulfonic acid is high enough in the non polar polymer matrix so that $-d(pK_a)/dT = (pK_a + 0.052\Delta S°)/T$ (taken from reference 13, pp 7) predicts a decrease in pK_a with temperature. For example, assuming a pK_a of 6.4 (toluenesulfonic acid in propylene carbonate taken from reference (*14*)) and a $\Delta S \sim -100$ $\text{Jdeg}^{-1}\text{mole}^{-1}$, $d(pK_a)/dT = -0.004$, increasing temperture by 100 deg will decrease the pK_a by ~0.4. A similar argument can be made for the dissociation from the BH^+ 'conjugate acid' of the resist polymer. In this instance, since no change of charge occurs, the entropy change is far less ~ -17 mJ/cm^2 giving, $-d(pK_a)/dT = (pK_a - 0.9)/T$, $d(pK_a)/dT = -0.018$. This predicts a much greater dependance on temperature, thus assuming that the pK_a of this conjugate acid is 6.4 then increasing the temperature by $100°C$ will decrease it ~1.8.

19. Nalamasu, O.; Vaidya, S.; Kometani, J.M.; Reichmanis, E.; Thompson, L.F., *Proc. Reg. Tech. Conf. on   Photopolymers*, Ellenville, NY Oct 28-30, **1991** 225.

20. Mc Kean, D. R.; Schaedeli, J.; MacDonald, S.A., In *Polymers in Microlithography* ACS   Symposium Series No 412; Reichmanis, E,, MacDonald, S.A., Iwayanagi, T. Eds   American Chemical Society; Washington DC 1989, 27.

21. There is also probably some residual spinning solvent, but the amount of this should also remain unchanged since the processing conditions for the different resists are the same.

22. Hinsberg, W.; MacDonald, S.A.; Clecak, N.; Snyder, C., *J. Photopolym. Sci. Tech*, **1993**, 6(4), 535.

23. Lowry, T.H. Richardson, K.S.; In *Mechanism and Theory in Organic Chemistry*; Harper and Row; New York, NY; 1976; pp149.

24. Ferguson, L.N.; In *Organic Molecular Structure*; Willard Grant Press; Boston, Mass.; 1975; pp 19.

25. Reference (*11*) pp 271.

26. Grant, D. J.W.; Higuchi, T., In *Solubility Behavior of Organic Compounds*; Editor,   Saunders Jr. W. H.; Techniques of Chemistry, Volume 21; Wiley Interscience; New York, N.Y.; 1990; pp 69.

27. Reference (*24*) pp 79.

28. Coates, D.; In *Liquid Crystals Applications and Uses*; Editor Bahadur, B.; World Scientific; New Jersey; 1990; Vol 1; pp 103.

29. Hansch, C., and Leo, A., In *Substituent Constants for Correlation Analysis in Chemistry and   Biology*; Wiley- Interscience; New York, 1979, pp1-69.

30. *Steric Effects in Organic Chemistry*, Newman, M. S.,   John Wiley and Son, Chapter   13 " Separation of Polar, Steric, and Resonance Effects in Reactivity," 1956, pp 556.

RECEIVED August 14, 1995

Chapter 7

Following the Acid: Effect of Acid Surface Depletion on Phenolic Polymers

James W. Thackeray, Mark D. Denison, Theodore H. Fedynyshyn, Doris Kang, and Roger Sinta

Shipley Company LLC, 455 Forest Street, Marlboro, MA 01752–4650

This paper reports on the diffusion coefficients for acid loss for triflic acid (TFA), toluenesulfonic acid (TSA), camphorsulfonic acid (CSA), and methanesulfonic acid (MSA) in an hydrogenated poly(p-vinyl)phenol matrix. A direct method for following acid content was derived from the TBPB spectrometric titration method. In this manner, the E_a for acid loss for TSA, MSA, and TFA was found to be 76, 87, and 106 kJ/mol, respectively. Using the E_a values, acid loss could be predicted over a wide range of processing conditions. It was found that MSA had an order of magnitude greater D than TSA and TFA, which renders MSA an impractical acid for use in chemically amplified resists. For TSA and TFA, D was found to be 1.2 x 10^{-5} μm^2/s and 4.8 X 10^{-6} μm^2/s, respectively, at 90°C. These values make these acids usable at 90°C, but TSA will be more problematic than TFA. All of these acids show sufficient acid loss at room temperature so that long term staging of exposed resists is not recommended. For DUV lithography with chemically amplified resists, we recommend linked PEB tracks directly to the exposure tool so that the maximum postexposure delay is 15 minutes. The one acid that shows virtually no acid loss even at 150°C was CSA. This acid shows no evidence of 't-tops' and may be the acid of choice for high thermal processing.

Chemically amplified resists have suffered from lack of environmental stability as evidenced by the formation of t-top structures at the surface of imaged positive resists, and by the formation of rounded images with negative resists.[1,2] The primary sources of these phenomena have been pinpointed to surface acid neutralization by airborne contaminants[3], and/or surface acid evaporation due to acid volatility under postexposure bake processing conditions.[4] In this paper, a sensitive method for determining acid concentration has been used in order to determine the fate of photogenerated acid under various process conditions.[5,6] First, the acid generation

0097–6156/95/0614–0110$12.00/0
© 1995 American Chemical Society

efficiency for each acid generator is determined in a poly(p-vinyl)phenol matrix. The amount of acid retained at room temperature at maximum exposure is used as a baseline acid concentration in the film. Then, the acid loss is monitored as a function of postexposure bake (PEB) temperature and PEB time. In this manner, the diffusion coefficient, D, for acid loss can be determined for a series of PEB temperatures. From these results, an Arrhenius plot of D vs. PEB temperature allows the determination of the activation energy for acid loss. The diffusion coefficient for acid loss is directly related to a number of factors: the relative volatility and strength of the photogenerated acid, the solvent type and the % solvent retained, and the relation between the T_g of the resist film and the softbake temperature during processing. Scheme I illustrates the possible reasons for acid loss. An earlier paper used a 'Sandwich blue' test where TBPB embedded in a polymer matrix is coated on an indicator quartz wafer which is placed on the resist during processing.[6] We elected not to use this method because the intimate contact of the indicator wafer to the resist could lead to diffusion due to intermixing. Instead, we remove the resist using THF after processing and add the extract to a known amount of TBPB in THF. This method circumvents any intermixing during processing.

The acid loss from the phenolic matrix can be associated with three possible sources: first, the solvent evaporation can act as a carrier of photogenerated acid. The relative solvent evaporation rate is determined by the solvent chosen and the softbake processing conditions of the resist film.[7] For solvent choice, only two solvents are currently recommended for microlithographic resists: propylene glycol monomethyl ether acetate (PGMEA), and ethyl lactate (EL). The relative solvent retention and evaporation rate of these two solvents should be critical to acid loss. If one can anneal the resist film above the polymer T_g, the solvent evaporation rate should be minimized and thus eliminated as a carrier of acid out of the film. The second source of acid loss is the intrinsic acid volatility out of the polymer film. Because of the large surface to volume ratio in 1 μm thick resist films, the ability of the acid to sublime or evaporate during heating will lead to significant acid loss. If the acid complexes to the phenolic matrix then the acid loss may be reduced. The third mechanism for acid loss is through neutralization by basic substances in the atmosphere such as ammonia, n-methylpyrrolidone, or tetramethylammonium hydroxide. The tetrabromophenol blue titration may not be able to monitor this type of acid loss if the ammonium salt formed also triggers bleaching at 620 nm. In this paper, an attempt will be made to sort out the relative acid loss of various sulfonic acids from a phenolic matrix. The acids studied in this work are methanesulfonic acid, camphorsulfonic acid, toluenesulfonic acid, and trifluoromethanesulfonic acid. A comparison of the fundamental acid loss measurements to lithographic results with PAGs that generate these acids will be made as well.

II. Experimental

A.   Materials.   Partially hydrogenated (~10%) polyvinylphenol (PVP) was obtained from Maruzen. Triarylsulfonium triflate (TASOTf), triarylsulfonium tosylate (TASOTs), triarylsulfonium camphorsulfonate (TASOCs) and tris(1,2,3-methanesulfonyloxy-)benzene (TMSB) were synthesized. Propylene glycol monomethyl ether acetate (PGMEA) and diacetone alcohol (DAA) were obtained from commercial vendors.

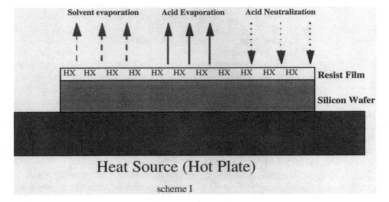

scheme I

Scheme I. The fate of the photogenerated acid upon postexposure bake heating is determined by three loss mechanisms: solvent evaporation rate, acid evaporation rate, and acid neutralization.

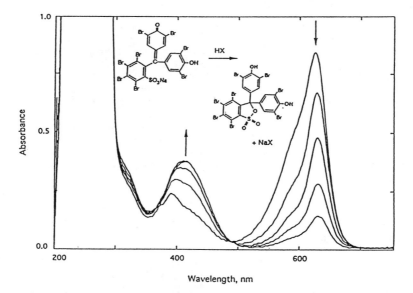

Figure 1. UV-Visible spectra of tetrabromophenol blue(TBPB) in THF as a function of trifluoromethanesulfonic acid concentration. The spectral bleaching at 620 nm is caused by successive addition of triflic acid to the solution in 0.06 μmol increments.

Tetrabromophenol Blue (TBPB) was obtained as the sodium salt from Aldrich. Tetrahydrofuran (THF) was obtained from Mallinckrodt or Aldrich and used as received.

B. Formulations and Coatings. Polymer/PAG solutions consisted of ~21% by weight PVP and 0.022 mmol PAG /g PVP in either PGMEA (for formulations containing TASOTf or TMSB) or 1:1 PGMEA/DAA (for formulations containing TASOTs). The solutions were passed through Gelman Acro 50 0.2 µm filters prior to use.

Polymer/PAG formulations were spin coated onto HMDS oven vapor primed 4 inch silicon wafers to a film thickness of ~9800 Å ± 25 Å as measured by a Prometrix SM300 system. The coated wafers were softbaked at 120 °C for 60 s, unless otherwise noted, and then flood exposed on a GCA ALS Laserstep DUV stepper (KrF laser, λ = 248.4 nm.) In the acid loss experiments, exposed wafers were postbaked at 120 °C, 150 °C, or 170 °C, as noted. In the acid generation experiments, no postbake was applied to the exposed wafers.

C. Acid Measurements. Each film to be analyzed was stripped from the Si wafer with THF and added to 2 mL of a stock solution of 0.1 mM TBPB in THF. The total volume of the solution was then raised to 10 mL by addition of THF, and the resulting solution was characterized on a Cary 3 UV-Vis spectrometer. The amount of acid in the film could be determined by monitoring the absorbance change of the TBPB indicator at 620 nm. Calibration curves showing a linear relationship between the amount of acid added to the indicator solution and the resulting absorbance change were used to interpolate the acid concentration in the films.

D. Lithographic Experiments. The resists used for this evaluation comprised a series of photoacid generators of varying anionic and cationic components in a polymeric matrix of poly (p-vinyl) phenol partially blocked with a proprietary blocking group. The molecules examined included a series of PAGs containing the triaryl sulfonium cation paired with three different specific counter anions, trifluoromethane sulfonate (TFA), p-toluene sulfonate (TSA), camphor sulfonate (CSA).

All of the above formulation matrices were generated using 5% w/w PAG as a function of polymer weight. The lithographic processing was identical for all of the evaluations unless otherwise noted. The resists were spin coated on HMDS oven vapor primed 4 inch silicon wafers to a film thickness of 7200Å ± 25Å as measured by a Prometrix SM300 system. The coated wafers were softbaked at 110°C for 60 s. Exposures were performed on a GCA ALS Laserstep DUV stepper (NA= 0.35, 248.4 nm KrF laser) and postbaked at 85°C for 60 seconds unless otherwise noted. The resists were developed with 0.21N developer using a 45 second double spray puddle process.

In order to minimize the effects of exposure to postexposure bake delay, all wafers were baked immediately following exposure, except during delay studies where exposure to bake delays were staged to yield 30 and 60 minute delays.

III. Results.

A. PAG Acid Generating Efficiency. Figure 1 illustrates the response of tetrabromophenol blue dye solutions to low trifluoromethanesulfonic acid

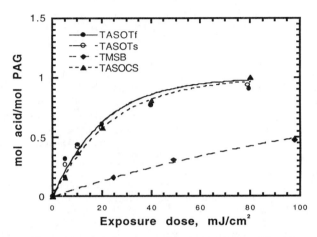

Figure 2. Acid generated vs. exposure for the PAGs used in this paper. The fitted lines are derived from equation (1).

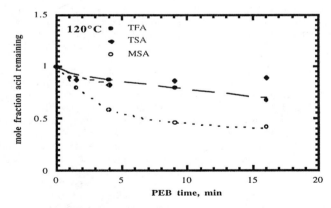

Figure 3. Plot of fraction of acid remaining in the resist film vs. PEB time under isothermal conditions, 120°C. Fitted curves are derived from equation (2).

concentrations. The dynamic sensitivity of this method monitors acid generation in resist films at normal resist thicknesses of 1 μm. All the acids studied in this paper show a linear response in bleaching at 620 nm with TBPB. The PAG acid generating efficiency can be determined using the titration method. Figure 2 illustrates the mole acid generated per mole of PAG vs the exposure dose. By fitting these plots to equation (1), the relative acid generating efficiency, α, can be determined:

(1) $\quad$ mol acid/mol PAG $= 1 - e^{-\alpha E}$

Table 1 indicates the relative acid generating efficiency, α, for the PAGs used in this study. The triarylsulfonium PAGs are the most efficient acid generators, with TMSB being the least efficient acid generator. The relative acid generating efficiency can not be totally explained by the relative extinction coefficient of the PAGs at the exposing wavelength, 248 nm. For instance, even though TMSB has an extinction coefficient 100 times less than TASOTf, TASOTf has only ten times the acid generating efficiency of TMSB. This result is due to the indirect sensitization of TMSB through the phenolic polymer matrix.[8]

B. Effect of Thermal Processing on Acid Loss. Utilizing the acid titration method, the amount of acid retained in the resist film could be monitored as a function of postexposure bake temperature and time. Each film was exposed such that the PAG was completely destroyed and the maximum acid concentration in the film was produced. In this manner, thermal production of acid was prevented. The measurement error for each data point was approximately 5%. The acids selected varied a great deal in pK_a values and relative melting point. The PEB temperatures used were sufficient to force acid evolution over a reasonable time observation such as ~20 minutes. Figure 3 shows the acid loss as a function of time for a PEB temperature of 120°C for methanesulfonic acid (MSA), trifluoromethanesulfonic acid(TFA), and toluenesulfonic acid (TSA). The relative acid retention is not surprising: MSA<TSA<TFA. Diffusion coefficients for each acid can be derived using the Crank-Nicholson equation in equation (2).[9] The stability of the calculation was checked via von Neumann stability analysis and it was found that Eq. (3) was satisfied such that the coefficients of the differential equations are varying at a sufficiently slow rate to be considered constant in space and time. Therefore, the concentration of acid can be determined from the previous time period's acid concentration for a given time period(Δt) and a given distance increment (Δx):

(2) $\quad C(x,t+1) = C(x,t) + D\Delta t/\Delta x^2[C(x-1,t) - 2C(x,t) + C(x+1,t)]$

(3) $\quad 2D\Delta t/\Delta x^2 \leq 1$

In this model, it is assumed that all acid molecules that reach the surface of the resist face a condition of zero acid present at the resist/air interface. It was also assumed that no acid loss occurs at the resist/Si interface. Figure 4 shows the acid loss vs. PEB time for 150°C PEB temperature for all the acids plus an even less volatile acid, camphorsulfonic acid (CSA). At 150°C, the fits of the modeled diffusion rates show lack of agreement after the initial PEB times, which indicates that the diffusion coefficient may be decreasing with time. Also, Figure 4 illustrates that CSA is retained in the resist film even at temperatures of 150°C. Figure 5 shows the acid loss vs. time for 170°C PEB temperature. Because of the extreme PEB temperature the longest time point for these acids

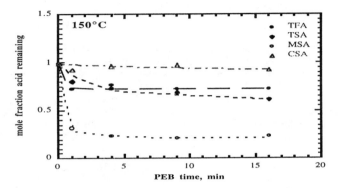

Figure 4. Plot of fraction of acid remaining in the resist film vs. PEB time under isothermal conditions, 150°C. Fitted curves are derived from equation (2).

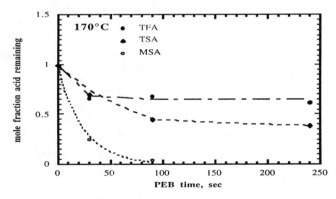

Figure 5. Plot of fraction of acid remaining in the resist film vs. PEB time under isothermal conditions, 170°C. Fitted curves are derived from equation (2).

was only 4 minutes. MSA was completely driven out of the film within 90 seconds.

Table 2 shows the calculated diffusion coefficients for these acids at the temperatures used in this study, 120°C, 150°C, and 170°C. In virtually all cases, our fits of D using the Crank-Nicholsen technique has led to large initial diffusion coefficients. We attribute the large values of D to the solvent evaporation occurring during PEB. In order to fit the acid loss vs. time plots, we had to make D a time-dependent variable which got smaller with bake time. However, for our overall analysis of D for the various acids we have chosen the highest apparent diffusion coefficients calculated. The lowest diffusion coefficient is exhibited for the photogenerated acid that has the lowest volatility, CSA. Within the experimental error of the titration, virtually all the CSA is retained at 150°C after 16 minutes of baking. This result shows that PAGs that generate CSA can be processed at high PEB temperatures without any acid loss. This result is impressive because the prebake condition was only 120°C for 1 minute, which means that some solvent loss during the PEB step was likely. For the other acids, PEB temperatures at or above 120°C cannot be used due to excessive acid loss from the resist film. The acid diffusion coefficients for each acid increase with PEB temperature, as expected. Utilizing this data, an Arrhenius plot of the acid loss vs. PEB temperature should provide a predictive model of acid loss at room temperature (i.e. during post-exposure delays) and at other PEB temperatures (i.e. 90°C, or 110°C).

C. Arrhenius Behavior for Acid Loss. The Arrhenius plot of the diffusion coefficient for acid loss vs. PEB temperature is shown in Figure 6. Table 3 shows the calculated preexponential factor, A, the activation energy for acid loss (E_a) for the three acids studied, and the predicted diffusion coefficients at lower processing temperatures of interest, 25°C, 90°C, and 110°C. The activation energy for acid loss is 87, 106, and 76 kJ/mol for MSA, TFA, and TSA, respectively. Overall, the diffusion coefficients measured for MSA are an order of magnitude greater than TFA and TSA, which eliminates MSA as a viable acid for chemically amplified resists that undergo thermal processing. In the cases of TFA and TSA, process conditions must be carefully selected in order to limit the diffusion of acid out of the resist. There are practical constraints to the size of the diffusion coefficient for acid loss at room temperature and at 90°C. In an earlier paper[9b], it was shown that when $D = 1 \times 10^{-7}$ $\mu m^2/s$ at room temperature, 55% of the acid would be lost from the uppermost 0.1 μm of resist film after a 90 minute postexposure delay. Therefore, we propose that D must be less than 1×10^{-7} $\mu m^2/s$ at room temperature and that the postexposure bake step must be linked to the exposure tool so that no acid diffusion at room temperature can occur. Similarly, it was shown that when $D = 1 \times 10^{-5}$ $\mu m^2/s$ at 90°C, that 53% of the acid would be lost from the uppermost 0.1 μm of resist film after only a 60 s PEB step. Table 4 shows the predicted mole fraction of acid remaining for these acids using the predicted diffusion coefficients in Table 3. At 25°C, the predicted moles acid remaining is 0.992 for TFA, which is probably an acceptable value. For MSA, the fraction of acid remaining is only 0.909, which means that the uppermost 0.1 μm is nearly completely depleted of acid content. The criticality of linking the exposure tool to the post-exposure bake track is apparent for these acid generators. We have also estimated the effect of these predicted diffusion coefficients at the PEB temperatures normally used for chemically amplified resists. For all of these acids except CSA, significant acid loss occurs during a normal 110°C PEB step for 2 minutes. At 90°C, only the MSA loses too much acid to be useful. Therefore, we propose D must be less than 5×10^{-6} $\mu m^2/s$ in

Table 1. Acid Generating efficiency of PAGs

Photoacid Generator	E(l*mol^{-1}*dm) @248 nm	α (cm^2/mJ)
Triarylsulfonium Triflate	10,900	0.062
Triarylsulfonium Toluenesulfonate	10,200	0.054
Triarylsulfonium Camphorsulfonate	9,600	0.043
Tris(methanesulfonyloxy) benzene	135	0.0070

Table 2. Diffusion Coefficients for Acids as a Function of PEB Temperature and Time

Acid	PEB Temperature (°C)	PEB time	D (μm^2/s)
TFA	120	0-16 min	6.5 x 10^{-5}
TSA	120	0-4 min	1.0 x 10^{-4}
MSA	120	0-4 min	4.1 x 10^{-4}
MSA	120	4-9 min	1.1 x 10^{-4}
MSA	120	9-16 min	3.0 x 10^{-5}
TFA	150	0-1 min	8.3 x 10^{-4}
TFA	150	1-16 min	0
TSA	150	0-4 min	2.4 x 10^{-4}
TSA	150	4-16 min	2.7 x 10^{-5}
MSA	150	0-1 min	3.2 x 10^{-3}
MSA	150	1-4 min	2.5 x 10^{-4}
MSA	150	4-9 min	5.0 x 10^{-5}
MSA	150	9-16 min	0
CSA	150	0-1 min	5.6 x 10^{-5}
TFA	170	0-30 s	2.4 x 10^{-3}
TFA	170	30-240 s	0
TSA	170	0-90 s	1.6 x 10^{-3}
TSA	170	90-240 s	1.8 x 10^{-4}
MSA	170	0-90 s	7.9 x 10^{-3}

Table 3. Activation Energy for acid loss and predicted diffusion coefficients for Acids at lower processing temperatures.

Acid Generated	$A(\mu m^2/s)$	$E_a(kJ/mol)$	D@25°C $(\mu m^2/s)$	D@90°C $(\mu m^2/s)$	D@110°C $(\mu m^2/s)$
MSA	1.4×10^8	87	9.2×10^{-8}	4.81×10^{-5}	2.2×10^{-4}
TFA	8.2×10^9	106	2.28×10^{-9}	4.78×10^{-6}	3.3×10^{-5}
TSA	1.05×10^6	76	4.86×10^{-8}	1.18×10^{-5}	4.4×10^{-5}

Table 4. Predicted acid loss at Lower Processing temperatures.

Acid	PEB Temperature (°C)	PEB time	mole fraction acid remaining
TFA	25	24 h	.992
TFA	90	2 min	.982
TFA	110	2 min	.939
TSA	25	24 h	.936
TSA	90	2 min	.967
TSA	110	2 min	.927
MSA	25	24 h	.909
MSA	90	2 min	.924
MSA	110	2 min	.825

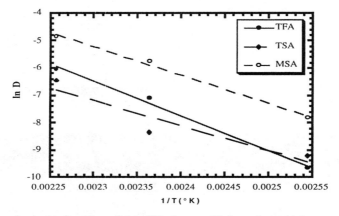

Figure 6. Arrhenius Plot of the diffusion coefficients for acid loss vs. PEB temperature.

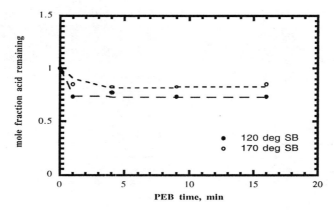

Figure 7. Plot of mole fraction acid remaining vs. PEB time under isothermal conditions, 150°C. The plot illustrates the effect of annealing the film at higher softbake in order to reduce acid diffusion.

order to retain sufficient acid for the deprotection or crosslinking reaction to occur effectively at the resist surface. A higher diffusion coefficient may be possible in resist systems that have a very fast reaction order whereby the acid reacts much faster than it diffuses.[10] Of the acids in this study, CSA, TFA, and TSA have sufficiently low enough acid loss diffusion coefficients for practical use in chemically amplified resists at 90°C. Only CSA may be usable at temperatures of 110°C or above.

D. Polymer Matrix Effects on Diffusion. In this study, hydrogenated poly(p-vinyl) phenol (M_w = 5000) has been the primary polymer matrix. The T_g of this polymer is 165°C. Also, this polymer has no thermally labile protecting groups. The softbake conditions used were 120°C for 1 minute, which means that the acid loss measurements under the PEB conditions in this study were at or substantially above the softbake conditions. Thus, the diffusion coefficients for acid loss may be aided by substantial solvent evaporation in the PEB step. Figure 7 shows the difference in acid retained when the polymer film is softbaked at a temperature, 170°C, significantly higher than the PEB conditions, 150°C, vs a softbake of 120°C. By annealing the resist film, acid diffusion can be substantially lowered. The TFA retained is only 0.72 at the lower softbake vs. 0.82 after the higher softbake. This effect has been reported by Ito et al. These results support the validity of Ito's high T_g resist design where the temperature of the blocking group decomposition should be higher than that of the polymer T_g.[11]

E. Effect of Acid Diffusion on Lithography. The diffusion coefficients derived under normal process conditions for these acids can be used to predict the relative acid retained under those conditions. In Figure 8, the predicted acid retained is inversely proportional to the rate of capping for the acids in this study. The rate of capping follows the relationship of TSA>TFA>CSA. The diffusion model predicts that under the process conditions for this experiment the relative acid retained is TSA = 0.962<TFA = 0.981<CSA = 1.00. Figure 9 illustrates the acid distribution vs. resist thickness for the acids used in this study. Even though relatively small amounts of acid are removed, there is a large effect on t-topping due to the fact that the uppermost 0.1 µm of the resist bears the brunt of acid loss.

IV. Conclusions. This paper has reported on the diffusion coefficients for acid loss for TFA, TSA, CSA, and MSA in an hydrogenated poly(p-vinyl)phenol matrix. A direct method for following acid content was derived from the TBPB spectrometric titration method. In this manner, the E_a for acid loss for TSA, MSA, and TFA was found to be 76, 87, and 106 kJ/mol, respectively. Using the E_a values, acid loss could be predicted over a wide range of processing conditions. It was found that MSA had an order of magnitude greater D than TSA and TFA, which renders MSA as an impractical acid for use in chemically amplified resists. For TSA and TFA, D was found to be 1.2 x 10^{-5} µm^2/s and 4.8 X 10^{-6} µm^2/s, respectively, at 90°C. These values make these acids usable at 90°C, but, TSA will be more problematic than TFA. All of these acids show sufficient acid loss at room temperature, that long term staging of exposed resists is not recommended. For DUV lithography with chemically amplified resists, we recommend linked PEB tracks directly to the exposure tool so that the maximum postexposure delay is 15 minutes. The one acid that shows virtually no acid loss even at 150°C was CSA. This acid shows no evidence of 't-tops' and may be the acid of choice for high thermal processing.

No Delay   0.5 hr. delay   1 hr. delay

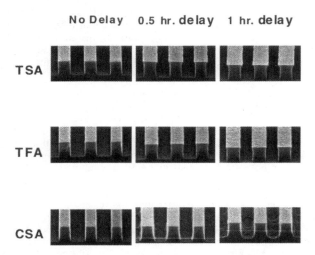

Figure 8. Effect of acid loss vs. acid type on resist profiles. Based on predicted D values, the relative acid retained after one hour delay for these acids is: TSA = 0.962<TFA = 0.981<CSA = 1.00.

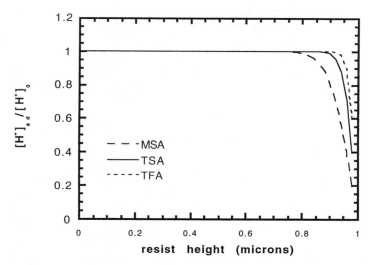

Figure 9. Acid depletion as a function of resist thickness for various acids assuming one hour delay at room temperature and a 90°C/1 minute PEB, using the extracted diffusion coefficients for each acid. $[H^+]_{ad}/[H^+]_0$ is the ratio of acid concentration after delay to the acid concentration immediately after exposure. This model does not account for absorbance of the resist film at the exposing wavelength.

V. Acknowledgments. The authors wish to acknowledge Tim Adams and Jim Cameron for synthesis of the triaryl sulfonium PAGs.

VI. References.

1. For reviews of chemically amplified resists see: (a.) A. A. Lamola, C. R. Szmanda, J. W. Thackeray, "Chemically Amplified Resists," **Solid State Technology**, <u>8</u>, 53 (1991); (b.) E. Reichmanis, F. M. Houlihan, O. Nalamasu, T. X. Neenan, **Chem. Mater.** <u>5</u>, 348 (1992).

2. (a.) H. Ito, C. G. Willson, Polym. Eng. Sci., 23, 1012 (1983); (b.) W. Brunsvold, W. Conley, J. Gelorme, R. Nunes, G. Spinello, K. Welsh, **Microlithography World**, <u>4</u>, 6 (1993).

3. S. A. McDonald, W. D. Hinsberg, H. R. Wendt, N. J. Clecak, C. G. Willson, **Chem. Mater.** <u>5</u>, 348 (1993).

4. J. W. Thackeray, T. H. Fedynyshyn, A. A. Lamola, R. D. Small, **J. Photopolymer Sci. Tech.**, <u>5(1)</u> , 215 (1992).

5. (a.) D. R. McKean, U. Schaedeli, P. H. Kasai, S. A. McDonald, **Poly. Mater. Sci. Eng.**, <u>61</u>, 81 (1989); (b.) G. L. Gaines Jr. **Anal. Chem.** <u>48</u>, 450 (1976).

6. H. Roschert, Ch. Eckes, H. Endo, Y. Kinoshita, T. Kudo, S. Masuda, H. Okazaki, M. Padmanaban, K.-J. Pryzbilla, W. Spiess, N. Suehiro, H. Wengenroth, G. Pawlowski, **Proc. SPIE**, *1925*, 14 (1993).

7. C. A. Mack, D. P. Dewitt, B. K. Tsai, G. Yetter, **Proc. SPIE**, <u>2195</u>, 584 (1994).

8. L. Schlegel, T. Ueno, H. Shiraishi, N. Hayashi, T. Iwayanagi, **Chem. Mater.**, 2, 299 (1990).

9. (a.) W. Press, S. Teukolsky, W. Vetterling, B. Flannery, *Numerical Recipes in Fortran*, 2nd Ed. (Cambridge University Press, Cambridge, Eng. 1992), pp. 827-842; (b.) T. H. Fedynyshyn, J. W.Thackeray, J. H. Georger, M. D. Denison, **J. Vac. Sci. Tech. B**, <u>12(6)</u>, 3888 (1994).

10. W.-S. Huang, R. Kwong, A. Katnani, M. Khojasteh, **Proc. SPIE**, <u>2195</u>, 37(1994).

11. H. Ito, W. P. England, N. J. Clecak, G. Breyta, H. Lee, D. Y. Yoon, R. Sooriyakumaran, W. D. Hinsberg, **Proc. SPIE**, <u>1925</u>, 65 (1993).

RECEIVED September 12, 1995

Chapter 8

Water-Soluble Onium Salts: New Class of Acid Generators for Chemical Amplification Positive Resists

Toshio Sakamizu, Hiroshi Shiraishi, and Takumi Ueno

Central Research Laboratory, Hitachi Ltd., Kokubunji, Tokyo 185, Japan

Alkyl-substituted-sulfonium salts were synthesized and investigated with a view to applying them as the acid generator in chemical amplification positive resists. It was found that the solubility in water of alkyl-substituted onium salts is high, while that of triaryl-substituted onium salt is quite low. This dissolution promotion ability of alkyl-substituted onium salts increase with the decreasing molecular size of the alkyl group. The quantum yield for acid generation from dialkylarylsulfonium salt was one order of magnitude larger than those of trialkyl-substituted onium salts. A difference was observed in acid-generation efficiency between electron-beam exposure and deep-UV exposure. We will discuss with this difference in terms of acid generation mechanism. Water-soluble onium salts were determined to be effective acid generators for electron-beam exposure: they can produce high resolution patterns (100-nm contact holes).

Diaryliodonium salts and triarylsulfonium salts, which have been employed as photoinitiators for cationic polymerization (*1*), are widely used as photoacid generators for chemical amplification resist systems (*2*). These arylonium salts used in the resist systems exhibit a strong dissolution inhibition capability in an aqueous base developer because these onium salts are little soluble in the developer (*3, 4*). Thus, the dissolution inhibition effect of the acid generator in positive resist lowers process latitude (*5*).

Chemical amplification resists utilizing various non-ionic acid-generators have been reported(*6-9*). One of the representatives of these compounds, tri(methanesulfonyloxy)benzene (MeSB) proposed by our group, showed a considerably lower dissolution inhibition effect compared with that of arylonium salts (*5*). We have already reported on a novolak resin-based chemical amplification positive resist using MeSB as an acid generator for electron-beam lithography (*10*). MeSB can yield methanesulfonic acid and phenol derivatives by the radiation-induced hydrolysis (*11*). It is expected that these products lead to dissolution promotion in the resist system. However, even though the dissolution inhibition effect of MeSB is small, the dissolution promotion of exposed area is not enough for the performance of the positive resist system because the decomposition degree of MeSB at a practical dose is only a few percent (*11*).

In a recent paper (*12*), it was found that trimethylsulfonium triflate (MES) shows dissolution promotion effect in the novolak resin for the aqueous-base development. Thus, we evaluated trimethylsulfonium triflate (MES) as an acid generator in a positive resist system. The use of MES results in a higher contrast positive tone in the exposure characteristic curve. From the above result, we considered that it need to understand the relation between chemical structure of alkylsulfonium salts and the performance as an acid generator in order to enhance the process latitude.

In this paper, we report on the difference in the dissolution promotion capability among the alkyl-substituted-sulfonium salts and the effect of the number of alkyl substituents on the resist sensitivity. Moreover, we discuss the investigation of the acid generation mechanism from these onium salts in the novolak resin matrix and make a comparison of acid generating efficiency from these compounds. We also describe the lithographic performance of a chemical amplification positive resist system using water-soluble onium salts.

Experimental

Materials preparation

The novolak resin was supplied from Hitachi Chemical Co. Commercially available aqueous-base developer NMD-3 (tetramethylammonium hydroxide: 2.38 wt%, Tokyo Ohka Co.) was used as the developer. The polymeric dissolution inhibitor, tetrahydropyranyl-protected poly(p-vinylphenol) (THP-M), was synthesized according to the previous paper (*10*). The inhibitor compound, bisphenol A protected with tert-butoxycarbonyl (tBOC-BA) (Figure 1) was synthesized as described in the literature (*13*). Triphenylsulfonium triflate (TPS) was synthesized according to the literature (*14*).

The procedures for syntheses of the sulfonium triflates (Figure 1) were adapted from literature preparations of sulfonium salts (*15, 16*). The details of the synthesis of trimethylsulfonium triflate (MES) is described below. Silver trifluoromethanesulfonate (25.2 g, 98.0 mmol) was added to a solution of 15.6 g (100 mmol) methyl iodide and 6.0 g (98.0 mmol) dimethyl sulfide in 200 mL of tetrahydrofuran. The reaction mixture was allowed to stir at room temperature for one day. Filtration to remove the yellow precipitate of AgI followed by removal of solvent in vacuo provided a white solid. The solid was recrystallized from 200 mL of ethanol. White crystals of the triflate were obtained (17.7 g, 80% yield). In a similar fashion, dimethylethylsulfonium triflate (DMS), dimethyl-n-butylsulfonium triflate (BMS), dimethylphenylsulfonium triflate (PMS), and diphenylmethylsulfonium triflate (DPS) were prepared. The analytical data for these sulfonium triflates are shown in Table I.

Characterization

Electron-beam exposure characteristics were measured by a scanning electron microscopy at 30 kV using a modified Hitachi S-570. Electron-beam lithography evaluation was conducted with a modified Hitachi HL-700 electron-beam lithography system at 50 kV. Deep-UV exposure at 250 nm was made with a Xe-Hg lamp through an interference filter. The light intensity was determined with a calorimetric flux detector thermopile (JASCO Inc.). The sample films were spin-coated from solutions in 2-methoxyethyl acetate onto silicon wafers and baked on a hot plate at various conditions. The film thickness was determined by using a profilometer, Alpha-step 200 (Tencor Instrument Co.).

The chemical structures of the synthesized compounds were identified from NMR spectra obtained with a Hitachi R-1900 FT-NMR system. The UV-absorption spectra were measured with a Hitachi U-3410 spectrophotometer. Mass spectra were measured by the field desorption (FD) technique using a Hitachi M-2000 mass spectrometer.

The FD spectra of the TPS, PMS, and MES are shown in Figure 2. The details of the fragmenting molecules and the explanation of these fragmentation are described below. A few fragment peaks were observed in the mass spectrum of TPS (Figure

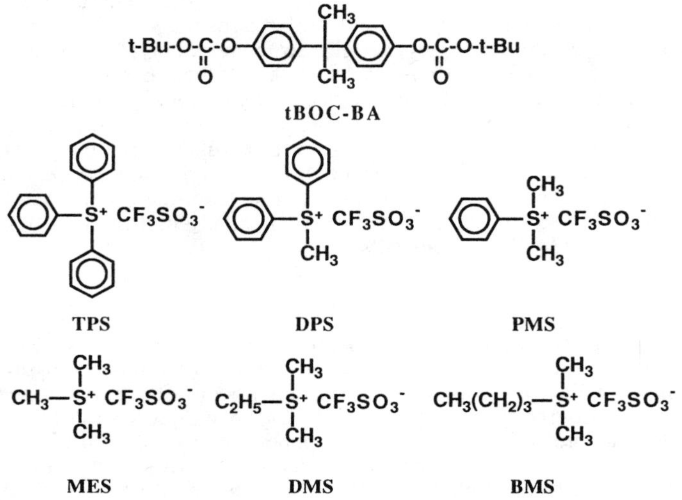

Figure 1. Chemical structures of tBOC-BA and acid generators.

Table I. Analytical data for sulfonium salts

Compound	Yield	^{1}H-NMR	mp	Solubility (g/100g of water)
TPS	75%	((CD3)2CO) δ 7.7-8.2 (m, 15H)	133-134°C	<1.0
DPS	69%	((CD3)2CO) δ 7.7-8.2 (m, 10H), 3.9 (t, 2H)	98-99°C	<3.0
PMS	73%	(D2O) δ 7.7-8.0 (m, 5H), 3.2 (s, 9H)	107-108°C	>10
MES	80%	(D2O) δ 2.9 (s, 9H)	206-207°C	>10
DMS	56%	(D2O) δ 2.9 (s, 6H),1.4 (t, 3H), 2.9 (q, 2H)	oily	>10
BMS	62%	(D2O) δ 2.9 (s, 6H), 0.9 (t, 3H), 1.5 (m, 2H), 1.8 (m, 2H), 3.3 (t, 2H)	55-56°C	>10

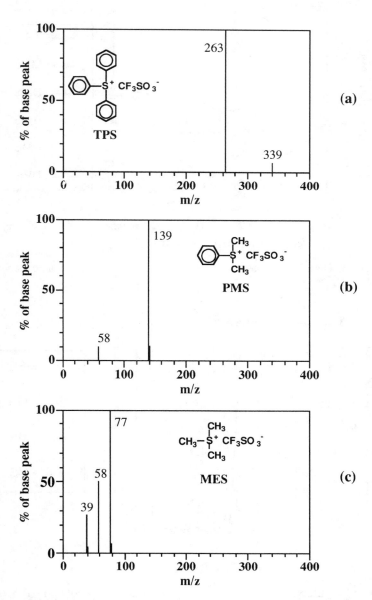

Figure 2. Mass spectra of sulfonium salts with field desorption ionization. (a) TPS, (b) PMS, (c) MES.

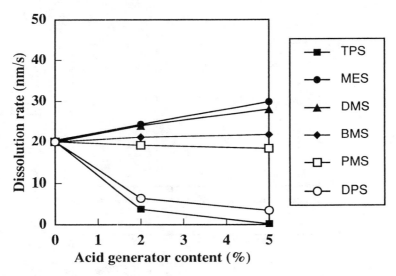

Figure 3. Dissolution rate of novolak resin as a function of acid generator content.

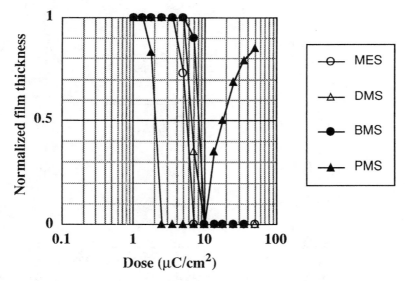

Figure 4. Electron-beam sensitivity curves of novolak/THP-M/acid generator=100/5.0/1.1 (mol. ratio). Prebaking: 100°C for 5 min. PEB: 100°C for 5 min. Film thickness: 520 nm. Development times, MES: 60 s, DMS: 60 s, BMS: 69 s, PMS: 76 s.

2(a)). The prominent peak at m/z 263 is the triphenylsulfonium ion. The only other peak at m/z 339 represents the biphenyldiphenylsulfonium rearrangement product (BPDPS). This fragmentation pattern for TPS may imply that a phenyl cation followed by the C-S cleavage can react with a triphenylsulfonium ion to give BPDPS and a proton. Figure 2(b) presents the mass spectrum of PMS. A prominent peak at m/z 139 is indicative of the dimethylphenylsulfonium ion. The only other peak at m/z 58 represents a rearranged cyclic ion C_2H_2S. This fragmentation pattern for PMS may imply that the formation of a cyclic ion with loss of large group followed by the C-S cleavage. However, no peak appeared beyond the dimethylphenylsulfonium peak, which means the absence of recombination between a phenyl cation and a dimethylphenylsulfonium ion. In the case of MES (Figure 2(c)), the prominent peak at m/z 77 is indicative of the trimethylsulfonium ion. The peaks at m/z 58 and m/z 39 correspond to the degradation ions of trimethylsulfonium ion, C_2H_2S and C_3H_3, respectively. The peak at m/z 58 represents the same fragment as shown in PMS. From these results, it seems that these three sulfonium salts (MES, PMS, and TPS) undergo the same cleavage pathway of the carbon-sulfur bond followed by the ionization of sulfur molecule.

Results and Discussion
Characteristics of water soluble sulfonium salts

Since the structure of acid generator has a strong influence on the dissolution rate of resist used for aqueous-base development, the dissolution rates of film composed of the novolak and onium salts (Figure 1) were measured as a function of the onium salt content. As shown in Figure 3, the dissolution rate of triphenylsulfonium salt (TPS) and diphenylmethylsulfonium salt (DPS) decreased sharply at relatively low contents (~2%), while those of the trialkylsulfonium salts MES, DMS, and BMS increased with increasing onium salt content (dissolution promotion effect). Dimethylphenylsulfonium salt (PMS) exhibited a slightly dissolution inhibition ability. These results indicate that the dissolution inhibition ability of sulfonium salt depends on the number of aryl groups bonded to the sulfur. In particular, it is clear that there is a considerable difference in dissolution inhibition ability between the dimethylphenylsulfonium salt (PMS) and diphenylmethylsulfonium salt (DPS). This phenomenon corresponds to the difference in the solubility in water between PMS and DPS (Table I). On the contrary, it turns out that the dissolution promotion ability increases with decreasing molecular size of the alkyl group.

From the above results, it is considered that the magnitude of the dissolution inhibition effect, that is, the solubility in water corresponds to the molecular size of the sulfonium salt. The difference in solubility of these onium salts in water is due to the interaction between the sulfonium ion and the electric dipole of water molecule. Therefore, in sulfonium salts, the steric bulk of the hydrophobic groups may prevent access of water molecules to the sulfonium cation site, so that a bulky onium salt may give a weak electrostatic interaction. Such an onium salt lowers solubility in water as well as an aqueous base developer.

In order to compare acid generation efficiency of these water soluble onium salts, electron-beam exposure characteristics of a model novolak resin-based positive resist were measured. Tetrahydropyranyl-protected polyvinylphenol (THP-M) was selected as a dissolution inhibitor in this experiment. The electron-beam exposure characteristic curves of the novolak/THP-M/acid generator=100/5.0/1.1 (mol. ratio) are shown in Figure 4. The development time of the resist films was determined depending on the dissolution inhibition ability of each onium salt, e.g., the development time of MES for 60 s and that of PMS for 76 s were used. As shown in this figure, complete positive tone of the resist using PMS as an acid generator was observed at about 2 $\mu C/cm^2$, whereas a negative tone was observed at doses higher than 10 $\mu C/cm^2$. In contrast, the resists using MES, DMS, and BMS gave sensitivity between 7 $\mu C/cm^2$ and 10 $\mu C/cm^2$ without negative tones up to 50 $\mu C/cm^2$.

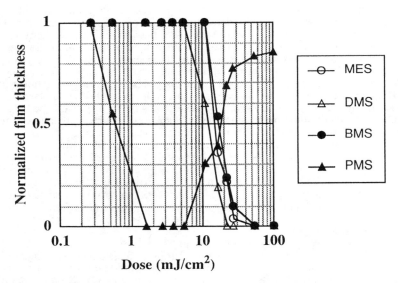

Figure 5. Deep-UV sensitivity curves of novolak/THP-M/acid generator=100/5.0/1.1 (mol. ratio).

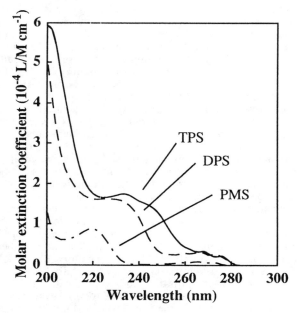

Figure 6. UV absorption spectra of TPS and DPS in acetonitrile and PMS in water.

According to these results, electron-beam sensitivity for PMS is about five times higher than those for MES, DMS, and BMS. It is also found that PMS produces negative tone behavior at about five times the dose required for a positive tone. This behavior may be ascribed to the inhibition effect of reaction products between intermediates formed by decomposition of PMS and the novolak resin matrix, though it needs further investigation.

We also measured the deep-UV exposure characteristics of the same resist composition as used in electron-beam experiment. The results are shown in Figure 5. A positive tone for PMS was observed at about 1 mJ/cm^2, whereas a negative tone was also observed at a doses over 5 mJ/cm^2. In the case of the onium salt MES, DMS, and BMS, positive tones were obtained at about 50 mJ/cm^2 without negative tones up to 100 mJ/cm^2. From this experiment, the deep-UV sensitivity for PMS is about one order of magnitude higher than those for the others. It is also found that PMS produces negative tone behavior at about five times the dose required for positive tone. This behavior is fair good correlation with the result of electron-beam exposure (the positive tone was observed at 2 μC/cm^2 and the negative tone was observed at 10μC/cm^2).

Since there is a large difference in the deep-UV sensitivity between the dimethylphenylsulfonium salt (PMS) and the trialkyl-substituted sulfonium salts, the effect of substituent group of sulfonium salt on UV absorption was investigated. The absorption spectra of these sulfonium salts in a solution (5x10^{-5} mol/L) were measured. The UV spectra of trimethylsulfonium salt (MES) and dimethylphenylsulfonium salt (PMS) were measured in water. The UV spectra for triphenylsulfonium salt (TPS) and diphenylmethylsulfonium salt (DPS) were measured in acetonitrile because they are little soluble in water as shown in Table I. The results are shown in Figure 6. It is clear from this figure that the absorbance in solution of PMS is considerably lower than that of TPS and DPS. Compared with the intensity of absorption for these onium salts, TPS exhibits a strong absorption at 248 nm (ε: 12900 L/mol cm^{-1}), and DPS has the extinction coefficient of 3700 L/mol cm^{-1} at 248 nm, while PMS shows very small absorption at 248 nm (ε: 300 L/mol cm^{-1}). On the other hand, a trialkylsulfonium salt such as MES has no absorption above 200 nm. It was confirmed by the experimental results that the intensity of absorption increases with an increase in the number of aryl groups bonded to the sulfur molecule.

From the deep-UV and electron-beam exposure experiment, it turns out that the acid generation ability of PMS was the highest among these water-soluble onium salts. However, it is to be pointed out that the dimethylphenylsulfonium salt and trialkylsulfonium salts can generate strong acid upon deep-UV exposure even if the absorption of dimethylphenylsulfonium salt is very small and the trialkylsulfonium salts have no absorption in the deep-UV region. This is mainly because the acid generation mechanism of these onium salts is based on the sensitization (electron-transfer process) from a strongly absorbing novolak resin matrix, as reported by Hacker et al. (*17, 18*). Furthermore, it is worth noting that the difference in sensitivity between PMS and the others is small in the case of electron-beam exposure compared with deep-UV exposure.

Radiation-induced acid formation from water soluble sulfonium salts

To be sure that the acid formation from these onium salt photosensitized by novolak resin, the UV spectra of film using cellulose acetate (CA) as a matrix polymer were measured before deep-UV exposure and after the exposure and PEB as studied in acid generation mechanism of MeSB (*19*). The model composition was CA/tBOC-BA/PMS=100/15/2 (wt ratio). Assuming that the onium salt in the film could give an acid by the exposure and the post-exposure baking (PEB), the inhibitor compound tBOC-BA would be deprotected to form bisphenol A. This

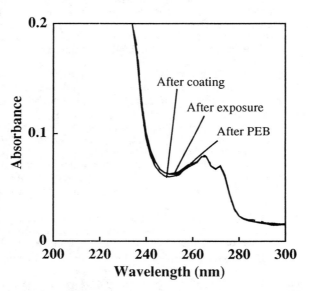

Figure 7. Spectral change in the resist CA/tBOC-BA/PMS=100/15/2 (wt ratio).
Exposure dose: 50 mJ/cm^2. PEB: 100°C for 5 min. Film thickness: 1.1 μm.

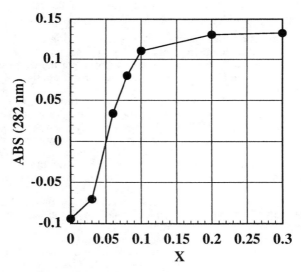

Figure 8. Increase in absorbance at 282 nm after baking at 100°C for 5 min.
with addition of acid. Novolak/tBOC-BA/trifluoromethanesulfonic acid=100/
12.8/X (mol. ratio).

deprotection of tBOC-BA (Figure 1) would lead to an increase in absorbance at 282 nm. Since CA does not absorb at the irradiation wavelength at 248 nm, this method is appropriate to confirm whether the sensitization by the novolak resin occurs or not. The UV spectra of the model film after spin-coating, and after exposure at 50 mJ/cm^2 and PEB at 100°C for 5 min are shown in Figure 7. Almost no change in the spectra of the film was observed, even at high exposure dose and PEB. On the contrary, the increase on absorbance at 282 nm was clearly observed for the resist composed of novolak, tBOC-BA, and PMS. These results support the sensitization mechanism by the novolak described above.

It is likely that the sensitivity of resists using water-soluble acid generators depends on the efficiency of the acid formation in the resist film. To investigate the difference in the sensitivity, we determine the quantum yield of acid generation and approach a relation between the quantum yield and substituent structure of the sulfonium salts. The method for the determination of quantum yield is essentially the same as the previous reported method (*13*). Post-exposure baking of resist film composed of novolak resin, tBOC-BA, and an acid generator leads to an increase in absorbance at 282 nm as described above. This increase, can be used as a detector of the deprotection degree of tBOC-BA. Therefore, the comparison of ΔAbs (282) of the film containing a sulfonium triflate with that of the film containing the triflic acid (TFA) can give the yield of generated acid in the film. Figure 8 shows the change in ΔAbs (282) as a function of TFA content in the film. The model composition was the novolak/tBOC-BA/TFA=100/12.8/X (mol ratio). As shown in this figure, for the film containing less than 0.1% TFA, the value of ΔAbs (282) increases with increasing TFA content. For higher TFA concentration more than 0.1%, the value of ΔAbs (282) was almost constant. Figure 9 shows the change in ΔAbs (282) as a function of the exposure dose. The model composition was the novolak/tBOC-BA/sulfonium salt=100/12.8/1.1 (mol ratio). It turned out that the value of ΔAbs (282) for PMS increased strongly for lower exposure dose of less than 5 mJ/cm^2, whereas the maximum value in ΔAbs (282) of films containing MES, DMS, and BMS was observed at a relatively high exposure dose. The value for the quantum yield of these onium salts are summarized in Table II. It is clear from this result that the quantum yield of aryl-substituted-sulfonium salt (PMS) is one order of magnitude larger than those of alkyl-substituted-sulfonium salts.

On the basis of these results, we can come to the following conclusion. The difference in quantum yields between aryl-substituted and alkyl-substituted onium salts can be ascribed to the difference in absorption coefficient at exposure wavelength 248 nm. The aryl-substituted onium salts can cause direct photolysis to yield an acid followed by absorption of exposed light, whereas the alkyl-substituted onium salts yield an acid by the sensitization (electron-transfer process).

The question which we have to consider next is that the difference in sensitivity between PMS and the others is quite small in the case of electron-beam exposure compared to the difference when deep-UV exposure is used. In the case of the irradiation with charged particles, such as electron-beam exposure, high-energy electrons produce electronic excitations as well as ionizations of resist components to lose their energy through numbers of such chemical events. It was reported that the acid formation from arylonium salts exposed to electron-beam is based on the reaction between proton adducts of phenolic matrix resin and the dissociated anions of the onium salts which are caused by scavenging electrons due to ionizations of base polymer (*20*). Based on the literature it is considered that the difference in sensitivity may be ascribed to the difference in dissociated anion formation efficiency of aryl-and alkyl-substituted sulfonium salts. However, it is difficult to confirm the reaction mechanism, since there is a possibility that the reaction induced by electron beam exposure is more complicated, such as direct radiolysis of onium salts, sensitization from excited novolak which is formed by direct excitation and/or recombination of novolak ion and electron, etc. Since water-soluble onium salts can

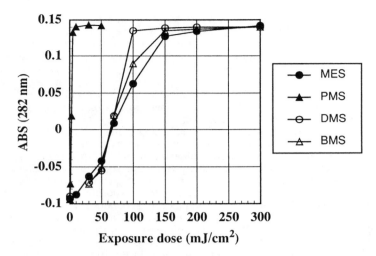

Figure 9. Change in absorbance after PEB as a function of exposure dose. Novolak/tBOC-BA/sulfonium salt=100/12.8/1.1 (mol. ratio).

Table II. Quantum yield for acid generation from sulfonium salts

Sulfonium salt	Quantum yield Φ
PMS	0.056
MES	0.0022
DMS	0.0024
BMS	0.0028

give the high sensitivity below 10 $\mu C/cm^2$, the use of these onium salts having dissolution promotion effect are advantageous to the irradiation with electron-beam rather than deep-UV.

Lithographic performance

As water-soluble onium salts can have dissolution promotion ability, it is expected that incorporation of these onium salts enhances the dissolution rate contrast of the resist system. The preliminary patterning of the resist was carried out with the electron-beam lithography system at 50 kV. The onium salt PMS shows high sensitivity, accompanying negative tone behavior. The onium salt MES, which showed the largest dissolution-promotion effect, was selected as an acid generator for our experimental positive resist system. A typical electron-beam delineated pattern of the resist using trialkylsulfonium triflate are shown in Figure 10. High-

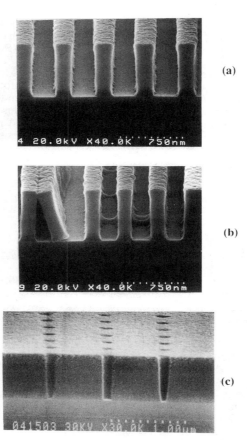

Figure 10. Scanning electron micrographs of (a) 0.25-μm line and space pattern, (b) 0.2-μm line and space pattern, and (c) 100-nm contact holes obtained by electron-beam exposure at 50 kV. Process conditions: (a) and (b) film thickness 0.7 μm, exposure dose 7.1 μC/cm^2, prebaking 120°C for 3 min, PEB 100°C for 2 min; (c) film thickness 0.7 μm, exposure dose 13 μC/cm^2, prebaking 120°C for 90 s, PEB 100°C for 2 min.

resolution patterns (0.25-μm, 0.2-μm line and space) were achieved at a dose of 7.1 μC/cm^2. We also obtained a very fine patterns (100-nm contact holes) at a dose of 13 μC/cm^2.

Conclusion

The performance of alkyl-substituted-sulfonium salts as an acid generator were investigated in terms of their dissolution promotion capability and the sensitivity of resists composed of novolak, THP-M, and one of these salts as the acid generator. It was found that the dissolution promotion ability increases with decreasing molecular size of the alkyl group. The dimethylphenylsulfonium salt (PMS) had the highest sensitivity among of these sulfonium salts for both deep-UV and electron-beam irradiation. However, the difference in sensitivity between PMS and the others was small for electron-beam exposure compared to the difference for deep-UV exposure. In electron-beam exposure, the amount of generated acid from the trialkylsulfonium salts due to electron-beam exposure may be considerably close to that from PMS. Therefore, water-soluble trialkylsulfonium salts have good potential as an acid generator for application in chemical amplification positive resists for electron-beam lithography.

Acknowledgments

The authors would like to thank Yoshihiro Ohta and Isao Sakama for their assistance in the electron-beam delineating experiments. They also thank Nobuaki Hayashi for his useful advice on the synthesis of sulfonium salts.

References

1. Crivello, J. V. Adv. Polym. Sci. **1984**, 62, 1.
2. Ito, H.; Willson, C. G. In Polymers in Electronics; Davidson, T., Ed.; ACS Symposium Series 242: Washington, DC, 1984; p 11.
3. Ito, H.; Flores, E.; Renaldo, A. F. J. Electrochem. Soc. **1988**,135, 2328.
4. Schlegel, L.; Ueno, T.; Shiraishi, H.; Hayashi, N.; Hesp, S.; Iwayanagi,T. Jpn. J. Appl. Phys.**1989**, 28, 2114.
5. Ueno, T.; Shiraishi, H.; Schlegel, L.; Hayashi, N.; Iwayanagi, T. In Polymers for Microelectronics-Science and Technology; Tabata, T.; Mita, I.; Nonogaki, S.; Horie, K.; Tagawa, S., Ed.; Kodansha: Tokyo, 1990; p 413.
6. Doessel, K. F.; Huber, H. L.; Oertel, H. Microelectron. Eng. **1986**, 5, 97.
7. Reichmanis, E.; Houlihan, F. M.; Nalamasu, O.; Neenan, T. X. Chem. Mater. **1991**, 3, 394.
8. Barra,M.; Calabrese, G. S.; Allen, M. T.; Redmond, R. W.; Sinta, R.; Lamola, A. A.; Small, R. D., Jr.; Scaiano, J. C. Chem. Mater. **1991**, 3, 610.
9. Brunsvold, W.; Montgomery, W.; Hwang, B. Proc. SPIE **1991**,1466, 368.
10. Shiraishi, H.; Hayashi, N.; Ueno, T.; Sakamizu, T.; Murai, F. J. Vac. Sci. Technol. **1991**, B9, 3343.
11. Sakamizu, T.; Yamaguchi, H.; Shiraishi, H.; Murai, F.; Ueno, T. J. Vac. Sci. Technol. **1993**, B11, 2812.
12. Shiraishi, H.; Yoshimura, T.; Sakamizu, T.; Ueno, T.; Okazaki, S. J. Vac. Sci. Technol. **1994**, B12, 3895.
13. Ueno, T.; Schlegel, L,; Hayashi, N.; Shiraishi, H.; Iwayanagi, T. Polym. Eng. Sci. **1992**, 32, 1511.
14. Crivello J. V.; Lam, J. H. J. Org. Chem. **1978**, 43, 305.
15. Kevill, D. N.; Anderson, S. W. J. Am. Chem. Soc. **1986**, 108, 1579.
16. Beak, P.; Sullivan, T. A. J. Am. Chem. Soc. **1982**, 104, 4450.
17. Dektar, J. L.; Hacker, N. P. J. Am. Chem. Soc. **1990**, 112, 6004.
18. Hacker, N. P.; Welsh, K. M. Proc. SPIE **1991**,1466, 384.
19. Schlegel, L.; Ueno, T.; Shiraishi, H.; Hayashi, N.; Iwayanagi,T. Chem. Mater. **1990**, 2, 299.
20. Kozawa, T.; Yoshida, Y.; Uesaka, M.; Tagawa, S. Jpn. J. Appl. Phys. **1992**, 31, 4301.

RECEIVED July 7, 1995

Chapter 9

Photoacid and Photobase Generators: Arylmethyl Sulfones and Benzhydrylammonium Salts

J. E. Hanson[1], K. H. Jensen[1], N. Gargiulo[1], D. Motta, D. A. Pingor[1], Anthony E. Novembre[2], David A. Mixon[2], J. M. Kometani[2], and C. Knurek[2]

[1]Department of Chemistry, Seton Hall University, South Orange, NJ 07079–2694
[2]AT&T Bell Laboratories, 600 Mountain Avenue, Murray Hill, NJ 07974

Studies of photoacid and photobase generators of interest for microlithography are reported. The efficiency of arylmethyl sulfone photoacid generators was studied as a function of internal hydrogen abstraction and polar substituents. In solution, benzyl phenyl sulfone derivatives with a methyl group as a hydrogen donor substituent on the benzyl side do not show an enhanced quantum yield. This contrasts with their behavior in polymer films. The difference can be explained by reference to the viscosity of the medium and the lifetime of the caged radical pair. A methyl substituent at the 2 position on the phenyl side, however, enhances the quantum yield both in solution and in polymer films. For this case, diffusion of the radical pair does not separate the abstracting sulfonyl radical from the hydrogen donor methyl group. Polar effects were studied by examining a series of benzyl phenyl sulfones with electron donor and acceptor substituents at the 4 and 4' positions. These produced only weak effects on photoacid generating efficiency as determined by deprotection of poly(t-butoxycarbonyloxystyrene - sulfur dioxide) films. On the benzyl side, electron donors (less electronegative substituents) appear to increase efficiency. On the phenyl side, the opposite is true: electron acceptors (more electronegative substituents) appear to increase efficiency. The effects are small and in some cases are masked by other influences. A new photobase generating molecule is trimethylbenzhydrylammonium iodide. Photolysis of this substance in the ultraviolet leads to formation of trimethylamine. This is the first example of a photogenerator of tertiary amines. Photogeneration of acid and base catalysts is important to the improvement of chemically amplified photoresists and other imaging technologies.

Photochemically generated catalysts and initiators have a long history, with numerous examples from photography and other imaging techniques.(1) The recent invention and development of chemically amplified photoresists(2-5) has produced a new demand for these materials, and so there has been an increase in reported

0097–6156/95/0614–0137$12.00/0

studies, particularly for photoacid generators.(6) There are many important design considerations for photocatalyst generators, which apply to both photoacid and photobase generators. These include high sensitivity (photospeed or quantum yield), good chemical stability for long term storage, good thermal stability for processing considerations, strength of the acid or base catalyst, low cost, and low toxicity. The number of design variables is sufficiently large that superior materials with regard to one or two variables may not be superior with regard to others. The importance of each variable may differ for different applications. We have previously reported on arylmethyl sulfones as photoacid generators.(7-10) These materials have certain advantages in their high chemical and thermal stability, low cost, and low toxicity. Their major disadvantage has been relatively low sensitivity, related to the photochemical quantum yield and the acidity of the photogenerated acid, most likely a sulfinic acid. We report here new studies on a variety of arylmethyl sulfone derivatives, which address the issue of quantum yield. We also report on a new family of photobase generators we have recently developed.(11)   These are quaternary ammonium salts of benzhydrylamines, photogenerators of tertiary amines. Previously reported amine photobase generators have been capable of photogenerating primary and secondary amines, but not tertiary amines.(12-15)

Arylmethyl Sulfone Photoacid Generators.

The published photochemistry of arylmethyl sulfones is dominated by photodesulfonylation, often related to the synthesis of strained hydrocarbons such as cyclophanes.(16) This photochemistry is typically carried out in poor hydrogen donor solvents such as benzene. In good hydrogen donor solvents, such as 2-propanol or ethanol, a different photochemistry can predominate. Langler *et al* observed yields of sulfinic acid products as high as 54% when irradiating arylmethyl sulfones in 2-propanol.(18) The literature of this alternate photochemistry of arylmethyl sulfones is rather sparse(17-18), and was unknown to us when we made our first observations of arylmethyl sulfones as photoacid generators.(7) This earlier work did not directly inspire our development of arylmethyl sulfone photoacid generators, but it has provided a mechanistic framework in which to understand their photochemistry, as shown in Scheme 1. The initial photochemical event is carbon-sulfur bond cleavage to give an arylmethyl radical and a sulfonyl radical. The path taken by the sulfonyl radical is the critical step, as the productive chemistry (for a photoacid generator) is hydrogen abstraction to give a sulfinic acid, but unproductive recombination and desulfonylation pathways are also available. (Under certain

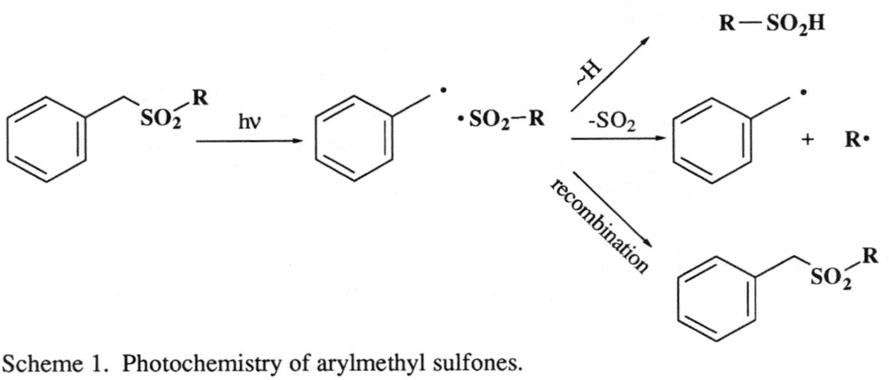

Scheme 1. Photochemistry of arylmethyl sulfones.

conditions, the SO_2 lost on desulfonylation might combine with water to give sulfurous acid, H_2SO_3, providing another source of photogenerated acid.) Since our first report, we have studied several series of sulfones, investigating the influence of structure on the efficiency of photoacid generation and the sensitivity of resist formulations.(7-10) Here we report on two distinct series of arylmethyl sulfones which were designed to reveal different structural effects.

The first series of arylmethyl sulfones was designed to investigate the effect of internal hydrogen donors. These compounds are shown in Scheme 2. An enhancement of quantum yield from an internal hydrogen donor had been observed in our earlier work for 2-methylbenzyl phenyl sulfone.(7-10) The series examined here all share the benzyl phenyl sulfone skeleton with methyl substituents at different positions. Our initial studies of resist formulations based on these photoacid generators has been published.(10) These earlier results, using 1.4 nm X-ray exposures in polymer films, suggested that the quantum yields of the derivatives with the methyl (Me) substituent on the benzyl fragment are ordered 2-Me > 4-Me ≈ 3-Me > unsubstituted benzyl phenyl sulfone. The derivative with the methyl group at the 2 position on the phenyl side, benzyl-2-methylphenyl sulfone, was found to have the highest efficiency in these studies, suggesting it has a higher quantum yield than any of the benzyl substituted derivatives. We have now measured solution quantum yields of acid generation for these molecules at 248 nm. Three solvents were employed: 2-propanol (a good hydrogen donor), glycerol (a similar solvent of much higher viscosity), and Freon-113 (1,1,2-trichlorotrifluoroethane, which lacks hydrogens and therefore cannot be a hydrogen donor). The measured quantum yields are listed in Table I.

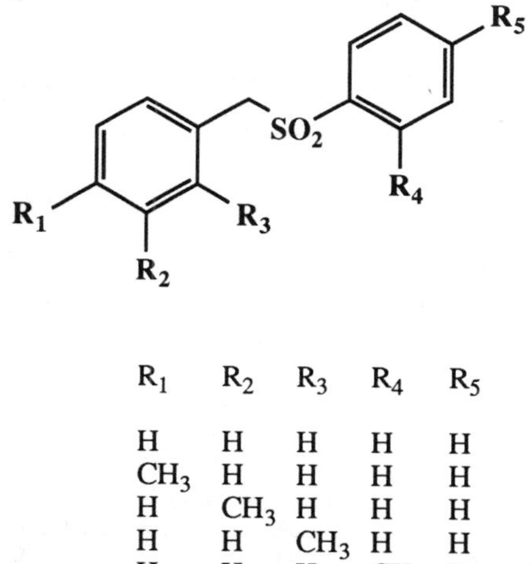

R_1	R_2	R_3	R_4	R_5
H	H	H	H	H
CH_3	H	H	H	H
H	CH_3	H	H	H
H	H	CH_3	H	H
H	H	H	CH_3	H
H	H	H	H	CH_3

Scheme 2. Methyl derivatives of benzyl phenyl sulfone.

In 2-propanol, benzyl-2-methylphenyl sulfone again stands out from the rest. The other sulfones' quantum yields all fall in a narrow range from 0.08 to 0.13. These values are probably the same within experimental error, although the standard deviation of these measurements is typically ±0.01 (average of four or five measurements). Benzyl-2-methylphenyl sulfone has a significantly higher Φ_{H^+} of

0.28. Our interpretation of these results and the differences between polymer film and solution results is based primarily on viscosity. In solution, low viscosity allows rapid escape of the photgenerated radical pair from the solvent cage. Recombination and "intra-pair" abstraction are minimized, and any abstraction that occurs is primarily intermolecular. Polymers are essentially "infinite viscosity" solvents below their T_g, so the radicals remain in the solvent cage for a much longer time. This increases recombination, which results in overall lower quantum yields for acid formation in polymers. "Intra-pair" abstraction also becomes important for a caged radical pair, so a hydrogen donor substituent on the other radical partner becomes significant for abstraction. So while the unsubstituted and benzyl substituted sulfones are essentially identical in their solution behavior, differences are observed in resist films.(10) Benzyl-2-methylphenyl sulfone is unique among this set of molecules, as the H-donor methyl group necessarily remains in proximity to the sulfonyl radical even after cage escape, and abstraction can be truly intramolecular. Glycerol, a relatively high viscosity solvent, was employed as an intermediate step between 2-propanol and polymer films. The measurements in glycerol are more difficult to make so that the results should not be overinterpreted, but the observed trends are suggestive. All of the Φ_{H^+} values decrease, as expected since recombination increases in the higher viscosity solvent. The decline in quantum yields suggests that the actual quantum yields in "infinite viscosity" polymer films are at best 5 - 10% of their value in 2-propanol.

Table I. Quantum Yields of Acid Generation for Methyl Derivatives of Benzyl Phenyl Sulfone

Compound	Φ_{H^+} 2-Propanol 1×10^{-3} M	Φ_{H^+} Glycerol 1×10^{-3} M	Φ_{H^+} Freon-113 5×10^{-4} M
Benzyl Phenyl Sulfone	0.12 ± 0.01	--	0.03 ± 0.02
2-Methylbenzyl Phenyl Sulfone	0.08 ± 0.01	0.05 ± 0.02	0.02 ± 0.02
3-Methylbenzyl Phenyl Sulfone	0.09 ± 0.01	0.05 ± 0.02	0.04 ± 0.02
4-Methylbenzyl Phenyl Sulfone	0.08 ± 0.01	0.02 ± 0.02	0.03 ± 0.02
Benzyl 2-Methylphenyl Sulfone	0.28 ± 0.03	0.14 ± 0.02	0.15 ± 0.02
Benzyl 4-Methylphenyl Sulfone	0.13 ± 0.01	--	0.10 ± 0.02

We recognized that intermolecular abstraction in 2-propanol need not necessarily be from solvent. The arylmethyl sulfones all have benzyl hydrogens and may serve as intermolecular as well as internal donors. To investigate this, the photochemistry in Freon-113 was studied. The sulfones are not highly soluble in this solvent, limiting the experiments that could be done. All of the sulfones were studied at 5×10^{-4} M, and all of the quantum yields decreased. The quantum yield in Freon-113 is expected to represent a bimolecular event involving two molecules of sulfone: one as the source of the sulfonyl radical and one as the hydrogen donor. Again, all of the benzyl substituted sulfones gave essentially identical results, although the numbers are low enough that error might mask any weak effects. Benzyl-2-methylphenyl sulfone was considerably higher, as expected, since it should retain a strong internal hydrogen abstraction component. Surprisingly, benzyl-4-methylphenyl sulfone also had a considerably higher Φ_{H^+}. As evidence for intermolecular abstraction from other molecules of sulfone, a significant dependence of quantum yield on concentration was noticed. For example, 4-methylbenzyl phenyl sulfone gave a quantum yield of 0.01 when the concentration was cut in half to 2.5 x 10-4 M, compared to 0.04 at 5 x 10-4 M. The low solubility and low quantum yields in Freon-113, coupled with the inherent errors and difficulties in making quantum yield measurements make such measurements difficult, however. However, the fairly large difference between the quantum yields in 2-propanol (~0.1) and in Freon-113 (~0.02) suggests that the chief abstraction route in 2-propanol is indeed abstraction from the solvent.

The other series of sulfones we have synthesized are also based on the benzyl phenyl sulfone skeleton, but these compounds carry electron donor or acceptor groups

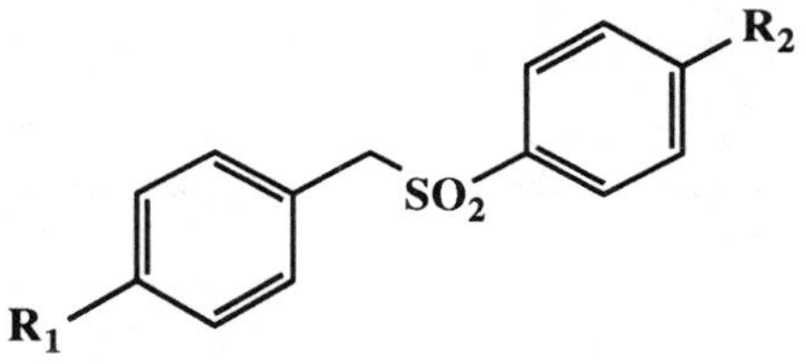

Series 1: R_1 = H, CH$_3$, OCH$_3$, Cl, F; R_2 = H

Series 2: R_1 = H; R_2 = H, CH$_3$, OCH$_3$, Br, Cl, F

Scheme 3. Benzyl phenyl sulfone derivatives used in study of polar effects.

at the 4 position of either the benzyl or phenyl group, as shown in Scheme 3. The primary purpose of these compounds is to investigate polar effects on the photochemistry. Such polar effects are not expected to be large, but have been observed in other radical reactions.(19-25) Polar effects could occur in the photohomolysis step due to stabilization or destabilization of the photogenerated radicals. For substituents on the phenyl side, polar effects could alter the reactivity of the sulfonyl radical toward hydrogen abstraction and desulfonylation as well. We have used lithographic evaluations to obtain the trends in the photoactivity of these compounds. This data is reported in Table II. The results were obtained for photoresists composed of .075 molar equivalents of the sulfone photoacid generators per t-BOC group in a 3:1 t-BOC-styrene/sulfur dioxide copolymer. The processing conditions are described in the experimental section. For these t-BOC-styrene/sulfur dioxide copolymer based photoresists, solubility in aqueous base developers usually becomes significant at ~90% deprotection. Minimum dissolution doses (D$_s$) are reported for each sulfone, and should represent the radiation dose necessary to generate sufficient acid to result in 90% deprotection. With two exceptions, these show little variation, confirming the expected small size of any polar effects. A more sensitive measure of the acid generating efficiency is the percent deprotection at an intermediate dose.(10) In Table II, percent deprotection is also reported at 11 mJ/cm^2. In analyzing this data, several of the sulfones appear to behave anomalously. In several studies of polar effects on radical reactions the unsubstituted derivative is unusually unreactive.(19-25) This appears to be the case here as well, where the parent benzyl phenyl sulfone is almost the least efficient sulfone. The methoxy derivatives appear to be unusually efficient. This may represent a difference in mechanism for the interaction of the methoxy derivatives with the radical center (resonance vs. inductive), or it could represent another instance of internal hydrogen donation as described above, with the methoxy groups as internal donors in place of the methyls. The phenyl substituted bromo derivative is unusually inefficient. This has been observed for other bromo substituted sulfones and may represent alternate pathways available to these compounds, such as homolysis of the carbon - bromine bond. In any event, it seems wise to exclude these anomalous values (parent, methoxy, bromo) until further data can be gathered. The remaining derivatives are the methyl, chloro, and fluoro for both benzyl and phenyl substitution. The results do not correlate well with any of the widely used substituent constants (σ_{para}, σ_{meta}, σ^+, or $\sigma\cdot$). They do establish trends based roughly on the electronegativity of the substituents. On the benzyl side, more electronegative (inductively withdrawing) groups decrease the efficiency of acid generation and therefore the percent deprotection: methyl is the most efficient, fluoro the least. On the phenyl side, more electronegative groups increase the efficiency of acid generation and therefore the percent deprotection: fluoro is the most efficient, methyl the least. The effects are at best moderate in size, as expected. The very small data set makes intensive analysis suspect, therefore we are currently synthesizing more derivatives to increase the size of the data set.

Table II. Polar Effects in Arylmethyl Sulfone Photoacid Generators

Compound	D_s (mJ/cm^2)	% deprotection at 11 mJ/cm^2
Unsubstituted		
Benzyl Phenyl Sulfone	20	51
Benzyl substituted		
4-Methoxybenzyl Phenyl Sulfone	17	79
4-Methylbenzyl Phenyl Sulfone	20	58
4-Chlorobenzyl Phenyl Sulfone	20	57
4-Fluorobenzyl Phenyl Sulfone	20	53
Phenyl substituted		
Benzyl 4-Methoxyphenyl Sulfone	8	99
Benzyl 4-Methylphenyl Sulfone	17	62
Benzyl 4-Bromophenyl Sulfone	26	39
Benzyl 4-Chlorophenyl Sulfone	17	66
Benzyl 4-Fluorophenyl Sulfone	17	70

Quaternary Ammonium Salt Photobase Generators.

Photochemical base generation has not been widely investigated. Tertiary amines are in some instances better basic catalysts than primary and secondary amines, as these may be incompatible with certain resist chemistries.(27-28) Photogeneration of tertiary amines also appears to be a relatively difficult problem, however. Our approach has been to build on the work of Kochi in photosolvolysis of benzylammonium salts(29) and Steenken in the photolysis of diphenyl- and triphenylmethane derivatives.(30-32) Conceptually, the photolysis of a quaternary ammonium salt where one substituent forms a stable cation should result in heterolysis to give a tertiary amine and a relatively stable cation. The cation, however, must not be able to eliminate H^+ to give an alkene, and the system must contain a nucleophile capable of competing with the tertiary amine to prevent recombination. This nucleophile also should not be protic (i.e., alcohols), since the product of reaction with the cation will subsequently eliminate H^+ and protonate the photogenerated amine. Ideally, the nucleophilic agent should be the counterion of the original quaternary ammonium salt. Counterions such as cyanide, thiocyanate, and iodide would appear to best meet this set of criteria. A photobase generator system that follows this pattern is trimethylbenzhydrylammonium iodide, as shown in Scheme 4.

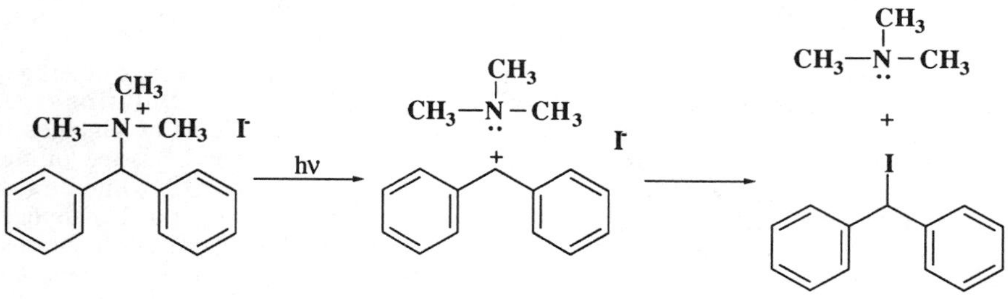

Scheme 4. Photochemistry of trimethylbenzhydrylammonium iodide.

The synthesis of trimethylbenzhydrylammonium iodide was accomplished by exhaustive methylation of benzhydrylamine with methyl iodide.(11, 33-35) Other counterions can be exchanged for the iodide (for example, we have exchanged for a completely nonnucleophilic counterion by reaction with silver trifluoromethane-sulfonate). Unfortunately, few other alkyl groups can be added to benzyhydrylamine by this technique: reactions with ethyl iodide, propyl iodide, and butyl iodide all fail. A molecular mechanics study suggested that there is considerable steric hindrance around the nitrogen lone pair of benzhydrylamine. The most stable conformation appears to be one where the amine lone pair is "guarded" by ortho hydrogens on the two phenyl rings, as shown in the Newman projection in Scheme 5. This apparently limits the nucleophilicity of the lone pair such that it can only react with relatively unhindered electrophilic centers. Even ethyl iodide appears to be too hindered.(36) While trimethylamine might be a useful catalyst due to its expected high mobility in polymer systems, volatility could be a problem since the boiling point is only 2.9°C. The ability to generate various amines with different mobilities, volatilities, and base strengths would be preferable. We have recently begun to experiment with the aminofluorene system, where the extra ring fusion forces near coplanarity and reduces the steric demand around the nitrogen. This amine is also easily exhaustively methylated to the trimethylfluorenylammonium salt, and we have been able to alkylate amines using 9-bromofluorene as well. This system promises to allow variation in the substitution of both the tertiary amine and the stabilized cation. This should provide a greater measure of control over the materials properties of these photobase generators.

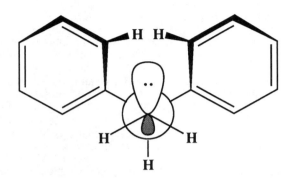

Scheme 5. Newman projection of calculated favored conformation of benzhydrylamine. The nitrogen is in the foreground (with lone pair).

We have demonstrated the solution photochemistry of the quaternary ammonium photobase generators using NMR.(11) Solutions of trimethylbenzhydrylammonium iodide in deuterated acetonitrile show changes in peak heights with UV exposure. The peak due to the methyl groups of the ammonium salt and the peak due to the benzhydryl proton decrease, while a peak due to free trimethylamine and a peak due to the benzhydryl proton of benzhydryl iodide grow in. For irradiation at 254 nm in a Hanovia photochemical apparatus, photochemical conversion is essentially complete in 60 minutes. Similar changes are observed for the trimethylfluorenylammonium iodide. Figure 1 shows the photolysis of the benzhydryl derivative. The thermal properties of these compounds are quite good, as both quaternary ammonium iodides have decomposition temperatures above 185°C. We are beginning to explore the use of these photogenerated amines in photoresist and polymer curing applications, with several promising results.(37) A number of organic transformations are catalyzed by tertiary amines, so the ability to generate these catalysts should open new avenues for resist chemistry.

Conclusions.

Advances in the photogeneration of catalysts are important in the development of new photoresists. Photoacid generation is a relatively mature field, and we are working on understanding the details of the photochemistry of the arylmethyl sulfones. We have established the importance of internal hydrogen donors, particularly in viscous solvents such as polymer films. Polar effects appear to be much less important, although there are measurable differences that correlate with substituent electronegativity. Photobase generation is far less developed, and we have reported a new type of photobase generator, where photochemical dissociation of a quaternary ammonium salt gives a tertiary amine. This appears to be the first tertiary amine photogenerator, which should open new possibilities for resist chemistry.

a) Before photolysis.

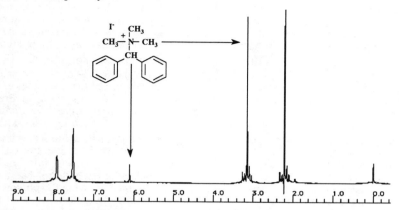

b) After 60 minutes photolysis at 254 nm.

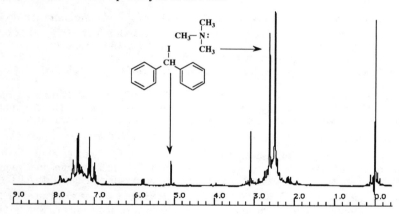

Figure 1. Photolysis of trimethylbenzhydrylammonium iodide in CD₃CN.

Experimental Section.

Materials Synthesis and Characterization. Arylmethyl sulfones were synthesized from the arylmethyl halides and the sodium sulfinates by the method of Baldwin and Robinson.(38) The arylmethyl halides and most of the sulfinates are available commercially. Some sodium sulfinates were obtained by reduction of the commercially available sulfonyl chlorides.(39) All sulfones were purified by recrystallization from toluene before use. The t-BOC-styrene/sulfur dioxide copolymer was prepared and characterized as previously described.(40) Trimethylbenzhydrylammonium iodide was prepared by exhaustive methylation of benzhydrylamine with methyl iodide and sodium carbonate in methanol as described elsewhere.(11)

Photochemistry: Quantum Yields. Commercially available spectroscopy grade solvents were used for the quantum yield determinations. Absorbance measurements were made on a Hewlett-Packard 8452A diode array spectrometer. A stock solution of the indicator bromophenol blue sodium salt was prepared in 2-propanol and used for all experiments. A calibration curve was established by treating 5.00 ml aliquots of this stock solution with 0-100 μl of 0.001 M HCl in methanol, diluting with 2-propanol, glycerol, or Freon-113 as appropriate such that each calibration sample had a total volume of 6.00 ml, and measuring the absorbance of these samples at 610 nm. Sulfone samples in the appropriate solvent (2-propanol, glycerol, or Freon-113) were prepared and their absorbance measured. The samples were then irradiated for 1 to 60 seconds at 248 nm using a Lambda-Physik KrF excimer laser with an intensity of 1.5 to 2.0 mJ/cm^2-sec. Malachite green leucocyanide solutions irradiated simultaneously were employed as an actinometer, following the method of Calvert and Rechen.(41) Following irradiation, 1.00 ml of each sulfone solution was combined with 5.00 ml of the bromophenol blue stock solution, and the absorbance at 610 nm measured. The quantum yield can be determined knowing the photon flux from the actinometry, the absorbance of the sample, and the concentration of acid produced from the calibrated indicator absorbance. Reported quantum yields are the average of two to four measurements.

Photochemistry: NMR Experiments. Trimethylbenzhydrylammonium iodide was dissolved in acetonitrile-d_3. The solution was divided among four to six quartz NMR tubes (Wilmad). These were irradiated in a Hanovia photochemical reactor using 254 nm lamps. The tubes were removed from the reactor at time intervals from 0 to 60 minutes and their NMR spectra obtained on a General Electric QE-300 FT-NMR.

X-Ray Lithography. Resist solutions (15 w/v%) were prepared using 3:1 t-BOC-styrene/sulfur dioxide copolymer in ethyl-3-ethoxypropanoate. The arylmethyl sulfones were added such that the ratio of t-BOC groups to sulfone was 100:7.5 (7.5 mol%). The solutions were filtered through 0.2 μm Millipore filters, then used to coat nominal 1.0 μm films on hexamethyldisilazane primed 5 inch (130 mm) silicon substrates. The wafers were subsequently baked at 105°C for 2 minutes on a vacuum hotplate. Film thickness was measured using a Nanometrics Nanospec/AFT film thickness gauge. X-ray (λ = 0.8 to 2.0 nm centered at 1.4 nm) exposures in 1.00 atmosphere helium were performed using a Hampshire Instruments Series 5000 point source proximity print stepper(42), operating with a pulse rate of 1.0 Hz and a flux of 0.8 - 1.2 mJ/cm^2/pulse. An open frame polysilicon mask (1.0 μm thick) was used in all exposures. Resist films were post-exposure baked on a vacuum hot plate for 140°C for 2.5 minutes. Wafers were customarily split in two, with one half developed and one half used for infrared measurements. Development was performed by immersion in 0.17 N tetramethylammonium hydroxide for 30 seconds followed by a de-ionized water immersion rinse for 30 seconds. The minimum exposure dose required to develop a approximately 5x5 mm area was reported as the measure of resist sensitivity D_s. Infrared analysis was performed on a Mattson Instruments Galaxy Series Model 8020 dual beam FT-IR spectrometer. Deprotection of the t-BOC-styrene/sulfur dioxide copolymer resist material is accompanied by disappearance of the carbonyl band at 1760 cm^{-1}. The extent of deprotection at a given exposure dose was determined by comparison of the integrated area of the carbonyl band for an exposed section of the resist film versus the integrated area of the carbonyl band for an unexposed section.

Acknowledgments. Discussions of this work with G.N.Taylor and E. Reichmanis have been useful. Acknowledgment is made to the donors of the Petroleum Research Fund, administered by the ACS; Research Corporation's Partners in Science Program; and the Seton Hall University Research Council for partial support of this research.

References

1. A.Reiser, "Photoreactive Polymers", Wiley & Sons, New York: 1989.

2. H.Ito, C.G.Willson *Polym. Eng. Sci.* **1983**, *23*, 1012.

3. C.G. Willson, H.Ito, J.M.J.Frechet, T.G.Tessier, F.M.Houlihan *J. Electrochem. Soc.* **1986**, *133*, 181.

4. E.Reichmanis, F.M. Houlihan, O.Nalamasu, T.X.Neenan *Chem. Mater.* **1991**, *3*, 394.

5. S.A.MacDonald, C.G. Willson, J.M.J.Frechet *Acc. Chem. Res.* **1994**, *27*, 151.

6. E.Reichmanis, A.E.Novembre *Ann. Rev. Mater. Sci.* **1993**, *23*, 11.

7. A.E.Novembre, W.W.Tai, J.M.Kometani, J.E.Hanson, O.Nalamasu, G.N.Taylor, E.Reichmanis, L.F.THompson *Chem. Mater.* **1992**, *4*, 278.

8. A.E.Novembre, J.E.Hanson, J.M.Kometani, W.W.Tai, E.Reichmanis, L.F. Thompson, R.J.West *Polym. Eng. Sci.* **1992**, *32*, 1476.

9., A.E. Novembre, J.E.Hanson, J.M.Kometani, W.W.Tai, E.Reichmanis, L.F. Thompson *Irradiation of Polymeric Materials,* ACS Symposium Series #527, 1993.

10. J.E.Hanson, D.A.Pingor, A.E. Novembre, D.A.Mixon, M.P.Bohrer, J.M. Kometani, W.W.Tai *Polym. for Adv. Technol.* **1994**, *5*, 49.

11. K.H.Jensen, J.E.Hanson submitted for publication.

12. J.F.Cameron, J.M.J.Frechet *J. Org. Chem.* **1990**, *55*, 5919.

13. Cameron, J.F.; Frechet, J.M.J. *J. Am. Chem. Soc.* **1991**, *113*, 4303.

14. J.M.J. Frechet *Pure and Appl. Chem.* **1992**, *64*, 1239.

15. K.A.Graziano, S.D.Thompson, M.R.Winkle *SPIE Advances in Resist Technology and Processing* **1991**, *1466*, 75.

16. R.S.Givens, R.J.Olsen, P.L.Wylie *J. Org. Chem.* **1979**, *44*, 1608.

17. C.L.McIntosh, P.de Mayo, R.W.Yip *Tet. Lett.* **1967**, 37.

18. R.F. Langler, Z.A.Marini, J.A.Pincock *Can. J. Chem.* **1978**, *56*, 903.

19. J.R.Shelton, C.K.Liang, P. Kovacic *J. Am. Chem. Soc.* **1968**, *90*, 354.

20. A.A.Zavitsas, J.A.Pinto *J. Am. Chem. Soc.* **1972**, *94*, 7390.

21. J.P.Engstrom, J.C.DuBose *J. Org. Chem.* **1973**, *38*, 3817.

22. J.R.Shelton, C.K.Liang *J. Org. Chem.* **1973**, *38*, 2301.

23. W.H.Davis, W.A.Pryor *J. Am. Chem. Soc.* **1977**, *99*, 6365.

24. T.H.Fisher, A.W.Meierhoefer *J. Org. Chem.* **1978**, *43*, 220.

25. T.H.Fisher, A.W.Meierhoefer *J. Org. Chem.* **1978**, *43*, 224.

26. O.Nalamasu, A.E.Novembre, J.M.Kometani, J.E.Hanson *J. Photopolym. Sci. Tech.* **1993**, *6*, 457.

27. T.Sakurai, S.Kojima, H.Inoue Bull. Chem. Soc. Jpn. 1990, 63, 3141.

28. S.E.Drewes, G.H.P.Roos *Tetrahedron* **1988**, *44*, 4653.
29. M.A.Ratcliff, Jr.; J.K.Kochi *J. Org. Chem.* **1971**, *36*, 3112.
30. R.A.McClelland, V.M.Kanagasabapathy, N.S.Banait, S.Steenken, *J. Am. Chem. Soc.* **1989**, *111*, 3966.
31. R.A.McClelland, N. Mathivanan, S.Steenken *J. Am. Chem. Soc.* **1990**, *112*, 4859.
32. J. Bartl, S.Steenken, H.Mayr, R.A.McClelland *J. Am. Chem. Soc.* **1990**, *112*, 6918.
33. Sommelet, M. *Comptes Rendus* **1925**, *180*, 76.
34. Battersby, A.R.; Binks, R. *J. Chem. Soc.* **1958**, 4333.
35. G.Wittig *Bull. Soc. Chim. France* **1971**, 1921.
36. We are also currently working on reductive amination routes through benzyhydrylamine, a route recommended by Dr. R.O.Hutchins. R.O.Hutchins, W.Y.Su *Tet. Lett.* **1984**, *25*, 695.
37. Dr. Y. Wei, personal communication.
38. W.A.Baldwin, R.Robinson *J. Chem. Soc.* **1932**, 1445.
39. P.Oxley, M.W.Partridge, T.D.Robson, W.F.Short *J. Chem. Soc.* **1946**, 763.
40. R.S.Kanga, J.M.Kometani, E.Reichmanis, J.E.Hanson, O.Nalamasu, L.F. Thompson, S.A.Heffner, W.W.Tai, P.Trevor *Chem. Mater.* **1991**, *3*, 660.
41. J.G.Calvert, J.L.Rechen *J. Am. Chem. Soc.* **1952**, *74*, 2101.
42. D.W.Peters, B.J.Dardinski, D.R.Kelly *J. Vac. Sci. Technol.* **1990**, *B8*, 1624

RECEIVED July 7, 1995

Chapter 10

Functional Imaging with Chemically Amplified Resists

Alexander M. Vekselman, Chunhao Zhang, and Graham D. Darling[1]

Department of Chemistry, McGill University, 801 Sherbrooke West, Montreal, Province of Quebec H3A 2K6, Canada

The same dramatic photosensitivity shown by films of chemically amplified resists that permits their patterned removal ("relief development", typical of microlithography), can also instead allow their further imagewise chemical modification ("functional development"), such as through exposure-controlled sorption of various species from contacting solutions or vapors. For example, radiation-defined deprotection of nonpolar poly(di-*t*-butyl fumarate-*co*-styrene) produced a pattern of polar and reactive carboxylic acid and anhydride moieties. Conditions were found for only these exposed areas of the resist material to take up Ca(II), Ni(II), Co(II), Pb(II) or some ammonium ions from the corresponding aqueous solutions, without being dissolved. Several organic dyes were also placed into either exposed areas from water/alcohol solutions, or into unexposed areas from hexane/toluene solutions. Modes and mechanisms are discussed in terms of resist, solute and solvent properties.

Since the early 1980's, sensitive Chemically Amplified (CA) resists such as based on P(*p*-TBOCST) (i.e. poly(*p*-*t*-butyloxycarbonyloxystyrene)) have been designed and tested to rapidly produce fine patterns of contrasting materials, such as in the manufacture of high-density DRAM devices and microprocessor chips by microlithography [1]. The traditional role for a resist in such applications has been to capture a projected radiation pattern as a "relief image" of removed and remaining areas of resist film (Figure 1), which is then translated into an underlying inorganic substrate through an etching process [2]. However, microlithography-like techniques can also be used for making fine-scale optical waveguides, couplers, and recording media, or array sensors, displays and supports for solid-phase (bio)chemical analysis or synthesis, or other useful patterned structures. Here, it would be often better not to remove, from the intermediate "latent image", either the exposed or unexposed areas of resist material ("relief development"), but instead only to further modify one or the other ("functional development"). Possible physical, chemical or biological properties that could thus be

[1]Corresponding author

0097–6156/95/0614–0149$12.00/0
© 1995 American Chemical Society

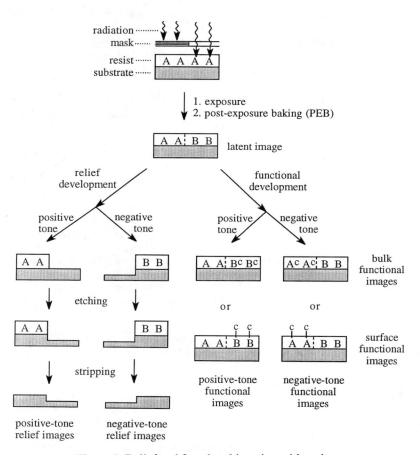

Figure 1. Relief and functional imaging with resists.

altered in selected patterns include density, refractive index (n), color, fluorescence, hyperpolarizability, reflectivity, heat capacity, dielectric constant, adhesiveness, liquid crystallinity, electrical or thermal conductivity; together with wettability, permeability, sorption, catalytic activity or chemical reactivity towards various liquid or gas-phase species; along with dissolution and etch resistance that would be relevant to later relief development steps as well.

Upon or following exposure, the polymers within CA or other resist materials can become altered in three general ways: i) polymerization or crosslinking, that increases average molecular weight; ii) scission or depolymerization, that decreases average molecular weight; iii) functional group transformation, that does not affect the lengths of polymer chains, but drastically alters their polarity, solubility and reactivity. This third mode, as exemplified in P(p-TBOCST) [3], is not only the most attractive for relief development because of the lack of swelling in undissolved material and the possibility of dual-tone imaging, but is also the most adaptable towards functional development. Indeed, both solubility and many other immediate or potential properties (including, tendency to further chemical modification) are critically dependent on the exposure-controlled functional composition of such resists, through highly cooperative behavior of functional groups.

Unlike relief development, a functional development step by definition does not remove any of the solid polymer film, and can be a heterogeneous reaction that produces a stratified product. Patterned modification exclusively at a surface has been done to adsorb catalysts for further metallization [4,5], or biomolecules for DNA sequencing [6]; non-photochemical surface imaging has also been reported by a stamping technique [7]. In near-surface imaging, only the upper 10-100 nm of material is transformed due to low penetration of the initial imaging radiation (ex. low-energy electron or ion beams, or UV in relatively opaque materials [8]), or due to low diffusion of chemical species during a later functional development step (ex. in the "diffusion enhanced silylated resist" = DESIRE processes [9,10]). Modifying species can also enter and diffuse deeper into the bulk of the polymer. The rate and extent of such uptake can depend on the polymer-solution interface, or on bulk polymer-polymer (ex. crosslinking), polymer-solvent (ex. plasticization), and polymer-solute interactions, covalent or not, often involving polymer functional groups [5,11-15]. The understanding and design of functional development would benefit from the study of relevant chemical and physico-chemical phenomena, similar to what has been done on airborne contamination of resists [16,17], and miscibility of their components [18].

One general way of modifying a thin film of material is by controlled uptake of ions or molecules from a contacting liquid or gas. We discuss here the functional development, through exposure-controlled sorption of various inorganic and organic chemical species, of a model CA resist based on poly(di-t-butyl fumarate-co-styrene) (PDBFS).

Experimental

Instruments.   Instruments for spin-coating, photo-exposure, and baking were described in earlier publications [11]. The pH of aqueous solutions was measured with a Cole-Palmer Digi-Sense 5938-00 pH-meter with 5992-40 combination electrode, which had been calibrated using commercial buffer solutions (Caledon Inc.) of 0.05 M potassium hydrogen phthalate (pH 4.00) and 0.01 M sodium borate (pH 9.22). FTIR spectra were obtained with a Brucker IF-48 spectrophotometer with a microscope accessory. UV-VIS spectra were done with a Shimadzu Spectronic-210UV spectrophotometer. Fluorescence measurements were made in a Spex model F112 spectrofluorimeter with a 450 W xenon lamp, with the emission detector at 90° to the incident beam for solution measurement, and at 22.5° for solid films. Oxygen Reactive Ion Etching (O_2-RIE) was performed in a custom-built Large-volume Microwave Plasma (LMP) apparatus [19].

Chemicals.   Hexane, toluene, methanol, *n*- and *iso*-propanol solvents were "all glass distilled" grade from OmniSolv Inc.; pyridine from BDH Inc.; *t*-butanol from Baker Inc.; and the dye Rhodamine-6G (R6G) from Aldrich Inc.. The preparation of 4-(5-N,N-dimethylaminonaphthalenesulfonylamido)-1-methylpiperazine (DSMP) was previously described [11]. 1,10-Diaminodecane, 2- and 4-aminopyridine and 2,6-diaminopyridine were obtained from Aldrich Inc.; 30% poly(ethyleneimine)/H_2O (M_w = 10K g/mol) from PolyScience Inc.. Metal salts of >99% purity were used as received: $CuCl_2$, $NiCl_2$, $CoCl_2$, $AlCl_3$, $FeCl_2$ and $FeCl_3$ from Alfa Inc.; $Pb(OAc)_2$ and $PdCl_2$ from Aldrich Inc. Each of these was dissolved in H_2O to ca. 0.1 M, except for $PdCl_2$ which was almost insoluble. $Ca(OH)_2$ was prepared [20] from calcium oxide (Anachemia Inc.), and saturated solutions were freshly filtered before use. 0.1 M HCl/H_2O and 5% NH_3/H_2O were prepared by dilution of the commercial concentrates (Caledon Inc. and BDH Inc., respectively).

Preparation and Exposure of Resist.   Di-*t*-butyl fumarate was prepared from fumaryl chloride and potassium t-butoxide, then polymerized with styrene in toluene solution with 2,2'-azobis(isobutyronitrile) (AIBN). The resulting PDBFS contained ca. 45 mol % of fumarate units, with M_w = 15K g/mol by GPC in chloroform, and [η] = 0.08 in chloroform [11]. Unless otherwise stated, samples for lithographic evaluation were prepared by dissolving 50 mg of PDBFS, together with 5 mg of either triphenylsulfonium or 4-(phenylthio)phenyldiphenylsulfonium hexafluoroantimonates [21] as Photo-Acid Generator (PAG), into 250-350 mg of propylene glycol methyl ether acetate, filtering this solution (0.2 μm) and spin-coating it at 1000-1200 rpm onto a silicon wafer, then performing a Post-Apply Bake (PAB) at 130 °C for 60 s, to give a 0.8-0.9 μm film that was optically clear. This was later irradiated with 0-100 mJ/cm^2 of deep- (254 nm) or mid-near-UV (ca. 20 % of 313 nm and 80 % of 365 nm) through an Optoline-Fluroware density photoresist step table REK/73 with resolution to 1 μm, or while half-shading the sample with an opaque object, before being subjected to a Post-Exposure Bake (PEB) at 135 °C for 60 s.

Functional Development with Metal Ions.   30 mL of solution containing metal ions was placed in a beaker with a pH electrode and continuous magnetic stirring. As needed, drops of 0.1 N HCl/H_2O or 5% NH_3/H_2O were then added to adjust the pH of the medium. Silicon wafers containing irradiated-baked resist were introduced for 10-180 s, then quickly rinsed for 30 s with distilled water, and air-dried before FTIR and other evaluation.

Reactive Ion Etching.   Samples were prepared and exposed in mid-near-UV through the step tablet (0-100 mJ/cm^2), dipped in saturated $Ca(OH)_2/H_2O$ for 90-180 s, dried, then later subjected to O_2-RIE in the LMP apparatus under conditions of 10 sccm O_2 flow, 50 mtorr pressure, 120 W power, and -500 V bias.

Refractometry.   Refractive index *n* of a thin resist film on a slide of fused silica was measured by a known wave-guiding technique [22], in which a laser beam was coupled by a prism into the resist layer, then decoupled ca. 1 cm further on by another prism. The relationship between coupling and decoupling angles and output light intensity gives both average n and film thickness. As the technique requires multi-mode wave-guiding, relatively thick films (1.1-2.0 μm) were prepared using a more concentrated resist solution and lower spin speed.

Functional Development with Amines and Dyes.   After standard processing (PAB, exposure and PEB), resist samples were immersed into 0.1-1 wt % solutions of amino-compounds or fluorescent dyes for 30 s, air-dried and evaluated. Modification from the gas phase was performed by suspending resist-coated wafers for a measured time over the liquid in a covered beaker partly filled with dilute aqueous ammonia or pure pyridine.

Results and Discussion

Poly(Di-*t*-Butyl Fumarate-co-Styrene) Resist.   The evolution of resists can be described as a progressive adjusting of materials properties towards simultaneously meeting all the requirements of the lithographic process, which are many and often conflicting: shelf-life vs. sensitivity, transparency at the short wavelengths needed for highest resolutions vs. plasma resistance requiring aromatic groups, ease of annealing vs. thermostability, etc. Many CA resists have been inspired by the use of protecting groups in synthetic organic chemistry, in which "capped" carbonate, ester or ether groups can undergo acid-catalyzed "deprotection" reactions to become much more polar carboxyl or hydroxyl groups. Carboxylic acids would be particularly attractive groups to permit both relief and functional development, via deprotonation and/or ion exchange. Such are generated from co/terpolymers of *t*-butyl with other alkyl acrylates undergoing thermo-acidolytic deprotection with photogenerated acid, permitting their relief development with aqueous base [23]. These aliphatic polymers also possess excellent UV transparency down to 190 nm but, without silicon or other plasma-resistant moieties, their films cannot later withstand conditions of the substrate etching step. Alternatively, polymers of protected vinyl-benzoic acid sufficiently resist such etching, however, conjugation between the carbonyl and aromatic ring results in strong optical absorbance below 320 nm [24]. A further improvement has been the copolymer of *t*-butyl methacrylate with α-methylstyrene, that combined etch-resistance with acceptable deep-UV transparency [25]. However, with only one ionizable group per 12 carbons, even the largely-deprotected matrix proved hard to dissolve in aqueous base without the addition of cosolvents such as *iso*-propanol.

The idea to copolymerize di-*t*-butyl fumarate with styrene for new CA resists appeared almost simultaneously in a Japanese patent [26], and in reports from our own [11,27] and Crivello's groups [28] (similar structure with additional α-cyano fumarate group was reported even earlier by Ito et al. [29]). Within the resulting PDBFS copolymer, each fumarate contributes a non-conjugated, UV-transparent unit with two acid-labile functionalities, while the styrene unit confers processability and plasma resistance. Excellent photolithographic properties have been reported [11,26-28], along with modest deep-UV absorbency (0.17-0.18 μm⁻¹), elevated glass transition temperature (T_g) both before (134 °C) and after deprotection (199 °C), and high thermostability (>300 °C). After deprotection, the copolymer could be removed easily by aqueous base alone, obviously due to the dianionic fumarate unit. According to developing conditions, PDBFS/onium salt combinations could give either positive- or negative-tone micron-scale relief images, with high photosensitivity of 14-40 mJ/cm^2

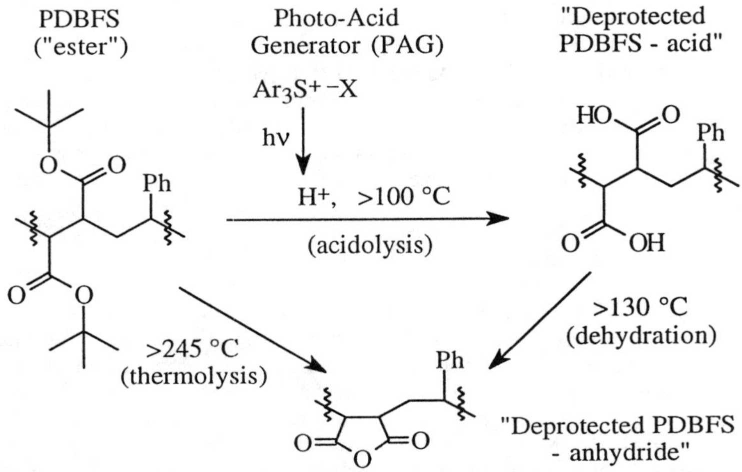

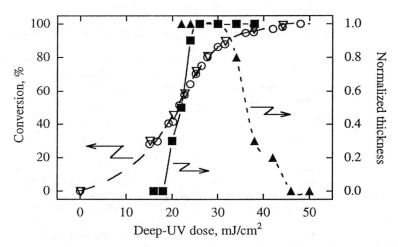

Scheme I. Reactions in PDBFS/onium salt resist.

Figure 2. Acid-catalyzed ester-to-acid conversion and developing of PDBFS resist vs. deep-UV dose (254 nm). Symbols represent normalized intensity of 1145 cm-1 (O) and 2977 cm-1 (∇) peaks, as well as normalized thickness after developing in toluene (■) and aqueous tetramethylammonium hydroxide (▲).

[11]. Versions of higher fumarate content and a more lipophilic PAG showed even better sensitivity of 2-4 mJ/cm^2 [28]. For this study, we chose both a high UV dose (100 mJ/cm^2) and PAG loading (10 wt %), as well as ample PAB and PEB (130-135 °C for 60 s), so that results of the new methods of development would not be affected by slight variations in the other processing steps.

Many general principles and development techniques worked out for PDBFS could be expected to apply to the other carboxyl-based resists [23-25] as well.

PEB Reactions and Relief Development of PDBFS. Our previous study of *t*-butyl elimination from PDBFS showed clean thermolysis at 245 °C in the absence of strong acid, and at ca. 100 °C in its presence [11]. However, FTIR showed that, at higher PEB temperatures, the appearance of free -COOH moieties was often followed by their conversion to anhydride (Scheme I). Comparison with IR spectra of commercial poly(styrene-*co*-maleic anhydride), and ease of dissolution in some solvents, suggested the occurrence of mostly intramolecular 5-member ring formation, without intermolecular anhydride crosslinking. It was roughly found that the ratios between ester, acid and anhydride groups could be controlled by UV dose (up to ca. 35 mJ/cm^2) and PEB temperature (variations in PEB times beyond 30 seconds had no great effect). Similar ester -> acid -> anhydride transformations have have also been studied in other resists [29,30].

Carboxylic acids and their ester, anhydride and anion derivatives each show characteristic strong absorption peaks in the carbonyl (1850-1400 cm^{-1}) and C-O (1000-1200 cm^{-1}) regions of their FTIR spectra; O-H of the free acids also absorb above 2900 cm^{-1}. The relative quantities of *t*-butyl ester groups in solid PDBFS films were measured by the intensities of the strong, sharp and isolated absorbance peaks from O-CMe$_3$ (1145 cm^{-1}) and CMe$_2$CH$_2$-H (2977 cm^{-1}) beyond a baseline defined by a spectrum of the same sample after obviously complete *t*-butyl cleavage. Comparison was also made between these peaks and that from RC$_6$H$_4$-H bend (703 cm^{-1}) in the same spectrum, as a reference that remained unaffected by chemical changes elsewhere in the polymer. The step tablet conveniently provided a wide range of UV dosage onto different areas of the same resist film. The silicon wafer substrates were effectively transparent to mid-IR radiation, so that FTIR could be conveniently performed on differently-exposed areas of film using an IR microscope in transmittance mode.

Whatever mode of its development, an effective resist must show high contrast for sharp demarcation between exposed and unexposed areas. For reasonable concentrations of PAG, the amount of strong acid produced in a CA resist increases only in gentle linear fashion with the UV dose supplied. Even after subsequent PEB treatment of a PDBFS/onium salt resist, the relationship between the content of its remaining *t*-butyl ester groups, and its previous UV dosage, typically appeared as a still shallow sigmoid curve (Figure 2). The slow start here of ester cleavage at lower doses may be due to basic contaminants, such as perhaps this polymer's -CN end groups from the AIBN [31], that deactivated the first few units of acid that had been photogenerated. By 15 mJ/cm^2, increasing exposure had a stronger effect on further alteration of material, possibly because the new -COOH groups assisted in further *t*-butyl cleavage (autocatalysis). Beyond 30 mJ/cm^2, the curve leveled off again, perhaps because some ester groups were in regions of low onium salt or acid concentration, either through bulk heterogeneity (i.e. microphase separation), or uneven vertical distribution (i.e. relating to the less-soluble "skin" often seen in positive-tone relief development of CA resists [32]).

Even the steepest slope of this "% ester vs. dose" curve was rather low. Similar low contrast between more- and less-exposed areas would also be expected for material properties which vary rather linearly with material composition, such as density, refractive index, or dielectric constant. However, for other material properties or processes that depend more on composition being above or below a critical threshold, contrast may be much higher. Thus, film dissolution of PDBFS/onium salt resist varied between nil and complete within a range of 7 mJ/cm^2 of UV energy and 40 mol % t-butyl ester content for negative-tone development with organic solvent, and 15 mJ/cm^2 and 10 mol % for positive-tone with aqueous base (Figure 2). Indeed, the complete mechanism of film dissolution involves consecutive steps whose thermodynamics and/or kinetics can be highly cooperative: (i) wetting - solvent or other solution components must penetrate the liquid/solid interface; (ii) permeation - these must then diffuse into the film to surround a given chain; (iii) (optional) reaction - polymer may have to be transformed, for example by deprotonation, to a more solvent-compatible form; (iv) solvation - chains are completely separated from each other and transported away into the liquid phase.

Even without eventual dissolution, the further alteration of a CA resist material during liquid functional development would also require at least the wetting then penetration of the resist, if not further reaction of its polymer, and thus would also be critically dependent on the composition of the latent image. Thus, even though a steadily increasing UV dose only gradually produces more acid and acid-cleaved polymer product, the onset of new or greatly altered materials properties upon functional development could also be remarkably sudden. This suggests that the same resist that shows high contrast in relief development would also be a good candidate for many forms of functional development as well.

Functional Development with Metal Ions: General.   Metal cations can be captured and held inside a polymer material by ion-ion or ion-dipole interactions, or may simply remain after swelling the polymer with solution then evaporating the solvent. Since an ion-exchange mechanism seemed to promise the best rate and capacity for metals loading, we explored the formation of insoluble polymer-supported metal carboxylates from the carboxylic acid moieties of deprotected PDBFS. Other candidates for good metal binding would be other carboxyl-based resists [23-25], protected phenolic resins (Novolac or hydroxystyrene) with possible formation of metal phenolates, or polymers containing other photo- or photocatalyst-deprotectable chelating groups.

Exposure-Controlled Sorption of Calcium Hydroxide.   As expected, beyond a critical UV dose (30 mJ/cm^2), aqueous solutions of either sodium or potassium hydroxides quickly dissolved exposure-deprotected areas of PDBFS resist, presumably due to formation of a water-soluble monocation-polycarboxylate polyelectrolyte. However, the same treated resist did not dissolve in even highly basic, saturated solutions of calcium hydroxide in water, though FTIR of the still-solid films (Figure 3) showed the appearance of peaks corresponding to carboxylate anion (ca. 1415 and 1565 cm^{-1}, sym and asym COO$^-$) and water (3000-3500 cm^{-1}) within the polymer matrix. Their increasing height with prolonged contact of Ca(OH)$_2$/H$_2$O correlated with further shrinking of remaining carboxylic acid and anhydride peaks (1700-1850 cm^{-1}). The broad shapes of these peaks suggested the formation of a variety of carboxylate-water-metal ion complexes in different microenvironments.

Since no other cation is present at such high pH, it is obvious that at least one Ca^{++} is being taken up into the polymer for every two COO^- being formed, to give molar proportions of styrene:fumarate:calcium of 55:45:45 (calculated 14 wt % calcium) from the completely deesterified PDBFS resist. Each divalent cation would thus ionically link two carboxylate groups, either intramolecularly (forming at least a 7-membered ring – not particularly favored), or intermolecularly to crosslink the polymer and keep it from dissolving. The literature reports instances of similar polycarboxylic acids being precipitated through ionic crosslinking by polyvalent cations: depending on pH, gels formed this way contained tetra- or hexadentate complexes, with up to 1/3 of the carboxylic groups still remaining un-ionized [33,34].

In our thin films, rough kinetics of $Ca(OH)_2/H_2O$ loading (see Figure 4) showed an initial delay without detectable formation of carboxylate anion, probably for the material surface to become sufficiently ionized and wetted, and possibly relating to the annoying "skin" often seen in positive-tone relief imaging [32]. Once begun, the reaction proceeded until, within a time comparable to or shorter than the induction period, all carboxylic acid and anhydride groups made available by UV/PEB had been converted to carboxylate, without hydrolysis of any more t-butyl ester groups. Even with this crosslinking species, the rate of material alteration was probably more limited by propagation of the reaction front than by ion movement into and through the now-hydrophilic polymer matrix, as has been suggested for other, even oligomeric crosslinkers entering similar maleic anhydride copolymers [12]. More complete initial deesterification shortened both these times and thus the total time for maximum functional development (180 s for 30 mJ/cm^2, to 10 s for 60 mJ/cm^2), and also increased the ultimate ion-exchange capacity of the film. As with relief development of the same resist, this abrupt infiltration of solute beyond critical exposure and developing times benefited the contrast of the resulting functional images (maximum slopes of curves A to D, Figure 4).

Exposure-Controlled Sorption of Other Metal Ions. 0.1 M aqueous solutions of $CuCl_2$, $NiCl_2$, $CoCl_2$ or $Pb(OAc)_2$ did not wet a film of PDBFS resist, even after its complete photo-deprotection. These solutions were neutral or slightly acidic (pH 7 to 4): it was expected that more basic conditions would be needed to deprotonate the polymer films for better wetting and influx of metal cations. Precipitation of metal oxides with rising pH was often largely forestalled by using NH_3 as base, since this is also a good ligand for many cations. Actually, aqueous ammonia alone was an effective solvent for fully-deprotected PDBFS (ex. for relief development). However, many of its mixtures with crosslinking polyvalent metal ions (likely containing NH_3, NH_4^+, M^{n+}, $(NH_3)_mM^{n+}$, $M^{n+}(OH^-)_m$ and OH^- species) did not dissolve even fully-deprotected resist at higher pH, though FTIR peaks near 1550 and 1400 cm^{-1} showed increasing conversion to carboxylate anion with increasing contact time and pH (Figure 5). Ammonia or ammonium hydroxide probably forms ammonium carboxylate first, but then metal or metal-ammonia complex would displace the ammonium ion for a thermodynamically more favorable association (lattice energies for acetates: 725 kJ/mol for NH_4^+, but 2835 for Cu^{++} and 2247 for Pb^{++} [35]). With increasingly higher ammonium ion concentrations (as reflected by increasing pH), monocation carboxylate groups eventually predominate within the material, ultimately leading to its dissolution. Thus, at pH 6-7, Ni^{++} and Cu^{++} began to penetrate exposed areas, which dissolved at pH 10-10.5. At pH 8 Ni^{++} or Cu^{++} ammonia solution does not cause any IR-detectable alteration within unexposed areas. With Co^{++}, dissolution occurred at pH 9, and was accompanied by small transformation of unexposed areas of

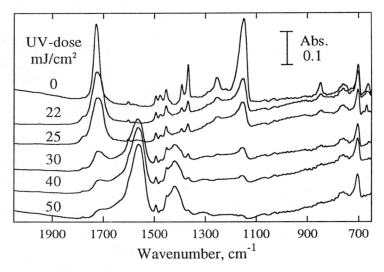

Figure 3. IR spectra of PDBFS resist after UV exposure at different doses, and 20 s treatment with aquaeous saturated Ca(OH)$_2$.

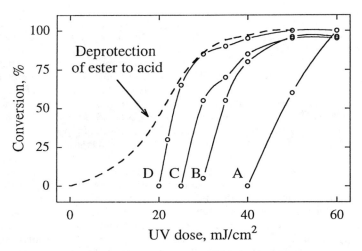

Figure 4. Conversion of -COOH into -COO⁻ after treating the deprotected resist with aquaeous saturated Ca(OH)$_2$ for 10 s (A), 20 s (B), 30 s (C) and 180 s (D).

resist (new FTIR peaks at 3500-3000, 1480 and 1380 cm^{-1}). Pb^{++} only slowly penetrated exposed areas even up to the point of dissolution at pH 10, while clearly causing hydrolysis of t-butyl ester groups in unexposed areas at all pH (new FTIR peaks at 1540 and 1390 cm^{-1}). Overall then, pH 8 seemed best for exposure-selective binding of these metal ions, though contrast was poor for Pb^{++}. Brief attempts to introduce FTIR-detectable quantities of Fe^{++}, Fe^{3+}, Al^{3+} or Pd^{++} failed, due to their precipitation in basic solutions.

Metal Ion-Developed Resist: Reactive Ion Etching. Mainly silicon [10,12-14], and some other elements that form refractive oxides (boron [36], tin [37], germanium [38], nickel and zirconium [5]), are used in microlithography in "dry development" processes, in which their compounds are selectively introduced either in exposed or unexposed areas of resist. The stable oxide layer produced under plasma bombardment protects polymer below. This technique gives excellent wall profile and high aspect (depth/width) ratio [9].

Preliminary tests showed that introducing of Ca^{++} into 0.8 μm films of exposed resist did indeed reduce the rate of its O$_2$-RIE plasma etching from 7.0 (ester without Ca) down to 1.5 nm/s (carboxylic Ca salt), to create 1 μm-resolution relief images. Sensitivities were 40 mJ/cm^2 for 90 or 120 s of immersion in Ca(OH)$_2$/H$_2$O, and 25 mJ/cm^2 for 180 s, similar to sensitivities associated with wet development (Figure 4).

Metal Ion-Developed Resist: Refractive Index. A waveguide consists of a channel of material with high n, surrounded by one with lower n. The value of n depends on material composition, being higher in materials containing more polarizable atoms (such as heavier atoms or ions) or groups (dipoles, delocalized charges). Exposure/PEB, then development with inorganic cations, each significantly and progressively increased the refractive index of PDBFS resist, as measured in different regions of the same film (Figure 6), through the formation of increasingly polarizable functional groups: RCOOR' < RCOOH (or (RCO)$_2$O) < (RCOO$^-$)$_n$M^{n+}. In principle, patterns of sharply- or gradually- varying refractive index could thus be generated, to 1 μm resolution and below.

Functional Development with Amines. Amines would generally be expected to form ionic or covalent bonds with carboxylic acid and anhydride groups, and none with t-butyl esters. Many biomolecules contain such groups which could be used to immobilize them in exposed areas of PDBFS resist.

Functional development with volatile amines proved possible from the gas phase. As shown by FTIR (Figure 7), deprotected resist (curve B) easily absorbed vapours of ammonia within 120 s to form the same carboxylates seen in liquid development (curve E: 1708 cm^{-1} replaced by 3200, 1560 and 1450 cm^{-1}). Though pyridine was also taken up in these areas, reaching a maximum at one hour with no further change in next 2 hours, polymeric carboxyls were not deprotonated by this weaker base (curves C and D, respectively; note new pyridine peaks at 1596, 1435, 1010, 750 and 700 cm^{-1}, and no peaks of carboxylate).

Depending on their concentrations, aqueous solutions of molecules with only one strongly basic amino group either did not wet the surface of photoexposed PDBFS resist, or dissolved it away completely [39]. Thus, ammonia, 2-amino- and 4-amino-pyridine at concentrations above 0.5 wt % acted as relief developers. Carboxylic amides were not detectably formed under these conditions. However, di- and poly-amines at concentrations above ca. 0.5 wt % were able to ionically crosslink the

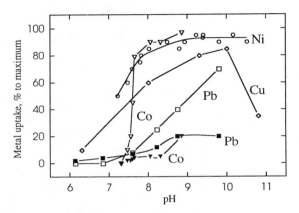

Figure 5.Uptake of different metal ions by the resist from aqueous ammonia solutions at different pH after 30 s. Hollow and filled dots represent uptake by -COOH and -COOR areas, respectively.

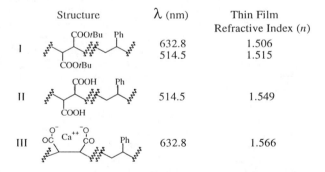

	Structure	λ (nm)	Thin Film Refractive Index (n)
I		632.8	1.506
		514.5	1.515
II		514.5	1.549
III		632.8	1.566

Possible micron scale features

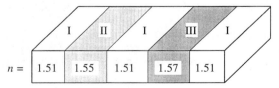

Figure 6. Refractive indices of unexposed, deprotected and Ca^{2+}-developped resist thin films, and possible patterned features.

polymer as it was ionized, preventing its dissolution as they penetrated. Thus, FTIR spectra of deprotected resist treated with 2,6-diaminopyridine, 1,10-diaminodecane or poly(ethyleneimine) showed formation of carboxylate anion (same peaks as from metal polycarboxylates), and also amide (3100 and 1640-1660 cm^{-1}), presumably by reaction with the minor anhydride component of the polymer, of diamine percolating in advance of water.

No FTIR or other evidence was seen of any of these amines being absorbed by unexposed resist under these conditions, meaning good contrast for this kind of functional development.

Functional Development with Dyes.   Development methods based on weak solute-polymer interactions were investigated with selected UV-VIS absorbent and fluorescent dyes (Table I). Even a minimal uptake of such compounds would allow direct visualization of a binding pattern. Moreover, such dyes are often able to supply interesting information, through shifts in their spectra, about their general environment and specific interactions.

Although some of the chosen dyes dissolved in water, no sorption was evident from purely aqueous solutions (Table I): drops rolled off both exposed and unexposed areas of resist, indicating its low wettability. However, the addition of different alcohols dramatically increased wettability of areas of exposed resist. Thus, solutions of methanol:water above 1:10 v:v wetted these areas of film but, beyond 8-12:10 v:v, swelled and even dissolved them to form positive-tone relief images. No *t*-butanol:water mixture dissolved any portion of the resist material, and one of 5:10 v:v seemed a good general mixture to introduce dyes deep into deprotected resist (Figure 8, curves A-B), so that prolonged water rinsing could not entirely remove them (curve C). UV-VIS spectroscopies could follow the uptake of dye into the film (curves D, E), and also subsequent drying of the film (curve F).

Though hexane wets the unexposed areas of resist, and not the exposed, it does not seem to penetrate beyond the surface (a dye applied from hexane was leached out later very quickly with the same solvent), and is a poor solvent for polar compounds such as R6G. Toluene is such a good solvent for neutral organic compounds that it acts as a negative-tone relief developer of the resist. Whereas 3-4:10 v:v toluene:hexane also dissolves unexposed areas of resist , a 1:10 v:v mixture only slightly swells them, though still enough to carry a dye deep into the film and trap it on drying so that hexane cannot wash it away later.

More detail report on selective binding with organic molecules will be published elsewhere.

Conclusion

As the dimensions of electrical, optical, bio- or chemo- devices and their features continue to shrink, demand will continue to grow for micron-scale patternable systems. One way to create such structures is by relief imaging, and filling spaces with inert or contrasting material. Another way is to area-selectively modify resists to active materials with valuable properties.

A variety of experimental approaches were shown for selective binding of organic and inorganic molecules into chemically amplified resists. Area-selective binding in positive or negative tone, from solution or vapor, and into bulk of a film or onto its surface, were successfully realized by choosing appropriate polymer structures, binding species, and solvents and other conditions.

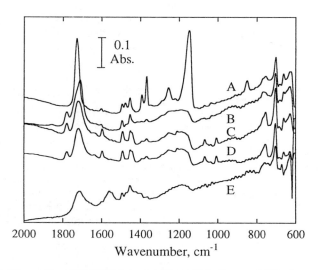

Figure 7. IR spectra of PDBFS based-resist, (A) before and (B) after deprotection, and after treatment with (C) vapours of ammonia for 2 min, or (D) pyridine for 60 min or (E) 180 min.

Table I. Functional development of PDFS resist with fluorescent dyes

Dyes	Solvent(s)	v:v	Effect on unexposed resist	Effect on exposed resist
R6G	H_2O		-	-
	MeOH:H_2O	1:10	-	+ (no swelling)
	tBuOH:H_2O	5:10	-	+++ (no swelling)
	MeOH:H_2O	10:10	-	+++ (partial swelling)
	MeOH:H_2O	20:10	-	(dissolves)
DSMP	H_2O		-	-
	MeOH:H_2O	10:10	-	+++ (partial swelling)
	Toluene		(dissolves)	-
	Hexane		+ (no swelling)	-
	Hexane:Toluene	1:10	+++ (slight swelling)	-

(+++) extensive sorption; (+) slight sorption; (-) no wetting

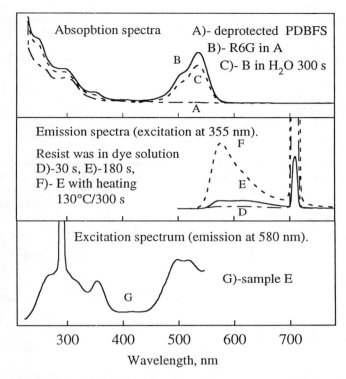

Figure 8. UV-vis absorbance and fluorescence spectra of R6G in deprotected PDBFS resist.

Acknowledgments

We thank L. Martinu and I. Sapieha (Ecole Politechnique, Montreal) for plasma etching experiments, and T. Kanigan (McGill University, Montreal) for refractive index measurements.

Literature Cited

(1)   Reichmanis, E.; Houlihan, F. M.; Nalamasu, O.; Neenan, T. X. *In: Polymers for Microelectronics*; Thompson, L. F.; Willson, C. G.; Tagawa, S., Eds.; ACS Symp. Ser. 537; American Chemical Society: Washington, DC, USA, **1994**, pp 2-24.

(2) Thompson, L. F. *In: Introduction to microlithography*; 2-nd ed.; Larry, F.; Thompson, C.; Grant, W.; Murrae, J. B., Eds.; American Chemical Society: Washington, DC, **1994**, pp 1-18.

(3) Frechet, J. M. J.; Eichler, E.; Willson, C. G.; Ito, H. *Polymer* **1983**, *24*, 995-1000.

(4) Dressick, W. G.; Dulcey, C. S.; Georger, J. H.; Calvert, J. M. *Chem. Mater.* **1993**, *3*, 148-150.

(5) Schilling, M. L., et al. *Macromolecules* **1995**, *28*, 110-115.

(6) Borman, S. *Chemical and Engineering News* **1994**, *72(23)*, 24-25.

(7) Gorman, C. B.; Biebuyck, H. A.; Whitesides, G. M. *Chem. Mater.* **1995**, *7*, 252-254.

(8) Shirai, M.; Sumino, T.; Tsunooka, M. *In: Polymeric Material for Microelectronic Application*; Ito, H.; Tagawa, S.; Horie, K., Eds.; ACS Symp. Ser. 579; American Chemical Society: Washington, DC, **1994**, pp 185-200.

(9) Baik, K.; Van den hove, L. *In: Polymeric Material for Microelectronic Application*; Ito, H.; Tagawa, S.; Horie, K., Eds.; ACS Symp. Ser. 579; American Chemical Society: Washington, DC, **1994**, pp 201-218.

(10) Coopmans, F.; Roland, B. *Proc. SPIE* **1986**, *633*, 34-41.

(11) Zhang, C.; Darling, G. D.; Vekselman, A. M. *Chem. Mater.* **1995**, *7*, 850-855.

(12) Sebald, M.; Ahne, H.; Leuschner, R.; Sezi, R. *Polymers for Advanced Technologies* **1994**, *5*, 41-48.

(13) MacDonald, S. A.; Schlosser, H.; Ito, H.; Clecak, N. J.; Willson, C. G. *Chem. Mater.* **1991**, *3*, 435-442.

(14) MacDonald, S. A.; Schlosser, H.; Clecak, N. J.; Willson, C. G. *Chem. Mater.* **1992**, *4*, 1364-1368.

(15) Allcock, H. R.; Nelson, C. J.; Coggio, W. D. *Chem. Mater.* **1994**, *6*, 516-524.

(16) Ito, H., et al. *J. Photopolym. Sci. Technol* **1993**, *6*, 547-562.

(17) Hinsberg, W. D.; MacDonald, S. A.; Clecak, N. J.; Snyder, C. D. *Chem. Mater* **1994**, *6*, 481-488.

(18) Reichmanis, E.; Galvin, M. E.; Uhrich, K. E.; Mirau, P.; Heffner, S. A. *In: Polymeric Material for Microelectronic Application*; Ito, H.; Tagawa, S.; Horie, K., Eds.; ACS Symp. Ser. 579; American Chemical Society: Washington, DC, **1994**, pp 52-69.

(19) Cutee, O. M.; Clemberg-Sapieha, J. E.; Martinu, L.; Wertheimer, M. R. *Thin Solid Films* **1990**, *193/194*, 155-163.

(20) Perrin, D. D.; Armarego, W. L. F. *Purification of Laboratory Chemicals;* 3-rd ed.; Pergamon Press: Oxford, England, **1988**, p 318.

(21) Crivello, J. V. *In: Initiators - Poly-reactions - Optical activity*; Advances in polymer science 62; Spriner-Verlag: Berlin, **1982**, pp 1-48.

(22) Swalen, J. D.; Tacke, M.; Santo, R.; Fisher, J. *Optics Communications* **1976**, *18*, 387-390.

(23) Allen, R. D.; Wallraff, G. M.; Hinsberg, W. D.; Simpson, L. L.; Kunz, R. R. *In: Polymers for Microelectronics*; Thompson, L. F.; Willson, C. G.; Tagawa, S., Eds.; ACS Symp. Ser. 537; American Chemical Society: Washington, DC, **1994**, pp 165-177.

(24) Ito, H.; Willson, C. G.; Frechet, J. M. J. *Proc. SPIE* **1987**, *771*, 24-31.

(25) Ito, H.; Ueda, M.; Ebina, M. *In: Polymers for Microlithography*; Reichmanis, E.; MacDonald, S. A.; Iwayanagi, T., Eds.; ACS Symp. Ser. 412; American Chemical Society: Washington, DC, USA, **1989**, pp 57-73.

(26) Murata, M.; Yamachika, M.; Yumoto, Y.; Miura, T. German Patent DE 4229 816 A1, **1993**

(27) Darling, G. D., et al. *Proc. 3-rd Pacific Polymer Conference*. Pol. Div. Royal Australian Chemical Institute: Gold Coast, Australia, **1993**, 399-400.

(28) Crivello, J. V.; Shim, S. Y. *J. Polym. Sci. Part A-Polymer Chemistry* **1995**, *33*, 513-523.

(29) Ito, H.; Padias, A. B.; Hall, H. K. J. *J. Polym. Sci. Part A. Polym. Chem.* **1989**, *27*, 2871-2881.

(30) Ito, H.; Ueda, M. *Macromolecules* **1988**, *21*, 1475-1482.

(31) Ito, H. *Jpn. J. Appl. Phys., Part 1* **1992**, *31*, 4273-4282.

(32) MacDonald, S. A., et al. *Chem. Mater* **1993**, *5*, 348-356.

(33) Sileo, E. E.; Morando, P. J.; Baumgartner, E. C.; Blesa, M. A. *Thermichimica Acta* **1991**, *184*, 295-303.

(34) Allan, J. R.; Bonner, J. G.; Gerrard, D. L.; Birnie, J. *Thermochimica Acta* **1991**, *185*, 295-302.

(35) *CRC Handbook of Chemistry and Physics*; 68-th ed.; Weast, R. C., Ed.; CRC Press, Inc.: Boca Raton, FL, USA, **1987**, p D-101.

(36) Talor, G. N.; Stillwagon, L. E.; Venkatesan, T. *J. Electrochem. Soc.* **1984**, *131*, 1664-1670.

(37) Nalamasu, O.; Baiocchi, F. A.; Taylor, G. N. *In: Polymers in microlithography*; Reicmainis, E.; MacDonald, S. A.; Iwayanagi, T., Eds.; ACS Symp. Ser. 412; American Chemical Society: Washington, DC, USA, **1989**, pp 189-209.

(38) Yoshida, Y.; Fujioka, H.; Nakajima, H.; Kishimura, S.; Nagata, H. *J. Photopolym. Sci. Technol.* **1991**, *4*, 497-507.

(39) Vekselman, A. M.; Zhang, C.; Darling, G. D. *Proc. 10th International Conference on Photopolymers*. Willson, C. G., Ed. SPE: Ellenville, NY, USA, **1994**, 116-127.

RECEIVED July 20, 1995

Chapter 11

Hydrogen Bonding in Sulfone- and *N*-Methylmaleimide-Containing Resist Polymers with Hydroxystyrene and Acetoxystyrene

Two-Dimensional NMR Studies

Sharon A. Heffner[1], Mary E. Galvin[1], Elsa Reichmanis[1], Linda Gerena[2], and Peter A. Mirau[1]

[1]AT&T Bell Laboratories, 600 Mountain Avenue, Murray Hill, NJ 07974
[2]Chemistry Department, Mount Holyoke College, Carr Laboratory Park Street, South Hadley, MA 01075

High resolution solution state NMR has been used to study structure/property relationships in hydroxystyrene and acetoxystyrene resist polymers containing either sulfone or N-methyl maleimide. The properties of resists, including the differential solubility of UV-exposed and nonexposed materials and the glass transition temperature, depend on the local molecular interactions. We have studied hydrogen bonding in the polymer component of selected resists by two dimensional NMR, where hydrogen bond donors and acceptors are identified by the appearance of off-diagonal intensity in the two dimensional nuclear Overhauser effect spectra. The results show that weak intermolecular hydrogen bonds are found in hydroxystyrene/acetoxystyrene/sulfone terpolymers where hydrogen bonding cross peaks are only observed at concentrations above 20 wt%, while hydrogen bonding cross peaks are observed at much lower concentrations in polymers that do not contain sulfone. Stronger intrachain hydrogen bonding is observed in polymers containing N-methyl maleimide. These results show that the materials with the best lithographic performance are those which do not form strong intra- or intermolecular hydrogen bonds.

Polymers are a key element in the development of resist technology used in the production of integrated circuits. The sensitivity and contrast in chemically amplified resists are dependent on the difference in solubility between the parent resist matrix, e.g., poly(t-butoxycabonyloxystyrene-co-sulfone) and the deprotected resists containing, for example, hydroxystyrene (*1-3*). Resists with improved imaging characteristics and lower weight loss have been engineered by introducing a variety of *para* and *meta* substituted styrene derivatives (*4*) into poly(t-butoxycabonyloxystyrene-co-sulfone) copolymers, or by replacing the sulfone monomers with maleimide or N-methyl maleimide (Galvin, R. E., Reichmanis, E. R., and Nalamasu, O., Bell Laboratories, unpublished data). However, little information is available on the

0097–6156/95/0614–0166$12.00/0

molecular level properties of the monomers or their functional groups. Intra- and intermolecular hydrogen bonding is potentially an important factor affecting the behavior of these materials because the UV exposed and postexposure baked materials contain high concentrations of hydrogen bond donors and acceptors, and the state of these polar groups (free or hydrogen bonded) can have a large impact on the aqueous base solubility and the diffusion and reactivity of small molecules in the resist matrix. Hydrogen bonding is most frequently studied by infrared spectroscopy, since the frequency and width of the carbonyl absorption is sensitive to the extent and strength of the hydrogen bonding (5). However, molecular level assignment of the donors and acceptors is difficult to achieve by this method when there are many possible donor and acceptor groups.

In these studies we have used high resolution solution NMR methods to measure intra- and interchain hydrogen bonding in hydroxystyrene, acetoxystyrene, N-methyl maleimide and sulfone-containing terpolymers, copolymers and mixtures. The hydroxystyrene materials were chosen for this investigation because this is the unit that is generated upon acid induced t-butoxycarbonyl deprotection and is responsible for allowing aqueous base dissolution of the exposed and post-exposure baked regions of the resists. We have used two dimensional NMR as a function of concentration to identify hydrogen bond donors and acceptors and to measure the effect of polymer microstructure on the hydrogen bonding interactions. The results show that the presence of sulfone, which is required for high contrast and sensitivity, leads to a large decrease in both intra- and intermolecular hydrogen bonding interactions between the hydroxystyrene donors and the acetoxystyrene acceptors. The introduction of N-methyl maleimide into the resist polymers leads to a material in which the hydroxyl groups generated by the acid catalyzed deprotection are strongly associated by intramolecular hydrogen bond formation. These results suggest a correlation between the molecular level properties and the imaging characteristics of the resist matrix.

Methods and Materials

N-methyl maleimide and deuterated solvents were purchased from Aldrich Chemical Co. and used as received. *Tert*-butoxycarbonyloxystyrene and acetoxystyrene were purchased from Hoechst-Celanese, Inc. and used as received. Polymerizations were conducted in cyclohexanone at concentrations of 2.5 moles of monomer per litre of solvent at 60°C with the free radial initiator azo-bis-isobutylnitrile. Dodecyl thiol was added to the polymerizations to control the molecular weight.

The polymers poly(*para*-hydroxystyrene-co-*para*-acetoxystyrene), poly(*para*-hydroxystyrene-co-*para*-acetoxystyrene-co-sulfone), poly(*para*-hydroxystyrene-co-*meta*-acetoxystyrene-co-sulfone), poly(*para*-hydroxystyrene-co-sulfone), and poly(*para*-acetoxystyrene) were prepared as previously described (4). In most polymers there was an approximately equal amount of hydroxystyrene and acetoxystyrene, and for the sulfone containing polymers the total ratio of styrene to sulfone was 3:1. Poly(*para*-hydroxystyrene) (MW=30,000) was obtained from Aldrich.

The model polymers for the UV-exposed and deprotected material were obtained by cleaving the *t*-butoxycarbonyl (t-BOC) moiety of the poly(tbutoxycarbonylstyrene-co-sulfone) parent polymer by heating in a rotary evaporator under vacuum at 170°C. A Perkin Elmer TGA-7 system thermal analyzer was used to insure that all t-BOC groups had been thermally removed. Heating rates of 10 °C/min were used for the thermal analysis.

The miscibility of the model resist polymers with the probe polymers poly(p-hydroxystyrene) and poly(p-acetoxystyrene) was evaluated by the optical clarity of solution cast films. In a typical experiment 2 mL of 3 wt% solutions of the probe polymer and the resist in tetrahydrofuran were mixed for several hours. Films were cast on a microscope slide under a nitrogen atmosphere overnight and annealed at 110°C for 4 h under vacuum to remove all traces of the solvent.

The high resolution solution NMR spectra were acquired at 500 MHz on a JEOL GX-500 spectrometer. The 90° pulse widths were 20 μs and the sweep widths were set to 7 kHz. The two dimensional nuclear Overhauser effect (NOESY) spectra were obtained with the $(90°$-t_1-$90°$-τ_m-$90°$-$t_2)$ pulse sequence in the phase sensitive mode (6). In a typical experiment 256 complex t_1 points and 512 complex t_2 points were acquired with a mixing time of 0.5 s and a recycle delay time of 4 s. The data were processed with 5 Hz line broadening in each dimension.

Results

Hydrogen bonding is a strong interaction that can alter the bulk properties of polymers. Such interactions can affect the processing of resist formulations by altering the local polarity, the glass transition temperature, the mobility of photoacid generators, and the aqueous base solubility. We have studied hydrogen bonding in model resists by comparing the solution NMR properties of several copolymers, terpolymers and blends. These polymers were chosen as model systems to understand the properties of exposed and deprotected resist matrix resins in terms of the local, molecular interactions. It might be expected that hydrogen bonding would affect the behavior of these materials since they contain a high concentration of hydrogen bond donors and acceptors. High resolution NMR in solution is used to measure the strength of the intra- and intermolecular hydrogen bond formation and to identify the donor and acceptor groups.

We have studied hydrogen bonding in sulfone (SO_2) and N-methyl maleimide (NMM) resist polymers containing $para$-hydroxystyrene (pOHSty), $para$-acetoxystyrene (pAcSty), and $meta$-acetoxystyrene (mAcSty) (Scheme I) by blend formation and by NMR spectroscopy. Most polymers do not form miscible blends because molecular mixing of the chains is not entropically favored in high molecular weight polymers ($7, 8$). As a result, films cast from most binary mixtures are phase separated and appear cloudy because the length scale of the phase separation is on the order of the wavelength of visible light. Miscibility in polymer blends can be promoted by introducing low levels (<5%) of hydrogen bond donors on one chain and acceptors on the other (9). In such cases the films are miscible, homogeneous and optically clear. In the first part of this study we measured the tendency of the resist polymers to form intermolecular hydrogen bonds by casting films of them with probe polymers containing hydrogen bond donors or acceptors. In a second set of experiments we used two dimensional nuclear Overhauser effect spectroscopy

(NOESY) (*10, 11*) to identify donors and acceptors, to measure the interaction strength, and to distinguish intra- from intermolecular hydrogen bonding.

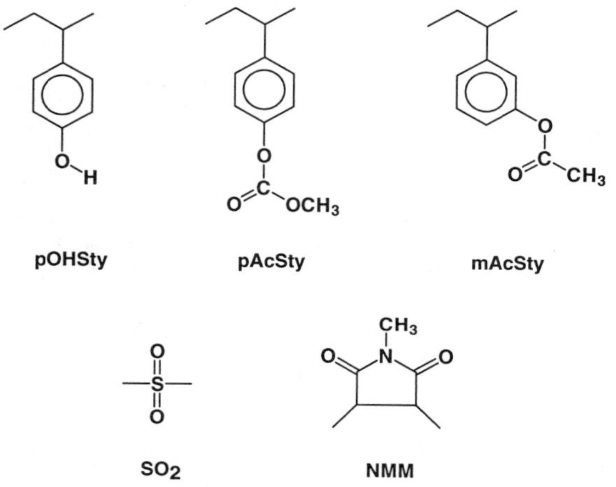

pOHSty **pAcSty** **mAcSty**

SO₂ **NMM**

Scheme I

The homopolymers *p*-hydroxystyrene and *p*-acetoxystyrene were used as probes to study intermolecular hydrogen bond formation (Scheme II). Poly(*p*-hydroxystyrene) has a tendency to self associate and can act as either a hydrogen bond donor or acceptor, while poly(*p*-acetoxystyrene) can only act as an intermolecular hydrogen bond acceptor. Hydrogen bonding has been extensively investigated in blends of poly(*p*-hydroxystyrene) (*5, 9*) but not poly(*p*-acetoxystyrene).

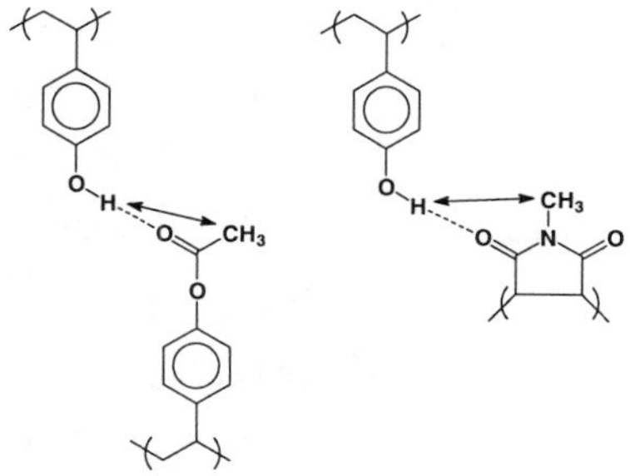

Scheme II

The results of the blend miscibility studies are shown in Table I. The control experiments show that optically clear films are obtained from mixtures of the donor and acceptor probe polymers, demonstrating that the chains have sufficient conformational flexibility to orient the hydrogen bond donors and acceptors in such a way as to promote strong intermolecular interactions. Miscible blends are also

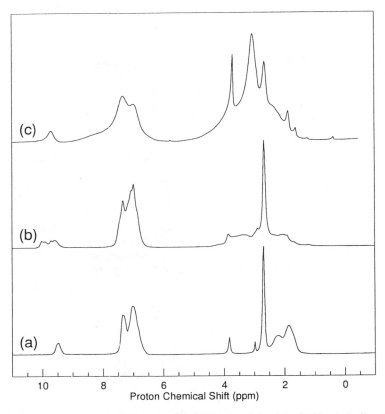

Figure 1. The high resolution proton NMR specta of the (a) pOHSty/pAcSty, (b) pOHSty/pAcSty/SO$_2$ and (c) pOHSty/pAcSty/NMM model resist polymers.

obtained with the donor and acceptor probe polymers and the *p*OHSty/*p*AcSty copolymer. These data demonstrate that there are a sufficient number of free donor and acceptor groups (>5%) (*9*) to stabilize the formation of a miscible polymer blend.

Table I. The miscibility[a] of poly(*p*-hydroxystrene) and poly(*p*-acetoxystyrene) with model resist copolymers and terpolymers

Polymer	p-OHSty	pAcSty
pOHSty	—	○
pAcSty	○	—
pOHSty/pAcSty	○	○
pOHSty/pAcSty/SO$_2$	○	○
pOHSty/pAcSty/NMM	●	●
pOHSty/Sty/NMM	●	●

a.   The symbols ○ and ● represent miscible and immiscible films.

Table I also shows that poly(*p*-hydroxystyrene) and poly(*p*-acetoxystyrene) form miscible blends with the *p*OHSty/*p*AcSty/SO$_2$ terpolymer while the terpolymers containing N-methyl maleimde monomers show dramatically different behavior. In contrast to the miscibility seen in the above systems, cloudy, phase separated films were obtained for all blends of the probe polymers with the NMM-containing materials. This result was unexpected due to the large number of potential hydrogen bond donor and acceptor groups in the model resist polymer. Immiscible films were also obtained for the NMM-containing polymers cast from dimethylsulfoxide and methyl ethyl ketone.

To gain an insight into the hydrogen bonding properties of these materials, we studied the resist polymer solutions by one and two dimensional NMR. We have previously used these methods to identify the interacting groups, to distinguish intra- from intermolecular association, and to estimate the interaction strength in mixtures of polymers that form miscible blends (*12-14*).

Figure 1a-c shows the high resolution proton NMR spectra of three examples of the model resist polymers examined in this study, *p*OHSty/*p*AcSty, *p*OHSty/*p*AcSty/SO$_2$ and *p*OHSty/*p*AcSty/NMM. The peaks of particular interest are the exchangeable hydroxyl protons appearing between 9 and 10 ppm, the aromatic protons at 6-8 ppm, and the N-methyl and acetoxy signals at 2.6 and 2.2 ppm that overlap with the signals from the main chain methylene and methine protons. Differences in the spectra can be attributed to differences in the chemical structure, the chain microstructure, and the chain dynamics. The differences in the chemical shift dispersion of the hydroxyl protons in the sulfone and N-methyl maleimide-containing polymers can be related to the difference in chain architecture between the sulfone and maleimide terpolymers. The environment of the hydroxyl protons in the maleimide polymers is expected to be more uniform as the maleimide and styrene units alternate (*15*), while there is a statistical distribution of styrene and sulfone units (*16*). The larger line widths observed in the NMM-containing

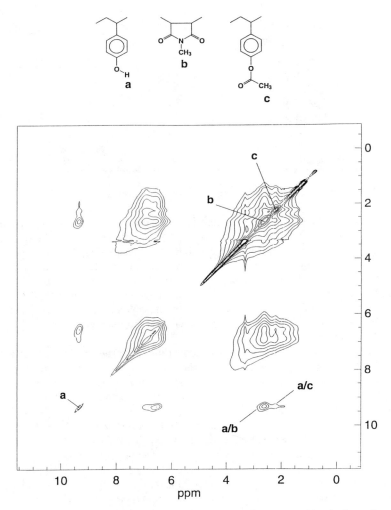

Figure 2. The 400 MHz 2D NOESY spectrum of the 20 wt% solution of pOHSty/pAcSty/NMM terpolymer obtained with a 0.5 s mixing time. The hydroxy-N-methyl cross peaks are labeled a/b and the hydroxy-acetoxy cross peaks are labeled a/c.

polymers may be due either to a difference in chain stiffness or hydrogen bonding (*vide infra*).

The hydroxyl protons of *para*-hydroxystyrene can be used as a molecular level probe of the hydrogen bonding in deprotected resists. The average distance between the hydroxyl protons of hydroxystyrene and the methyl protons in a nearby acetoxystyrene for random coil copolymers in solution is much larger than the 5 Å separations that can be measured by two dimensional nuclear Overhauser effect spectroscopy (*10, 11*). However, hydrogen bonding brings these groups into close contact (Scheme II) and cross peaks can be observed in the two dimensional spectra between the hydrogen bond donor protons and protons near the hydrogen bond acceptor.

Figure 2 shows the two dimensional NOESY spectrum for the *p*OHSty /*p*AcSty/NMM terpolymer at 20 wt% acquired at 25 °C in DMSO solution with a 0.5 s mixing time. In addition to the off-diagonal peaks expected from the chemical structure of these materials, such as between the OH and aromatic protons, cross peaks between the OH protons and the protons in the vicinity of the hydrogen bond acceptors are also observed. In this terpolymer the cross peaks are observed mainly between the OH protons and the N-methyl protons of N-methyl maleimide (cross peak a/b). Only small cross peaks are observed to the acetoxy protons (cross peak a/c), indicating that the carbonyl group of the N-methyl maleimide is the primary acceptor for the hydrogen bonds. Further evidence for the hydrogen bond formation between the OH protons and the N-methyl maleimide carbonyl are obtained from the observation of similar NOESY cross peaks in the spectrum of the *p*OHSty/Sty/NMM terpolymer that lacks the acetoxy group (not shown). Our previous studies have shown that such cross peaks are not expected from random coil polymers in the absence of specific interactions (*14*).

Figure 3 shows a comparable experiment for the *p*OHSty/*p*AcSty/SO$_2$ terpolymer obtained on a 20 wt% sample with a mixing time of 0.5 s. Under these conditions intermolecular cross peaks are observed, but they are much weaker than those observed for the *p*OHSty/*p*AcSty/NMM sample shown in Figure 2.

We have previously shown that the concentration dependence of the two dimensional NMR cross peaks can be used to estimate the relative interaction strength in weakly interacting polymer mixtures (*12, 14*). In weakly interacting polymer mixtures, such as polystyrene and poly(methyl vinyl ether) (*12*) or poly(vinyl chloride) and poly(methyl methacrylate) (*14*), intermolecular NOESY cross peaks are observed at concentrations above 25 wt%. In strongly interacting systems, such as the complex of poly(acrylic acid) and poly(ethylene oxide) (*13*) that is stabilized by hydrogen bond formation at low pH, these cross peaks are observed at concentrations in the range of 1 wt%.

We have measured the strength and mode, intra- vs intermolecular, of hydrogen bond formation in the model resist polymers from the concentration dependence of the NOESY cross peaks. As noted above, larger cross peaks are observed at lower concentrations for strongly interacting polymers. Intra- and intermolecular hydrogen bond formation can be distinguished since intermolecular association is expected to be concentration dependent while intramolecular association is not.

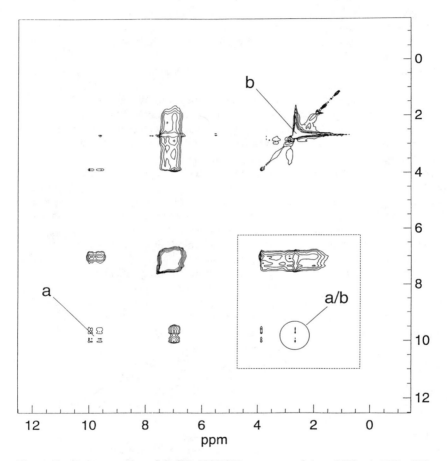

Figure 3. A contour plot of the 2D NOESY spectrum of the pOHSty/pAcSty/SO$_2$ terpolymer acquired with a 0.5 s mixing time. The hydroxy-acetoxy cross peaks are labeled a/b and the area enclosed in the box contains both the hydrogen bonding cross peaks and the intramolecular aromatic-main chain cross peaks that are used as an internal intensity calibration standard.

The concentration dependence of hydrogen bond formation has been studied by two dimensional NMR for the N-methyl maleimide- and sulfone-containing terpolymers, the pOHSty/pAcSty copolymer, and the solution mixture of poly(p-hydroxystryene) and poly(p-acetoxystrene). The results of these studies are compiled in Table II, which lists the minimum concentration that the hydrogen bonding cross peaks were observed and the concentration dependence.
Table II. The minimum polymer concentration ([C]$_{min}$)and concentration dependence (Δ[C]) for the observation of intermolecular hydrogen bonding cross peaks in the two dimensional NOESY spectra of model resist polymers.

Table II. The minimum concentration ([C]$_{min}$) and concentration dependence (Δ[C]) of the hydrogen bonded cross peaks detected by two dimensional NMR

Polymer	[C]$_{min}$ (wt%)	Δ[C]
pOHSty + pAcSty	2	+
pOHSty/pAcSty	<0.2	—
pOHSty/pAcSty/SO$_2$	20	+
pOHSty/mAcSty/SO$_2$	20	+
pOHSty/pAcSty/NMM	<0.2	—
pOHSty/Sty/NMM	<0.2	—

Typical data from these studies is illustrated in Figure 4, which shows the effect of concentration on the two dimensional NOESY cross peaks for the pOHSty/pAcSty/NMM terpolymer at 20 and 0.2 wt%. These stacked plots show the cross peaks between the aromatic and hydroxy protons in one dimension with the main chain, acetoxy and N-methyl protons along the other dimension. The results show that while the signal-to-noise ratio is lower for the sample at 0.2 wt%, the hydroxyl to N-methyl cross peaks (a/b) are clearly visible, and have not changed significantly in intensity relative to the intramolecular aromatic to main chain cross peaks. The cross peak connecting the hydroxyl and acetoxy proton resonances (a/c) is weaker than the hydroxyl-N-methyl cross peak in the 20 wt% sample and is missing from the spectrum of the 0.2 wt% sample. A similar concentration dependence is observed for the hydroxy-N-methyl cross peak in the pOHSty/Sty/NMM terpolymer that lacks the acetoxy group.

The concentration dependence of the hydrogen bonding cross peaks for the solution mixture of poly(p-hydroxy styrene) and poly(p-acetoxy styrene) illustrates the behavior expected for the intermolecular hydrogen bond formation for the hydroxy and acetoxy functional groups. Acetoxy-hydroxy cross peaks can only be observed when there is intermolecular hydrogen bond formation, and Table II shows that the cross peaks are concentration dependent and only observed at concentrations above 2 wt%. We consider this the standard behavior for intermolecular hydrogen bond formation in flexible polymers. Table II shows that cross peaks are observed at lower concentration in the pOHSty/pAcSty copolymers that are not concentration dependent. This behavior is considered typical for intramolecular hydrogen bond formation.

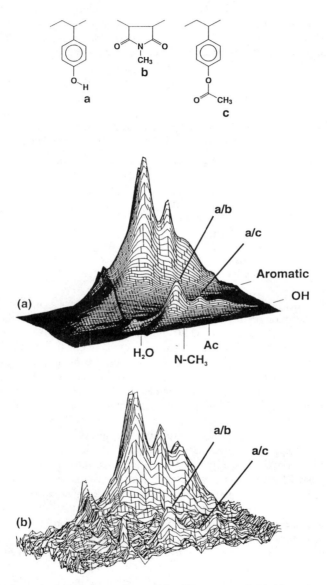

Figure 4. A stacked plot comparison of the 2D spectra of (a) 20 wt% and (b) 0.2 wt% NOESY spectra of the pOHSty/pAcSty/NMM terpolymer.

The behavior of the sulfone-containing polymer is quite different from the pOHSty/pAcSty copolymer, where concentration dependent cross peaks are only observed above 20 wt%. This shows that the introduction of sulfone into the main chain greatly decreases the hydrogen bonding ability of the hydroxy and acetoxy groups, and may be related to the favorable lithographic properties of these materials.

The behavior of the NMM-containing polymers differs markedly from the sulfone-containing materials, and the hydrogen bonding cross peaks for the hydroxyl and the carbonyl of the NMM-containing polymers is observed at low concentrations while the hydroxy-acetoxy cross peaks are observed only at very high concentrations. Thus, this polymer shows two modes of hydrogen bond formation, strong intramolecular association and weak intermolecular association. The pOHSty/Sty/NMM terpolymer also shows strong intramolecular hydrogen bond formation.

Discussion

The imaging characteristics of resist polymer formulations ultimately depends on the molecular level intra- and intermolecular interactions between the monomers. We studied these interactions in a polymer system that is a model for the UV-exposed and deprotected regions of resists in an attempt to understand the factors that control the difference in aqueous base solubility between the parent polymer and the hydroxystyrene containing material. The solubility of the exposed materials depends on several factors, including the competition between the polymer-polymer, solvent-solvent, and polymer-solvent interactions, and the rate of solvent ingress into the polymer matrix which is expected to strongly depend on the state (free or hydrogen bonded) of the polar functional groups. It has been demonstrated that the sensitivity, contrast, and weight loss of resist polymers is improved with the introduction of a variety of comonomers (*4*, Galvin, M.E., Reichmanis, E.R., Nalamasu, O., Bell Laboratories, unpublished data). The goal of our studies is to provide a molecular understanding of the relationship between copolymer structure and resist function.

In these studies we have used solution NMR methods to measure hydrogen bonding in model resist polymers. The primary tool in these studies is nuclear Overhauser effect spectroscopy (*10, 11*) which is used to map out the environment of hydrogen bond donors and acceptors. These NMR data can be used not only to identify donors and acceptors in complex materials, but also to evaluate the strength of hydrogen bonding interactions and to distinguish intra- from intermolecular association.

The high concentrations required for the observation of cross peaks in the sulfone-containing terpolymers shows that hydrogen bonding is very weak in these polymer systems, and that the hydrogen bonds formed at high concentrations result from intermolecular association. The observation that cross peaks become visible above 20 wt% is similar to our observations of intermolecular cross peaks in solution mixtures of poly(styrene) and poly(vinyl methyl ether) (*12, 17*) or poly(methyl methacrylate) and poly(vinyl chloride) (*14*). This result was surprising because the intermolecular associations in these systems arise from van der Waals interactions that are much weaker than those expected from hydrogen bonding polymers. These experiments demonstrate that one effect of introducing sulfone into the polymer backbone is to decrease the intra- and intermolecular interactions.   This effect is directly demonstrated by comparing the results for the sulfone containing terpolymer with the hydroxystyrene/acetoxystyrene copolymer, where hydrogen bonding cross peaks are observed at all concentrations studied.   This result suggests that the pOHSty/pAcSty copolymer can form both intra- and intermolecular hydrogen bonds. The incorporation of sulfone is known to increase chain stiffness in some copolymers (16). A possible explanation for the weak interactions in the sulfone containing terpolymers is that the decreased conformational flexibility in these materials makes it energetically unfavorable for these systems to adopt a geometry that promotes the formation of hydrogen bonds. The similar results obtained for the *para* and *meta* acetoxystyrene terpolymers suggests that a difference in hydrogen bonding between these materials is not responsible for the different lithographic properties (Reichmanis, E. R., Uhrich, K. E., Neenan, T. X., and Nalamasu, O., Bell Laboratories, unpublished data)

The observation that the hydrogen bond donating and accepting probe polymers are immiscible with the N-methyl maleimide copolymers was surprising given the high level of potential hydrogen bond donors and acceptors in the model resist polymers. This observation can be explained by the NMR results that show strong intramolecular hydrogen bonding in the pOHSty/pAcSty/NMM and pOHSty/STY/NMM terpolymers. This intramolecular association ties up a large number of the potential donors and acceptors and make these polymers immiscible with the probe donors and acceptors. This interaction may also account for the greater line widths in the maleimide polymers compared to the sulfone polymers, as strong intramolecular interactions are expected to decrease the chain flexibility, leading to greater line widths.

The results from these studies provide a fundamental insight into the function of the resist matrix resin that may be applied to the design of the next generation of these products. Among the key aspects of the hydroxystyrene/acetoxystyrene based materials is their dissolution in an aqueous base developer, and the conversion of acetoxystyrene to hydroxystyrene that increases the base solubility while minimizing the weight loss. The penetration of solvent into the resist matrix depends on the local polarity and, therefore, on the state of the hydrogen bond donors and acceptors. We suggest on the basis of these studies that good resist materials are those containing polar groups such as hydroxystyrene and acetoxystyrene that are not strongly hydrogen bonded. The free hydroxyl and carbonyl groups can interact with the aqueous solvent and promote the conversion of acetoxystyrene to hydroxystyrene and dissolution of the polymer (4).   This interpretation is supported by the observation that the hydroxystyrene/acetoxystyrene copolymer lacking sulfone is not soluble in the aqueous base developer and that acetoxystyrene is converted to hydroxystyrene with great difficulty in this system, which these NMR studies show is strongly hydrogen bonded. The suggestion that the imaging characteristics of hydroxystyrene based resists is correlated with the hydrogen bonding properties is further corroborated in our studies of resists containing N-methyl maleimide in place of sulfone, where poorer dissolution characteristics are correlated with the formation of strong intramolecular hydrogen bonds.   These studies are part of our continuing effort to gain a molecular level understanding of resist function to aid in the design of the next generation of resist materials.

References

1. Ito, H.; Wilson, C. G. *Polymers in Electronics*; Davidson, J.; ACS: Washington, D.C., 1984; pp 3.
2. Frechet, J. M. J.; Eichler, E.; Ito, H.; Wilson, C. G. *Polymer* **1980**, *24*, 995.
3. Ito, H.; Willson, C. G.; Frechet, J. M. J.; Farrall, M. J.; Eichler, E. *Macromolecules* **1983**, *16*, 1510.
4. Kometani, J. M.; Galvin, M. E.; Heffner, S. A.; Houlihan, F. M.; Nalamasu, O.; Chin, E.; Reichmanis, E. *Macromolecules* **1993**, *26*, 2165.
5. Coleman, M.; Lichkus, A.; Painter, P. *Macromolecules* **1989**, *22*, 586.
6. States, D.; Haberkorn, R.; Ruben, D. *J. Magn. Reson.* **1982**, *48*, 286.
7. Walsh, D.; Rostami, S. *Adv. Polym. Sci.* **1985**, *70*, 119.
8. Coleman, M. M.; Serman, C. J.; Bhagwagar, D. E.; Painter, P. C. *Polymer* **1990**, *31*, 1187.
9. Serman, C. J.; Xu, Y.; Painter, P. C.; Coleman, M. M. *Macromolecules* **1989**, *22*, 2015.
10. Jeneer, J.; Meier, B.; Bachmann, P.; Ernst, R. *J. Chem. Phys.* **1979**, *71*, 4546.
11. Ernst, R.; Bodenhausen, G.; Wokaun, A. *Principles of Nuclear Magnetic Resonance in One and Two Dimensions*; Clarendon Press: Oxford, 1987; pp 610 .
12. Mirau, P.; Tanaka, H.; Bovey, F. *Macromolecules* **1988**, *21*, 2929.
13. Mirau, P.; Heffner, S.; Koegler, G.; Bovey, F. *Poly. Int.* **1991**, *26*, 29.
14. Kogler, G.; Mirau, P. *Macromolecules* **1992**, *25*, 598.
15. Van Paesschen, G.; Timmerman, D. *Makromol. Chem.* **1964**, *78*, 112.
16. Cais, R. E.; Bovey, F. A. *Macromolecules* **1977**, *10*, 757.
17. Mirau, P.; Bovey, F. *Macromolecules* **1990**, *23*,4548.

RECEIVED July 20, 1995

Chapter 12

NMR Investigation of Miscibility in Novolac–Poly(2-methyl-1-pentene sulfone) Resists

Sharon A. Heffner, David A. Mixon, Anthony E. Novembre, and Peter A. Mirau

AT&T Bell Laboratories, 600 Mountain Avenue, Murray Hill, NJ 07974

The success of positive resists containing a blend of novolac and poly(olefin sulfones) (NPR) as effective materials for x-ray and electron beam lithography is critically dependent on the miscibility of the component polymers. Solution and solid state NMR techniques along with thermal analysis are used to gain a molecular level understanding of the factors influencing miscibility of the novolac and poly (2-methyl- 1- pentene sulfone) (PMPS) in the NPR system. The two dimensional nuclear Overhauser effect solution spectra of several novolacs and PMPS show interactions between the PMPS side chain and the novolac aromatic rings, suggesting that miscibility is related to these interactions. Solid state cross-polarization magic-angle spinning (CPMAS) experiments show mixing of the blend components on the length scale of less than 200Å. Possible improvements for the miscibility in the NPR system are discussed in light of these results.

The development of improved resist materials is critical for the successful manufacture of the next generation of integrated circuits using either x-ray or electron-beam lithographic technologies. These materials must not only have the desired properties in terms of contrast, resolution and sensitivity, but also must withstand the thermal stresses required in the manufacturing process.

Several years ago a positive resist formulation consisting of a blend of novolac and a poly(olefin sulfone) (NPR) was introduced as a potential electron beam resist for optical mask and direct write wafer fabrication.[1,2] NPR is a blend of novolac and poly (2-methyl-1-pentene sulfone) (PMPS), a radiation sensitive dissolution inhibitor. PMPS undergoes main chain scission and depolymerization under irradiation, rendering the exposed film regions soluble in aqueous base. NPR was originally considered promising since it provided vastly improved dry etching resistance characteristics and aqueous base solubility in comparison with the widely used poly(1-butene sulfone) electron-beam resist. More recently NPR represents an alternative path to radiation induced chemical amplification processes.

0097–6156/95/0614–0180$12.00/0

Amplification in NPR is achieved solely in the process exposure step and does not rely on the presence of a radiation sensitive acid catalyst. The lack of need for this acid catalyst removes all of the associated process variations widely reported for the chemically amplified resists. [3]

Although promising, problems were encountered in the processing of NPR, including the irreproducibility of the exposure sensitivity, development time and resolution. [4] These problems have been linked to immiscibility [5] and batch-to-batch variations in the novolacs and/or the PMPS. To understand the chemical factors affecting NPR performance, Tarascon and coworkers [5,6] investigated the NPR residue after exposure and development. They found that phase separation was observed in some samples, but that this problem can be minimized using lower molecular weight PMPS and/or novolac. However, this reduction in molecular weight does not overcome the need to further increase the phase stability of the resist. More recently the introduction of an interruptive spray/spin development step has improved the contrast obtained with NPR and produced 0.25 μm resolution using x-ray radiation. [7]

The miscibility in novolac:PMPS blends has not been extensively investigated. On the basis of the chemical structure of the component polymers it might be expected that hydrogen bonding would be an important factor. However, IR studies of novolac model compounds show much smaller changes in the hydroxyl bands in the IR spectra than are typically observed for hydrogen bonding compounds, and no changes in the spectra of the sulfone. [8] This study concluded that the hydrogen bonding must be very weak.

We have previously used two dimensional NMR methods to study intermolecular interactions in polymer blends. [9,10] These studies have contributed to our understanding of the forces that control blend formation because the signals can be assigned to individual main chain and side chain atoms, the interacting groups can be identified and the relative strength of the interactions measured. We have recently extended these studies to the solid state where we can measure the length scale of chain mixing. [11]

In this study, we use thermal analysis in combination with solution and solid state NMR methods to investigate NPR miscibility and to gain a molecular level understanding of the factors controlling the interchain mixing of the novolacs and PMPS. The goal is to understand which of several commercially available novolacs would make the best phase-compatible NPR system and to suggest other polymers that would enhance lithographic performance.

Methods and Materials

Materials: The PMPS was synthesized via free radical polymerization as described previously, [6] using toluene as the solvent, t-butyl hydroperoxide as the initiator and bromotrichloromethane to reduce the molecular weight by chain transfer. The intrinsic viscosity of the PMPS, as measured at 30°C in methyl ethyl ketone, is 5.3 cm^3/g. The novolac resins were obtained from Borden Chemical and OCG Microelectronic Materials Inc. The composition of the novolac samples is provided in Table I.

Table I. The thermal analysis of novolac and PMPS samples.

Sample	Type[a]	T_g (°C)	T_d^b (°C, under N_2)
SD126	m-cresol	79.2	310
427A	75/25 m-/p- cresol	99.1	340
CN-2	50/50 o-cresol/o-sec butyl phenol	78.5	350
UHT-1[c]	30/70 m-/p- cresol	75.5	340
BSD-8	70/30 m-/p- cresol	101.5	340
PMPS		85.2	220

[a] novolac synthesis monomer feed composition, mole basis (exclusive of fomaldehyde content)

[b] decomposition temperature; all novolac samples start losing weight between 200-225°C

[c] synthezised using chloracetaldehyde in place of formaldehyde

<u>Thermal Analysis:</u> Differential scanning calorimetry (DSC) and thermal gravimetric analysis (TGA) were obtained using a Perkin-Elmer DSC-7 and a TAC-7 thermal analysis controller. DSC samples were heated from -50°C to 200°C at a rate of 20 °C/min. TGA samples were heated from 50 °C to 400 °C at a rate of 10 °C/min.

<u>NMR:</u> The high resolution solution NMR spectra were acquired at 500 MHz on a JEOL GX-500 spectrometer. The 2D nuclear Overhauser effect (NOESY) spectra were obtained with the (90° - t_1 - 90°-t_m - 90° - t_2) pulse sequence in the phase sensitive mode. [12] In a typical experiment, 256 complex t_1 points and 512 complex t_2 points were acquired with a recycle delay time of 4 s and mixing times between 0.1 and 0.5 s. Solid state carbon spectra were acquired at 100.6 MHz on a Varian UNITY 400 spectrometer using cross polarization and magic angle spinning in a 5 mm probe (Doty Scientific). The proton and carbon 90° pulse widths were 5 μs and the spinning speeds were on the order of 3 kHz. The proton T_1 and $T_{1\rho}$ relaxation times were measured using published pulse sequences [13] and 8-10 delay times were measured for each experiment. Single exponential decay rates were observed in all cases.

Results

In addition to the lithographic properties, other materials properties must be considered in making a suitable resist material. For NPR, these additional important criteria include the miscibility of the component polymers and the thermal and phase stability during pre- or post-exposure baking.

The two important thermal properties for the component polymers are the glass transition temperatures (T_g) and the thermal degradation temperatures (T_d). The temperature required for lithographic processing is dependent on the T_g of the components and the boiling point of the film-casting solvent. Solvent is removed most efficiently by post-application baking the spin-cast films above the T_g , whereas the T_g and T_d of the polymers must be above those temperatures encountered in any subsequent pattern transfer steps.

The thermal analysis of several novolacs and PMPS are summarized in Table I. The novolacs exhibit Tg's in the range of 79-101°C and T_d's in the range of 310-350°C. The UHT-1 sample (70/30 *m-/p-*cresol) shows some weight loss (~8.5 wt%) between 100-200°C in N_2 that may be due to residual solvent. PMPS has a T_g of 85°C and a T_d above 200°C.

High resolution NMR in solution has been used to characterize the novolacs and the blends with PMPS. Figure 1 shows the 500 MHz proton NMR spectra of novolac BSD-8, PMPS, and the 40 wt% blend in dioxane-d_8. The resonance assignments are noted in the figure and were assigned using one and two dimensional NMR methods.

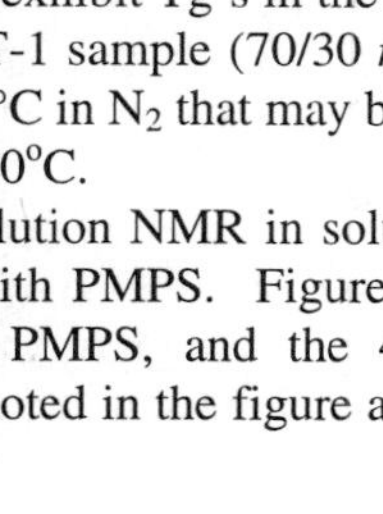

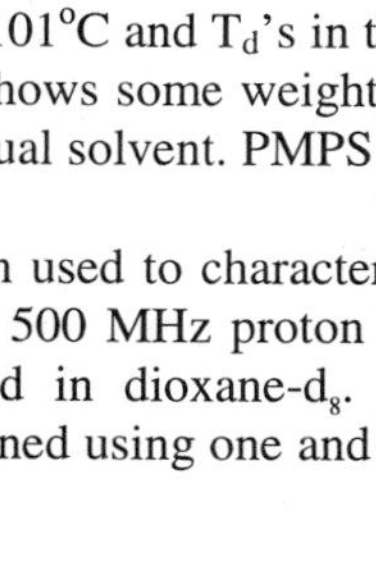

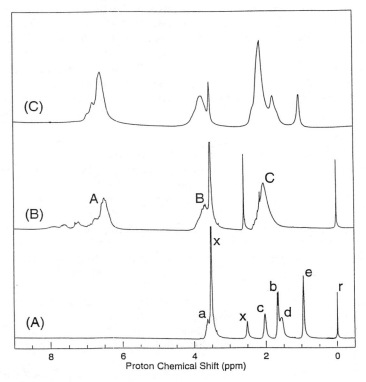

Figure 1. The 500 MHz proton NMR spectra of (A) PMPS, (B) novolac BSD-8, and (C) the 40 wt% novolac BSD-8:PMPS solution mixture in dioxane-d_8 at 21 °C. The peaks attributed to residual solvent are marked with an x and the reference compound with an r.

We have previously used 2D NOESY NMR methods to measure the strength and specificity of intermolecular interactions in solutions of polymers capable of forming miscible polymer blends.[9,10] The types of interactions are determined from the observation of intermolecular cross peaks, which can only arise between pairs of protons separated by less than 5 Å. The interacting groups can be identified from their chemical shifts, and the strength of the interactions measured by the concentration required for the observation of these cross peaks. In strongly interacting systems, such as the hydrogen bonding intermolecular complex formed by poly(acrylic acid) and poly(ethylene oxide),[10] cross peaks are detected at concentrations of 1-2 wt%, while the cross peaks for weakly interacting polymer pairs, such as polystyrene:poly(methyl vinyl ether)[14] or poly(methyl methacrylate):poly(vinyl chloride),[9] are observed only at concentrations above 25 wt%.

Figure 2 shows the 500 MHz 2D NOESY[15] spectrum for the 40 wt% 3:1 molar mixture of the novolac BSD-8 and PMPS obtained at 21 °C with a mixing time of 0.5 s. The intense peaks along the diagonal correspond to the signals observed in the 1D spectrum shown in Figure 1C, while the smaller off-diagonal peaks arise from through-space dipolar interactions between pairs of protons separated by less than 5 Å. The majority of the cross peaks arise between protons that are in close proximity as a result of the chemical structure of the polymers, such as the intense cross peaks connecting the novolac aromatic signals at 6.5 ppm with the methylenes at 3.6 ppm and the aromatic methyl protons at 2 ppm. Of particular interest for these studies are the cross peaks between the novolac aromatic signals and the methyl and methylene protons from the PMPS side chain that are enclosed in boxes in Figure 2. The unfortunate overlap between the novolac methylene and the aromatic methyl protons with the main chain methylene and methyl protons of PMPS makes it difficult to determine if there are other intermolecular interactions.

In an effort to characterize the strength of the intermolecular interaction, we have measured the NOESY cross peaks for the novolac:PMPS mixtures as a function of mixing time and concentration. Intermolecular cross peaks are observed with shorter mixing times (Figure 3), but they are substantially weaker than in spectra acquired with a longer mixing time (Figure 2). We also note that the intermolecular cross peaks in Figure 3 are much weaker than the intramolecular cross peaks for both the novolac and the PMPS. Taken together, the observations suggest that the novolac:PMPS intermolecular interactions are very weak.

The concentration at which intermolecular cross peaks are observed can also be a measure of the intermolecular interaction strength. We measured the 2D NOESY spectrum for the 20 wt% novolac BSD-8:PMPS mixture under conditions similar to those listed in Figure 2 and were unable to observe the intermolecular cross peaks observed in the 40 wt% samples, although the intramolecular cross peaks were of comparable magnitude (data not shown). These data are also consistent with weak intermolecular interactions.

To understand structure-property relationships in these novolac:PMPS blends, we measured the 2D NOESY spectra of PMPS mixed with the novolacs shown in

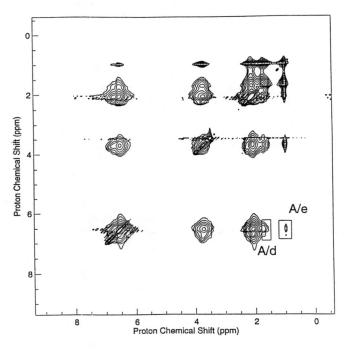

Figure 2. The 500 MHz 2D NOESY spectrum of the 40 wt% novolac BSD-8:PMPS solution mixture in dioxane-d$_8$ acquired at 21 $^\circ$C with a mixing time of 0.5 s. Intermolecular cross peaks are enclosed in the boxes.

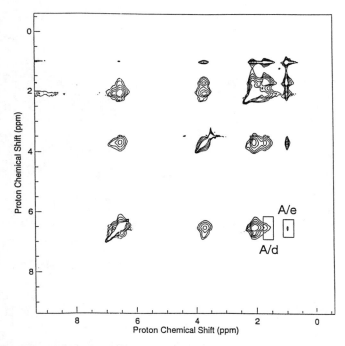

Figure 3. The 500 MHz 2D NOESY spectrum of the 40 wt% novolac BSD-8:PMPS solution mixture in dioxane-d_8 acquired at 21 °C with a mixing time of 0.3 s. Intermolecular cross peaks are enclosed in the boxes.

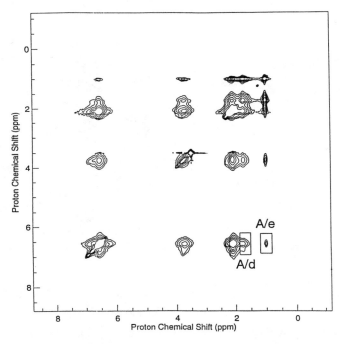

Figure 4. The 500 MHz 2D NOESY spectrum of the 40 wt% novolac 427A:PMPS solution mixture in dioxane-d$_8$ acquired at 21°C with a mixing time of 0.5s.

Table I. A typical example is shown in Figure 4, which shows the NOESY spectrum for the 40 wt% mixture of novolac 427A and PMPS obtained with a 0.5 s mixing time. Note, that as in Figure 2, intermolecular cross peaks are observed between the terminal methyl protons on the PMPS side chain and the novolac aromatic protons. Similar behavior was observed for the other novolacs under the same experimental conditions. Based on these observations, we suggest that the weak intermolecular interaction between the PMPS side chain and the novolac aromatic group is a feature common to all novolac:PMPS blends studied.

Reproducibility in blend formation in novolac:PMPS resins remains an important factor that limits the applicability of this technology. We have previously used our NMR methods to explore other blends, such as poly(vinyl chloride):poly(methyl methacrylate), [9] where considerable controversy exists in the literature about the compatibility of these materials. In that study we found that the intermolecular interactions were quite weak, but they were also sensitive to small concentrations of contaminants, such as water. To determine if the intermolecular novolac:PMPS complex was susceptible to such factors, we measured the 2D NOESY spectrum for the 40 wt% solution mixture in the presence of a small amount (3.8 wt%) of H_2O, and the results are shown in Figure 5. The presence of water has no measurable effect on the intermolecular novolac:PMPS cross peak.

The NOESY experiments demonstrate the existence of weak intermolecular association of these polymers in solution. The miscibility in the solid state may be examined using the solid state proton NMR relaxation times T_1 and $T_{1\rho}$ measured via the CPMAS spectrum. These relaxation times can provide an insight into the length scale of polymer mixing [13] since proton spin diffusion from one domain to another proceeds quite rapidly. In the case of phase separated materials, separate values for the relaxation times may be measured for each polymer. If the polymers are mixed on a 20 Å length scale an averaged value for the $T_{1\rho}$ will be measured, and if they are mixed on a length scale less than 200 Å then an averaged value for the T_1 will be measured. Such measurements of the phase separation length scale are only possible if the relaxation times for the two polymers differ by a factor of two.

Figure 6 shows the 100 MHz carbon spectrum of the BSD-8 novolac, PMPS, and the 10:1 molar blend cast from THF obtained with magic angle spinning and cross polarization. [13] The chemical shifts for the constituent polymers and the blend are similar to those observed in solution (data not shown). The observation that the spectrum for the blend appears as a composite of the blended polymer spectra is consistent with the conclusions from the NOESY analysis that there are weak intermolecular interactions between novolac and PMPS.

The proton T_1 and $T_{1\rho}$ relaxation times were measured and the results are compiled in Table II. The $T_{1\rho}$ relaxation times are typically two orders of magnitude shorter than the T_1's, and can therefore measure spin diffusion over a shorter length scale. However, as shown in Table II, the values for $T_{1\rho}$'s for the novolac BSD-8 and PMPS are similar to each other (7.1 and 5.5 ms) so it is not possible to determine if the polymers are mixed on a length scale of less than 20 Å using this approach. The T_1 values for the novolac and PMPS differ by an order of magnitude and the value for the blend (3.6 s) is close to the weighted average value of 4.3 s. These data show that the novolac:PMPS blend is mixed on a length scale of less than 200 Å.

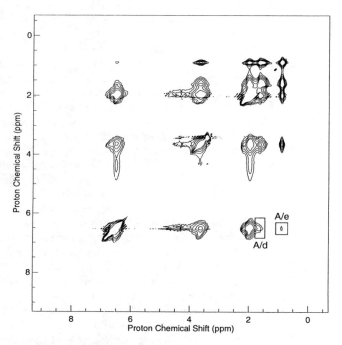

Figure 5. The 500 MHz 2D NOESY spectrum of the 40 wt% novolac BSD-8:PMPS solution mixture in dioxane-d$_8$ acquired at 21 °C with a mixing time of 0.5 s. The spectrum differs from Figure 2 in that the sample contains 3.8 wt% H$_2$O.

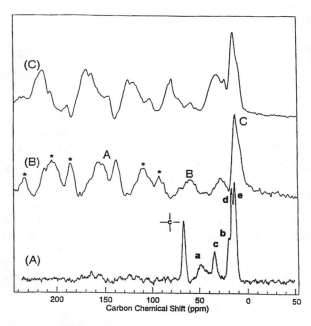

Figure 6. The 100 MHz solid state cross polarization magic angle spinning spectra of (A) PMPS,(B) novolac BSD-8, and (C) the 10:1 novolac BSD-8:PMPS blend cast from THF.

Table II. The solid state proton spin-lattice and rotating frame spin-lattice relaxation times for novolac, PMPS, and the novolac:PMPS blend (1:10, mole basis).

Sample	T_1 (s)	$T_{1\rho}$ (ms)
Novolac BSD-8	10.1	7.1
PMPS	0.7	5.5
Novolac:PMPS[a] (1:10)	3.6	8.4

[a] The 10:1 blend of novolac BSD-8 and PMPS was cast from THF.

Discussion

Resist systems based on a polymer blend of cresol novolac and poly(olefin sulfones) show promise for the manufacture of integrated circuits, but this potential has not been realized due to problems with the reproducibility of blend formation. In this study we have used thermal analysis and high resolution NMR to study intermolecular interactions in novolac and PMPS blends. Our goal is to compare the results for the NPR blends with the stronger and more weakly interacting blends that we have previously studied by these methods, and to suggest further studies for materials characterization and additional monomers that may be incorporated into the novolacs and/or PMPS to enhance the miscibility.

The NMR results show weak intermolecular interactions between the novolacs and the PMPS. As in blends such as poly(methyl methacrylate):poly(vinyl chloride) [9] or polystyrene:poly(methyl vinyl ether) [16] intermolecular cross peaks are observed only with long mixing times and at high solution concentrations. This is in contrast to other studies of intermolecular interactions in hydrogen bonded polymers where the cross peaks are observed at shorter mixing times and at lower concentrations. [10] We also observe that the presence of such intermolecular cross peaks does not seem to be strongly dependent on the chemical structure of the novolac resin, suggesting that the critical interacting groups are present in all novolacs studied. On the basis of small molecule studies it has been suggested that hydrogen bonding between the novolac and the sulfone in PMPS is very weak. [8] We suggest from these studies that intermolecular complex formation may also be stabilized by van der Waals interactions between the novolac aromatic ring and the PMPS side chain.

The results from these NMR studies show that the thermodynamics favor the intermolecular association for all of the novolac resins studied here. This observation by itself, however, does not necessarily ensure miscible blend formation, as the phase structure of the film is influenced by the casting or spinning solvent. For example, even though polystyrene and poly(methyl vinyl ether) are thermodynamically miscible, phase separated films are obtained when the films are cast from chloroform. [17] The miscibility of weakly interacting polymer pairs is particularly susceptible to such solvent effects.

Given the results of this study, we can suggest two possible courses of action that may lead to higher quality films. One is the careful study of blend formation as a function of solvent, temperature and composition. It is well known that both solvent and temperature can affect blend formation in weakly interacting polymer pairs. This is also true for the novolac:PMPS blends where it has been reported that miscible films are obtained from isoamyl acetate but not from some other solvents. [4] Closely related to such an observation is the possibility of a lower critical solution temperature (LCST) for this blend. Understanding such effects are critical for the novolac:PMPS blends because either the pre- or post-exposure bake could lead to phase separation in these materials. The goal of such studies would be to find the proper combination of temperature and solvent leading to the highest quality films. Initial steps in this area have already been taken with a study of the effect of temperature on the dissolution characteristics of NPR (unpublished).

A second possibility is to consider the synthesis of polymers related to PMPS that might be expected to have stronger intermolecular interactions, making the blend more stable. One obvious possibility is to add hydrogen bonding groups to the PMPS as in the work of Cheng et al [18] with the incorporation of methylallylethyl ether . Such hydrogen bonding groups must be carefully chosen, however, because blend formation is a careful balance of the PMPS:PMPS, novolac:novolac and novolac:PMPS interactions. Adding hydroxyl groups, for example, can increase both intra- and intermolecular hydrogen bond formation. On the basis of the results presented here, we suggest that a better polymer might be one containing methoxyl groups. Such groups can act as hydrogen bond acceptors but do not strongly self associate. We have also observed that methoxyl groups increase the stability of polymer blends through van der Waals interactions with aromatic groups. [16]

In summary, we have used a combination of thermal analysis and high resolution NMR to study the stability and intermolecular interactions between novolacs and PMPS, the components of the NPR resist formulation. The results show thermodynamic miscibility between the PMPS and all novolacs studied, and we have suggested experiments which may lead to enhanced miscibility and stability in this polymer blend.

References

1.   Bowden, M. J.; Allara, D. L.; Vroom, W. I.; Frackoviak, J.; Kelley, L. C. ; Falcone, D. R. *Polym. Electr.* **1984**.
2.   Thompson, L. F.; Bowden, M. J.; Fahrenholtz, S. R. ; Doerries, E. M. *J. Electrochem Soc.*, **1981**, 128, 1304.
3.   Reichmanis, R. ; Novembre, A. E. *Ann. Rev. Mat. Sci.* **1993** , 23, 11.
4.   Shiraishi, H.; Isobe, A.; Murai, F. ; Nonogaki, S. *Polym. Elect.* **1984**.
5.   Tarascon, R.; Frackoviak, J.; Reichmanis, E.; Thompson, L. *SPIE Proceedings*, **1987**, 771, 54.
6.   Frakoviak, J.; Tarascon, R.; Vaidya, S.; Reichmanis, *SPIE Proceedings*, **1987**, 771, 120.
7.   Novembre, A. E.; Kruck, C. S.; Kometani, J. M.; Kumar, U.; Mixon, D. A.; Alonzo, J. C. *Polym. Prep., ACS* **1994** , 942.
8.   Fahrenholtz, S. R. ; Kwei, T. K. *Macromolecules*, **1981**, 14, 1076.
9.   Kogler, G. ; Mirau, P. *Macromolecules* **1992** , 25, 598.
10. Mirau, P.; Heffner, S.; Koegler, G. ; Bovey, F. *Poly. Int.* **1991** , 26, 29.
11. White, J. L. ; Mirau, P. A. *Macromolecules* **1993**, 26, 3049.
12. States, D.; Haberkorn, R. ; Ruben, D. *J. Magn. Reson.* **1982** , 48, 286.
13. Schaefer, J.; Stejskal, E. O. ; Buchdahl, R. *Macromolecules* **1977** , 10, 384.
14. Mirau, P.; Tanaka, H. ; Bovey, F. *Macromolecules* **1988** , 21, 2929.
15. Jeneer, J.; Meier, B.; Bachmann, P. ; Ernst, R. *J. Chem. Phys.* **1979** , 71, 4546.
16. Mirau, P. ; Bovey, F. *Macromolecules* **1990** , 23, 4546.
17. Kwei, T.; Nishi, T. ; Roberts, R. *Macromolecules* **1974** , 7, 667.
18. Cheng, R.,Grant, B., Pederson, L., Willson, C.G., *US Patent* # 4,938,001, **1983**.

RECEIVED August 26, 1995

Chapter 13

Styrylmethylsulfonamides: Versatile Base-Solubilizing Components of Photoresist Resins

Thomas X. Neenan, E. A. Chandross, J. M. Kometani, and O. Nalamasu

AT&T Bell Laboratories, 600 Mountain Avenue, Murray Hill, NJ 07974

A series of styrylmethylsulfonamides have been prepared and used as components of chemically-amplified deep-UV photoresists based upon the acid catalysed removal of t-butoxycarbonyloxy (t-Boc) groups. The sulfonamide group is similar in acidity to the phenolic moiety commonly used in resists, but offers the advantages of much lower optical absorption and substantial resistance to chemical degradation. Reaction of p-chloromethylstyrene with N-acetyl methanesulfonamide, followed by removal of the N-acetyl group, yielded 4-amido-sulfonylmethylstyrene (3). The reaction of 4-bromo-benzylsulfonyl chloride with methylamine, followed by replacement of the bromide with ethylene by means of the Heck reaction, yielded N-[(4-ethenylphenyl)methyl]-sulfonamide (4). Polymerization of 3 or 4 with t-butoxycarbonyloxystyrene (TBS) yielded polymers with low optical density but having low glass transition temperatures (Tg). Polymerization of 3, TBS and SO_2 yielded terpolymers of higher Tg (>130 °C) which could be imaged with sub 0.30 μm resolution using a conventional photoacid generator such as 2-nitro-6-trifluoromethyl-benzenesulfonate. The terpolymers have good post-exposure bake stability without the need for a protective overcoat, and are attractive new materials for deep-UV lithography at 248 nm.

The development of high numerical aperture (NA) deep UV (DUV) exposure tools, which allow printing of sub 0.30 μm features, continues to drive the demand for high performance positive tone resists. The low transparency of conventional novalac/diazonaphthoquinone resists at 248 nm precludes their use for deep-UV imaging. Chemically amplified systems, based upon more transparent matrix resins are promising candidates for manufacturing scale deep-UV lithography (*1*), provided that several critical problems can be overcome. Chemically amplified systems often use copolymers of polyhydroxystyrene as a base resin, in which the hydroxyl groups have been masked by a protecting group, typically the t-butoxycarbonyl (t-Boc) group (*2-3*).

0097–6156/95/0614–0194$12.00/0

Though capable of being imaged with sub-0.30 μm resolution with appropriate photoacid generators (PAGs), typical t-Boc styrene resins (such as poly(4-acetoxystyrene-4-t-butoxycarbonyloxystyrene-sulfone) (PATBSS) or poly(4-t-butoxycarbonyloxystyrene-sulfone) (PTBSS) suffer excessive weight loss and shrinkage with loss of the t-Boc during exposure *(4)*. Typical shrinkages are of the order of 35%, with ~10 % thickness loss being desirable. Excessive thickness loss leads to serious processing problems such as loss of line-width control *(4-5)*. A solution to excessive thickness loss is to reduce the t-butoxycarbonyloxystyrene (TBS) content of the resins without affecting performance. PTBSS polymers have been prepared in which a portion of the t-Boc groups are chemically *(6)* or thermally *(7)* removed (to form hydroxystyrene moieties) prior to spinning and imaging of the polymers. A second approach, also leading to hydroxystyrene residues in the polymer, has been the preparation of matrix resins in which a proportion of the TBS has been replaced by trimethylsilyloxystyrene *(8)*. Upon polymerization and acidic workup, the trimethylsilyl group is removed to give a partially deprotected polymer.
An alternative approach to the use of these *inactive* matrix components is the use of intrinsically base soluble monomers which, copolymerized with TBS, yield polymers finely balanced between base solubility and insolubility. Removal of the t-Boc groups upon exposure and baking yields a polymer with high base solubility. Such a monomer must fulfill several requirements besides base solubility. It should be polar to ensure good adhesion to silicon, and be compatible towards co-polymerization with TBS. The resulting copolymers must have high enough T_g's (glass transition temperatures) to withstand pre- and post-baking temperatures, and be essentially transparent at 248 nm. The polymers must be stable to both acids and base, have long shelf lives in solution and be inexpensive to manufacture.

We describe here one such class of materials which show promise to fulfill these requirements. They are based on the weak acidity ($pK_a \sim 10$) of the sulfonamide group, which is the same as that of the phenol groups of novalac resins and hydroxy styrene polymers. Specifically, we have prepared and characterized a series of copolymers of TBS and styrylmethylsulfonamides and shown that these materials function as chemically amplified resists over a wide range of copolymer compositions. These materials have low optical densities at 248 nm, and adhere well to silicon. The copolymers are of relatively low molecular weight, and have T_g's that are too low to be useful as deep-UV matrix resins . However, the preparation of *ter*polymers of these sulfonamide monomers, TBS and sulfur dioxide furnishes materials with greatly improved T_g's. Lithographic evaluation indicates that this class of polymers offer excellent resolution (0.25 μm lines and spaces in a 0.8 μm thick film) with good process latitude.

Experimental Section.

Reagents. 4-Bromobenzyl bromide, methanesulfonyl chloride, triethylamine, acetonitrile, palladium acetate and tris(o-tolyl)phosphine were obtained from Aldrich and used without further purification except where noted.   2-Nitro-6-trifluoromethylbenzyl tosylate was prepared as described previously *(9-10)*. Ethylene (Matheson Gas Products) was used without purification.

Characterization. ^{1}H and ^{13}C NMR spectra were measured in CDCl$_3$ using a Bruker AM360 MHz spectrometer and using the solvent proton signal as reference. Fourier transform infrared spectrometry (FTIR) was performed on Mattson Instruments Galaxy Series 8020 FTIR spectrometer on NaCl discs or double polished silicon wafers. Size exclusion chromatography (SEC) was performed with a Waters Model

510 pump in conjunction with a Waters Model 410 differential refractometer detector and a Visotek Model 100 differential viscometer detector.

Thermal analysis data were obtained on a Perkin-Elmer TGA-7 thermogravimetric analyzer interfaced with a TAC 7 thermal analysis controller and a PE-7700 data station. TGA samples were heated at a rate of 10 °C/min. with a purified N_2 gas flow of 20 cm^3/min. DSC samples were heated at a rate of 20 °C/min.

Deep-UV Lithography. Resist solutions (11-15 wt% relative to solvent) were prepared by dissolving the polymers in propylene glycol monomethyl ether acetate (PGMEA) or in methyl-3-methoxypropionate (MMP). To these solutions was added 2-nitro-6-trifluoromethylbenzyl tosylate (*11*) (15 wt% relative to polymer) as the photoacid generator (PAG). The solutions were filtered through at least a 0.2 μm average pore size Millipore Teflon filters. Resist films of 0.6-0.8 μm thick were prepared by spin coating onto 4 inch silicon substrates (previously primed with hexamethyldisilazane (HMDS)) at spinning speeds ranging from 2000 to 2500 rpm. The films were baked after coating at 120 °C for 30 seconds on a vacuum hot plate. The resist coated substrates were then exposed by a Suss Model MA56A contact aligner equipped with a Lambda Physik excimer laser and also by a GCA Laser Step prototype deep UV exposure tool operating at 248 nm. After exposure, the wafers were baked for 30 seconds at 120 °C unless otherwise stated. The developer solution was OPD 262 (0.262 N tetramethylammonium hydroxide). Exposed and baked films were developed in the aqueous base solution for 1-2 minutes. Film thickness was measured on a Nanospec film thickness guage (Nanometrics, Inc.).

N-Acetyl-methanesulfonamide (7). A solution was prepared of methanesulfonamide (95.0 g, 1 mol) in pyridine (400 mL). The solution was cooled in ice and acetic anhydride (122.4 g, 1.2 mol) was added. The solution was allowed to warm to room temperature, and was then heated at reflux for 3 h. The solvent and excess acetic anhydride were removed on the rotary evaporator. Cooling in ice caused crystallization of the product. The white crystals were filtered, dried in air, and recrystallized twice from absolute ethanol. A total of 118 g (86 %) was recovered, m.p.101-102 °C. IR (KBr) 3129, 2907, 1681 (carbonyl), 1476, 1338, 1241, 1158, 977, 872 cm^{-1}. ^1H NMR (DMSO-d6) 11.72 (bs, 1H), 3.21 (s, 3H), 3.03 (s, 3H). Calculated for $C_3H_7NO_3S$, C, 26.28; H, 5.11; N, 10.22. Found C, 26.36; H, 5.22; N, 10.14.

N-Acetyl-methanesulfonamide, potassium salt (8). To **7** (40.0 g, 0.29 mmol) in hot ethanol (300 mL) was added a solution of potassium hydroxide (16.24 g, 0.29 mol) (in ethanol, 100 mL). An immediate white precipitate formed. The reaction mixture was cooled in ice, filtered, and the white crystalline potassium salt (46 g, 89%) washed with cold ethanol, dried in air and used without further purification.

Synthesis of compound 9. To a solution of N-acetyl-methanesulfonamide, potassium salt (**8**) in dry DMF (200 mL) was added chloromethylstyrene (**5**). The reaction mixture was heated under nitrogen for 16 h, cooled and poured into cold water (500 ml). The mixture was extracted with ethyl acetate (3 x 200 ml), the organic extracts were combined, washed with water, brine, and dried over $MgSO_4$. The solvent was removed under reduced pressure and the residue was recrystallized from methanol. Compound **9** was recovered as white needles. ^1H NMR (CDCl3) 7.35 (dd,

4H), 6.68 (q, 1H), 5.75(d, 1H), 5.27 (d, 1H), 4.99 (s, 2H), 3.13 (s, 3H), 2.35 (s, 3H).

4-Methylsulfonamidobenzylstyrene (3).

To a solution of **9** (10.0 g) in hot methanol (200 mL) was added a solution of KOH (2.0 g) in methanol (30 ml). The solution was stirred at reflux for three hours, filtered and made acidic with conc. HCl. The precipitated white solid was filtered, washed with water, dried and finally recrystallized from methanol. Monomer **3** was recovered as white needles, (4.13 g 76 %). M.p. 124-125 °C. ^{1}H NMR (CDCl$_3$) 7.42 (dd, 4H), 6.67 (q and bs, 2H CH= and NH), 5.64 (d, 1H), 5.23 (d, 1H), 4.34 (d, 2H), 2.82 (s, 3H); Anal. Calcd for C$_{10}$H$_{13}$NO$_2$S: C, 56.85; H, 6.20; N, 6.63; S 15.17. Found C, 56.71; H, 6.21; N, 6.55; S, 15.07.

4-Bromobenzylsulfonic acid, sodium salt (11).

To a saturated solution of sodium sulfite in water (205 g salt in 300 mL water) was added 4-bromobenzyl bromide (**10**) (20.0g, 0.54 mol). The mixture was heated at 75 °C for 6 hrs. The 4-bromobenzyl bromide initially melted and then dissolved to form a faintly cloudy solution. After 4 hrs of heating a white precipitate began to form. The solution was cooled to 5 °C, and filtered. The white crystalline solid (heavily contaminated with sodium sulfite) was recrystallized from warm water and dried in a vacuum oven over P$_2$O$_5$. The yield was 13.7g (63 %). ^{1}H NMR (DMSO-d$_6$) 6.93 (dd, J = 1.8 Hz, 4 H), 3.3 (s, 2H). This material was used without further purification in subsequent reactions.

4-Bromobenzylsulfonyl chloride (12).

An intimate mixture of 4-bromobenzylsulfonic acid, sodium salt (12.90 g, 43.6 mmol) phosphorus penta-chloride (10.8 g, 52 mmol), and phosphorus oxychloride (7.90 g, 52 mmol) was heated to 80 °C under an atmosphere of nitrogen. After 30 minutes a homogeneous liquid was formed. Heating was continued for a total of 5 hrs. The reaction mixture was cooled, diluted with diethyl ether and (CAUTION !) added to a large volume of ice. The ether layer was separated, washed with water and saturated brine solution, dried over magnesium sulfate and the solvent removed. 4-Bromobenzylsulfonyl chloride (13.3 g) (recovered as white needles) was used without further purification. A small quantity was recrystallized from hexanes for elemental analysis. M.p. 114-116 °C. ^{1}H NMR (CDCl$_3$) 7.42 (dd, J = 1.7 Hz, 4 H), 4.83 (s, 2 H). Anal. Calcd for C$_7$H$_6$BrClO$_2$S : C, 31.19; H, 2.24; S, 11.89. Found: C, 31.23; H, 2.41; S, 11.73.

Methanesulfonamide,N-[(4-bromophenyl)methyl]-(13).

Methylamine (~ 10 mL) was condensed into a flask containing methylene chloride (100 mL), and equipped with a Dry Ice condenser. The flask was cooled to 0 °C with an ice bath. To this solution was added crude **12** (13.1 g, 48.3 mmol) rapidly as a solution in methylene chloride (50 ml). An exothermic reaction took place. The solution was stirred at 0 °C for two hours, the Dry Ice condenser and the ice bath were removed, and the excess methylamine allowed to evaporate into the fume hood. The methylene chloride was removed at reduced pressure, and the residue flushed through a short plug of silica gel, using ethyl acetate as the eluting solvent. Removal of the ethyl acetate followed by recrystallization from methanol furnished **13** as white needles (9.6 g, 74 %). M.p. 125-127 °C. ^{1}H NMR (CDCl$_3$) 7.34 (dd, 4H), 5.25 (t, 1H, NH), 4.34 (d, 2H), 2.87 (t, 3H). Anal. Calcd for C$_8$H$_{10}$BrNO$_2$S: C, 36.51; H, 3.83; N, 5.33. Found C, 36.44; H, 3.89; N, 5.11.

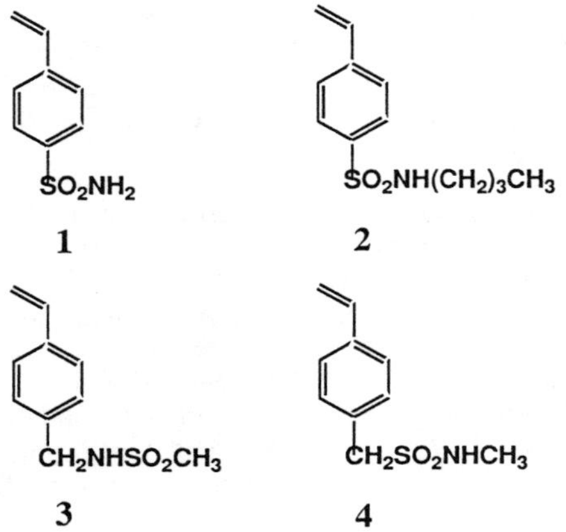

Fig. 1. Structures of styrylsulfonamides **1-4**.

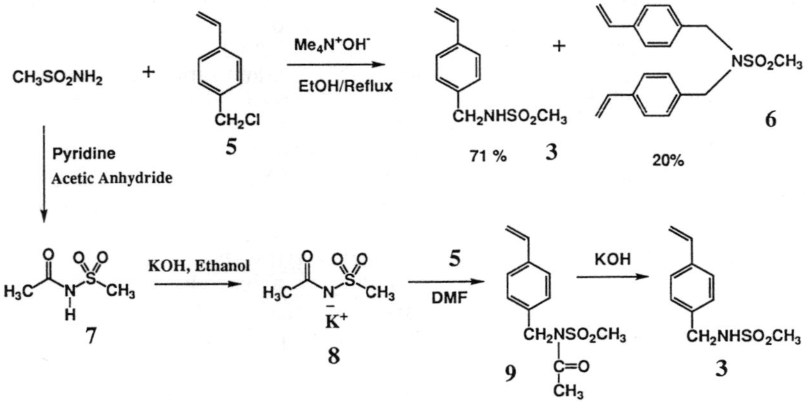

Scheme I. Synthesis of monomer **3**.

Methanesulfonamide,N-[(4-ethenylphenyl)methyl]- (4). A 300 mL Teflon lined bomb equipped with a Teflon coated magnetic stirring bar was charged with **13** (6.8 g, 26 mmol), palladium acetate (57 mg, 1 mol% with respect to the bromide), tris(o-tolyl)phosphine (155 mg, 2 mol% with respect to bromide), triethylamine (10 mL) and acetonitrile (10 mL). The bomb was sealed, connected to an ethylene cylinder and the air flushed from the bomb with ethylene. The bomb was pressurized to 100 psi with ethylene, sealed and placed in a thermostatted oil bath at 95 °C. The bomb was heated for 16 hours while being magnetically stirred. The bomb was cooled and vented into the hood (CAUTION !). Diethyl ether (100 ml) was added to the bomb, causing precipitation of the triethylamine hydrobromide as a white crystalline solid. The contents of the bomb were poured through a Buchner funnel, and the salts were washed with a further 50 mL of diethyl ether. The solvent was removed from the combined filtrates on a rotary evaporator, the residue was dissolved in the minimum amount of ethyl acetate and the solution passed through a short column of silica gel (50 g packed with the same solvent in a two inch diameter column). Removal of the solvent yielded **4** as a white solid (4.13 g, 76 %). M.p. 103-105 °C. ^{1}H NMR (CDCl$_3$) 7.44 (dd, 4 H), 6.72 (dd, 1 H), 5.82 (d, 1 H), 5.33 (d, 1 H), 4.23 (bs, 3 H, NH and CH$_2$) 2.68 (s, 3H).

Synthesis of Polymers. Preparation of 14, the nominally 50/50 copolymer between 3 and TBS. A solution was prepared of **3** (5.27 g, 25 mmol) and TBS (5.50 g, 25 mmol) in dry THF (100 mL) in a heavy wall polymerization tube. To this solution was added AIBN (164 mg, 2 mmol, 2 mol% with respect to total monomer). After degassing by a series of three freeze pump thaw cycles, the tube was sealed and placed in an oil bath whose temperature was kept between 68-72 °C. The polymerization reaction was heated for 22 h, cooled and vented. The polymer was recovered by precipitation in hexanes, and purified by dissolution in the minimun amount of THF followed by precipitation into methanol. The product was dried in a vacuum oven at 35 °C overnight. Yield 5.91 g (55%).

Results and Discussion.

Materials Preparation. Styrylsulfonamides are attractive components of lithographic matrix resins because of their base solubility. The commercial availability of sodium 4-styryl sulfonate prompted us to prepare 4-styrylsulfonamide and it's n-butyl analog (Fig 1, **1** and **2**) and to examine their copolymers with TBS. Typically, a lithographically useful polymer needs to have an optical density of ~<0.2 absorbance units/μm before the addition of photoacid generators. A 1 μm film of a 50/50 copolymer of TBS and **1** had an optical density at 248 nm of 0.76, while a 1 μm thick film of a 50/50 co-polymer of **2** and TBS had an absorbance of 0.82. In order to reduce the absorbance of these systems, it is necessary that the sulfonamide group be decoupled electronically from the aryl ring by, for example, a methylene linkage. Two such monomers, **3** and **4** are shown in Fig. 1. The simplest route to monomer **3**, reaction of the potassium salt of methanesulfonamide with chloromethylstyrene (**5**) yields primarily the bis-alkylated (**6**) product unless a large excess of the potassium salt is used. To circumvent this difficulty, an alternative synthesis was developed in which methanesulfonamide was first acetylated followed by conversion of the monoacetylated product (**7**) to its potassium salt (Scheme I). Reaction of the potassium salt with chloromethylstyrene and treatment of the acetylated styrenesulfonamide (**9**) with base led to monomer **3** in 51% overall yield. Scheme II shows our route to monomer **4**, in which the positions of the sulfur and nitrogen of **3** have been interchanged. The synthesis involved treating commercially available 4-bromobenzyl bromide (**10**) with an excess of sodium sulfite to form the corresponding benzenesulfonic acid, sodium salt (**11**) (*11*). Conversion of **11** to the sulfonyl chloride (**12**) and reaction of **12** with an excess of methylamine formed the methanesulfonamide derivative (**13**). Reaction of **13** with ethylene (100 psi) under Heck conditions (*12*) yielded **4**.

Polymerizations of TBS with **3** and **4** respectively were carried out in tetrahydrofuran (THF) solution, using AIBN as initiator. A series of co-polymers of **3** and TBS were prepared in which the composition of **3** was varied from 50 to 90% (Table 1, **14-17**) to determine the minimum amount of TBS needed to allow imaging. Due to the relatively low Tg of the resulting polymers (Table 1), a series of polymerizations were carried out between **3**, TBS and sulfur dioxide (Table 1, **18-20**). It has been previously shown that the incorporation of sulfur dioxide into TBS polymers raises the Tg and improves adhesion of the polymer to silicon substrates. The polymerizations involving SO_2 (Scheme III) were carried out in the absence of solvent as described earlier for other t-Boc-styrene-sulfone co-polymers (*13*).

Polymer Characterization. The polymers were isolated as white powders after purification by repeated precipitations and their physical properties are summarized in Table 1. The composition of the polymers was determined by a combination of the best fit between proton NMR integration, thermogravimetric and elemental analysis. Of specific interest was the ratio of sulfonamide ($NHSO_2 = 3$; $SO_2NH = 4$) to TBS. By all methods, the sulfonamide:TBS ratio was close to that of the monomer feed. In the NMR studies, integration of the sulfonamide methyl protons of **3** at 4.3 ppm relative to the butoxy protons at 1.55 ppm gave an approximate value for the ratio of the two monomers in the polymers. Elemental analysis was used to confirm these data. Molecular weights were determined by high pressure size exclusion chromatography. The molecular weights of the polymers of the series of co-polymers between **3** and TBS, (**14-17**) though relatively low, did not vary significantly with the polymer composition. However, the polydispersity and the yields increased slightly with increased t-Boc content. The terpolymers of **3**, TBS and sulfur dioxide (**18-20**) had significantly higher molecular weights, and the incorporation of sulfone into these polymers also increased the absorbance, an observation noted previously for other TBS-sulfone copolymers (*13*).

Thermal Analysis. Thermogravimetric analysis (TGA) of TBS-based polymers generally shows one major weight loss event between 100 and 200 ºC, corresponding to the thermal deprotection of the tert-butoxycarbonyloxy (t-Boc) group, with loss of carbon dioxide and isobutylene. For the homopolymer of TBS, the onset temperature for deprotection (T_O) is 188 ºC, while for PTBSS copolymers T_O ranges from 150 to 180 ºC, depending upon composition and molecular weight. For the series of co-polymers of between **3** and TBS, there is a clear relationship between TBS content and T_O, with higher TBS content leading to higher T_O temperatures. The onset of backbone decomposition of all eight polymers was in the range of 350-370 ºC, typical of TBS based polymers. Thermogravimetric analysis (TGA) was also useful in determining the compositions of the co-polymers. There is a clear distinction between the loss of the t-Boc group and the thermal decomposition of the polymers in the TGA curves. From weight loss after thermal deprotection of the t-Boc group, the copolymer composition was calculated to be essentially equivalent to the monomer feed ratios. The composition data for the series of sulfone polymers (**18-20**) shows that the synthesis of these terpolymers is relatively reproducible.

The glass transition temperatures of the polymers varied with composition, and again higher TBS content led to higher Tg's in the series **14-17**. It is surprising that a

Table 1. **Characterization Data for Polymers 14-21**

Polymer	Composition	Yield (%)	M_w ($\times 10^3$)	M_w/M_n	Optical Density Abs/μm	T_g (°C)	T_o (°C)	T_d (°C)
14	$P(NHSO_2)_{0.9}(t\text{-}Boc)_{0.1}$	52	37.8	1.42	0.09	90	148	348
15	$P(NHSO_2)_{0.75}(t\text{-}Boc)_{0.25}$	55	33.5	1.51	0.12	96	161	347
16	$P(NHSO_2)_{0.62}(t\text{-}Boc)_{0.37}$	61	36.5	1.58	0.06	106	167	351
17	$P(NHSO_2)_{0.5}(t\text{-}Boc)_{0.5}$	69	38.8	1.66	0.08	108	170	350
18	$P(NHSO_2)_{0.38}(t\text{-}Boc)_{0.32}(SO_2)_{0.30}$	52	127	2.04	0.19	128	154	379
19	$P(NHSO_2)_{0.36}(t\text{-}Boc)_{0.32}(SO_2)_{0.32}$	54	181	1.67	0.18	136	161	376
20	$P(NHSO_2)_{0.34}(t\text{-}Boc)_{0.35}(SO_2)_{0.31}$	81	83	1.95	0.17	137	156	356
21	$P(SO_2NH)_{0.5}(t\text{-}Boc)_{0.5}$	45	23.4	1.53	0.13	119	173	363

T_g = Glass transition temperature. T_o = Onset of t-Boc loss. T_d = Onset of backbone decomposition. The acronyms $NHSO_2$ and SO_2NH indicate polymers prepared from monomers **3** and **4** respectively.

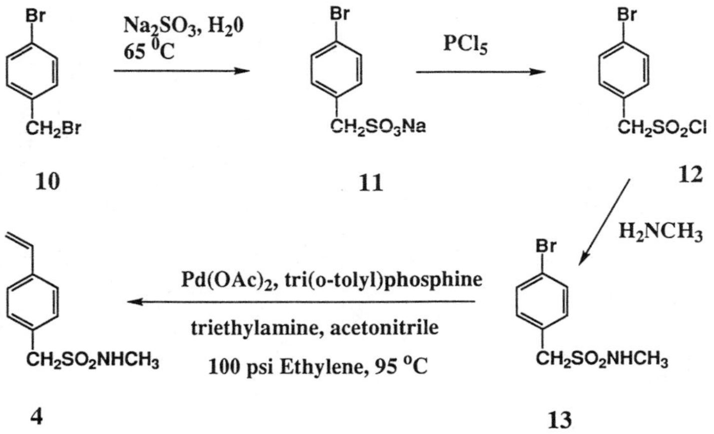

Scheme II. Synthesis of mononer 4.

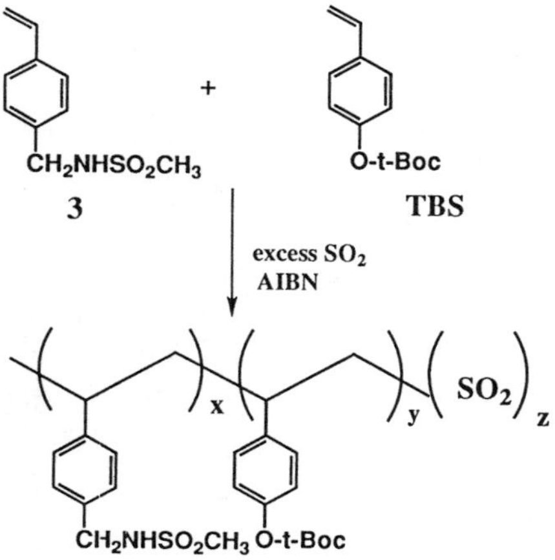

18-20

Scheme III. Synthesis of polymers 18-20.

molecule as polar as **3** should lead to such low T_g's. Low T_g values are of concern since they place limits on the pre and post bake temperatures, and decrease process latitude. Polymer **21** has a T_g ~10 °C higher than **17**, suggesting that polymers of higher T_g's are accessible, depending upon sulfonamide structure. Polymers **18-20** had the highest T_g 's, consistant with previous work (*14*) which has shown that inclusion of sulfones into t-Boc-styrene polymers raises the T_g by 30-40 °C.

Lithography.

Copolymers. Polymers **14-17** and **21** have low optical densities of 0.12-0.13 absorbance units/μm. The optical densities of the co-polymers of TBS and **3** are very similar over the whole composition range, while **21**, the co-polymer of **4** and TBS, has a slightly higher absorbance. In an initial lithographic survey, all resists were formulated with 2-nitro-6-trifluoromethylbenzyl tosylate as the photoacid generator. Solutions were prepared by dissolving the polymers in cyclohexanone to which was added 15 wt % PAG. Typical optical densities for a 1 μm film of these formulations were 0.33-0.36. Our first objective was to determine the maximum amount of a styrylsulfonamide that could be incorporated into a TBS copolymer and still form images. Although **14-17** all gave latent images upon exposure, attempts to develop **14** and **15**, the two polymers with highest **3** content, with aqueous TMAH were unsuccessful. Only the addition of ~20% isopropanol to the developer allowed image formation for these latter polymers. This data suggests that the upper limit of sulfonamide content that can still be developed by aqueous developer lies between 62 and 75 %. Clearing doses of **14-17** ranged from ~17 mJ/cm^2 for **14** to ~32 mJ/cm^2 for **17**. The higher sensitivities observed for **14** seems to confirm earlier observations that polar polymers form the most sensitive resists (*9-10*).

Terpolymers. Because of the problems encountered with the copolymers (low T_g and low base solubility) we focused on the lithographic development of the terpolymers **18-20**. In a typical experiment, polymer **18** was formulated with 2-nitro-6-trifluoromethylbenzyl tosylate as the photogenerator of acid. Resolution of 0.25 μm l/s (line/space) with vertical wall profiles was obtained in 0.80 μm thick films with a 0.53 NA stepper (Fig. 2). Isolated single lines were observed with the same resolution (Fig. 3). Additionally, 0.30 μm contact hole resolution could be achieved with overexposure. The relatively high resist absorbance (0.41 absorbance units/μm) at 248 nm did not adversely affect resolution. Fig 4 shows the focus latitude of 0.275 μm l/s bright field patterns. Linewidth variation of +/-10% and 10% profile erosion are the criteria used in determining depth of focus (DOF). Within these limits, polymer **18** had a DOF of 1.5, 1.0 and 0.8 μm for 0.35, 0.30 and 0.275 μm l/s respectively.

Pattern deterioration associated with time delay has been the bane of many chemically amplified resist systems. A resist needs to have pattern stability between exposure and post exposure bake of at least 30 minutes (and ideally > 60 minutes). A post exposure bake experiment on the resist formulated from **18**, was performed with the 0.37 NA stepper. After a 1 hour time delay after exposure, no shrinking or capping of the 0.35 μm lines could be detected. Additionally, no capping was observed when the patterns were significantly underexposed. The time delay stability between prebake and exposure is less critical, but a 30 minute delay between PB and exposure also showed no deterioration of 0.35 μm l/s patterns.

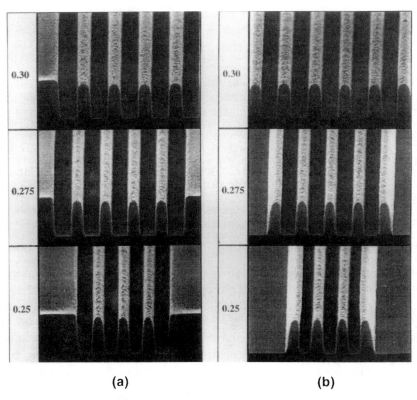

(a)   (b)

Fig. 2. Resolution of 0.3 to 0.25 µm l/s patterns for polymer **18** in (a) dark field and (b) bright field, showing wall profiles in 0.80 µm thick films with a 0.53 NA stepper.

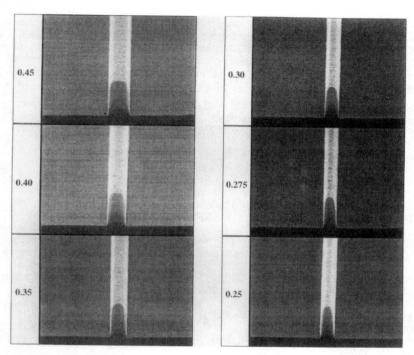

Fig. 3. Resolution of 0.45-0.25 μm isolated single lines in resist **18**.

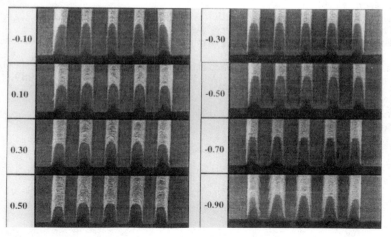

Fig. 4. Focus latitude for 0.275 μm l/s bright field patterns in resist **18**.

Conclusion.

We have prepared a new class of sulfonamide styrenes and shown that they may be co-polymerized with TBS to form polymers which, upon deprotection, yield base soluble materials. Formulation of these polymers with an acid generator allows imaging at 248 nm but the T_g's of these polymers are too low to allow them to function as practical resists. Terpolymers of the sulfonamide styrenes, TBS and sulfur dioxide have higher T_g's, have sensitivities of 27-40 mJ/cm^2 and may be imaged with resolutions of down to 0.25 µm l/s and 0.30 µm contact holes. The terpolymers have good post exposure bake stability without the use of an overcoat, and are promising new materials for deep-UV lithography at 248 nm.

Literature Cited.

1. Ito, H.; Willson C. G.; in "Polymers in Electronics", ACS Symposium Series no. 242, Davidson, T. Ed., ACS, Washington, D. C. 1984, pp 11-23.
2. Frechet, J. M. J.; Eichler, E.; Ito, H.; Willson, C. G. *Polymer*, **1980**, *24*, 995.
3. Ito, H.; Willson, C. G.; Frechet, J. M. J.; Farrall, M. J.; Eichler, E. *Macromolecules*, **1983**, *16*, 510.
4. Nalamasu, O.; Reichmanis, E.; Cheng, M.; Pol, V.; Kometani, J. M.; Houlihan, F. M.; Neenan, T. X.; Bohrer, M. P.; Mixon, D. A.; Thompson, L. F.; Takemoto, C. *Proceedings SPIE 1991*, *1466*, 2.
5. Nalamasu, O.; Kometani, J. M.; Cheng, M.; Timko, A. G.; Reichmanis, E.; Slater, S.; Blakeney, A., *J. Vac. Sci. Technol.*, **1992**, *B10*, 2536.
6. Mixon, D. A.; Bohrer, M.; Alonzo, J. C.; *Proceedings SPIE 1994*, *2195*, 297.
7. Merritt, D. P.; Moreau, W. M.; Woood, R. L. Canadian Patent Application, 2,001,384, 1989.
8. Uhrich, K. E.; Reichmanis, E.; Heffner, S. A.; Kometani, J. M., *Macromolecules*, **1994**, *27*, 4936.
9. Houlihan, F. M.; Chin, E.; Nalamasu, O.; Kometani, J. M.; Neenan,T, X.; Pangborn, A. *J. Photopolymer Sci. and Technol.* **1993**, *6(4)*, 515.
10. Houlihan, F. M.; Neenan, T, X.; Reichmanis, E.; Kometani, J. M.; Chin, T. *Chem. Mater.* **1991**, *5*, 2345
11. Lichtenberger, J.; Tritsch, P. *Bull. Chim. Soc. Fr.* **1961**, *78*, 363
12. Plevyak, J. E.; Heck, R. F. *J. Org. Chem.* **1978**, *43*, 2454.
13. Tarascon, R. G.; Reichmanis, E.; Houlihan, F. M.; Shugard, A.; Thompson, L. F. *Polym. Eng. Sci.* **1989**, *28*, 13.
14. Kanga, R. S.; Kometani, J. M.; Reichmanis, E.; Hanson, J. E.; Nalamasu, O.; Thompson, L. F.; Heffner, S. A.; Tai, W. W.; Trevor, P. *Chem. Mater.* **1989**, *3*, 660.

RECEIVED July 17, 1995

Chapter 14

4-Methanesulfonyloxystyrene

A Means of Improving the Properties of *tert*-Butoxycarbonyloxystyrene-Based Polymers for Chemically Amplified Deep-UV Resists

J. M. Kometani, F. M Houlihan, Sharon A. Heffner, E. Chin[1], and O. Nalamasu

AT&T Bell Laboratories, 600 Mountain Avenue, Murray Hill, NJ 07974

A series of novel 4-methanesulfonyloxystyrene/4-t-butoxycarbonyloxystyrene/sulfone ($MSS/TBS/SO_2$) terpolymers in conjunction with a 2-nitrobenzyl photo-acid generator (PAG) provided positive resists that were developable in aqueous tetramethylammonium hydroxide (TMAH) and gave lower volume loss than TBS/SO_2 copolymers. These polymer resists have low optical density (0.14 AU/μm) and volume loss of ~15% after post exposure bake (PEB). Even lower volume losses were achieved, while retaining aqueous base solubility, by introducing a base soluble unit, 4-hydroxystyrene (OHS), into $MSS/TBS/SO_2$ polymers. Similarly, by using 4-t-butoxycarbonyl-2,6-dinitrobenzyl tosylate (a dissolution inhibitor solubilizable by chemical amplification, DISCA PAG) as the photo-acid generator, it was possible to increase the MSS content in the $MSS/TBS/SO_2$ polymer and improve dissolution in aqueous TMAH. These resist systems exhibited low volume loss (5%) and low clearing doses (14 mJ/cm^2).

Volume loss due to acidolysis is detrimental for the development of positive tone deep UV (248 nm) chemically amplified (CAMP) resists formulated with 4-*t*-butoxycarbonyloxystyrene (TBS) based polymers. Volume loss has been minimized by replacing a fraction of the TBS units in the matrix resin, with a monomer unit such as acetoxystyrene (AS)(*1*) which is not subject to acidolysis, but to base hydrolysis. The 4-methanesulfonyloxystyrene (MSS) monomer was evaluated because the polar mesylate (Ms) protecting group is acid resistant, base labile, low in optical density, and thermally stable(*2*). We have designed and synthesized the copolymer 4-methanesulfonyloxystyrene/4-t-butoxycarbonyloxystyrene (MSS/TBS), and the terpolymer MSS/TBS/Sulfone (SO_2) for application as resist matrix resins(*3*).

---

[1]Current address: Affy Metrix, 3380 Central Expressway, Santa Clara, CA 95051

0097–6156/95/0614–0207$12.25/0

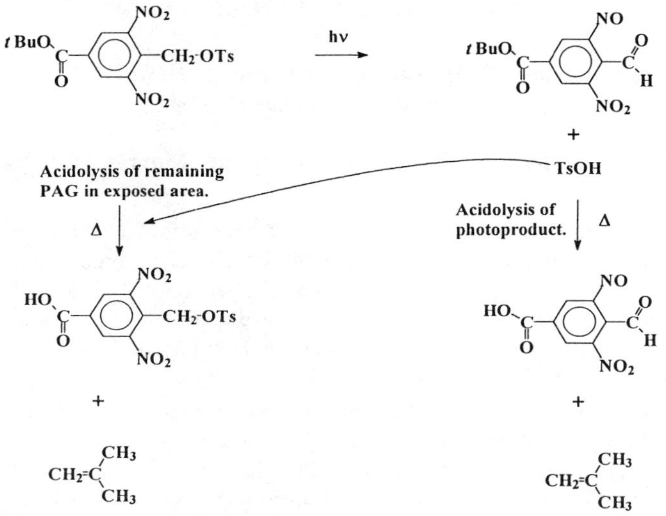

Scheme 1. Photo-generation of acid, followed by acidolysis of photoproduct and of the remaining PAG in exposed area.

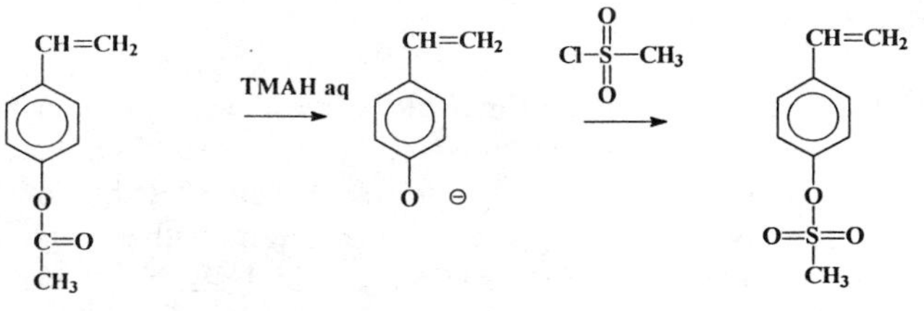

Scheme 2. Synthesis of 4-methanesulfonyloxystyrene.

A decrease of volume loss for MSS/TBS/SO$_2$ and MSS/TBS polymer resists was obtained by incorporating 4-hydroxystyrene (OHS) into the polymer resins. These new polymers MSS/TBS/OHS/SO$_2$ and MSS/TBS/OHS resulted in base developable resist systems which allowed for increased MSS loading.

Another approach for improving aqueous base solubility of low volume loss resists is to use a photo-acid generator (PAG) that enhances dissolution in exposed regions. 4-*t*-butoxycarbonyl-2,6-dinitrobenzyl tosylate acts as a PAG and a dissolution inhibitor solubilizable by chemical amplification (DISCA PAG)(*4*). Scheme 1 shows photo-generation of acid, followed by acidolytic removal of the *t*-butyl group, liberating the base soluble carboxylic acid moiety. DISCA PAG should not inhibit dissolution in the exposed regions, since both photoproduct and unreacted PAG are base soluble.

Experimental

Synthesis.

The DISCA PAG molecule, 4-*t*-butoxycarbonyl-2,6-dinitrobenzyl tosylate(*4*), 4-methoxy-2,6-dinitrobenzyl tosylate(*4*), 2-trifluoromethyl-6-nitrobenzyl sulfonate PAG's(*5,6*), and 4-trimethylsilyloxystyrene(*7*) were synthesized as previously described. TBS was obtained from Kodak. ACS was obtained from Hoechst-Celanese. Tetramethylammonium hydroxide (TMAH), 25% aqueous solution, was obtained from the Johnson Matthey Company. All other chemicals were obtained from the Aldrich Chemical Co. and used as received.

Synthesis of 4-methanesulfonyloxystyrene (Scheme 2). 4-Acetoxystyrene (100 g, 617 mmol) was added slowly to a cooled, stirred solution (480 mL) of 25 % TMAH (the temperature was maintained below 25°C during the addition). After 0.5-1 h of stirring, a clear, yellow solution resulted. This solution was cooled to 5°C and the methanesulfonyl chloride 79.5 g (694 mmol) was added with stirring. The reaction mixture was stirred for 0.5h until neutral pH was observed. The precipitated product was dissolved in a minimal amount of methylene chloride and recrystallized by the addition of hexane to remove excess methanesulfonyl chloride. 76.0g (62% yield) of pure(tlc, LC/UV) white, crystalline product, mp:57-58°C was recovered.

IR(cm^{-1}): 2980(CH stretch CH$_3$SO$_2$), 1500(CH$_3$ antisym deformation), 1385(SO$_2$, antisym. stretch), 1180(SO$_2$ sym stretch), 1150(C-O stretch), 980(antisym CH out of plane deformation CH=CH$_2$, 880(CH, out of plane deformation CH=CH$_2$), 850(CH out of plane deformation p-substituted benzene).

^{1}H NMR(CDCl$_3$, ppm): 3.14(s, 3H, CH$_3$SO$_2$), 5.31(d, J=11Hz, 1H, CH$_2$ vinyl), 5.74(d, J=18Hz, 2H, CH$_2$ vinyl), 6.70(d of d(11, 18Hz), 1H, CH vinyl), 7.24(d(9Hz), 2H, CH styrene), 7.44(d(9Hz), 2H, CH styrene).

Elemental Analysis: Calculated (C:54.53, H:5.09, S:16.17), Found(C:54.35, H:5.19, S:16.22).

Polymer Synthesis.

Copolymers and terpolymers containing MSS were synthesized by adapting procedures for the polymerization of ACS/TBS/SO$_2$(1). Table 1 summarizes the characterization data for selected polymers. Examples of each polymerization procedure are as follows: **MSS/TBS copolymer.** A solution of 12.0 g (54.4 mmol) TBS and 7.50 g (37.8 mmol) MSS dissolved in 20 mL of dry toluene was prepared and degassed under vacuum. Under argon atmosphere, 0.61 g (3.67 mmol) of azobis isobutyronitrile (AIBN) was added to the solution. The stirred solution was heated under argon for 17 h at 70°C. The polymer was precipitated twice in methanol yielding 16.0 g (83% yield) of white polymer. Mw: 69 K, Mw/Mn: 2.1, polymer composition MSS/TBS: 0.80/1.00. **MSS/TBS/SO$_2$ terpolymer.** In the synthesis of a sulfone polymer appropriate safety precautions were followed as described in the *Matheson Gas Data Book*(8). A mixture of 33.3 g (151 mmol) TBS and 20.0 g (101 mmol) MSS dissolved in toluene was degassed under vacuum. To this mixture was added to 0.83 g (5.0 mmol, 1/50 to monomer) AIBN in 30 mL of dry toluene and degassed at -78°C under vacuum. The reaction mixture was cooled to -78°C, evacuated, and 24.8 mL (631 mmol) of dry SO$_2$ was condensed into the reaction flask. The evacuated flask was heated to 65°C while stirring. After 3.5 h, the reaction mixture was cooled in a dry-ice/acetone bath, the flask was opened to allow excess SO$_2$ to escape. The reaction mixture was diluted with 160 mL of acetone and precipitated in 1,200 mL of methanol. After two more precipitations, 10.8 g of polymer (52% yield) was recovered. Mw: 126 K, Mw/Mn: 1.5; polymer composition MSS/TBS/SO$_2$: 0.67/1.00/ 0.64. **MSS/TBS/OHS terpolymer.** The synthesis was similar to that of the MSS/TBS copolymer except for the addition of the monomer 4-trimethylsilyloxystyrene as the precursor for OHS(7) in the terpolymer. The styrene/AIBN ratio was 50/1, and the reaction was done at 65°C for 6 hours. The precipitation in methanol allowed for the isolation of the trimethylsilyl derivatized polymer (49-78% yield) which was characterized by GPC. The OHS polymeric derivative was obtained almost quantitatively by reprecipitation of the trimethylsilyl derivatized material into methanol acidified with a small amount of HCl. The polymer was then reprecipitated into methanol for final purification. **MSS/TBS/OHS/SO$_2$ tetrapolymer.** The synthesis was similar to that previously described for the ACS/TBS/OHS/SO$_2$ tetrapolymer(7) substituting MSS for ACS. The polymerization was done at 65°C with a 50/1 styrene/AIBN ratio for 6 hours.

Thermal analysis.

Differential scanning calorimetry (DSC) data for the solid samples were obtained using a Perkin-Elmer DSC-7 differential scanning calorimeter interfaced with a TAC 7 thermal analysis controller and a PE-7700 data station. All samples were heated from 30 to 450°C at a heating rate of 10°C per minute. Samples of ~4 mg were encapsulated in aluminum pans. All measurements were obtained using ultra high purity N$_2$ with a flow rate of 20 cc per minute. The thermal analysis of the polymers as summarized in Table 1.

Lithography.

Lithographic samples were irradiated at 248 nm using a Lambda Physik excimer laser source and a Karl Süss Inc. model MA56A contact aligner. Scanning electron (SEM) cross-sections were obtained on a JEOL scanning electron microscope. Final exposure studies used in obtaining SEM cross sections were done using a Laserstep prototype deep-UV exposure tool (NA: 0.35, 5X optics) operating at 248 nm. All thickness measurements were obtained on a Nanospec film thickness gauge (Nanometrics, Inc.) or a Dektak model IIA profilometer. Solutions for lithographic study were prepared as outlined in Tables 2 and 3. Photoresists for lithographic experiments consisted of 13 wt % of polymer dissolved in 2-ethylethoxypropionate (EEP). To this was added 6 mole % of 2-nitrobenzyl sulfonate PAG or DISCA PAG relative to the polymer's t-BOC repeat units. Photoresists were spun (speed 2 K rpm) onto hexamethyldisilazane-primed oxidized silicon substrates and prebaked at 120°C for 30 s. A resist solution formulated with ACS/TBS/SO$_2$ was prepared in the same way and used as a reference material. Development was accomplished with a 0.26 N TMAH either in aqueous solution or in 10-20% isopropyl alcohol (IPA) aqueous solution.

Resist IR studies.

Resist solutions, described above, were used to prepare the samples for IR study. The films were prepared by spinning resist solutions onto double polished silicon substrates which were then exposed and processed. The IR spectra of the resist films were recorded on a Nicolet 510P after exposure and also after PEB.

Resist NMR studies.

The ^{13}C NMR spectra were recorded at 25°C on a JEOL GX-500 spectrometer at a resonance frequency of 125 MHz. Quantitative spectra were obtained with gated decoupling without the nuclear Overhauser effect and a pulse delay of 8 s. To simulate the dissolution of the deprotected polymer during the lithographic experiments 10 wt % samples were prepared with solutions composed of TMAH, IPA and D$_2$O. The exact proportions are specified in the text. Polymers for these studies were prepared by thermally deprotecting the t-BOC groups. Procedures for the synthesis of these materials follow: **MSS/OHS copolymer.** 3.0 g of the 0.80/1.00 MSS/TBS (Mw: 69 K, Mw/Mn: 2.1) were placed into a dry round bottomed flask equipped with a dry-ice condenser connected to an argon/vacuum manifold. The polymer sample was degassed 4 times under high vacuum. The polymer sample was heated under vacuum to 160°C. The sample was heated for 1 h; and the deprotection was monitored by the amount of isobutene condensing on the dry ice cold finger. TGA, FT-IR and ^{1}H NMR analysis of this polymer showed that all t-BOC groups had been removed. 2.0 g of the 0.80/1.00 MSS/OHS (91 % yield) were prepared. **MSS/OHS/SO$_2$ terpolymer.** 3.0 g of the 0.67/1.00/0.64 MSS/TBS/SO$_2$ (Mw: 126 K, Mw/Mn: 1.5) were heated, for 2 h at 150°C The procedure was similar to that

Table 1

Synthesis and Characterization of Polymers

Monomer Feed Ratio of monomers[a] MSS/TBS/OHS/SO$_2$	Polymer Composition- Ratio of monomers[b] MSS/TBS/OHS/SO$_2$	% yield polymer[c]	M_w (M_w/M_n)	TGA onset of decomposition (DSC, T_g) °C
1.00/1.00/0/0	0.90/1.00/0/0	82 %	69 K (2.1)	185 (127)
0.70/1.00/0/0	0.80/1.00/0/0	83 %	69 K (2.1)	183 (129)
1.00/0.67/0/0	1.00/0.60/0/0	80 %	58 K (2.0)	186 (128)
0.67/1.00/0/4.2	0.67/1.00/0/0.64	52 %	126 K (1.5)	174 (146)
1.00/1.00/0/5.0	1.00/0.97/0/0.66	62 %	143 K (1.5)	177(155)
1.00/1.00/0/5.0	1.00/0.97/0/0.50	60 %[d]	42 K (1.8)	174 (141)
1.00/0.87/1.00/0	1.00/0.87/1.00/0	72%	41 K (2.0)	158 (124)

0.90/1.00/0.81/0	0.90/1.00/0.81/0	65%	47 K (1.6)	153 (125)
1.00/0.50/0.84/0	1.00/0.50/0.77/0	56%	89 K (1.9)	154 (120)
1.00/0.87/0.63/5	1.00/0.86/0.62/0.80	53%[e]	110 K (1.5)	158 (135)
1.00/0.50/1.00/5	1.00/0.49/0.99/0.85	46%[f]	143 K (1.5)	148(135)
1.00/1.00/0.87/5	1.00/0.98/0.85/0.87	55%	75K (1.5)	156(133)

a) MSS = 4-methanesulfonyloxystyrene; TBS = 4-t-butoxycarbonyloxystyrene; OHS = 4-hydroxystyrene; SO_2 = sulfur dioxide

b) Polymer composition evaluated from ^1H NMR, elemental analysis (C:H:S), and TGA weight lost of *t*-BOC.

c) Reaction conditions are as outlined in the examples given in the experimental section.

d) Monomer/AIBN molar ratio 25/1.

e) Monomer/AIBN molar ratio 40/1

f) Monomer/AIBN molar ratio 50/1

Table 2

LithographicSensitivity of Resists Formulated with 4-Methanesulfonyloxystyrene Polymers. Copolymers and Terpolymers with 4-t-Butoxycarbonyloxystyrene and 4-Hydroxystyrene[a]

Polymer Composition MSS/TBS/OHS	M_w	PEB Temperature °C	PEB Time (s)	Volume loss (%)	TMAH normality, % IPA (develop time)	Clearing Dose mJ/cm² (Contrast)
0.9/1.0/0	62 K	120	10	16%	0.26 N, 30 % (10 s)	13 (>10)
0.9/1.0/0	62 K	120	10	16%	0.26 N, 20 % (10 s)	20 (>10)
0.9/1.0/0	62 K	120	10	16%	0.26 N, 10 % (10 s)	>60[b]
0.9/1.0/0	62 K	105	20	16%	0.26 N, 20 % (10 s)	37 (>10)
0.9/1.0/0	62 K	105	60	16%	0.26 N, 20% (10 s)	18 (>10)
1.0/0.6/0	58 K	120	10	15%	0.26 N, 20% (10 s)	15 (>10)
1.00/0.50/0.77	89 K	120	60	9%	0.26 N, 20% (10 s)	6(1)
0.90/1.00/0.81	47 K	120	60	8%	0.26 N, 20% (60 s)	2(4)
1.00/0.87/1.00	41 K	120	60	10%	0.26 N, 20% (10 s)	2(5)

a) Resists Formulated with 6 mole % PAG as described in the experimental section.

b) Maximum dose employed giving no development.

previously described for the preparation of MSS/OHS copolymer. TGA, FT-IR and ^{1}H NMR analysis of this polymer showed that all t-BOC groups had been removed. 2.1 g of the 0.67/1.00/0.64 MSS/OHS/SO$_2$ (94 % yield) were prepared.

Results and Discussions:

Terpolymers of MSS/TBS/SO$_2$ are desirable candidates for deep UV (248nm) resists because of their low optical density, high Tg's and acceptable decomposition temperature. It is essential to minimize the absorbance of the resist polymer at 248 nm (~0.1-0.2 AU/μm) to prevent it from interfering with the photolysis of the PAG. Both MSS/TBS/SO$_2$ terpolymers and MSS/TBS copolymers gave low absorbances (~0.14 AU/μm). Thermal analysis data of the MSS/TBS/SO$_2$ terpolymers in Table 1 indicated onset temperatures for t-BOC loss from 174-177°C. By comparison the onset temperatures for the MSS/TBS copolymers ranges from 183-186°C. The Tg's for the MSS terpolymers range from 141-155°C, somewhat higher than that observed for the copolymers, 127-129°C. The incorporation of SO$_2$ into the MSS/TBS increased the Tg and slightly lowered the onset temperature for removal of t-BOC group.

The mechanism of MSS/TBS/SO$_2$ and MSS/TBS resists during lithographic processing, delineated by the NMR data, indicate that the Ms group in the resist is not deprotected during acidolytic-thermal processing, thereby minimizing volume loss (15%). The NMR data also indicate that the Ms group in the terpolymer and copolymer undergo minimal base catalyzed hydrolytic cleavage with an aqueous TMAH developer as shown in Figures 1 and 2.

Table 2 shows that exposed MSS/TBS copolymer resists were not developable in 0.26 N TMAH without the addition of 20-30% v/v IPA. Unlike the MSS/TBS copolymer, the MSS/TBS/SO$_2$ (1.00/0.67/0.64) terpolymer resist was developable without the IPA additive (clearing dose: 32mJ/cm^2, PEB: 20 s at 120°C) and did not show fading of the latent images(*3*) or loss of solubility in aqueous base when increasing PEB time to 3 min. This is attributed to the incorporation of SO$_2$ units which improved both the hydrophilicity and Tg of the polymer. The addition of 10% IPA improved the clearing dose of the MSS/TBS/SO$_2$ resist from 32 to 22 mJ/cm^2 (Table 3), and by increasing the PEB to 1 minute improved the clearing dose to 16 mJ/cm^2.

The SEM cross section micrograph in Figure 3 indicates that the (1.00/0.67/0.65) MSS/TBS/SO$_2$ terpolymer resist is capable of producing 0.4 μm lines and spaces. The sloped profile of the resist processed with normal development was unexpected because of its exceptionally low absorbance (0.14AU/μm). A probable explanation is that the pendant Ms group induces dissolution inhibition. The dissolution inhibition was overcome by the addition of IPA to the developer as shown by the vertical wall profiles in Figure 3B. Although IPA enhanced the effectiveness of the developer, its use in manufacturing processes is environmentally undesirable.

To further improve developability and volume loss of these resists, we investigated the addition of the base soluble OHS units into both MSS/TBS and MSS/TBS/SO$_2$ polymer framework. The MSS/TBS/OHS polymers has low optical

Table 3

LithographicSensitivity of Resists Formulated 4-Methanesulfonyloxystyrene Polymers. Terpolymers and Quaternary Polymers with , 4-t-Butoxycarbonyloxystyrene 4-Hydroxystyrene ˈand Sulfur dioxide[a].

Polymer Composition MSS/TBS/OHS/SO$_2$	M$_w$	PEB Temperature °C	PEB Time s	Volume loss %	TMAH normality, % IPA (develop time)	Clearing Dose (mJ/cm^2) (Contrast)
0.67/1.00/0/0.64	126 K	120	20	16%	0.26 N, 0% (45 s)	32 (4)
0.67/1.00/0/0.64	126 K	120	20	16%	0.26 N, 10% (45 s)	22 (>10)
0.67/1.00/0/0.64	126 K	120	60	16%	0.26N, 0% (45 s)	16 (>10)
0.67/1.00/0/0.64	126 K	120	90	16%	0.26 N, 0% (45 s)	14 (>10)
0.67/1.00/0/0.64	126 K	120	60	16%	0.26 N, 20% (10 s)	12(>10)
1.00/0.97/0/0.66	143 K	120	60	15%	0.25 N, 0% (600 s)	300 (1)
1.00/0.97/0/0.66	143 K	120	60	15%	0.26 N, 20% (30 s)	19 (>10)

1.00/0.97/0/0.50	42 K	120	60	16%	0.26 N, 20% (5 s)	12 (>10)
1.00/0.86/0.62/0.80	110 K	120	60	10%	0.26 N (120 s)	22 (6)
1.00/0.86/0.62/0.80	110 K	120	120	10%	0.26 N (120 s)	20(5)
1.00/0.86/0.62/0.80	110 K	120	180	10%	0.26 N (120 s)	20(3)
1.00/0.49/0.99/0.85	143 K	120	60	5%	0.26 N (120 s)	60 (0.8)
1.00/0.49/0.99/0.85	143 K	120	60	5%	0.26 N(180s)	16(3)
1.00/0.98/0.85/0.87	75 K	120	60	10%	0.26 N(120s)	14 (>10)

a) Resists Formulated with 6 mole % PAG as described in the experimental section.

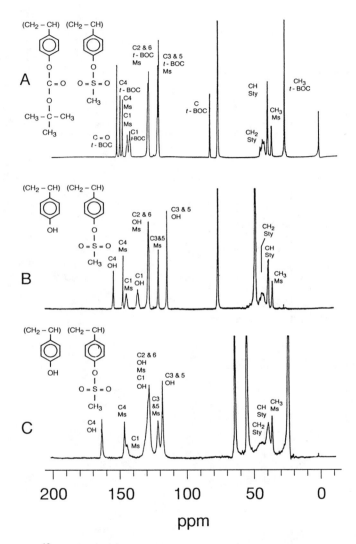

Figure 1. A. ^{13}C NMR spectrum of MSS/TBS (1.0/0.8) copolymer dissolved in CDCl$_3$ (77.0 ppm). **B.** ^{13}C NMR spectrum of 1.0/0.8 MSS/OHS copolymer dissolved in CDCl$_3$ with added methanol (49.0 ppm). **C.** ^{13}C NMR spectrum of 1.0/0.8 MSS/OHS copolymer dissolved in a solution consisting of 0.46 N TMAH solution containing 20% IPA and left overnight. (TMAH, 54.5 ppm; IPA, 24.2 and 62.9 ppm).

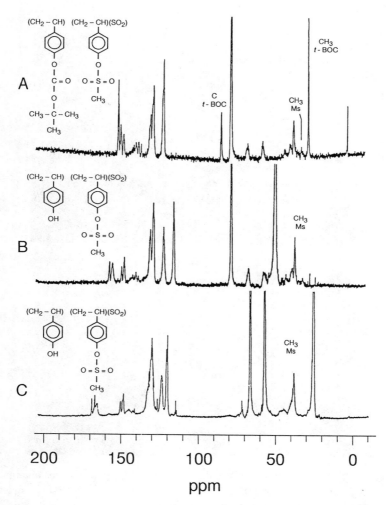

Figure 2. A. ^{13}C NMR spectrum of MSS/TBS/SO$_2$ (1.00/0.66/64) terpolymer dissolved in CDCl$_3$. **B.** ^{13}C NMR spectrum of MSS/OHS/SO$_2$ (/0.66/1.00/0.64) terpolymer dissolved in CDCl$_3$ with added methanol. **C.** ^{13}C NMR spectrum of MSS/OHS/SO$_2$ (0.66/1.00/0.64) terpolymer dissolved in a solution consisting of 0.46 N TMAH solution containing 20% IPA and left overnight.

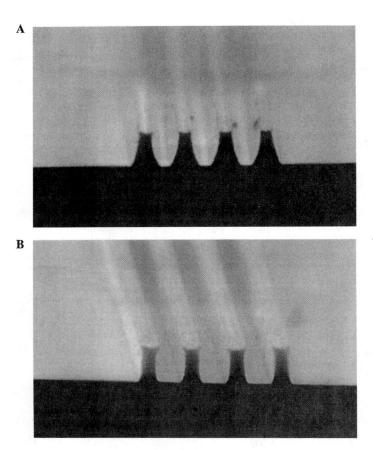

Figure 3. SEM cross section micrographs of 0.40 μm lines and spaces obtained with resists formulated from 0.67/1.00/0.53 MSS/TBS/SO$_2$ terpolymer and 6 mole % PAG relative to the polymer. The PEB of 60 s at 115°C. **A.** The D$_p$ was 34 mJ/cm^2 and development in 0.26 N TMAH. **B.** The D$_p$ was 20 mJ/cm^2 and development in 0.26 N TMAH with added IPA (10%v/v.).

density at 248 nm (0.14-0.20 AU/μm). The optical density was not appreciably affected by increasing the MSS and OHS content. The TGA onset for MSS/TBS/OHS polymers (153-156°C) is somewhat lower than for the MSS/TBS materials (Table 2). Unexposed MSS/TBS/OHS resists do not show volume loss nor decrease in C=O concentration for t-BOC after 3 min PEB at 120°C. These new resists (Table 2) exhibit very low clearing dose (D_p's) (2-6 mJ/cm^2) and low volume losses (7-9%). Resists formulated with these materials required development with IPA doped TMAH and have lower contrasts (Table 2).

The MSS/TBS/OHS/SO$_2$ polymer resists (Table 3) were developable in aqueous TMAH and resulted in low volume loss (5-10%). The TGA onset for the deprotection of t-BOC of the MSS/TBS/OHS/SO$_2$ polymers was lower (148-158°C) than that of the materials without OHS units (Table 3). Similar to MSS/TBS/OHS resists, MSS/TBS/OHS/SO$_2$ resists do not undergo volume loss nor deprotection of t-BOC (148°C) during PEB at 120°C. The exception is the resist containing the polymer having the highest OHS/TBS (~2/1) ratio which showed volume loss (~6%) and decrease in the intensity (~14%) of the t-BOC C=O stretching band (1759 cm^{-1}) after PEB of 3 min. The lower thermolysis temperature of t-BOC is due to an increase number of vicinal TBS and OHS moieties(9). The unexposed areas of the resist formulated with OHS/TBS ratio (~2/1) polymer did not show significant degradation after PEB of 1 min at 120°C. This material also had the highest (MSS+OHS)/TBS ratio and predictably gave the lowest volume loss (5%) of all the MSS/TBS/OHS/SO$_2$ formulations (Table 3). However, because of the high content of the dissolution inhibiting MSS unit, the resist took the longest to develop (3 min). The tetrapolymer resists with lower MSS+OHS/TBS ratio showed higher volume loss (9-10%), but developed more quickly (2 min). The absorbance at 248 nm of the MSS/TBS/OHS/SO$_2$ polymer matrix is higher than the MSS/TBS/OHS system and ranged from 0.31-0.48 AU/μm increasing with the OHS/SO$_2$ ratio. One possible explanation is an increase in quinone formation due to increasing sulfone-benzylic moieties. Increasing donor-acceptor interactions between sulfone and OHS causing absorbance band shifts, is another possibility. However, despite this high absorbance, resists formulated with these materials still show good imaging properties with aqueous TMAH development (Figure 4).

We also investigated an approach to increase solubility of MSS/TBS/SO$_2$ in aqueous TMAH by using a PAG which does not impart dissolution inhibition in the exposed areas. Figure 5 shows a conventional strong acid PAG (PAG I). Figure 6 illustrates the dramatic difference in D_p with aqueous TMAH and IPA doped TMAH development of the high MSS content MSS/TBS/SO$_2$ polymer resist formulated with PAG I. This resist shows very strong dissolution inhibition without the use of IPA, manifested by a ~290 mJ/cm^2 difference in the D_p's as seen in Table 4. In contrast, the resist formulated with a DISCA PAG (Figure 5) gives a low D_p (~30 mJ/cm^2) with TMAH developer (Figure 7). Only a small decrease in D_p (<5 mJ/cm^2) is seen when IPA is added. To ensure that this effect was due to the acidolytic cleavage of the t-butyl group of the DISCA PAG (Scheme 1), PAG III, 4-methoxy-2,6-dinitrobenzyl, was employed for comparison (Figure 5). Figure 7 shows that resists formulated with PAG III have both poor D_p's and contrasts unless IPA is added to the

A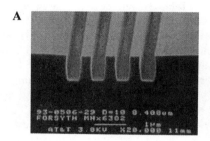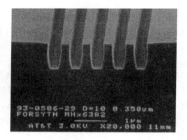

B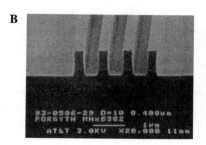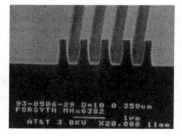

Figure 4. SEM cross-section micrographs of 0.40 μm and 0.35 μm lines and spaces obtained with a resist formulated from 1.00/0.98/0.85/1.70 MSS/TBS/OHS/SO$_2$ tertrapolymer and 6 mole % PAG relative to the polymer. The D$_p$ was 14mJ/cm^2 for a 60 s at 120°C and 2 min development in 0.26 N TMAH. **A.** Dark Field SEM cross section micrographs of 0.40 μm and 0.35 μm lines and spaces. **B.** Bright Field SEM cross section micrographs of 0.40 μm and 0.35 μm lines and spaces.

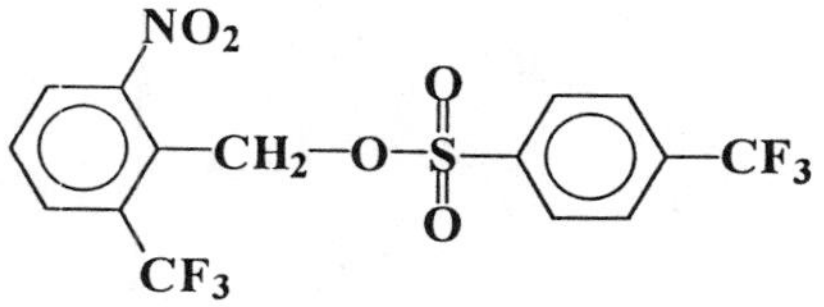

EXAMPLE OF STRONG ACID PAG

PAG I

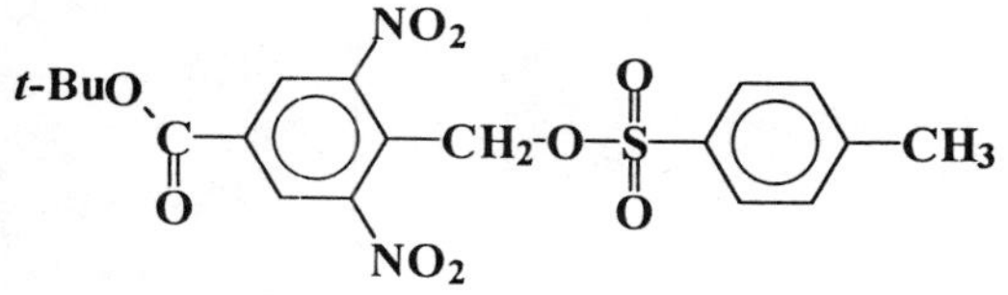

DISCA/PAG

PAG II

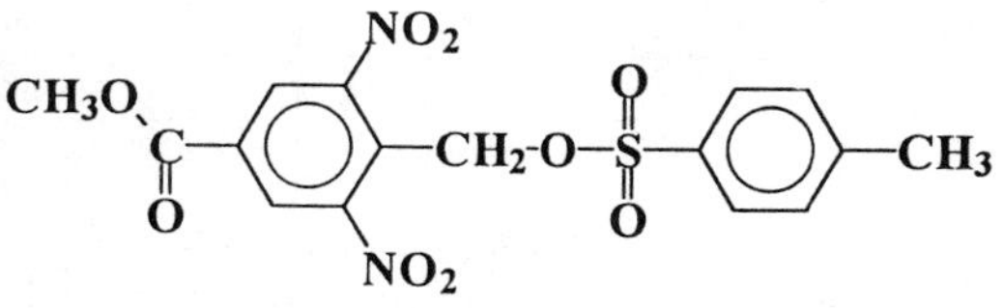

MODEL PAG

PAG III

Figure 5. Chemical structures: PAG I, 2-trifluoromethyl-2,6-dinitrobenzyl 4-trifluoromethylbenzenesulfonate, DISCA PAG II, 4-t-butoxycarbonyl-2,6-dinitrobenzyl tosylate; and the model PAG III, 4-methoxycarbonyl-2,6-dinitrobenzyl tosylates.

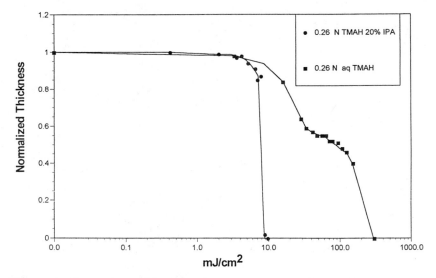

Figure 6. Comparison of the exposure response curves for a resist formulated with $0.84/1.00/0.47$ MSS/TBS/SO$_2$ terpolymer and 6 mole % of strong acid PAG I under different development conditions. The curves were obtained using a 1 min PEB at 120°C using either a 10 second development in 0.26 N TMAH with 20% v/v IPA or a 5 min development in 0.26 N TMAH.

Table 4

Lithographic Comparison of Resists Formulated with an MSS/TBS/SO$_2$ Terpolymer[a] and Either a Strong Acid Pag PAG I, the DISCA/ PAG PAG II Tosylate or the Model PAG III

Photo-acid Generator[b]	Development[c] solvent TMAH/H$_2$O/IPA	Development time	Clearing Dose[d] mJ/cm^2	Contrast γ
PAG I	0.26 N	600s	300	1.2
PAG I	0.26 N 20% IPA	10 s	9	>10
DISCA/PAG PAG II	0.26 N	1 min	34	>10
DISCA/PAG PAG II	0.26 N 20% IPA	10 s	30	>10
PAG III	0.26 N	3 min	120	1.3
PAG III	0.26 N 20% IPA	10 s	30	>10

a) Resist formulated with MSS/TBS/SO$_2$ (1.00/0.97/0.66) 143 K and 6 mole % PAG.

b) For the structure assignment of these materials see Figure 5.

c) For information regarding composition and processing of the resists solutions please refer to the   experimental section.

d) Clearing dose for large features.

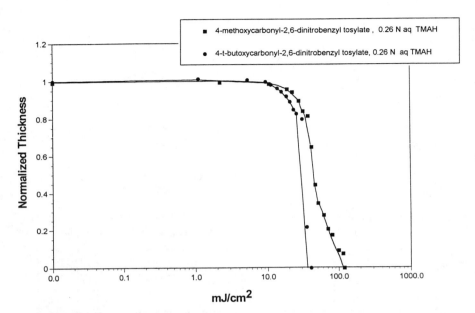

Figure 7. Comparison of the exposure-response curve for resists formulated with 0.84/1.00/0.47 MSS/TBS/SO$_2$ terpolymer and 6 mole % of DISCA PAG II, or the model compound, 4-methoxy-2,6-dinitrobenzyl tosylate, PAG III. The curves were obtained by using a PEB of 120°C for a min followed by either development with 0.26 N TMAH (3 min for III, 1 min for II) or 0.26 N TMAH with 20% v/v IPA (10 s for II and III).

developer (Table 4). PAG I appears to significantly slow the dissolution rate of exposed regions of the high MSS content resist in aqueous TMAH, resulting in a high D$_p$ and low contrast material. The DISCA PAG yields a more sensitive resist despite the fact that it releases a weaker acid. Since PAG I contains CF$_3$ which has a large π hydrophobicity parameter(*10*) (π = 0.88) it is understandable that it would act as a dissolution inhibitor. In contrast (Scheme 1), the DISCA PAG and its photoproducts (π COOH: -0.32) after exposure and PEB, will be much more hydrophilic. Since more than 90% of the PAG is not decomposed during normal exposure, the dissolution rates, in this high MSS loaded system, are largely a function of the hydrophobicity of the PAG.

Conclusions

The incorporation of MSS into *t*-BOC based polymers for chemically amplified deep UV resists resulted in MSS/TBS/SO$_2$ (0.67/1.00/0.64) terpolymer which provided a resist with low optical density (0.14 AU/μm) at 248 nm) and improved volume loss (~15%) after PEB. The resist has good tolerance to harsh PEB conditions and is readily developed in aqueous TMAH. The aqueous base solubility and imaging characteristics of MSS based resists was improved by incorporating OHS units on the polymer or by using a DISCA PAG. Such approaches have allowed us to formulate resists with high resolution (0.35 μm lines and space), low volume loss (5%), good D$_p$ (14 mJ/cm^2) while maintaining good dissolution in aqueous base.

Acknowledgments

The authors would like to thank Elsa Reichmanis and Kathryn E. Uhrich for helpful discussions and Wai W. Tai for the molecular weight analyses.

REFERENCES

1.   Kometani, J. M.; Galvin, M. E.; Heffner, S. A.; Houlihan, F. M.; Nalamasu, O.; Chin, E.; Reichmanis, E., *Macromolecules*.**1993,** *2*, 2165.
2.   Greene, T. W. *Protective Groups in Organic Chemistry*, Wiley-Interscience, New York, NY, **1981,** 307.
3.   Kometani, J. M., Houlihan, F. M., Heffner, S. A., Chin, E., Nalamasu, O., Tai, W. W., Reichmanis, E., *Proc. 10th International Conf. Photopolymers*, **1994,** 146-156.
4.   Houlihan, F. M., Chin, E., Nalamasu, O., Kometani, J. M., *Polymers for Microelectronics* , ACS Symposium Series **537,** ACS, Washington, D. C., **1992,** 25-39.
5.   Houlihan, F. M.; Neenan, T. X.; Reichmanis, E.; Kometani, J. M.; Chin, E. *Chem. Mater,.***1991,** *3*, 462.
6.   Houlihan, F. M.; Neenan, T. X.; Reichmanis, E.,   U.S. Patent 5 5,200,544, **1993**.
7.   Uhrich, K. E.; Reichmanis, E.; Heffner, S. A.: Kometani, J. M.; Nalamasu, O.; Baiocchi, F. *Chem. Mater.*, **1994,** *6*, No. 3, 287-294.
8.   Braker,W., Mossman, A. L., *Matheson Gas Data Book*, 6th Edition, Matheson Product, Lyndhurst, NJ, **1980,** 641-648.
9.   Ito, H., *J. Polym. Sci., Polym. Chem. Ed.*, **1986,** 24, 2971.
10.   Hansch, C; Leo, A. *Substituent Constants for Correlation Analysis in Chemistry and Biology*, Wiley-Interscience, New York, NY, **1979,** 13-17, 69.

RECEIVED July 17, 1995

Chapter 15

Dienone–Phenol Rearrangement Reaction: Design Pathway for Chemically Amplified Photoresists

Ying Jiang[1], John Maher, and David Bassett

Technical Center, Union Carbide Chemical & Plastics, Co., Inc., 1100–1200 Kanahwa Turnpike, South Charleston, WV 25303

A new design of chemically amplified Deep-UV photoresists based on the dienone-phenol rearrangement reaction is described. Dienone compounds in which a cyclohexadienone has two substitute groups in the 4-position undergo 1,2 migration of one of these groups with acid treatment. Since the acid catalyzed rearrangement reaction generates phenol derivatives, a chemically amplified deep UV photoresist system can be designed based on this chemistry by using photo-acid generators and polymers bearing dienone pendant groups. Improved sensitivity of the system can be achieved as the result of aromatization as its driving force of the rearrangement reaction. As the chemical transformation is not a deprotection process, no volatile compounds are generated and the film loss which is often observed in the chemically amplified photoresist systems is greatly limited. The resists were formulated with a photo-acid generator such as triphenylsulfonium hexafluoroantimonate and a polymer such as poly(methyl cyclohexadienone methacrylate). Positive tone image could be resolved by exposing the resist film in deep UV, post-baking and developing in a tetramethylammonium hydroxide solution.

Chemical amplification based on photo-acid catalyzed deprotection of polymers has been successfully employed in the design of highly sensitive resist systems for use in short wavelength lithographic technology (*1-5*). Most of these polymers are based on carbonate- or ether-protected poly(vinylphenols) or ester-protected polymeric acids (*6-8*). Recently, we reported a new chemically amplified resist system based on phenoxyethyl protection of poly(vinylphenol) (*9*). The chemistry design of this system provides a highly sensitive system and very limited film loss because one of its deprotection products is non-volatile phenol. The polarity change from a nonpolar to a polar state as the result of the deprotection described in the above systems provides positive images with aqueous base development. Recently, several chemically amplified resist systems with reverse polarity change based on acid-catalyzed dehydration have also been reported (*10*).

We have been interested in utilizing acid-catalyzed rearrangement reactions to design chemically amplified resist materials using photoacid generators. The acid-

---

[1]Current address: U.S. Surgical Corporation, 195 McDermott Road, North Haven, CT 06479

catalyzed rearrangement reactions are attractive as a versatile design pathway for chemically amplified photoresists because they do not produce any volatile chemicals as the result of the chemical transformation which is usually seen in most chemically amplified photoresist systems and the property change is built in the polymers and achieved without changing chemical composition of the polymers. Use of acid-catalyzed rearrangement reactions to design resist systems has been applied by several researchers in the field. Ito and co-workers described a polymer system based on Pinacol rearrangement reaction, in which dehydration changed the solubility behavior of the polymer (*10*). Frechet and Willson reported a resist system based on Claisen rearrangement (*11-13*). In this paper we report a new design of a chemically amplified deep UV photoresist system based on the acid-catalyzed dienone-phenol rearrangement reaction. Dienone compounds, cyclohexadienone derivatives bearing two substitute groups in the 4-position, undergo 1,2 migration of one of theses groups with an acid-catalyst treatment (*14-17*). Since the acid-catalyzed rearrangement products are phenol derivatives, the dienone-phenol rearrangement reaction can be used to design chemically amplified and aqueous-base developable deep UV photoresists using photo-acid generators. While the focus of discussion of this report is placed on the design of the system and its chemistry, efforts towards using this material to formulate deep UV photoresists are a subject for future investigation. The syntheses of the monomer and polymers and their characterization are also reported.

EXPERIMENTAL

Materials and Instrumentation

Methyllithium solutions in ether, pyridine, and N,N'-dimethyl aminopyridine were purchased from Aldrich and used without purification. 1,4-Benzoquinone was crystallized from petroleum ether and then from benzene. Methacryloyl chloride was freshly distilled before use.

NMR spectra were recorded in deuterated chloroform. FT-IR spectra were recorded on a BioRad FTS-40 Fourier-Transform Infrared Spectroscometer. Films were spun over silicon wafers using a manual spinner (Solitec, Inc., Model 5100). Irradiation was done using a Xenon deep UV lamp (ORIEL, Model 66142) which was powered with an arc power supply (ORIEL Model 8530).

Preparation of 4-hydroxy-4-methyl cyclohexadien-1-one

A 3000 mL three-necked flask equipped with a mechanical stirrer was dried carefully with heating and then purged with dry nitrogen. After the flask was cooled to room temperature, it was placed in a dry-ice-acetone cooling bath at -78°C. A solution of 46.5 g of 1,4-benzoquinone in 2500 mL dry ether was added into the flask and the solution was stirred under dry nitrogen. When the temperature of the solution reached -70°C, some solid fell out the solution. Methyllithium solution (400mL, 1.4 M in ether) was added slowly over a period of three hours and then the reaction was kept stirring in the cooling bath for additional 2 hours. The reaction was stopped by adding an aqueous solution of 39 g of NH_4Cl in 200 mL of deionized water. A temperature increase was observed during this addition. The phases were separated and the ether phase was collected. The aqueous phase was extracted with ethyl acetate three times and the combined organic phases were dried with magnesium sulfate, filtered. The filtrate was evaporated to dryness under vacume using a rotoevaporator. After purification by silica gel column chromatography using a mixture of hexane/ethyl acetate (80/20), white crystals were obtained which, characterized by both FT-IR and NMR, was the desired product. The product had a melting point of 77°C.

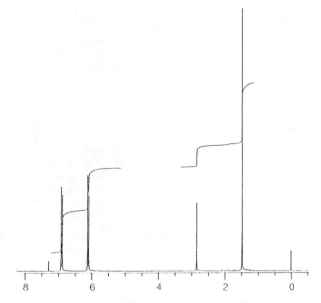

Figure 1.   **Proton NMR spectrum of 4-hydroxy-4-methyl cyclohexadien-1-one**

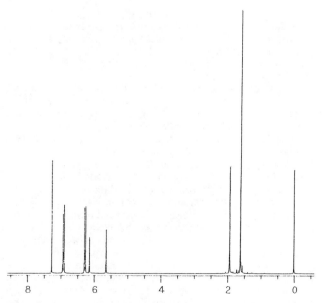

Figure 2.   **Proton NMR spectrum of 4-(4-methyl cyclohexadien -1-one) methacrylate**

Preparation of 4-(4-methyl cyclohexadien-1-one) methacrylate

A dried 100 mL flask was charged with 3.16 g of 4-hydroxy-4-methyl cyclohexadien-1-one prepared as above and 5 g of freshly distilled methacryloyl chloride in 20 mL of dry tetrahydrofuran (THF) at room temperature. To the solution were then added 3.5 g of pyridine and 0.34 g of N,N'-dimethylaminopyridine (DMAP). The reaction was stirred at room temperature for 16 hours. TLC analysis indicated that the reaction was not complete and therefore, the reaction mixture was gently refluxed for additional 4 hours. After the mixture was cooled to room temperature, the insoluble white solid was filtered off and the filtrate was collected. The solid was washed three times with THF and then the THF washes were combined with the filtrate. The combined THF solutions were washed with an aqueous NH$_4$Cl solution and water, separated and dried over magnesium sulfate and finally the solvent was removed under reduced pressure. The crude product was further purified by silica gel chromatography. The product was obtained in 58% yield (m.p.= 71-72°C).

Preparation of Poly[4-(4-methyl cyclohexadien-1-one) methacrylate]

In a typical experiment, 2 g of the monomer, 4-(4-methyl cyclohexadien-1-one) methacrylate prepared as above, was dissolved in 2 mL of benzene in a three-necked and dry flask equipped with a condenser under nitrogen. After 10 mg of 2,2'-azobis(isobutyronitrile) (AIBN) was added, the solution was degassed with nitrogen for 30 minutes. The flask was then placed in an oil bath at 85°C with stirring for 24 hours. The solution, which became very viscous because of the polymer formation and partial evaporation of the solvent, was precipitated in methanol and the white solid was filtered and dried in vacuum. The solid was proven to be the desired polymer after characterization by FT-IR and NMR.

RESULTS AND DISCUSSION

Synthesis

The synthetic pathway to 4-(4-methyl cyclohexadien-1-one) methacrylate and its free radical polymerization is illustrated in the **scheme 1**.

Benzoquinone was first reacted with one mole of methyllithium under carefully controlled condition and only the mono-addition product was observed by TLC. The lithium salt intermediate was neutralized to form the hydroxy-cyclohaxadienone product which, after purification by silica gel chromatography, was a stable crystalline solid. The hydroxy-methyl cyclohexadienone was then reacted with methacryloyl chloride to attach the cyclohexadienone form to the polymerizable methacrylate functionality. It was found that using a mixture of pyridine and DMAP as the acid scavenger gave the most satisfactory results. A reaction to use the lithium salt intermediate directly reacting with methacryloyl chloride was attempted. A low reaction yield was obtained and the reaction mixture was difficult to purify.

Figure 1 is the proton NMR spectrum of 4-hydroxy-4-methyl cyclohexadien-1-one. The methyl group is a singlet located at about 1.5 ppm and another singlet at 2.8 ppm can be assigned to the hydroxy group at the 4-position. The cyclohexadienone protons are doublets located in the region of from 6.2 to 6.8 ppm. The proton NMR spectrum of the monomer, 4-(4-methyl cyclohexadien-1-one) methacrylate, is shown in **Figure 2**. A singlet located at 1.61 ppm is the methyl group of the cyclohexadienone and the other methyl group, a singlet located at 1.93 ppm, can be assigned to the ester methyl group. The protons of the cyclohexadienone, as doublets, are located at 6.24 to 6.28 and 6.89 to 6.93 ppm respectively. Two protons of the methacryl double bond are seen at 5.63 and 6.13

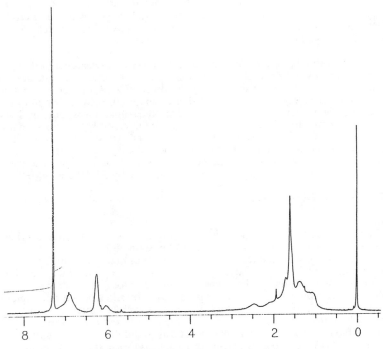

Figure 3.   **Proton NMR spectrum of poly[4-(4-methyl
cyclohexadien-1-one) methacrylate]**

ppm. FT-IR spectrum of the monomer shows a vinyl absorption which disappears after polymerization.

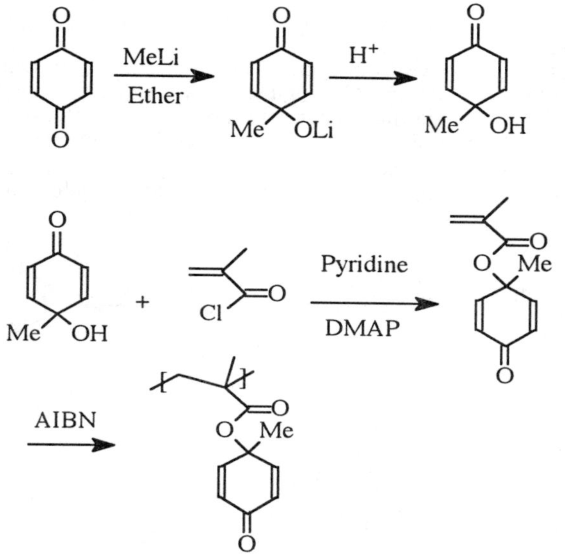

Scheme 1. Synthesis of 4(4-methyl cyclohexadien-1-one) methacrylate and its polymerization

The monomer, 4-(4-methyl cyclohexadien-1-one) methacrylate, was easily polymerized under free radical conditions using AIBN as the initiator to afford the desired polymers. **Table 1** summarizes some results of the free radical polymerization. As shown in the table, aromatic solvents such as toluene or benzene produced satisfactory results. High molecular weight polymers were obtained using these aromatic solvents with yields over 90%. Not surprisingly, chlorohydrocarbons such as methylene chloride generated lower molecular weight polymers due to the fact that these solvents are free radical chain transfer agents.

Table 1. Free Radical Polymerization of 4-(4-methyl cyclohexadien-1-one) methacrylate[a]

AIBN (mg)	Solvent	Yield%	MW[b]	Mw/Mn
20	Benzene	93.2	112,000	2.39
10	Benzene	91.5	152,000	2.45
10	CH$_2$Cl$_2$	80.9	67,500	3.10
10	Toluene	95.2	125,000	2.42

[a] 2 G of the monomer was used in 2 mL of the solvents. All reactions were carried out under nitrogen at 85°C for 24 hours. [b] GPC data with polystyrene standard.

Figure 3 is the proton NMR spectrum of poly[4-(4-methyl cyclohexadienone) methacrylate]. The protons of the cyclohexadienone are located in the region of 6.25 to 6.70 ppm. While the methyl groups' chemical shifts in the polymers do not change much, the protons of the double bond from methacrylate disappear after polymerization. New broader peaks located in the region of 1.0 to 2.5 ppm can be assigned to the polymer backbone' protons.

Chemistry of the Dienone-phenol Rearrangement and Its Application to Design of Resist Materials

The formation of phenols from cyclohexadienones in the presence of an acid-catalyst is a well known reaction. Although some of the mechanistic details remain unclear, it is believed that the first step involves protonation of the carbonyl oxygen to form a cyclohexadienyl cation. The second step is the migration of a group to a neighboring carbon and the subsequent repulsion of a proton produces stable phenol products. This general rearrangement sequence is illustrated below in **Scheme 2**.

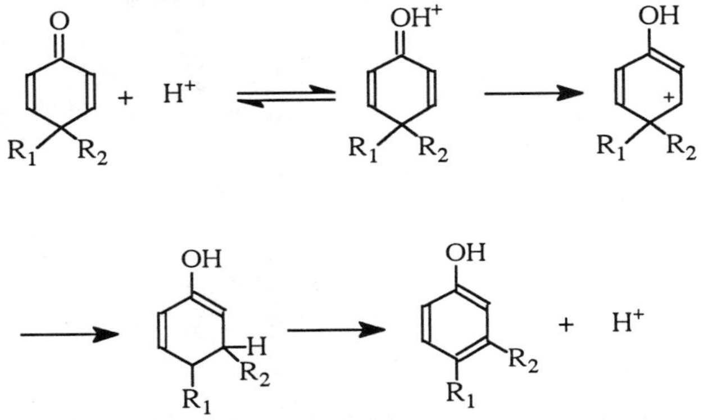

Scheme 2. Acid-catalyzed Dalkyldienone-phenol Rearrangement

In our polymer system, however, an ester substitute group in the 4-position should be a better model for the rearrangement mechanism because the linkage of the cyclohexadienone moiety is attached to the polymer backbone by an ester linkage. Although the rearrangement reaction mechanism for 4-ester substituted cyclohexadienone is less understood, some studies (18-21) suggested that the acetoxy group migrates more readily than alkyl groups because of the formation of the bridged acetoxy-cation which was assumed to be the intermediate with a very low energy pathway. **Scheme 3** is an illustration of this rearrangement reaction using 4- alkyl-4-acetoxy cyclohexadienone as the model compound. In this reaction, a ring position adjacent to the acetoxy group is unoccupied and the acetoxy group can easily migrate to it.

Since the acid-catalyzed rearrangement products are phenol derivatives, the reaction is useful to design chemically amplified and aqueous base developable resist materials using photo-acid generators. Therefore, our approach is to attach the cyclohexadienone to a polymer backbone, such as a methacrylate polymer as illustrated in the scheme 1. If formulated with a photo-acid generator, the photochemically generated acid catalyzes the rearrangement of the organic soluble polymer to an aqueous base soluble phenolic polymer. In this way, dual-tone images can be developed depending on developing conditions.

The photoresist system based on the acid-catalyzed dienone-phenol rearrangement reaction shows several advantages as the result of its novel design of chemistry. Because the driving force in the overall rearrangement reaction is creation of an aromatic system, the reaction is extremely sensitive to acid catalysts. Therefore, the resist systems based on this type of rearrangement should demonstrate high sensitivities. Another advantage of the system is that it does not have any mass loss as the result of the chemical transformation which is commonly seen in most chemically amplified resist systems based on the deprotection chemistry and therefore, the film loss in this system is greatly limited.

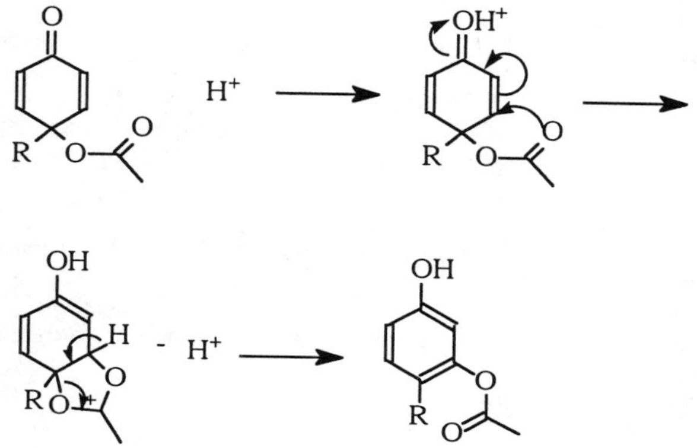

Scheme 3. Acid-catalyzed 4-Alkyl-4-acetoxycyclohexadienone Rearrangement Reaction

Imaging Experiments

The preliminary imaging tests were carried out using triphenylsulfonium hexafluoroantimonate as a photo-acid generator. The resist was formulated in diglyme with 5 wt% of the photo-acid generator and a small amount of a surfactant. The film was prepared by spin-coating on a silicon wafer and then exposed to UV radiation at 254 nm through a mask in a contact printing process. A positive image was resolved after development in a commercial aqueous base developer, KTI 934. Although the system demonstrated a high sensitivity as expected, the quality of the image was not completely satisfactory. It was concluded that further study of formulation and processing was needed to produce high quality images.

CONCLUSION

The acid-catalyzed dienone-phenol rearrangement is a promising tool for designing chemically amplified resist materials. This can be accomplished by anchoring the cyclohexadienone onto a polymer backbone and formulating the polymer with photo-acid generators. A variety of polymers containing a pendant cyclohexadienone group can be easily synthesized. The novel design of the chemistry provides high sensitivities of the resist systems as the result of aromatization and very limited film loss since the chemical transformation, a simple rearrangement reaction, does not induce any mass loss.

ACKNOWLEDGMENTS

The authors would like to thank the members of the Photoresist R&D group of Union Carbide Chemicals and Plastics Company, Inc..

REFERENCES

1. Ito, H.; Willson, C. G.; Frechet, J. M. J.; *Digest of Technical Papers of 1982 Symposium on VLSI Technology*, **1982**, 86.
2. Frechet, J. M. J.; Ito, H.; Willson, C. G.; *Proc. Microcircuit Engineering*, **1982**, 260.
3. Frechet, J. M. J.; Eichler, E.; Willson, C. G.; Ito, H.; *Polymer,* **1983**, 24, 995.
4. Ito, H.; Willson, C. G.; Frechet, J. M. J.; U.S. Patent No. 4,491,628, **1985.**
5. Houlihan, F. M.; Reichmanis, E.; Thompson, L. F.; Eur. Patent Appl. No. 89301556.0, **1989.**
6. Ito, H.; Willson, C. G.; Frechet, J. M. J.; *Proc. SPIE*, **1987**, 771, 24.
7. Colon, D. A.; Crivello, J. V.; Lee, J. L.; O'Brien, M. J.; *Macromolecules*, **1989**, 22, 509.
8. Ito, H.; Ueda, M.; Ebina, M.; in " *Polymers in Microlithograph*", Reichmanis, E.; MacDonald, S. A.; Iwayanagi, T.; Eds.; ACS Symposium Series No. 412, ACS, Washington, D.C., 57.
9. Jiang, Y.; Bassett, D. R.; in " *Polymers for Microelectronics*", Thompson, L. F.; Willson, C. G.; Tagawa, S.; Eds.; ACS Symposium Series No. 537, ACS San Francisco, CA, **1992**, 40.
10. Ito, H.; Maekawa, Y.; Sooriyakumaran, R.; Mash, E. A.; in " *Polymers for Microelectronics*", Thompson, L. F.; Willson, C. G.; Tagawa, S.; Eds.; ACS Symposium Series No. 537, ACS San Francisco, CA, **1992**, 64.
11. Frechet, J. M. J.; 2nd SPSJ International Conference, Tokyo, **1986**, #2CIL4.
12. Frechet, J. M. J.; Kallman, N.; Kryczka, B.; Eichler, E.; Houlihon, F. M.; Willson, C. G.; Polymer Bulletin, **1988**, 20, 4.
13. Frechet, J. M. J.; Eichler, E.; Gauthier, S.; Kryczka, B.; Willson, C. G.; *"Radiation Chemistry of High Technology Polymers*", Reichmanis, E.; O'Donnell, J.; Eds.; ACS Symposium Series No. 381, **1989**, 155.
14 Miller, B.; *Accounts of Chem. Research*, **1975**, 8, 245.
15. Vitullo, V. P.; *J . Organic Chemistry*, **1969**, 34, 224.
16. Vitullo, V. P.; *J . Organic Chemistry*, **1970**, 11, 224.
17. Swenton, J. S.; *Accounts of Chemical Research*, **1983**, 16, 74.
18. Metlesics, W.; Wessely, F.; Budzikiewicz, H.; *Tetrahedron*, **1959**, 6, 345.
19. Budzikiewicz, H.; Metlesics, W.; *J. Org. Chem.,* **1959**, 24, 1125.
20. Takacs, F.; *Monatsh*, **1964**, 95, 961.
21. Wessely, F.; Zbiral, E.; Sturm, H.; *Ber.,* **1960**, 93, 2840

RECEIVED July 7, 1995

NOVEL CHEMISTRIES AND APPROACHES FOR SUB-0.25-μm IMAGING

Elsa Reichmanis

Progress in VLSI device design and manufacture continues to demand increasingly smaller and more precise device features. Today, 248 nm (KrF) photolithography is beginning to be introduced into manufacture for the production of devices with feature sizes on the order of 0.25 µm. As this technology matures, device feature dimensions will continue to shrink and new lithographic strategies will be required. Concomitant with the development of new lithographic techniques is the development of new resist materials and processes. One attractive approach to extending optical lithography beyond the 0.25 µm generation of devices is the use of ArF excimer laser (193 nm) lithography. The design of a traditional solution-developed single layer resist chemistry for 193 nm, however, represents a significant challenge for the synthetic chemist. As for conventional and 248 nm photolithography, the polymer resins used for resists must i) exhibit solubility in solvents that allow the coating of uniform defect-free films, ii) be sufficiently thermally stable to withstand the temperatures and conditions used with standard device processes, iii) exhibit no flow, and limited erosion during pattern transfer of the resist image into the device substrate, iv) possess a reactive functionality that will facilitate pattern differentiation after irradiation, and v) have absorption characteristics that will permit uniform imaging through the thickness of the resist film. It is precisely this latter criterion in conjunction with the requirement that the resist exhibit only limited erosion during pattern transfer that represents a challenge. Several approaches have been documented, ranging from the use of polycyclic aromatic rings to shift the absorption spectra of the materials to allow imaging at 193 nm, to the incorporation of alicyclic substituents to effect the same result. Incorporation of silicon into the polymer has also been demonstrated to effect low absorption and oxygen reactive ion etching resistance.

0097–6156/95/0614–0237$12.00/0

An alternative to the traditional solution developed chemistries is the use of dry development techniques employing reactive ion etching pattern transfer. The driving force that is behind the development of these schemes is simply the demand for improved resolution which requires imaging features with increasingly higher aspect ratios and smaller linewidth variations over steep substrate topography. A number of schemes have been proposed to address this issue, namely; the use of polymeric planarizing layers, anti-reflection coatings, and contrast enhancement materials.

Notable are the dry-developed resist chemistries, many of which are associated with top-surface imaging mechanisms. Simplification of both tri- and bi-level resist processes can be achieved by incorporating all of the desirable features of these multilevel technologies into a single layer of resist. Incorporation of species other than ions in the first 200-300Å of the resist can be accomplished through gas-phase functionalization processes. For this process to work effectively, it is desirable to have an organic polymer that both contains a reactive functionality and is sufficiently absorbent to limit deposition of the irradiation dose to the topmost part of the resist film. The high absorption limits radiation induced modification of the polymer to the topmost region of the film. Treatment of the exposed film to vapors of a silylating agent such as HMDS provides for the selective reaction with the silicon rich reagent. The silylating agent is thus covalently bound throughout the depth of the exposed film. An oxygen etch atmosphere further develops the pattern to the substrate.

Each of the alternative resist materials and processes described above need to be further explored and evaluated for their applicability to VLSI device manufacturing. The future of microlithography is bright and contains many challenges in the areas of resist research and associated processing. There is no doubt that within the decade, many new materials will be commonplace within the manufacturing environment.

RECEIVED August 21, 1995

Chapter 16

Single-Layer Resist for ArF Excimer Laser Exposure Containing Aromatic Compounds

Tohru Ushirogouchi, Takuya Naito, Koji Asakawa, Naomi Shida, Makoto Nakase, and Tsukasa Tada

Materials and Devices Research Laboratories, Toshiba Research and Development Center, 1, Komukai Toshiba-cho, Saiwai-ku, Kawasaki 210, Japan

Abstract: Aromatic compounds have been considered as indispensable materials for resist, since the aromatic backbone has high thermal stability, high etching resistance to plasmas and high photo–efficiency. The aromatic phenolic moiety also has good solubility characteristics in alkaline developers. However, few papers have reported the application of aromatic compounds as resists for ArF excimer laser exposure, since the conventional aromatic compounds have strong absorption at 193 nm. Using MO calculation, the authors tried to find an effective modification method for obtaining aromatic compounds transparent to the ArF excimer laser. The calculated absorption maximum of series of aromatic compounds were found to be significantly red shifted upon conjugation of the aromatic ring, such as in polycyclic aromatic compounds. This expectation was confirmed with spectral experiments. We tried to prepare a novel resist for the ArF laser, consisting of aromatic compounds, and acceptably fine pattern with 0.17 μm size was obtained with up to 30 wt % of aromatic compounds. The aromatic phenolic moiety of the polymer in the resist was also found to effect the efficiency of photo–acid generation in the polymer film.

Continuing reduction of pattern geometries in order to increase the packing density of semiconductor devices has led to shorter wavelength light exposure systems, since the resolving power of an optical image is proportional to the wavelength of the light source. Since the exposure environment has strong absorption due to absorption by oxygen at wavelengths up to 180 nm light, the ArF excimer laser (193 nm wavelength) exposure system is regarded as the limit of photo–exposure systems suitable for mass production of semiconductor devices. However, in this wavelength range, conventional components used in resist materials, such as photosensitive components and polymers, usually have extremely low transparency. Thus, the development of ArF excimer laser resist materials is a key technology for 193 nm lithography. Some papers have reported novel dry–etch resistant acrylic co–polymers having aliphatic poly–rings in the side chains [1-3]. This idea extended the possibility of single–layer resist processing in 193 nm lithography; however, the alkaline solubility achieved with this kind of the polymer was relatively

0097–6156/95/0614–0239$12.00/0

lower than phenol–based polymer, therefore, organic solvents have been introduced in the developer.

Few papers, however, have reported the application of aromatic compounds as resists for ArF excimer laser exposure. This is because of extremely strong absorption (about 30 /μm) near 193 nm, which is due to the $\pi - \pi^*$ electronic transition of the aromatic ring. The authors believe that aromatic compounds are indispensable materials for resists for achieving high solubility in alkaline developers, high thermal stability, high etching resistance to plasmas, and high photo–efficiency, and by means of molecular orbital (MO) calculations[4], have tried to find an effective modification method for obtaining aromatic compounds transparent to ArF excimer laser irradiation. This paper reports a study of transparency of aromatic species using MO calculations, and aromatic materials (polymers, photo–acid generators or dissolution inhibitors) suitable for ArF laser exposure, and also demonstrates single–layer resist performance of a novel resist using aromatic components. The authors also discuss advantages of aromatic polymers for 193 nm resist by comparison with non–aromatic resists, then consider a modification method enhancing transparency of the aromatic compound to 193 nm wavelength UV.

EXPERIMENTAL SECTION

EMG–160MSC type ArF excimer laser (Lambda Physik) was used for the contact exposure. Exposure energy of the laser measured with a TPM–310 power meter (GENTECH), was 0.25–0.4 mJ/pulse. A prototype ArF excimer laser exposure system, equipped with a 0.55 numerical aperture (NA) projection lens and 0.7 σ illumination system (Nikon), was used for resist micro–patterning. The absorption spectra of resist materials in the vacuum ultraviolet (VUV) region were measured with a VTMS–502 VUV spectrophotometer system (Acton Research Co.). Polyvinyl (acetal) (70% acetal capped, Sekisui Kagaku Co.) was effectively used as a transparent binder polymer, and 1–mmol/g sample (in the case of the polymer, the sample corresponding to 1–mmol of monomer unit) was incorporated in to it in the spectral measurement of VUV. Photo acid generators (PAGs) chemical species (triphenyl sulfonium triflate (TPSTF), NAT–105 [137867–61–9], TAZ–106 [69432–40–2], NAI–105[85342–62–7]) were obtained from Midori Co. Organic alkaline developer (eluted AD–10, Tama Kagaku Kogyo Co.: an aqueous solution of tetramethyl ammonium hydroxide) was used as a developer. An epoxy resin with a naphthalene backbone (EP–1) was synthesized by the reaction of epichlorohydrin with 1,5–dihydroxy naphthalene using NaOH. A co–polymer of naphthyl (methacrylate), methacrylic acid, tert–butyl methacrylate and methyl methacrylate was synthesized according to the previously reported method[9]. All the other chemicals were obtained from Aldrich and Poly Sciences Inc.. Photo–acid generation efficiency of PAG in polymers was measured as follows; 5 wt% of PAG was dissolved in a polymer solution, then spin–coated to a thickness of 1 μm on Si wafer. After exposing the film, 2.5 cm^2 of exposed area is dissolved to a solution of 0.036 mmol/l tetrabromophenolblue (TBPB) indicator in cyclohexanone. Then, the absorption changes at 627 nm (which corresponds to the amount of the acid generated) of the indicator solution was measured with a UV–310PC (Shimazu) UV spectral measurement system. Acid–catalyzed conversion of the poly (tert–butyl methacrylate) to poly (acrylic acid) was measured by IR absorption change at 2940 cm^{-1} (corresponds to C–H vibration for tert–butyl groups) and 1715 cm^{-1} (corresponds to C=O vibration for carboxylic acid) by employing microscopic FTIR measurement system (Nippon Bunko Co.).

RESULTS AND DISCUSSION

Study of Absorbance of Aromatic Compounds at 193 nm light using MO calculation

Configuration Interaction approaches (CI), which consist of modeling excited states as combinations of multiple substitutions of Hartree–Fock ground states, are qualitatively predictive methods for the excitation energy and its oscillation strength of the molecules. The excitation energies (corresponding to the maximum absorption wavelength of the molecules (λ_{max})) are calculated using the CI method in MOPAC[5] (semi–empirical MO calculation program, 8 orbitals near HOMO–LUMO boundary levels with all configurations of electrons taken into consideration). CNDO/S[6] (semi–empirical spectra calculation program using the CI method with singly excited configurations) was used for excitation energy simulation including oscillation–strength evaluation. ZINDO[7] was also used for spectral simulation of the molecules.

The absorption near 193 nm of benzene derivatives is the third absorption band from longer wavelength side of the UV region and is extremely strong compared with the other two forbidden transition absorptions. This absorption is considered to arise from degenerate, allowed 1st $\pi - \pi^*$ electron transitions. If the transition energy can be changed by substituent effects in the benzene ring, the transparency of the aromatic compound might be increased. Therefore, the excitation energies for the third absorption band of model aromatic compounds having different types of substituents were calculated using the CI method in MOPAC and CNDO/S. In each calculation method, the geometry applied to the CI calculation had been optimized with MOPAC using the AM1 Hamiltonian. The AM1 Hamiltonian was effectively used for calculation in MOPAC, since this approximation was found to result the best precision.

Table 1 shows the results of calculated λ_{max}. The tendencies of the shift in λ_{max} were thought to be similar between two calculation methods, but the results obtained with MOPAC were slightly shorter than those for CNDO/S, in spite of the MOPAC method included the doubly–excited states, which usually interact with ground states and are therefore thought to lower the energy of the ground state. This may due to the difference of empirical parameters used in two calculation methods. MOPAC was suitable for calculation of absolute value of λ_{max}, since this program was found to result better precision than CNDO/S.

The introduction of typical functional groups (electron–withdrawing Cl and NO_2, or electron donating OMe) on the benzene ring had only a small effect, resulting in a small decrease of the $\pi - \pi^*$ transition energy (red shift of λ_{max}), whereas substituent groups having double bonds conjugated with the benzene ring indicated significant decrease of the transition energy implying decrease of absorption near 193 nm. The calculated transition energies were found to decrease with increasing length of conjugation of double bonds. Multi–membered aromatic rings, such as naphthalene, are effective in lowering the transition energy, since they have rigid molecular structures to align the double bonds in a plane. Figure 1 shows simulated spectra of typical aromatic backbones calculated by ZINDO. Benzene was simulated as having a very strong absorption at 193 nm, whereas the spectrum simulated for naphthalene clearly indicated a significant decrease of absorption at 193 nm. Anthracene also has a similar spectral window at 193 nm. From these results, the authors predicted that aromatic compounds transparent to 193 nm light will be based on polycyclic compounds. We named this novel technique of creating a spectral window at 193 nm with a polycyclic compound, as the " Absorption band shift method".

Transparent resist materials with Aromatic Rings for ArF Excimer Laser Exposure

In order to confirm the above expectations, VUV absorption spectra of the aromatic compounds have been measured. Figure 2 shows a typical example of the comparison of VUV spectra of polycyclic compounds in a transparent polymer. As shown in Figure 2, naphthalene and anthracene derivatives were more transparent than benzene derivatives. Anthracene derivatives, however, were found to have other small absorption bands near 193 nm, that are thought to be the vibronic allowed transition bands. We therefore tried to look for resist materials in naphthalene derivatives. The compounds with naphthalene backbones tested for resist materials were as follows; PAGs: NAT-105 (onium salt known as i-line PAG[8]), TAZ-106 (chlorine-substituted triazine for i-line PAG), NAI-105 (naphthalimide triflate, non-ionic PAG). Acid-sensitive materials; 2-Npt-tBoc (tert-butoxy carbonyloxy naphthalene), NCR-1 (2-naphthaldehyde) (acid catalyzed cross-linker with naphthol backbone). Polymers; NV-1 (2-naphthol novolak), PNMA (poly (naphthyl methacrylate)[9]), EP-1 (epoxy resin synthesized with 1,5-dihydroxy naphthalene and epichlorohydrin.). Figure 3 shows the chemical structures of the above components. Table 2 summarizes the calculated λ_{max} employing the C.I. method in MOPAC (approximations with a one monomer unit model terminated by hydrogens were used in polymer calculations instead of calculating for the huge polymer molecule) and the experimentally obtained λ_{max} for the above components. The calculated values with MOPAC were found to be in fairly good agreement with the experimental values. Table 2 also includes experimental values of the absorbance of the aromatic derivatives at 193 nm (this was measured as a transparent polymer matrix containing 1 mmol/g of the aromatic derivatives, which corresponds to very high concentration of aromatic compounds, such as 0.389g/1g for species having Mw=389.). As shown in Table 2, naphthalene compounds with low molecular weight are characterized as materials very transparent to ArF excimer laser exposure. The naphthalene polymer species, however, were found to have relatively high absorption, revealed by the broadening of the absorption band. This broadening was thought to occur with absorbed energy distribution due to the non-uniformity of naphthalene configuration in the polymer (this will be discussed in later section). Example of the typical VUV absorption spectrum of NAT-105 (1.4 wt%) with comparative data are shown in Figure 4.

Novel Resists for ArF Excimer Laser Exposure Containing Aromatic Compounds.

We have examined the above resist components and prepared some effective compositions. The following materials have been precisely examined as a prototype resists for ArF excimer laser exposure.

PAGs: We have already reported the photo-efficiency of the PAGs mentioned above. The photo-efficiency of NAT-105 was estimated by acid titration, and found to have the highest efficiency compared with other aromatic PAGs. We therefore selected NAT-105 as the best PAG for ArF excimer laser exposure. TAZ-106 was also employed for negative resist composition (because of the difficulty of the titration method in the photo-efficiency measurement; no efficiency data for TAZ-106 was available except transparency data[4]) .

Dissolution inhibitor: 2-Npt-Boc was used as a dissolution inhibitor. It works as a sensitizer rather than dissolution inhibitor in chemically amplified resist system[9].

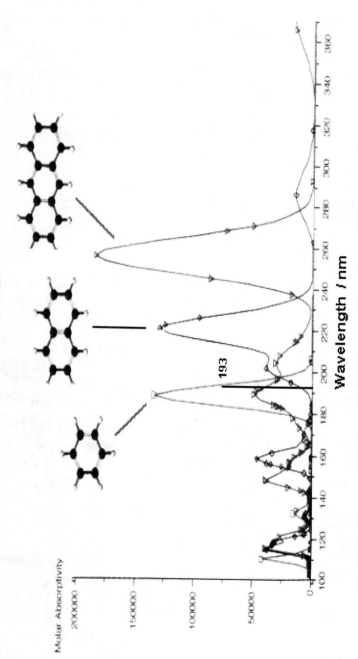

Figure 1. Simulated spectra for polycyclic aromatic compounds using ZINDO.

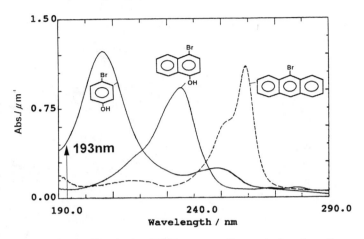

Figure 2. Experimentally obtained UV spectra for typical polycyclic aromatic compounds.

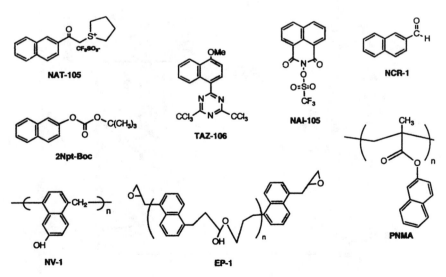

Figure 3. Structure of resist materials with naphthalene rings.

Table 1. Calculated λ_{max} for model aromatic compounds using CI methods in CNDO/S and MOPAC.

Compounds	Length of Conjugation	Calculated Absorption maximum (nm) [CNDO/S]	Calculated Absorption maximum (nm) [MOPAC]
Benzene	3	225	197
Chlorobenzene	3	229	203
Nitrobenzene	3	217	206
Methoxy benzene	3	224	205
Styrene	4	228	226, 215
Phenyl butadiene	5	262	231, 226
Naphthalene	5	259	225

Table 2. Calculated λ_{max} using MOPAC and experimentally obtained λ_{max} of the naphthalene containing resist materials .

Materials	Calculated λ max (nm)	Measured λ max (nm)	Measured Abs. at 193 nm
TPSP	205	197	9.83
NAT-105	251	251	1.64
TAZ-106	232	229	2.9
NAI-105	224	232	5.0
2Npt-Boc	222	222	0.52
NCR-1	223	228	0.97
NV-1	226	210, 240	3.2
PNMA	219	222	2.5
EP-1	230	222, 248	1.17

*1: Concentration in the film was 1 m mol (monomer unit for the polymer) / 1g solid.

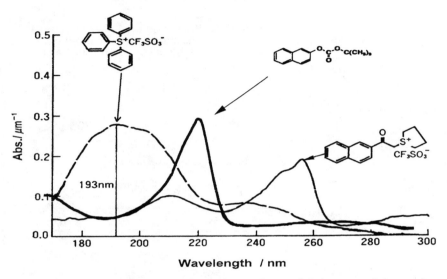

Figure 4. Typical VUV absorption spectra for naphthalene-containing resist materials.

Table 3. Resist composition and sensitivity of ALR-1 and ALR-2.

Resist Name	Polymer	PAG (wt%)	Inhibitor (wt%)	Sensitivity (mJ/cm^2)
ALR-1 (positive)	NATM	NAT105 (2)	2-Npt-Boc (20)	150
ALR-2 (negative)	40 wt% EP-1 in poly (vinylacetal)	TAZ-106 (5)	-	210

Although the solubility of acrylic co–polymers in the alkaline developer tends to decrease with increasing etching resistance, this kind of sensitizer (material enhancing the dissolution rate in exposed area) helps to increase the alkaline solubility effectively without loss of etching resistance.

Acid sensitive polymers: EP–1 was insoluble in the alkaline developer, therefore, it was developed by organic solvents such as methyl (isobutyl) ketone. Acceptably high (about 2.0) absorbance were achieved with the polymer mixture of EP–1 homo–polymer and transparent polymer up to 30 wt% of EP–1. It was possible to use this polymer as an acid–curable, transparent polymer. The copolymer of naphthyl (methacrylate), methacrylic acid, tert–butyl methacrylate and methyl methacrylate (NTAM) is used as an acid–sensitive co–polymer for positive tone 193 nm resist[9].

Compositions of novel prototype single layer resists containing aromatic compounds are shown in Table 3. The total amount of naphthalene rings in both polymers was about 30 wt%. Table 3 also contains typical sensitivity data for these resists, and it should be noted that both resists have good sensitivity to ArF excimer laser exposure in spite of the 30 wt % of aromatic composition. Pattern formation with these resists was also carried out using the contact exposure method. Acceptable patterning was obtained in the positive tone resist, ALR–1,but ALR–2 showed poor patterning performance, due to swelling of the fabricated patterns during development. By optimizing the co–polymerization ratio of the monomer and content of the components of ALR–1 (naphthyl (methacrylate) : methacrylic acid : tert–butyl methacrylate : methyl methacrylate = 5 : 20 : 50 : 25), patterns with 0.175–0.17 μm feature size were successfully fabricated using an NA 0.55 prototype ArF excimer laser lens. The patterns are shown in Figure 5. We also found here a problem of absorption enhancement in the 193–nm region especially in this prototype polymer (10 mol% introduction of naphthyl(methacrylate) to copolymer caused Abs.=0.9/μm), according to the increase in naphthalene content in the polymer. Therefore, sufficient etching rate could not be obtained [9)10)] (about 1.73 times of Novolac etching rate for CF_4 plasma) by the polymer with 5 mol % introduction of naphthyl (methacryrate). However, it was striking fact that acceptable fine patterns were obtained even in 30 wt % of total aromatic compounds of the resist composition. This means the aromatic component is no longer the minor part of the ArF excimer laser resist. We think, at least, aromatic PAGs and aromatic inhibitors are materials imperative for 193– nm resist.

Properties of the acrylic–based 193–nm single layer resist.

In this section, we would like to discuss the acrylic polymer species for 193–nm resist. Acrylic polymers have been commonly employed in published ArF prototype resists, as it is in our prototype resist. In developing the ALR–1 resist, we saw several problems for acrylic–based alkaline–soluble polymers. For example, the dissolution rate of acrylic–based polymers is very fast in conventional concentrations of developer. This may due to the relatively higher acidity of carboxylic acids in comparison with the phenolic moiety. Therefore, eluted developers must be used for acrylic–based resist.

Figure 6 shows plots of the dissolution rate of poly (tert–butyl (methacrylate)) (PTBM) vs. deprotection % of tert–butyl group, which has measured with FTIR. Great increase in dissolution rate, corresponding to the increase in amount of deprotection were observed when the polymer was developed with conventional 2.38% tetramethyl (ammonium) hydroxide aqueous solution (TMAH). This step–function–like large dissolution rate change usually reveals the difficulty in controlling the synthesis of a polymer having appropriate dissolution rate or difficulty of micropatterning. In contrast, eluted developer, such as 0.436% TMAH soln.,

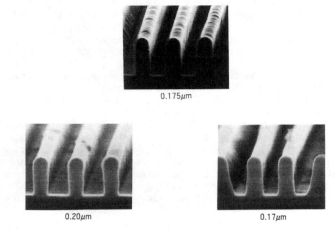

Figure 5. The fabricated patterns of ALR-1 with the prototype ArF projection system.

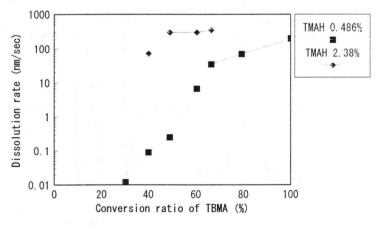

Figure 6. Relationship Between Dissolution Rate and Conversion Ratio of TBMA.

seems to indicate better dissolution, similar to that of a phenol–based resist. In general, the eluted developer loses basicity easily, which therefore leads to low stability of developer and large change of pattern size, depending on the density of the patterns. A problem was also found in "adhesion" of the pattern in acrylic–based resist. Figure 7 shows the typical example of pattern collapse after development. This may occur due to peeling of the pattern by shrinkage of the polymer during PEB. Figure 8 is a plot of PTBM thickness vs. deprotection % of the tert–butyl group. This large shrinkage seems to be a big problem for micro–patterning.

Another interesting phenomenon was difference in sensitivity of resists, depending on polymer matrix. Naphthalene–containing PAGs usually have sensitivity to i–line exposure, therefore, if the polymers are transparent to i–line, the photo–acid generation efficiency of the PAG can be estimated with removal of electron transfer from the polymer matrix to PAGs. Table 4 is the comparison of the sensitivity of the chemically amplified resist with the phenol moiety and the acrylic–based resist (co–polymer of tert–butyl (methacrylate) and methyl (methacrylate)). A sensitivity about a order of the magnitude lower was observed for the system without phenol, in spite of the same amount of PAG. In this case, the sensitivity of the resist is thought to be determined by a combination of the efficiency of photo acid generation, the mobility of the acid in the polymer and the rate of the deprotection reaction. We therefore directly measured the amount of acid generated in these polymers by titration to remove the combination due to mobility of acid and the deprotection reaction. Figure 9 shows the plot of difference in 627–nm absorption of the titration solution vs. absorbed i–line exposure dose. The slope of the curve at origin is directly indicates the efficiency of photo–acid generation. A 3–times–higher photo–efficiency was observed in the phenolic polymer. From these results, the phenolic–polymer is considered to be a very effective media for photo–acid generation in a chemically amplified resist.

Disadvantages of acrylic polymers mentioned in this section can occur in all aliphatic 193–nm resists. We reported that the full substitution of naphthol containing polymers led to relatively strong absorption compared with acrylic–based aliphatic polymers. However, the authors believe that the partial substitution of acrylic–polymer by the naphthol moiety is the solution for the problems mentioned above, and also reproduces "phenol chemistry " in the 193–nm resist.

How do we enhance the transparency at 193 nm for the naphthalene–containing polymer?

We reported that the naphthalene–containing polymers have relatively high absorption revealed by the broadening of the absorption band. As mentioned in 3–2, the broadening of absorption occurs due to absorbed energy distribution. This is the "polymer effect", which has been commonly observed in polymer spectroscopy. Although the broadening of the absorption band in polymer species seems a natural phenomenon, our study has been extended to control this phenomenon with molecular structures. In this case, we also started from the simulations of polymer spectra.

In order to evaluate the influence on λ_{max} of the distance between naphthalene rings, a 2–naphthalenes–molecule system was also calculated. We used CNDO/S–CI for this calculation, since this method can treat huge molecular systems. The calculations were tested under conditions in which the two naphthalenes approach perpendicular to each other or parallel to each other. In both cases, the energy of the 2–naphthalenes system started to increace from a distance of 7 Å . Figure 10 shows the calculated values of λ_{max} according to the distance between the 2–naphthalenes molecules. A blue shift of about 5 nm was calculated at distance of 5 Å , and 4 Å distance resulted 10–nm shift, that strongly influenced the absorption minimum near

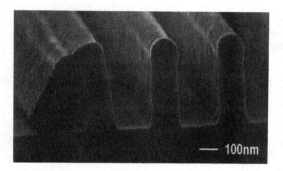

Figure 7. Typical example of the pattern collapse after development.

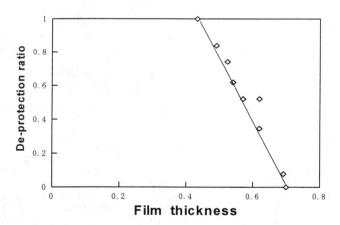

Figure 8. Relationship between film thickness and de-protection % in TBMA.
FTIR spectroscopy was used for de-protection % determination.

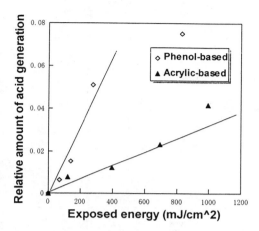

Figure 9. Difference between photo-acid generation in the polymers.

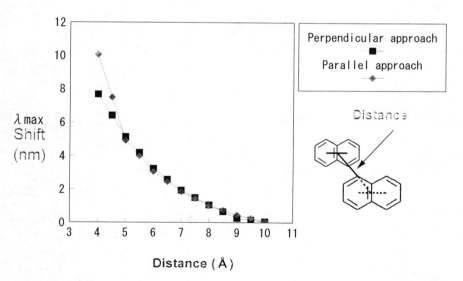

Figure 10. Influence on maximum absorption wavelength of the distance between two naphthalene molecules.

Polymer	PAG	Sensitivity to i-line exposure (mJ/cm^2)
Phenol-based	NAT-105 (5wt%)	48
Acrylic- based	NAT-105 (5wt%)	350

Table 4. Comparison of the i-line sensitivity of chemically amplified resist between a phenol-based system and an acrylic-based system.

193 nm. We think 6–7 Å is sufficiently far to result in negligible interaction of electrons between the two naphthalenes. Therefore, the three dimensional structure of the polymer appears to be an important factor in reducing 193 nm absorption of polymer materials.

The relatively low absorption in the EP-1 polymer may arise from this structural effect in the polymer species, since the two long epoxy groups perform the role of separator of the naphthalene residues. Figure 11 is a comparison of the VUV spectra for EP-1 and NV-1. A difference of about 10 nm in maxima seems to be fatal for NV-1 in creating spectral window at 193 nm. Our study is continuing now and is at the phase of synthesizing a highly transparent naphthol polymer using the separation effect noted before.

CONCLUSIONS

A novel single layer resist containing aromatic compounds have been developed employing the "absorption band shift method", which creates spectral windows at 193 nm by extended conjugation with polycyclic aromatic rings. This was expected from the molecular orbital calculations. By developing ArF resist materials with containing naphthalene, the fine patterns with 0.17 μm L/S could be obtained in spite of 30% content of naphthalene compounds. Polymer materials with naphthalene rings lead to relatively lower transparency, but molecular orbital calculations showed a method for obtaining transparent naphthalene polymers for the ArF excimer laser.

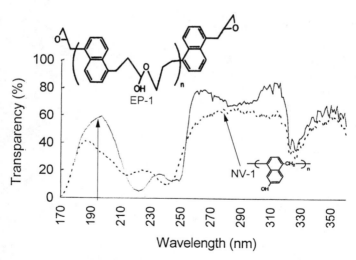

Figure 11. Typical examples of VUV spectra for naphthalene-containing polymers.

Solid line : EP-1, Doted line : NV-1.

Acknowledgment

The authors thank Nikon Co. for exposing resist on their prototype ArF excimer laser exposure equipment.

Literature Cited

1) M. Takahashi et al., J. Photopolym. Sci. Technol. **1994**,7(1),31.
2) Robert D. Allen et al., J. Photopolym. Sci. Technol. **1994**,7(3),507.
3) Kaichiro Nakano et al., SPIE, **1994**,2195,194.
4) T. Ushirogouch, N. Kihara, S. Saito, T. Naito, K. Asakawa, T. Tada and M. Nakase, SPIE , **1994**,2195,205.
5) J. J. P. Stewart et al., Research Lab. U.S.Air Force Academy, MOPAC (QCPE#455): Colorado, **1988**.
6) J. Del Bene and H. H. Jaffe, J. Chem. Phys. ,**1968**,48,1807.
7) Anderson W. P. ,Edwards W. D. and Zenner M. C. ,Inorg. Chem., **1986**, 2728.
8) N. Hayashi et al. , Jpn. J. Appl. Phys. ,**1990**,29,2632.
9) N. Naito, K. Asakawa, N. Shida, T. Ushirogouchi and M. Nakase, Jap. J. Appl. Physics. , **1994**,33,7028.
10) M.Nakase, T. Naito, K. Asakawa, A. Hongu, N. Shida and T. Ushirogouchi.,SPIE, **1995**, in press.

RECEIVED August 14, 1995

Chapter 17

Design Considerations for 193-nm Positive Resists

Robert D. Allen[1], I- Y. Wan[1], Gregory M. Wallraff[1],
Richard A. DiPietro[1], Donald C. Hofer[1], and Roderick R. Kunz[2]

[1]IBM Almaden Research Center, 650 Harry Road,
San Jose, CA 95120–6099
[2]Lincoln Laboratory, Massachusetts Institute of Technology,
Lexington, MA 02173–9108

Our approach to the design of positive, single layer resists for 193 nm lithography will be discussed. Phenolic resins, the archetype in positive photoresist materials, cannot be used at this wavelength due to optical opacity. Acrylic polymers combine the required optical transparency at 193 nm with easily tailored properties. With a design based on *methacrylate terpolymers*, we have recently developed a high resolution positive resist for 193 nm lithography with good imaging at both 193 and 248 nm. Our work has centered on gaining further insight into methacrylate polymer structure/property relationships, improving the imaging performance and finally increasing the etch resistance. Towards that end, we have employed a class of *dissolution inhibitors* for 193 nm resists that are combined with methacrylate polymers to provide 3-component resists. A family of 5B-steroid dissolution inhibitors that also increase etch resistance will be described. Imaging and etch performance of these resists will be disclosed, with particular emphasis on the impact of these steroid dissolution inhibitors on the thermal properties of the resist. These methacrylate chemically amplified resists show resolution capability below 0.25 micron, etch rates 20% higher than novolak resins, and dual wavelength (193/248 nm) imaging.

The explosive growth in performance of semiconductor devices has been fueled by advances in microlithography and photoresist technology. The current generation of advanced microprocessors and DRAM memory chips have critical dimensions approaching 0.5 microns and are printed using novolak-based mid-UV photoresists. Next generation devices will be produced with optical lithography at shorter wavelengths (ca. 250 nm, deep UV) combined with newer (chemically

0097–6156/95/0614–0255$12.00/0
© 1995 American Chemical Society

amplified) photoresists. The technology path toward device generations beyond 0.25 micron is currently the subject of much discussion. ArF excimer (193-nm) lithography is a viable approach to extend optical lithography beyond 0.25 microns, but the resist technology develped for traditional DUV lithography is problematical at this 'deeper' UV wavelength. We will discuss the materials issues involved in the design of a positive resists for 193 nm lithography with regard to optical properties, resolution, photospeed and etch resistance.

The design of positive (single-layer) resists for 193 nm lithography is a significant challenge. This emerging field of photoresist research has recently been reviewed. [1] The imaging chemistry is quite similar to that practiced in traditional DUV lithography: photogeneration of a strong acid followed by acid catalyzed deprotection to render the exposed regions of the film soluble in aqueous base. The differences in resist design between traditional (248 nm) and 193 nm lithography are related to matters of optical transparency. Traditional DUV resists are based on hydroxystyrene polymers, phenolic resins with much improved optical properties at 248 nm than the structurally similar novolak resins. Hydroxystyrene polymers are extremely opaque at 193 nm, however. In fact, these resins are ideally suited for top-surface imaging at 193 nm. [2]

New polymer materials are required for 193 nm (single layer) resists, with high optical transparency at the exposure wavelength combined with properties that hydroxystyrene polymers (and few others) possess; 1) hydrophilicity (for good positive-tone development characteristics); 2) high T_g (130-170 °C), for good thermal properties and the latitude to perform higher post-expose bakes; 3) aromatic rings in high concentration (for good etch resistance), and 4) an easily blocked hydroxyl group (for incorporation of acid cleavable functionality).

These four taken-for-granted characteristics that are present *by default* in DUV (248 nm) resists need to be painstakingly designed into single-layer resists for 193 nm lithography, where phenolic resins cannot be used. In light of their excellent optical transparency (see Figure 1), methacrylate polymers are (to date) the new paradigm for 193 nm resist design. In this case, gaining high resolution imaging can be accomplished,[3] as can etch resistance.[4] Building a resist with both excellent etch resistance *combined* with good imaging characteristics is the challenge. This paper discusses our resist design with special emphasis on modification of a high resolution first generation 193 nm resist (Version 1) to gain enhanced etch resistance. Control and balancing of the separate factors of T_g, hydrophilicity, etch resistance, and imaging properties (simultaneously) will be discussed.

EXPERIMENTAL

Methacrylate polymers described here for use in 193 nm resists were prepared by free radical solution polymerization. Molecular weights were controlled by inititator structure and concentration, polymerization solvent and temperature, and through the use of chain-regulating additives.[5] Conversions of monomer to polymer are dependent on polymerization conditions, but were typically between 80-100%. Polymers were isolated by precipitation into hydrocarbon solvents, filtered and dried at elevated temperatures for 24 hours.

For example, methyl methacrylate (MMA) (200 grams, 2 moles), t-butyl methacrylate (TBMA) (100 grams, 0.70 moles) and methacrylic acid (50 grams, 0.58 moles) were charged to a 2 liter round bottom flask with a magnetic stir bar. Unstablized (inhibitor-free) tetrahydrofuran was added as the polymerization solvent (1200 grams). Finally, the polymerization initiator bis-azoisobutyronitrile (AIBN) was added (1.5 grams, 0.009 moles). The polymerization reactor was fitted with a reflux condenser. The polymerization mixture was repeatedly degassed with nitrogen/vacuum cycles. The reaction was carried out by heating to reflux (67 °C) and was allowed to proceed for 24 hours under a nitrogen blanket using a Firestone valve. The polymerization mixture was then cooled to room temperature, at which time the viscous polymer solution was diluted with approximately 400 grams of THF. The polymer was isolated by precipitation into hexane, via dropwise addition, into a large excess of the rapidly stirred non-solvent. The precipitated polymer was filtered, washed, and dried in a vacuum oven at elevated temperature for 1-2 days. Collected yield of 90% was found (310 grams of polymer was produced) with a molecular weight (M_w) of 75,000 g/mole, a polydispersity of 2.2 and a T_g = 150 °C.

Resists solutions are prepared by dissolving the polymer of interest into propylene glycol methylether acetate (PGMEA), then adding the other components of the formulation (dissolution inhibitors, photoacid generators). The PAG used in all resist compositions reported here was bis-(t-butyl phenyl)iodonium triflate (TBIT) (Figure 2). Loading of TBIT in the resist was typically between 1 and 2 wt% (vs. resist solids). Resists were processed as follows: resist solutions were spin-coated onto HMDS-primed wafers, post-apply baked (PAB) above 100 °C for 1 min, exposed at 193 nm (SVGL Micrascan 193 prototype, NA= 0.50) or DUV (GCA XLS "lotus" lens, NA= 0.48), post-expose baked (PEB) above 100 °C for 1 minute, developed in 0.01-0.05N TMAH (tetramethylammonium hydroxide) (Version 1), and in 0.02-0.13N TMAH (Version 1.5 and Version 2). Exposed wafers were immersion developed for 20-60 seconds.

RESULTS AND DISCUSSION

We use a building block approach in resist design, employing different monomers to tailor polymer properties. The use of methacrylate monomers facilitates this approach, as these are easily incorporated into the polymer because the copolymerization characteristics are largely unaffected by the structure of the ester group. Using this methodology, we developed a high speed, aqueous developing positive tone resist several years ago for direct-write printed circuit board lithography.[6] The approach taken for materials design used acid-labile methacrylate polymers as a polymer platform. This is also a class of materials with the required optical transparency at 193 nm and with an easily tailored structure. These versatile

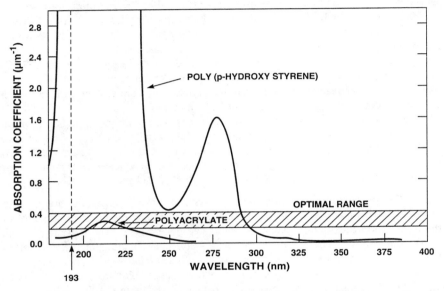

Figure 1. Optical absorbance spectrum of poly(hydroxystyrene) vs. acrylic polymer.

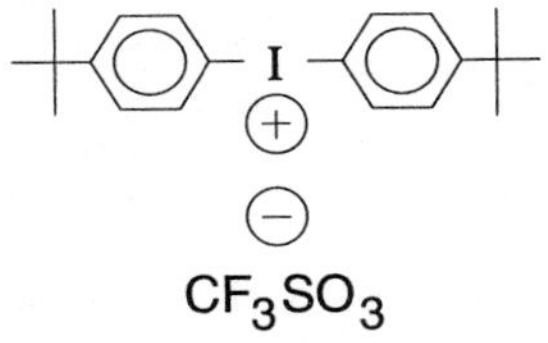

Figure 2. Structure of TBIT photoacid generator.

acrylic polymers are methacrylate terpolymers of methyl methacrylate (MMA), t-butyl methacrylate (TBMA) and methacrylic acid (MAA) (Figure 3).

We first designed a methacrylate polymer-based 193 nm positive resist with excellent optical transparency, aqueous-development, high T_g, good adhesion and excellent imaging properties (photospeed, resolution, environmental stability).[3] This resist (Version 1) lacks the required etch resistance for most semiconductor processes, but offers a vehicle for learning structure/property relationships of methacrylate-based chemically amplified resists.

Version 1 Resist. The Version 1 resist is comprised of two components, an iodonium triflate onium salt (TBIT), (Figure 2), and the methacrylate terpolymer (Figure 3) which has been discussed previously.[3] Each monomer serves a separate function in the terpolymer. T-butyl methacrylate (TBMA) provides an acid cleavable side group which is responsible for creating a radiation induced solubility change. Methyl methacrylate (MMA) promotes hydrophilicity for photoinitiator solubility and positive tone development characteristics, while also improving adhesion and mechanical properties, and minimizing shrinkage after expose/bake. Methacrylic acid (MAA) controls aqueous development kinetics. This polymer is prepared in a single step from readily available, inexpensive components. By selecting the terpolymer composition and molecular weight, imaging properties (including dissolution properties, photospeed, contrast) can be altered to a significant extent. For example, Figure 4 shows the dissolution rate of the spin-coated film of neat terpolymer as a function of methacrylic acid concentration. The non-linear response is important for high contrast imaging behavior. The MAA concentration in a resist formulation is selected to exhibit highly non-linear dissolution response to dose. For example, as shown in the figure, an MAA concentration of 18% would yield a resist with relatively low unexposed dissolution rate. Very low exposure doses to produce MAA concentrations only slightly higher (e.g., 20%) would yield little change in dissolution rate, but moderate doses which could convert a substantial fraction of t-butyl ester into carboxylic acid would produce dramatic dissolution acceleration.

Version 1A resist was exercised extensively on the SVGL Micrascan 193 prototype. In fact, the resist was integral to the optical characterization of the tool and accelerated optimization of the prototype's optical characteristics. Highest resolution obtained at 193 nm for Version 1A was 0.22 micron in 0.75 microns of resist.

An advanced version of this resist (Version 1B) resulted from optimization of the polymer composition, molecular weight, PAG loading, process temperatures and developer concentration. Version 1B has principally been exercised on the GCA (XLS) 248 nm stepper at MIT Lincoln Laboratory. The resist is linear to 0.30 microns at a resist thickness of 0.75-0.85 microns with good process latitude and is able to resolve features beyond the limit of this exposure tool (down to 0.25 micron) with a slight overexposure. Figures 5 demonstrates the imaging quality of Version 1B exposed with this 248 nm stepper at dimensions of 0.30 microns, pictured through a dose range of 5.2-6.4 mJ/cm². This resist can resolve features as small as 0.25

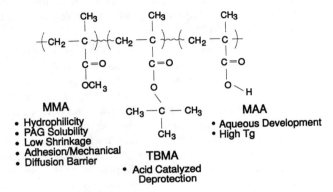

Figure 3. Structure of MMA-TBMA-MAA methacrylate terpolymer.

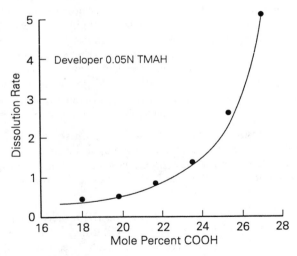

Figure 4. Aqueous base dissolution rate vs. methacrylic acid concentration for a series of methacrylate terpolymers.

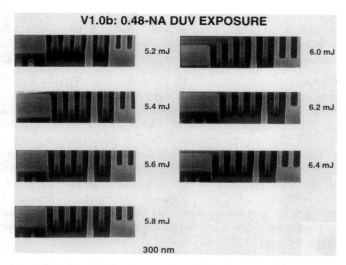

Figure 5. Imaging results of Version 1B resist.

microns in 0.8 microns of resist, through a (slightly overexposed) dose range of 7.6-8.6 mJ/cm^2.

These 'Version 1' resists have etch properties surprisingly similar to conventional DUV resists (e.g. APEX-E) in CF$_4$ based (oxide) etch recipes.   More aggressive etch chemistries (e.g. aluminum and polysilicon etching) demand substantial increases in etch resistance. The approach currently being used to impart etch resistance without degrading the optical transparency of 193 nm single layer resists involves incorporation of alicyclic (aliphatic/cyclic) compounds into the polymer structure.   Methacrylate polymers with pendent alicyclics with a rich structural variety include adamantyl, [4] norbornyl, [6] tricyclodecanyl [7] and adamantanemethyl[8] have been used.   Alicyclics have been identified which also undergo acid-catalyzed deprotection.[8]   Aromatic compounds with enhanced optical transparency at 193 nm have been identified and incorporated into a resist.[9]

Etch Resistant Versions (1.5 and 2.0). The quest for a high performance 193 nm positive chemically amplified (CA) resist with etch resistance adequate for production lithography is a significant challenge.   We examined the structure/property relationships of methacrylate polymers with increased etch resistance over the Version 1 resist. Modifications which improve etch resistance often negatively impact aqueous solubility and polymer thermal properties. Several approaches will be discussed which attempt to address competing considerations in positive resist design.

Our exploration of alicyclic incorporation into Version 1 methacrylate polymers to increase etch resistance has met with only limited success. Alicyclic methacrylate monomers are excellent for imparting etch resistance; 50 mole percent of an alicyclic methacrylate reduced the etch rate in chlorine plasma by a factor of 2[8] (2-2.5 times faster than novolak for non-alicyclic methacrylate polymers to 1.2-1.4 times faster for a 50% copolymer of adamantane methylmethacrylate-co-methyl methacrylate). This increased etch resistance comes at the expense of imaging performance (in our hands). Some unintended consequences of alicyclic incorporation are strong increase in T_g, runaway hydrophobicity, poor adhesion, and embrittlement. The delicate *balances* required for a high performance positive resist are difficult to achieve when etch resistance is provided only through alicyclic methacrylate copolymers.

We developed an alternative strategy to upgrade etch resistance.[10] This design involves the introduction of 3-component resists: *methacrylate polymer* slightly modified from Version 1, *alicyclic dissolution inhibitor* compound and finally a *photoacid generator*. After evaluating large numbers of dissolution inhibitor compounds, it became apparent that a class of compounds was available (from natural sources) with very desirable properties:

   1) high solubility in resist and PGMEA
   2) strong dissolution inhibition
   3) high exposed dissolution rate
   4) 193 nm transparency
   5) moderating influence on T_g
   6) etch resistance
   7) good thermal stability (> 200 °C)

Bile acid esters (5B-Steroids) were used in early (pre-chemically amplified) DUV resists by Reichmanis and co-workers in the early 1980's.[11] Photo-induced deprotection of o-nitro benzyl esters created carboxylic acids in the exposed portion of the film. Extensive studies of substituent effects demonstrated the possibility of good dissolution inhibition, but photospeeds were quite slow. O'Brien and co-workers examined t-butyl cholate as a dissolution inhibitor for novolak resins, an example of a three-component chemically amplified resist.[12] The central idea in this work was the search for an inhibitor with little or no absorbance at 248 nm, so as to afford a dilution in the optical density of novolak resins at this DUV wavelength.

We were intrigued at the possibility of a dissolution inhibitor with *combined* etch resistance from this steroid family of alicyclic compounds. We prepared a variety of these compounds (see Figure 6) and investigated structure/property relationships

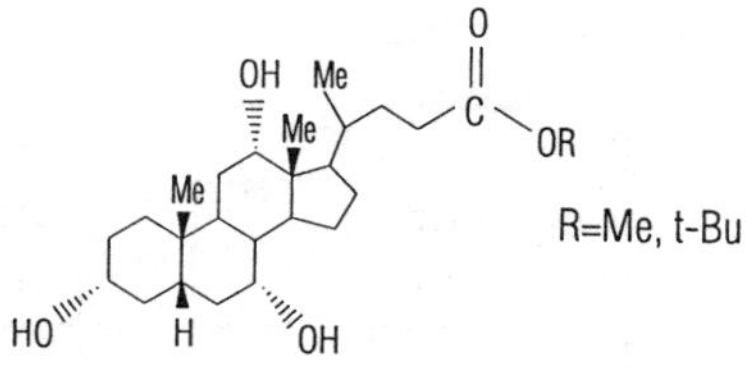

Cholate Esters (ME-1; TB-1)

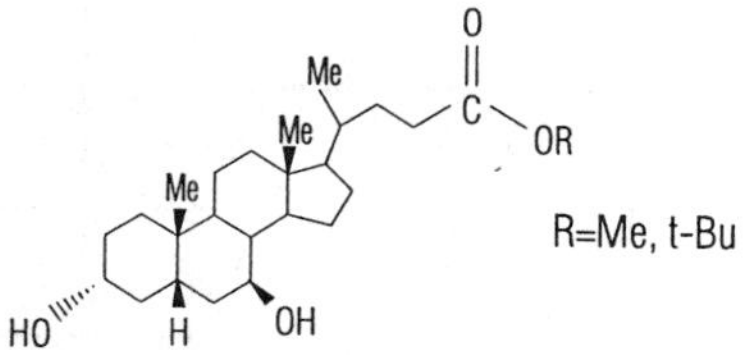

R=Me, t-Bu

Ursocholate Esters (ME-2)

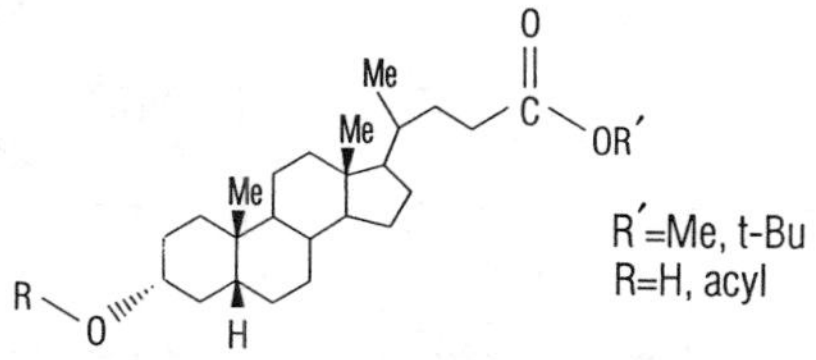

R=Me, t-Bu

Lithocholate Esters (ME-3; TB-3; TB-4; TB-5)

R'=Me, t-Bu
R=H, acyl

Figure 6. Steroid dissolution inhibitor structures.

with attention paid to solubility, dissolution inhibition in methacrylate terpolymers, dissolution promotion after exposure and imaging performance.   The impact of manipulation of structure in the 5B-steroid family on dissolution properties was extreme (see Table I). Both passive (methyl esters) and active (t-butyl esters) steroids were examined in an aqueous base soluble methcrylate polymer used in our version 1 resist. Films were developed in 0.1 N TMAH before and after exposure to 25 mJ/cm^2 of 254 nm filtered light.   Unexposed and exposed dissolution rates were measured, and inhibition (decrease in dissolution rate by adding 25% inhibitor) and the dissolution rate ratio (R/R$_0$) were calculated. Methyl esters of cholic, ursocholic and lithocholic acid demonstrated a strong substituent dependence on dissolution properties.   The cholate ester compound is by far the least efficient inhibitor, with a percent inhibition of only 5 %. Simply removing one hydroxyl group improves the inhibition by a factor of 5 (methyl ursocholate) (ME-2) and further removal of the second hydroxyl to yield methyl lithocholate (ME-3) improves the inhibition by a further factor of 3, to 83.5%.

Table I. Alicyclic Etch/Dissolution Inhibitors

	Dissolution Inhibitor	Unexposed Dissolution Rate(μ/min.)	Percent Inhibition	Exposed Dissolution Rate(μ/min.)	R/R(0)
passive	none (K2)	4.14	0	26.0	6.3
	ME-1	3.91	5.5	17.5	4.5
	ME-2	3.00	27.5	11.4	3.8
	ME-3	0.68	83.5	8.6	12.6
active	TB-1	2.6	38	23.0	8.8
	TB-3	0.84	80.0	27.0	32.3
	TB-4	0.53	87.2	25.0	47.2
	TB-5	0.25	94	35.0	140

Developer, 0.10N TMAH   Expose dose, 25 mJ/cm^2, 254 nm
25% Inhibitor loading

Acid-cleavable "active" (t-butyl) esters were prepared. The substitution trends are similar to the passive compounds. Tert-butyl cholate is a very poor inhibitor, while t-butyl lithocholate is a much better inhibitor (see Table I). The exposed dissolution rate is much less sensitive to the alicyclic substitution pattern (not the case in the "passive" compounds). As a result, R/R_0 is ca. 4 times higher for t-butyl lithocholate (TB-3) than for t-butyl cholate (TB-1).

Further enhancements in dissolution inhibitor were realized through hydroxyl substitution of t-butyl lithocholate. Two compounds are shown in the table (TB-4 and TB-5), which have substantially improved dissolution inhibition properties. These compounds inhibit the dissolution rate of the Version 1 terpolymer by close to 90% and maintain a high exposed dissolution rate. TB-5 is highly soluble in methacrylate polymers and in PGMEA. Loadings as high as 50% (wt) were achieved without detectable phase separation with this compound.

Mixtures of the steroids with the Version 1 methacrylate terpolymers affords (*Version 1.5*) resists with relatively good imaging properties and enhanced etch resistance. Figure 7 shows 0.35 micron images printed in the Version 1.5 resist as a function of PEB temperature. Note that a high PEB (130 °C) causes degradation in the imaging resolution, while a much lower PEB (104 °C) provides better imaging quality.

The requirement for sharply lower PEB in this Version (1.5) of the resist appears to be a function of the T_g of the steroid/methacrylate mixture. The thermal properties of these mixtures are dominated by the plasticizing effect of the steroid. For example, TB-5 is slow to crystallize, melts at ca. 50 °C, then shows a strong glass transition below -10 °C (see Figure 8). At the relatively high steroid loadings employed, significant T_g suppression (plasticization) is quite likely. In our experience, post-expose baking at or above T_g when using a triflic acid generator causes image degradation apparently due to acid diffusion. In order to gain meaningful reduction in etch rates by increasing the steroid loading, *overplasticization* of the resist occurs. If one could raise the T_g of the resin, and its intrinsic etch resistance, the steroid could be added in somewhat lower concentration. This T_g increase would allow for increased loading of the steroid before overplasticization becomes a problem.

EFFECT OF DISSOLUTION INHIBITOR ON GLASS TRANSITION

RESIST: V1.5 – POLYMER / INHIBITOR/PAG, 100 / 33 / 2.5
POLYMER T_g ~165°

PROCESS: 1. 130°C PAB FOR 1 min
2. EXPOSE – (193-nm, 0.5 NA)
3. POST-EXPOSURE BAKE
4. DEVELOP IN 0.05 N TMAH, 20 s

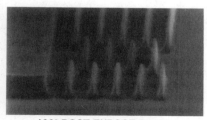 

130° POST-EXPOSE BAKE   105° **POST-EXPOSE BAKE**

Figure 7. Imaging of Version 1.5 three-component resist.

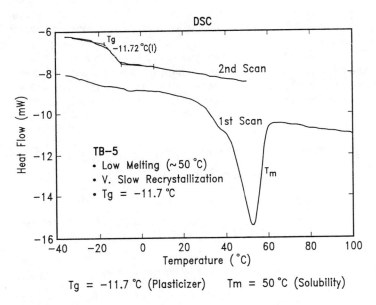

Figure 8. DSC thermogram of a pure steroid dissolution inhibitor.

Version 2 resist was born of this concept. The steroid dissolution inhibitor TB-5 is added to a *tetrapolymer* of isobornyl methacrylate (IBMA) (or adamantanemethyl methacrylate) and MMA-TBMA-MAA (Figure 9).[7] The IBMA in this case raises T_g (to over 200 °C) *and* imparts increased etch resistance. Figure 10 shows the impact of the steroid on the thermal properties of this high T_g tetrapolymer, through the use of thermomechanical analysis (TMA) on spin-coated resist formulations. Note the strong plasticization of this resist as a function of steroid loading. Note also the strong impact of PAB temperature on resist thermal properties.

The increased Tg of the IBMA tetrapolymer/TB-5 mixture allows for an increase in PEB temperature and a corresponding increase in imaging quality. High quality 0.25 micron features have been printed in 0.6 microns of Version 2 resist, exposed with the SVGL 193 nm step-and-scan prototype. Figure 11 shows 248 nm imaging results at 0.5 micron resolution in this Version 2 resist. Etch resistance of Version 2 resist is far better than Version 1 or Version 1.5. The presence of alicyclics in *both* the polymer and dsissolution inhibitor produced chlorine etch rates only slightly faster than novolak resins. Version 2 resist has achieved etch rates as low as 1.2 times that of novolak, in a resist formulation with good imaging quality. This is perhaps the first example of the *combination* of etch resistance and imaging quality in a non-phenolic, 193 nm-transparent photoresist.

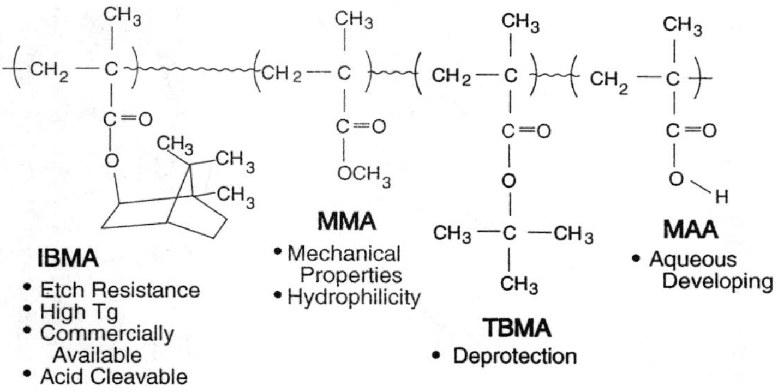

Figure 9. Methacrylate tetrapolymer structure used in Version 2 resist.

SUMMARY

The evolution from a high quality imaging resist with little etch resistance (Version 1), to a resist which *combines* imaging quality with etch resistance (Version 2) was described. Integral to this transformation was the introduction of a three component resist. Alicyclic dissolution inhibitors from the 5*B*-steroid family, when combined with methacrylate tetrapolymers and an iodonium triflate photoacid generator, form three-component resists with the appropriate balance of

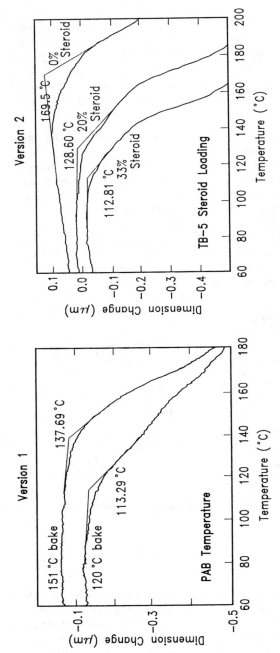

Figure 10. Thermomechanical analysis (TMA) of methacrylate tetrapolymer/steroid mixtures.

RESIST: V2.0 – THREE-COMPONENT RESIST WITH IMPROVED ETCH RESISTANCE
POLYMER – 16% WEIGHT ISOBORNYL MOIETY
INHIBITOR – CHOLIC ACID ESTER DERIVATIVE
PHOTOACID – BIS (t-BUTYL PHENYL) IODONIUM TRIFLATE
TOTAL RESIST 32% WEIGHT ALICYCLIC CARBON

ETCH RESISTANCE: 1.2 × NOVOLAC IN HIGH-DENSITY Cl_2 PLASMA (Helicon)
IMAGING: 0.48 NA DUV STEPPER

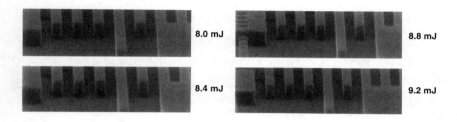

500-nm FEATURES

Figure 11. Imaging results of Version 2 resist exposed at 248 nm.

hydrophilicity, glass transition, alicyclic carbon, acid cleavable protecting groups and transparency at 193 nm. A high speed, aqueous developing 193 nm single layer resist with a combination of chlorine etch resistance approaching novolak resin and good imaging quality resulted from this approach.

ACKNOWLEDGMENTS

The authors thank Hoa Truong and Monica Barney from the IBM Almaden Research Center for help with material characterization and polymer synthesis, respectively, and Deanna Downs from MIT Lincoln Laboratory for help with photoresist formulation and processing. Dr. Roger Sinta, Shipley Company, is gratefully acknowleged for his description of the TMA analysis of resist films. This work was supported by the Advanced Lithography Program of the Advanced Research Projects Agency.

LITERATURE CITED

1. R. D. Allen, G. M. Wallraff, D. C. Hofer, R. R. Kunz, S. C. Palmateer and M. W. Horn, *Microlithography World*, 21, Summer 1995.

2. M.A. Hartney, R. R. Kunz, D. J. Ehrlich, and D. C. Shaver, *Proc. SPIE*, **1262**, 119 (1990).

3. R. R. Kunz, R. D. Allen, W. D. Hinsberg and G. M. Wallraff, *Proc. SPIE*, **1925**, 167(1993); R. D. Allen, et al. *J. Photopolym. Sci. Tech.*, **6(4)** 575 (1993); R. D. Allen, G. M. Wallraff, W. D. Hinsberg, L. L. Simpson, and R. R. Kunz, In "Polymers for Microelectronics", ACS Symposium Series **537**, Thompson, L. F. , Willson, C. G. and Tagawa, S. Eds., ACS, Washington, D.C., 1994, pp. 165-177.

4. Y. Kaimoto, K. Nozaki, S. Takechi and N. Abe, *Proc. SPIE*, **1672**, 66 (1992).

5. G. Odian, *"Principles of Polymerization"*, Third Edition, Wiley, New York, 1991.

6. M. Endo, et al., IEDM Tech. Digest, 45, December (1992).

7. K. Nakano, K. Maeda, S. Iwasa, J. Yano, Y. Ogura, E. Hasagawa, *Proc. SPIE*, **2195**, 194 (1994).

8. R. D. Allen, G. M. Wallraff, R. A. DiPietro, D. C. Hofer, and R. R. Kunz, *J. Photopolym. Sci. Technol.*, **7(3)**, 507(1994).

9. T. Ushirogouchi, N. Kihara, S. Saito, T. Naito, K. Asakawa, T. Tada, and M. Nakase, *Proc. SPIE*, **2195**, 205(1994).

10. R. D. Allen, G. M. Wallraff, R. A. DiPietro, D. C. Hofer, and R. R. Kunz, *Proc. SPIE*, **2438** (1995) in press.

11. E. Reichmanis, C. W. Wilkins, D. A. Price, E. A. Chandross, *J. Electochem. Soc.* **130**, 1433 (1983); E. Reichmanis, et al., *J. Polym. Sci. Polym. Chem. Ed.* **21**, 1075 (1983).

12. M. J. O'Brien, *J. Polym. Eng. Sci.* **29**, 846 (1989).

RECEIVED September 15, 1995

Chapter 18

Top-Surface Imaged Resists for 193-nm Lithography

Roderick R. Kunz, Susan C. Palmateer, Mark W. Horn, Anthony R. Forte, and Mordechai Rothschild

Lincoln Laboratory, Massachusetts Institute of Technology, Lexington, MA 02173–9108

We have optimized a positive-tone silylation process using polyvinylphenol resist and dimethylsilyldimethylamine as the silylating agent. Imaging quality and process latitude have been evaluated at 193 nm using a 0.5-NA SVGL prototype exposure system. A low-temperature dry etch process was developed that produces vertical resist profiles resulting in large exposure and defocus latitudes, linearity of gratings down to 0.175 μm, and resolution of 0.15-μm gratings and isolated lines.

Optical lithography using 193-nm radiation (ArF excimer laser) is a leading candidate for manufacturing integrated circuits with sub-0.25-μm features, such as the 1-Gb DRAM. For this technology to be realized photoresists sensitive to 193-nm radiation must be developed. At longer wavelengths, single-layer bulk photoresists have been and continue to be the dominant technology, largely because of their proven track record and mature commercial infrastructure. However, the single-layer bulk approach has become more complex and expensive as the exposure wavelength decreases to 248 and 193 nm. This trend stems from the need for acid-catalyzed resist chemistries and for antireflective coatings. Furthermore, at the shorter exposure wavelengths the thickness of the resist approaches the total focal plane depth of the exposure optics, thereby reducing the process depth of focus to values which are impractically small in a manufacturing environment.

Top-surface imaged (TSI) resists have been developed as an alternative to bulk imaged resists (1). These TSI resist processes operate via area-selective in-diffusion of a silyl amine into a phenolic polymer to form a silyl ether. Once the silicon has been selectively incorporated, the latent image is developed in an anisotropic oxygen plasma etch. At 193 nm, the simplest TSI resist scheme is based on photocrosslinking of the polymer, followed by selective silylation of the unexposed areas (2). Although chemical amplification schemes offer higher sensitivity (3) and may ultimately be the process of choice, the single-component direct crosslinking approach provides a simple platform for initial process evaluation of TSI at 193 nm. In this paper we report on recent lithographic results obtained by imaging with a prototype large-field step-and-scan exposure system, which is operational in Lincoln Laboratory's class 10 cleanroom.

0097–6156/95/0614–0271$12.00/0
© 1995 American Chemical Society

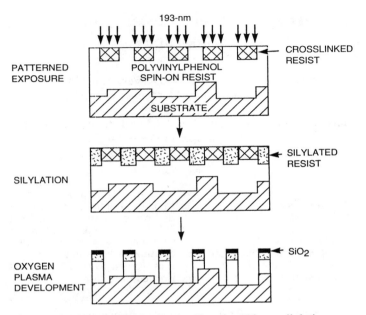

Figure 1   Schematic process flow for 193-nm silylation.

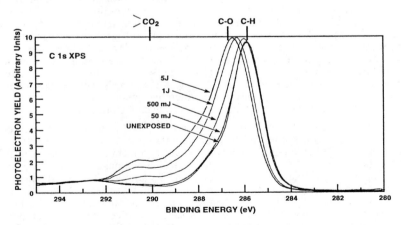

Figure 2   Carbon 1s x-ray photoelectron spectra for PVP irradiated at 193-nm in air for the indicated doses.

Experimental

The positive-tone 193-nm silylation resist process is illustrated schematically in Figure 1. A polyvinylphenol (PVP) resin of molecular weight M_w of 38,000 g/mol is selectively crosslinked by exposure on a 0.5-NA step-and-scan 193-nm optical lithography system. The sample is then treated with a silicon-containing vapor, dimethylsilyldimethylamine (DMSDMA), at a temperature of 90°C and a pressure of 25 Torr for 30 to 75 s. The uncrosslinked PVP areas incorporate a controlled amount of silicon, whereas the crosslinked areas exhibit near-zero silicon incorporation. The wafers are then dry developed in a high-ion-density helicon plasma etcher in an oxygen-based plasma (4,5). The oxygen reacts with the silylated areas to form SiO_2, which acts as an etch mask in the unexposed areas.

The silylation resist process has been evaluated as two separate steps, the silylating agent diffusion/reaction and the oxygen plasma etching. We have performed scanning electron microscope (SEM) metrology of the silylated profiles, including exposure-defocus matrices for the silylation step alone. This approach treats the silylation as a step which can be optimized independently from the exposure and etching.

In parallel, the etching step has been studied and optimized with respect to resist profile, etch rate, uniformity across the wafer, and selectivity between exposed and unexposed areas. First, unpatterned wafers consisting of a 0.50- to 0.76-µm-thick PVP film with a 100-nm-thick silylated layer were used to optimize the rf source power, chuck power, O_2 flow rate, temperature and pressure. An imaging interferometry system was used for these measurements, resulting in greatly reduced process development times (6). Then, patterned trilayers consisting of a 1.1-µm-thick hard-baked resist planarizing layer, a 0.2-µm-thick SiO_2 hardmask, and a 0.3-µm-thick resist imaging layer were used to optimize the resist profile independently of the silylation mask shape. Finally, patterned silylation wafers were etched to evaluate pattern transfer into resist and process latitudes for the silylation process. Lithographic results were all obtained by performing SEM measurements. An Amray SEM equipped with a WICS measurement system for top-down critical dimension (CD) measurements was used.

Results and Discussion

Resist and Silylating Agent Chemistry. Characteristics of 193-nm-based TSI resist processes differ from those at other wavelengths. For example, neat PVP used in these studies has an absorption coefficient α of roughly 25 µm^{-1} at 193 nm, limiting the crosslinking to the top ~50 nm, in contrast with α ~0.5 µm^{-1} at 248 nm. In addition, direct photochemical crosslinking at levels sufficient to inhibit silylation occurs at 193 nm for exposure doses of only 50 mJ/cm^2, whereas doses in excess of 500 mJ/cm^2 are necessary at 248 nm. As a result, efficient crosslinking at 248 nm can be accomplished only through addition of sensitizers and crosslinkers, and true TSI is only achieved through addition of a dye (7). These factors are important at 193 nm only when high photospeed (<20 mJ/cm^2) is desired. For unsensitized PVP, we believe the crosslinking mechanism to proceed via free-radical intermediates analogous to those present during UV-induced crosslinking of polystyrene (8). Little or no contribution to crosslinking is achieved through oxidative mechanisms involving quinone derivatives, as indicated by the carbon 1s XPS spectra of PVP irradiated at doses of 50, 500, and 2000 mJ/cm^2, shown in Figure 2.

Based on initial materials characterization we chose to use DMSDMA to silylate the PVP because of its high vapor pressure and small molar volume, which allows it to diffuse rapidly at low temperatures. This is useful for our single-component high-molecular-weight PVP resist which has a higher glass transition temperature (T_g), ~180°C, than many multiple-component chemically amplified TSI resists. One effect of this high T_g is slower diffusion rates which can be minimized by going to lower-molar-volume silylating agents. The larger-molecular-weight analogs trimethylsilyldimethylamine (TMSDMA) and trimethylsilyldiethylamine (TMSDEA) were also used to silylate PVP (9,10). The absolute diffusion/reaction rate is significantly faster for the smaller-molecule DMSDMA, while the silylation contrast for all three reagents are comparable. Because of these considerations we have optimized our process with DMSDMA, and the results reported below have been obtained with this silylating agent.

Silylation Mask Shape. For a positive-tone process, the silylation mask shape is governed by the silylating diffusion agent through the crosslinked latent image in the exposed resist film. For PVP and DMSDMA the diffusion process and its dependence on feature size and type have been studied in detail previously (11). Figure 3 shows SEM images of plasma-stained samples of the silylation mask shape for 0.25-μm gratings as a function of exposure energy and defocus, when the open field silylation is 200 nm deep. Note the relatively weak dependence of the silylated profile linewidth on defocus, i.e., only at the extremes of defocus (greater than ±0.8 μm) does the profile linewidth change dramatically. However, the profile angle, which is critical in determining how much dimensional loss occurs during the dry development step (12), does become somewhat narrower at modest (±0.5 μm) defocus (13). Not only does this behavior place greater demands on the dry development step to realize the maximum depth of focus, but we have found the profile angle to be profoundly affected by silylation conditions as well. As such, well-optimized dry development conditions are essential in realizing the maximum process window for a given set of silylation conditions. MOSES (Modeling of Surface Exposed Systems), an empirical model developed at Lincoln Laboratory, has been used to simulate silylation mask shape and its response to both defocus and the dry development conditions, to predict linewidth control and process latitude (13). Results of the simulations, together with the experimentally measured profiles, have been used to qualitatively optimize the silylation process latitude.

Dry Development. The following trends have been observed in the O_2-plasma etching of PVP, using the helicon etcher. The etch rate increases with increasing source power from ~0.95 μm/min at 1 kW to 1.4 μm/min at 2.5 kW. The etch rate increases with chuck power from 1.1 μm/min at 20 W to 1.7 μm/min at 90 W. The etch rate also increases with flow rate (from 0.85 μm/min at 25 sccm to 1.3 μm/min at 100 sccm). The etch rate increases from 1.4 to 1.6 μm/min when the wafer temperature is reduced from 30 to -70°C. This is consistent with etchants adsorbing to the resist: adsorption of reactive species increases at lower temperatures and enhances the ion-assisted etching rate (14,15). The etch rate is nearly independent of pressure between 1 and 6 mTorr. To achieve lower pressures the flow rate had to be decreased to 50 sccm, and these experiments indicated that from 1 to 0.5 mTorr the etch rate decreases from 1.05 to 0.90 μm/min. Our results also indicate that the etch rate varies with the amount of material being removed, consistent with results reported by Jurgensen *et al.* (16). Under identical plasma conditions, the etch rate decreases with increasing wafer size, from 1.65 μm/min for 3-in.-diameter wafers, to 1.35 μm/min for 4-in.-diameter wafers, to 1.05 μm/min for 6-in.-diameter wafers. The dependence of etch rate on flow rate and wafer size indicates that loading effects are important in this helicon etcher.

Selectivity between the silylated and unsilylated resist is the variable which most dramatically affects exposure, focus and etch process latitude as well as feature linearity. We have found that the chuck power is the etching parameter that produces the largest change in selectivity, from ~160:1 at 20 W to less than 20:1 at 90 W. Selectivity is nearly independent of pressure (2 to 6 mTorr) and wafer temperature (30 to -70°C) at 2-kW source power, 50- to 75-W chuck power and 100-sccm O_2. Again we note a change at low pressure where the selectivity decreases from 30:1 at 2 mTorr to 20:1 at 0.5 mTorr, all other variables remaining the same. The selectivity at a given chuck power can be increased by increasing the flow rate or the source power. To obtain acceptable linewidth control with the 193-nm silylation resist, MOSES modeling indicates that a selectivity greater than 20:1 is required. This requirement can be satisfied by lowering the chuck power. However, a lower limit on chuck power is imposed by the need to maintain a low lateral etch rate, for profile control (see below). Thus, an operating window on the chuck power is defined. For typical conditions of 2-kW source power, 2-mTorr pressure, 100-sccm O_2, and -70°C, it is 35 to 75 W. Under these optimized process conditions blanket wafer etch rate nonuniformity is less than ±3% over the central 85 mm of a 100-mm wafer.

In addition to adequate selectivity, it is necessary to optimize the resist profile. We used patterned trilayers with a SiO_2 hardmask to evaluate the effect of etch process parameters on resist profile under idealized silylation mask shape. A high degree of anisotropy is obtained by increasing the chuck power and decreasing the wafer temperature (14,15). Decreasing the wafer temperature from 30 to -70°C practically eliminates the isotropic etch component, while the vertical etch rate increases by about 15% (Figure 4). An alternative way to reduce the lateral etch rate is to use a sidewall passivating gas such as SO_2. However, the lateral etch rate is not eliminated but only reduced to ~20 nm/min, a value which causes a greater than 10% variation in 0.18-μm linewidths (17).

Lithographic Results. The overall lithographic performance of the silylation process depends on both the silylation step (i.e., the silylation mask shape) and the dry development step (i.e., selectivity, vertical/lateral etch rates and amount of overetch). One of the main criteria for good performance is linearity, i.e., the ability to print a wide range of feature sizes and types within prescribed tolerances (e.g., ±10% of linewidth). Figure 5 demonstrates the excellent linearity obtained with 193-nm TSI, using the process conditions listed in Table I for a 100% overetch and a 60-s silylation time (open field silylation depth 200 nm). For a wafer temperature of -70°C, linearity is maintained down to 0.20 μm for both a 25 and 100% overetch. For an etch temperature of 30°C, linearity is maintained down to 0.20 μm for a 25% overetch, but only to 0.30 μm for a 100% overetch. This is due to an increased isotropic etching component at 30°C. The nominal best dose for a 100% overetch is 20% lower than for a 25% overetch in order to enable more silylation which compensates for the linewidth erosion occurring during longer etch.

The preceding linearity data are for a fixed silylation time (60 s). Dense feature linearity was extended down to 0.175 μm for a 30-s silylation time using the process conditions in Table I and a 25% overetch. For longer silylation times (60 to 75 s) feature linearity could not be maintained below 0.25-μm features for these process conditions.

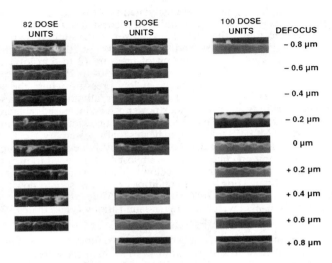

Figure 3   Scanning electron micrographs showing the variation in the profile of silylated areas with changes in exposure dose and defocus. The features were nominally 0.25-μm equal lines and spaces. Each 193-nm dose unit corresponds to ~1 mJ/cm². Silylation was performed with 25-Torr DMSDMA at 90°C for 60 s.

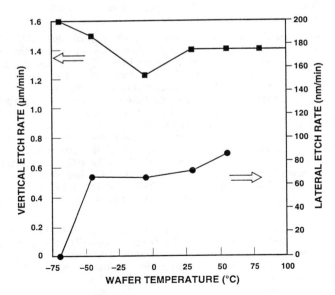

Figure 4   Vertical and lateral resist etch rates vs actual wafer temperature measured for 0.70-μm-wide isolated lines patterned in trilayers.

Table I   Silylation Process Parameters

Parameter	Conditions
Substrate	HMDS-treated Si (100) 100-mm-diameter wafers
Resist	0.50- to 0.76-µm PVP
Post-apply bake	120°C, 60 s
Exposure	SVGL 193-nm step and scan 0.5 NA
Silylation	DMSDMA 90°C/25 Torr/30-75 s
Dry development	2-kW source power, 75-W chuck power, 2-mTorr pressure, 100-sccm O$_2$, -70°C, helicon etcher
Resist etch rate	1.6 µm/min
Selectivity	27:1 open field

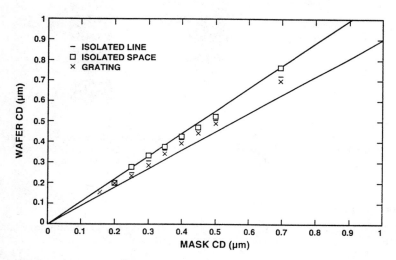

Figure 5   Silylation process linearity at best dose (100 dose units ~100 mJ/cm^2). Resist 0.76-µm-thick PVP silylated with DMSDMA at 90°C and 25 Torr for 60 s was etched under optimized conditions (see Table I) with a 100% overetch. The grating linearity is extended to 0.175 µm for a 30-s silylation time and 25% overetch. The two solid lines are the acceptable performance, i.e., ±10% deviations from the nominal feature size.

In addition to linearity, lithographic performance is also judged by the extent of the exposure and defocus process latitudes. Figure 6 is a graphic representation of the exposure-dose matrix for 0.20-, 0.175- and 0.15-μm gratings for a 30-s silylation time and a 25% overetch. At best dose the depth of focus is 1.6, 1.0, and 1.0 μm, respectively. However, this maximum depth of focus is obtained at higher doses as the feature size decreases. Such behavior is in qualitative agreement with the aerial image and corresponding exposure-defocus plots (18). Figure 7 shows SEMs of 0.20-μm silylated resist features, whereas Figure 8 shows 0.15-μm isolated and dense features.

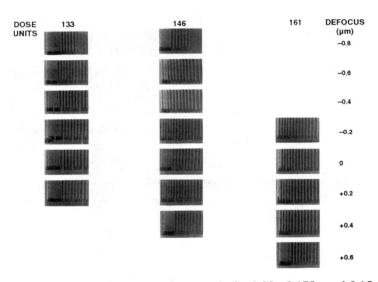

Figure 6   Experimental exposure-dose matrix for 0.20-, 0.175-, and 0.15-μm gratings, (±10% CD). Each 193-nm dose unit corresponds to ~1 mJ/cm^2. Resist 0.50-μm-thick PVP silylated with DMSDMA at 90°C and 25 Torr for 30 s was etched under optimized conditions (see Table I) with a 25% overetch.

Conclusions

The large process latitudes, linearity, depth of focus, and exposure latitude obtained using silylation at 193 nm make it an extremely attractive process for sub-0.25-μm lithography. By understanding both the silylation and dry development steps and their interplay, the process can be further optimized for different feature sizes and types. Other concerns regarding implementation of TSI in manufacturing are related to its perceived higher cost than that of single-layer resist. Although outside the scope of this paper, it is worth noting that a simple cost of ownership calculation demonstrates that the TSI approach is competitive with the single-layer approach (assuming the yields are equivalent) and that cost should not be the deterring factor in implementing TSI for 248-nm imaging (19).

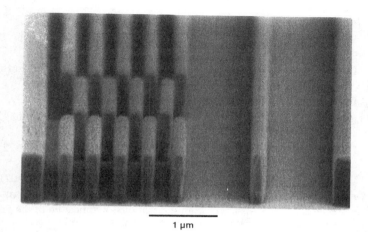

1 μm

Figure 7   Scanning electron micrograph of 0.20-μm silylated resist features. Resist 0.76-μm-thick PVP silylated with DMSDMA at 90°C and 25 Torr for 60 s was etched under optimized conditions (see Table I) with a 100% overetch.

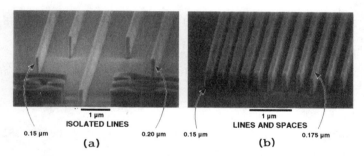

Figure 8   Scanning electron micrograph of 0.15-μm (a) isolated and (b) dense features. The doses used for these two features were not the same.

Acknowledgments

We would like to acknowledge the help of Lynn Eriksen on silylation, Scott Doran for aerial image simulations, and Cheryl Graves for SEM work. This work was supported by the Advanced Lithography Program of the Advanced Research Projects Agency. Opinions, interpretations, conclusions, and recommendations are those of the author and are not necessarily endorsed by the United States Air Force.

Literature Cited

1.   Wolf, T. M.; Taylor, G. N.; Venkatesan, T.; and Kraetsch, R. R., *J. Electrochem. Soc.* **131**, 1664 (1984).

2.   Hartney, M. A.; Rothschild, M; Kunz, R. R.; Ehrlich, D. J.; and D. C. Shaver, *J. Vac. Sci. Technol. B* **8**, 1476 (1990).

3.   Hartney, M. A. and Thackeray, J. W., *SPIE Proc.* **1672**, 486 (1992).

4.   Benjamin, N.; Chapman, B.; and Boswell, R., *SPIE Proc.* **1392**, 95 (1990).

5.   Horn, M. W.; Hartney, M. A.; and Kunz, R. R., *SPIE Proc.* **1672**, 448 (1992).

6.   Low Entropy Systems, Inc., 83A Monroe Street, Somerville, MA 02143.

7.   Linehan, L.; Conley, W.; Stewart, K.; Wood, R.; Muller, K. P.; and LaTulipe, D., in Extended Abstracts, 10th International Symposium of the Society of Plastics Engineers, Inc., Mid-Hudson Section, Ellenville, NY, October 31 – 2 November 1994.

8.   Ranby, B. and Rabek, J. F., in Photodegradation Photo-oxidation and Photostabilization of Polymers: Principles and Applications, Wiley-Interscience, New York, 1975.

9.   Hartney, M. A.; Johnson; D. W.; and Spencer, A. C., *SPIE Proc.* **1466**, 238 (1991).

10.   Dao, T. T.; Spence, C. A.; and Hess, D. W., *SPIE Proc.* **1455**, 257 (1991).

11.   Hartney, M. A., *J. Vac. Sci. Technol. B* **11**, 681 (1993).

12.   Goethals, A. M.; Baik, K. H.; Ronse, K.; Vertommen,J.; and Van den hove, L, *SPIE Proc.* **2195**, 394 (1994).

13.   Kunz, R. R.; Hartney, M. A.; and Otten, Jr., R. W., *SPIE Proc.* **1927**, 464 (1993).

14.   Tachi, S.; Tsujimoto, K.; and Okudaira, S., *Appl. Phys. Lett.* **52**, 616 (1988).

15.   Bensaoula, A.; Ignativ, A.; Strozier, J.; and Wolfe, J. C., *Appl. Phys. Lett.* **49**, 1663 (1986).

16.   Jurgensen, C. W.; Hutton, R. S.; and Taylor, G. N., *J. Vac. Sci. Technol. B* **10**, 2542 (1992).

17.   Joubert, O.; Martinet, C.; and Pelletier, J., *SPIE Proc.* **1803**, 130 (1992).

18.   Palmateer, S. C.; Kunz, R. R.; Horn, M. W.; Forte, A. R.; and Rothschild, M., *SPIE Proc.* **2438** (1995), in press.

19.   LaTulipe, J. C.; Simons, J. P.; and Seeger, D. E., *SPIE Proc.* **2195**, 372 (1994).

RECEIVED August 14, 1995

Chapter 19

Silicon-Containing Block Copolymer Resist Materials

Allen H. Gabor and Christopher K. Ober

Department of Materials Science and Engineering, 214 Bard Hall, Cornell University, Ithaca, NY 14853

Polymer chains of controlled block and graft architecture have been investigated for use as microlithographic resists. For the last ten years, such research has focused on polymers having a silicon-containing block. These materials offer highly desirable properties, including excellent oxygen reactive ion etch resistance and good thermal stability due to their microphase separated structure. We review the use of silicon-containing block copolymers as resist materials and update the reader on our own research in this developing area. This will include a discussion of our most recent block copolymer resists which are developable in environmentally friendly supercritical CO_2.

Materials designed for use as resists-of-the-future must meet stringent requirements. In addition to resolving features less than 0.3 μm, the resist must stand up to both dry and wet etches, develop in environmentally friendly solvents, adhere well to the wafer, have high contrast and be highly sensitive to the exposing radiation. We believe that block copolymers will allow further improvements in resists so that these demands can be met. Although, block copolymer resists have been prepared and studied for the last two decades, there seems to be a general lack of awareness of this literature. Some of this is due to the limited readership of work "published" only as patents or in conference proceedings.

This paper consists of three sections. In the first section we briefly review polymer phase behavior so that the reader understands what parameters need to be considered when designing block copolymer resists. In the second section we review past research on silicon-containing block copolymers designed to be microlithographic resists. We also review silicon-containing graft copolymer resists which, while not block copolymers, do have a controlled architecture. (By 'controlled architecture' we mean that the placement of the monomer units making up the polymer chain is not random but instead is regulated by the polymer chemist. The polymer chemist, for example, could prepare a graft copolymer by copolymerizing a monomer with a macromonomer or alternatively could prepare a block copolymer through a sequential addition of different monomers to a reactor.) In the final section we present a detailed account of our own research on block copolymers for 193 nm resist materials. Often, the superior properties of polymers with controlled

0097–6156/95/0614–0281$12.00/0

architecture are caused by the microphase separated structures that these types of polymers form. We therefore begin by reviewing the phase behavior of polymers.

Section I: Phase Behavior of Polymers

Different polymers are rarely miscible with each other and thus when blending two polymers, phase separation almost always occurs (1). Unlike small molecules, which have a high entropic driving force for mixing, large linear chain molecules have a low entropic driving force for mixing (2). The enthalpic forces generally favor phase separation since the interaction energy is usually lower when two identical molecules are neighbors, than when two different molecules are neighbors. The morphology, which results from the phase separation of two polymers, is dependent on kinetic factors. In order to lower the free energy, the mixture decreases the area of interface between the two phases (3). This results in materials with phases of macroscopic dimensions and in many cases with disappointing properties (4). For example, the strength of the interface between the phases can be weak due to the absence of chains bridging the interface.

In contrast, many of the spectacular properties of block copolymers result from the microphase separated structures that they form (5). Block copolymers consist of at least two different polymer segments, covalently connected together. A block copolymer with two blocks is termed a diblock copolymer. While the blocks often phase separate due to a low entropy of mixing, they are not able to macrophase separate due to the covalent bonds connecting them. The connections between the blocks will cause the mixture to have a larger interfacial area than the corresponding homopolymer mixture. The interfaces of the block copolymers, will be mechanically strong due to the chains of the block copolymers crossing them (6). Although the morphology of block copolymers is commonly described as microphase separated, nanophase separated is actually more descriptive since the size of the phases is roughly equal to the radius of gyration of the blocks, typically 5-30 nm (7).

A diblock copolymer, that does microphase separate, can have either an ordered or disordered microstructure depending on the magnitude of the enthalpic driving force for phase separation. In order for block copolymers to form an ordered microphase separated structure, the enthalpic driving force must be higher than the minimal enthalpic driving force needed for a mixture of corresponding homopolymers to phase separate. This is because there is an entropic penalty associated with the requirement that the joints between the two blocks be positioned at an interface (8). Also, the compositional sharpness of the interface increases with the enthalpic driving force for phase separation. The enthalpic driving force increases as χN increases where χ is the Flory Huggins parameter and N is the number of monomeric repeat units of the chain (9). Thus, one way to increase the driving force is to increase the length of the block copolymer. The Flory Huggins parameter, χ, relates the interaction energy cost associated with having the same or different nearest neighbors. For symmetric diblock copolymers (block copolymers which have two distinct blocks each occupying the same volume), when χN is much greater than 10, the interfaces between the two phases are narrow and compositionally sharp and the microstructure is ordered as shown in Figure 1 (10). As χN decreases towards 10, the width of the interface is increased but the microphase separated structure is still ordered. At approximately $\chi N=10$, there is an order-disorder transition (ODT). Block copolymers with values of χN greater than ~2 but less than 10 still microphase separate, but the structures formed do not have long range order. For χN less than ~2, the two blocks are miscible with each other and the material is homogeneous.

When the enthalpic driving force is large (χN greater than 10), different ordered microstructures can be obtained depending on the length ratio of the two blocks. Figure 2 shows the structures predicted theoretically for a diblock copolymer (9). The symmetric diblock copolymer forms lamellae where the thickness of each lamella is on the order of the radius of gyration of the polymer. Diblock copolymers with asymmetric blocks have different equilibrium structures. At very asymmetric block ratios, the block copolymers form spheres of the minority phase in a continuous matrix of the majority phase. The structures shown in Figure 2 are at equilibrium because eliminating additional interfaces actually increases the free energy since such elimination would require the polymer chains to stretch further out of their random coil configuration (11).

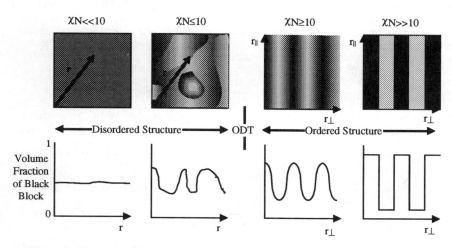

Figure 1. A symmetric block copolymer with both a black and a white block will phase separate when the enthalpic driving force (χN) is large enough. The top pictures show an aerial view of the monomeric-unit composition. The bottom graphs show the volume fraction of black block as a function of r for the disordered structures and r perpendicular to the lamellae for the ordered structures (adapted from 10).

Increasing length of black block relative to white block

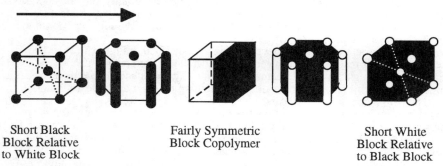

| Short Black Block Relative to White Block | Fairly Symmetric Block Copolymer | Short White Block Relative to Black Block |

Figure 2. Microphase separated structures predicted by theory for a diblock copolymer. In this figure one block is color coded black while the other block is coded white (adapted from 9).

The different microstructures formed by block copolymers act as micro-composites. For example, poly(isoprene-*b*-styrene) which has a microstructure of spheres of poly(isoprene) in a continuous matrix of poly(styrene) has two distinct glass transition temperatures (Tg's): a low Tg for the isoprene phase and a high Tg for the styrene phase. However, the material would not become rubbery until the temperature was raised above the Tg of the continuous poly(styrene) phase. Thus, a material can have the high modulus and thermal stability of poly(styrene) but take advantage of toughening by poly(isoprene). At the other extreme of monomeric composition, very different properties can be observed. For example a poly(styrene-*b*-isoprene-*b*-styrene) which has a rubbery continuous phase of poly(isoprene) can have properties similar to a rubber with chemical crosslinks. The behavior occurs because the high Tg poly(styrene) phase acts as a physical crosslink for the poly(isoprene) phase. However, this block copolymer rubber is thermoplastic since when heated above the Tg of the poly(styrene) it can be reshaped (5). When block copolymers are properly designed and applied they offer the undiminished desirable properties of each phase and minimize the undesirable properties of each phase. The superior properties of block copolymers are the reason that several different groups have investigated block copolymers as resists.

Section II: Review of Silicon-Containing Copolymer Resists With Controlled Architecture

Copolymers with controlled architecture have been sporadically investigated as resist materials for the last 20 years. They were and still are studied because of their promise of offering superior properties compared to corresponding random copolymers, i.e. copolymers having similar molecular weight and chemical composition to that of the block copolymer, but with the monomer units in a random sequence. During the last ten years most research on block copolymer resists has focused on preparing and using them as the imageable polymer of bilevel resist schemes.

Bilevel Resist Scheme. Bilevel resist schemes are effective in the intermediate to advanced stages of lithography where a single-level resist might not cover the circuitry previously deposited on the silicon wafer. The bilevel resist scheme involves planarizing the surface of a wafer with a nonimageable, organic layer and imaging the desired pattern with a thin, top-coated resist (12). There are other lithographic advantages to the bilevel resist system (13). Due to the thinness of the imageable layer, optical lithography can be performed with minimal linewidth irregularities associated with the problem of limited depth of focus. Also, when exposing an electron beam (e-beam) resist, fewer backscattered electrons "expose" the resist, since many are stopped in the planarizing layer. Finally, the bilayer system offers greater ease in obtaining high aspect ratio features (the height of a feature divided by the width) since a planarizing layer with thermal stability superior to that of the resist can be used.

The lithographic steps for bilevel imaging are outlined in Figure 3 (14). The resist must offer resistance to oxygen RIE so that the imaged pattern can be transferred through the organic planarizing layer. The resistance to the oxygen RIE in most cases is achieved by using a silicon-containing polymer as the resist material. Resists with higher silicon-concentrations generally have increased oxygen RIE resistance. When the etch rate of the silicon-containing resist is at least 20 times slower than that of the organic planarizing layer, reproducible pattern transfer can be achieved (15). Thus, there is a minimal concentration of silicon necessary for the resist to have adequate oxygen RIE resistance. In an oxygen RIE, the top surface of the silicon-containing polymer is oxidized. The oxidized surface was the subject of many investigations, and it was shown that the oxidized top surface was similar in chemical bonding to silica (15,16). After a silica layer forms, the film is etched by sputtering which is a much slower process than the chemical etching of the unprotected organic planarizing layer (17).

Despite all their benefits, silicon-containing resists are not used commercially. One problem associated with siloxane based polymers is their low Tg, which leads to dimensional instability of the imaged features (18). Traditionally the Tg of silicon-containing resists was raised by randomly copolymerizing with a "high Tg" monomer. This method is a compromise in that the dimensional stability and oxygen RIE resistance are intermediary between those of the two parent homopolymers. In contrast properly designed block copolymers are able to be both dimensionally stable and have superb oxygen RIE resistance. Another problem with silicon-containing resists is their extreme hydrophobicity which makes aqueous base developability difficult (14). In our research, preliminary evidence points to the possibility of using aqueous base developer for block copolymer resists with high silicon-content (19).

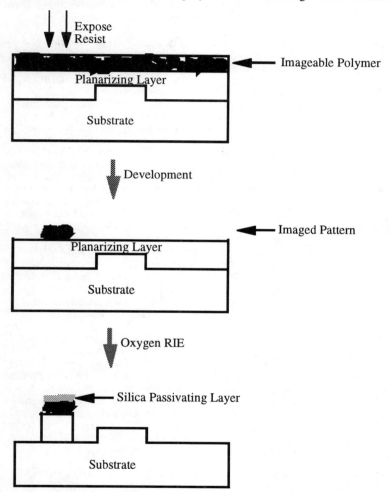

Figure 3. Bilevel resist scheme involves imaging a pattern into a top-coated resist and then transferring the pattern through the planarizing layer.

First Use of Silicon-Containing Block Copolymers as Resist Materials. Hartney et al. were the first to report using silicon-containing block copolymers as resist materials (20-22). They demonstrated that two different polymers, which can not be used as a polymer blend because of macroscopic phase separation, can be used as a resist when prepared in a block copolymer configuration. They prepared diblock-triblock mixtures of poly(*p*-methylstyrene-*b*-dimethylsiloxane) and poly(*p*-methylstyrene-*b*-dimethylsiloxane-*b*-*p*-methylstyrene) which they subsequently chloromethylated. Table I shows the chemical structure for this and the other polymers discussed in this review. The chloromethylated polymer had a slightly larger polydispersity than the unmodified mixture of poly(p-methylstyrene-b-dimethylsiloxane) and poly(p-methylstyrene-b-dimethylsiloxane-b-p-methylstyrene). Two different block ratios were synthesized. The first (CPMS/DMS 1) had a relatively short poly(dimethylsiloxane) [PDMS] block and a silicon-concentration of 4.0 wt %. The second (CPMS/DMS 2) had a longer PDMS block and a silicon-concentration of 15.5 wt %. CPMS/DMS 1 and CPMS/DMS 2 were effective e-beam resists with sensitivities of 2.2 and 0.9 μC/cm^2 and contrasts of 1.4 and 1.3 respectively. With CPMS/DMS 2, 0.75 μm lines were imaged.

The sensitivity of the CPMS/DMS 2 was actually slightly better than that of a chloromethylated poly(p-methylstyrene) homopolymer of similar molecular weight to the block copolymer. In comparison, the sensitivity of a random copolymer consisting of two different monomer units, is usually intermediate to the sensitivity of the two parent polymers. The high sensitivity was surprising since poly(dimethylsiloxane) is a low sensitivity e-beam resist. Several factors could be involved in the high sensitivity of this block copolymer resist. Because the two blocks are each in their own distinct phase, the block with a high sensitivity is not diluted by the block with a lower sensitivity. Brewer had found in 1977 a similar improvement in e-beam sensitivity for styrene-diene block copolymers (23). Hartney et al. proposed, as an explanation for the improved sensitivity, that in the microphase separated structure the PDMS microphase acted as an effective crosslink. While this is probably not true for all block copolymer resists, we believe that for crosslinking resists, a second phase can act to effectively decrease the dose needed if the developer used has a greater affinity for the block of higher sensitivity. Thus, we believe that if a developer with a greater affinity for the PDMS was used, the sensitivity improvement caused by the PDMS "crosslinks" would not occur.

Hartney et al. demonstrated that the block copolymers could be mixed with chloromethylated poly(p-methylstyrene) homopolymer to modify the oxygen RIE resistance. The CPMS/DMS 2 had the slowest etch rate, etching 13 times slower than a hard baked novolak. As the percent silicon was decreased, by adding homopolymer, the etch rate increased. In their study, the weight percent silicon needed to be greater than 14 to obtain an etch selectivity ratio higher than 10.

First Use of Silicon-Containing Graft Copolymers as Resist Materials. Bowden et al. in 1987 published their work on graft copolymers having a poly(methyl methacrylate) [PMMA] main-chain with PDMS grafts (24). PMMA is positive tone while PDMS is negative tone when exposed to e-beam. When exposed to e-beam the resist was negative tone. However, the resist had poor contrast and even in the best cases 40% of the resist film was removed. The authors attributed this to the slightly higher sensitivity of PDMS. The resist was negative tone, when exposed to e-beam, even when the PDMS was the minority phase. When exposed to deep UV, the resists were low sensitivity (2700 mJ/cm^2) positive tone resists. The oxygen RIE resistance of the graft copolymers was intermediate to that of the parent homopolymers.

Block Copolymers with Deep UV Imageable Blocks. In 1987 Crivello received a patent for a novel method of preparing block copolymers containing both PDMS blocks and imageable blocks (25). He prepared several block copolymers having various imageable blocks using a novel free radical initiator which had PDMS units attached. Among the imageable blocks he prepared were poly(t-BOC styrene) which is the basis for the current generation of 248 nm resists and poly(tert-butyl methacrylate) [P(t-BMA)] which is being investigated for use as a 193 nm chemically amplified resist (see Table I). The block copolymers had approximately 10 wt % Si. Crivello was able to image 2 μm features with these resists, using 250 nm, contact printing. The resists were developed with aqueous base even though the copolymers had hydrophobic PDMS blocks ranging from 1400 to 4500 g/mol. One problem with the PDMS radical initiators that Crivello used, and with the resulting polymers, is that they contain $\equiv$Si-OR bonds (R is either an organic group or an organic polymer). The

≡Si-OR bond can be hydrolytically cleaved, as will be discussed in the ≡Si-OR bond stability section.

Block Copolymer-Homopolymer Resist Mixtures. In 1989 Allen and MacDonald investigated the oxygen RIE resistance of block copolymer-homopolymer mixtures (26). Before their work, researchers believed it was necessary to have greater than 8 wt % silicon in the resist to obtain selectivities acceptable for reproducible pattern transfer. Allen and MacDonald proposed that the silicon-concentration only needed to be greater than 8 wt % in the top surface of the resist, and that this could be achieved using a surface active, microphase separated block copolymer.

They designed and synthesized the triblock copolymer: poly(2,6-dimethyl-1,4-phenylene oxide-*b*-dimethylsiloxane-*b*-2,6-dimethyl-1,4-phenylene oxide) [PXE-*b*-PDMS-*b*-PXE] (see Table I) (27). Because PDMS has a lower surface energy than poly(2,6-dimethyl-1,4-phenylene oxide) [PXE], it segregates to the surface. PXE is known to be miscible with poly(styrene) (28), and the block copolymers they prepared were compatible with poly(styrene) homopolymer. One problem with PXE-*b*-PDMS-*b*-PXE, which is otherwise well designed, is the presence of ≡Si-OR bonds connecting the PXE and PDMS blocks, which results from the coupling chemistry used. As will be discussed in the ≡Si-OR bond stability section, this bond is susceptible to hydrolysis, and we therefore believe that the block copolymers can have a short shelf-life.

Allen and MacDonald's proposal, that uncompromised oxygen etch resistance could be obtained by using silicon-containing block copolymer resists, did not follow smoothly from the previously performed research. As discussed in the subsections reviewing the earlier studies of Bowden et al. (24) and of Hartney et al.(21), silicon-containing surface active block copolymers had been found to have etch resistance intermediate to that of the parent homopolymers. Indeed Allen and MacDonald measured poor etch resistance with their block copolymers, one of which had 4 wt % silicon and one of which had 5 wt % silicon. Thus, the idea turned into two questions: 1.) 'Why is the etch resistance of the surface active block copolymers not enhanced?' 2.) 'How thick does the top surface need to be, to obtain the desired etch resistance?' They asked the latter question since it was known that the etch rate decreases as the silica-like layer gradually builds up during the oxygen RIE. The thickness of the silicon-containing layer therefore needs to be at least thick enough to allow the protecting silica film to form. There are two ways to increase the thickness of the silicon-containing surface layer. The first is to increase the molecular weight of the PDMS block. The second is to swell the silicon-containing surface layer by adding PDMS homopolymer. Allen and MacDonald were not able to increase the molecular weight of the PDMS block due to limitations with their synthetic method. However, they were able to swell the silicon-containing surface layer by adding PDMS homopolymer of a molecular weight less than that of the PDMS block. They found that as long as r was less than 0.4, where r is the ratio of unbound to bound PDMS, homopolymer could be blended with the block copolymer without macroscopic phase separation occurring.

Adding the PDMS homopolymer had a dramatic effect on the oxygen RIE rates of the polymer blends. The etch rate of the triblock which had 5 wt % silicon was only two times slower than the etch rate of poly(styrene). This same block copolymer, when blended with PDMS homopolymer (r=0.2, a total of 6 wt % silicon) was not observed to etch during the RIE treatment. A blend that consisted of a 1:1 mixture of the triblock copolymer and poly(styrene) and PDMS homopolymer (r=0.2, a total of 3 wt % silicon) had an etch rate 80 times slower than poly(styrene) homopolymer. Allen and MacDonald studied the surface composition of their blends using variable angle XPS. Based on the XPS results they estimated that the silicon-enriched surface layer should be at least 5 nm thick and preferably greater than 10 nm to obtain good etch resistance. The dramatic increases in etch resistance were maximized by using oxygen RIE conditions that gave optimal selectivity. Furthermore, the block copolymer-homopolymer systems used by Allen and

Table I.   Polymers With Controlled Architecture Studied For Use As Microlithographic Resists

Block or Graft Copolymer Resists	Authors	Citation
(A)	Hartney, Novembre and Bates	(20-22)
(B)	Crivello	(25)
(B)	Crivello	(25)
(B, C)	Allen and MacDonald	(26,27)
(B, D) $R_1 = CH_3$ or other; $R_2 = H$ or OH	Jurek et al.	(29,30)
(B, D)	Jurek et al.	(29,30)

Table I. Continued

Structure	Reference	
$-(CH_2-\underset{\underset{C=O}{\overset{CH_3}{\mid}}}{C})_m$ ran $(CH_2-\underset{\underset{C=O}{\overset{CH_3}{\mid}}}{C})_n$ ran $(CH_2-\underset{\underset{C=O}{\overset{CH_3}{\mid}}}{C})_o$ — with side groups including $H-C-OH$, CH_3, $(CH_2)_3$, $CH_3-Si-CH_3$, and boxed R: $-CH=CH_2$, $:-CH=C\underset{}{\overset{H}{\mid}}$ (phenyl), $:$ (phenyl)$-N_3$	Tachibana et al.	(33)
$-(CH_2-\underset{\underset{CH_3}{\overset{CH_2}{\mid}}}{CH}-\underset{O}{\overset{O}{\parallel}}S)_x$ ran $(CH_2-\underset{\underset{CH_3-Si-CH_3}{\overset{(CH_2)_4}{\mid}}}{CH}-\underset{O}{\overset{O}{\parallel}}S)_y$	DeSimone et al.	(34,35)
$-(CH_2-CH)_n^b$ (phenyl) $[(CH_2-\underset{\underset{CH_2}{\overset{R_1}{\mid}}}{C})_{m1}$ ran $(CH_2-\underset{\underset{CH-R_2}{\overset{R_1}{\mid}}}{C})_{m2}]_o$ — R_1=H or CH$_3$; if R_1=CH$_3$ then R_2=H; if R_1=H then R_2=H or CH$_3$	Gabor et al.	(36-41)
$-(CH_2-\underset{\underset{C=O}{\overset{CH_3}{\mid}}}{C})_m^b$ $(CH_2-\underset{\underset{C=O}{\overset{CH_3}{\mid}}}{C})_n$ — with $(CH_2)_3$, $CH_3-Si-CH_3$, O, $CH_3-Si-CH_3$, CH_3 and CH_3-C-CH_3, CH_3	Gabor and Ober	(19,42)

(A) Is a mixture of diblock and triblock copolymers.

(B) Bond connecting the two blocks is a $\equiv$Si-OR bond which is susceptible to hydrolysis and condensation as discussed in the $\equiv$Si-OR bond stability section.

(C) Is not imageable but was used to study the oxygen RIE resistance of silicon-containing block copolymers.

(D) Synthetic procedure couples difunctional oligomers of dimethylsiloxane and multifunctional phenolic polymers together. Therefore branching between the phenolic polymers is possible.

MacDonald were model systems and were not imageable. Regardless, this work demonstrated that it is possible to obtain superior etch resistance, at low silicon-concentration, if favorable microphase separation of the block copolymers takes place. It also demonstrated that it should be possible to greatly enhance the etch resistance of more conventional resist materials by adding suitable block copolymers.

Phenolic Resin-Oligomeric Dimethylsiloxane (DMS) Resists. In 1989, Jurek et al. prepared graft copolymers consisting of an oligomeric DMS block and one of three different phenolic resins [poly(hydroxy styrene), 2-methyl resorcinol novolac and o-cresol novolac; see Table I] (29,30). The phenolic resins all function as resists when mixed with an appropriate amount of diazonaphthoquinone dissolution inhibitor. The graft copolymers were formed through a condensation reaction between a small fraction of the hydroxyl groups of the phenolic polymers and the dimethylamine groups of oligomeric DMS that had both chain-ends functionalized. Crosslinking of the mixture was minimized by keeping the ratio of oligomeric DMS blocks to phenolic blocks less than one. The $\equiv$Si-OR bonds which form during the condensation reaction are susceptible to hydrolysis, as will be discussed in the next section. Thus, the polymers may not be stable. This could perhaps explain some of the reported TEM micrographs which show rather large scale phase separation in which the DMS domains were approximately 20-200 nm (30). Such domains are much larger than we might expect for oligomeric DMS of molecular weight ranging from 510 to 4400 g/mol. Regardless, with these systems Jurek et al. were able to demonstrate several points about silicon-containing, phase separated resists. The resist with the best oxygen RIE resistance, was found to etch 36 times slower than a hard baked novolak while for the poorest resist the ratio was reduced to 13. The resist with the highest silicon-content at the air-resist surface had the slowest oxygen RIE rate. All the polymers were soluble in aqueous base. However, when prepared with diazonaphthoquinone dissolution inhibitor many of the formulated resists had unsatisfactory performance, due to pinholes forming in the unexposed regions during development. This was attributed to phase separation (30). The better resist formulations were able to resolve 0.5 micron line-space patterns which were transferred through an organic planarizing layer using oxygen RIE.

$\equiv$Si-OR bond stability. The $\equiv$Si-OR bond is susceptible to hydrolysis in mildly acidic or basic conditions, and its stability decreases further if the pH departs greatly from neutral (31,32). The incorporation of $\equiv$Si-OR bonds (designated 'B' in Table I) into the back-bone of the block copolymer chain is a weakness with several of the early block copolymer resist studies. When the $\equiv$Si-OR bond connects two blocks, degradation will result in a mixture of homopolymers. Of the pioneering studies investigating block copolymer resists with $\equiv$Si-OR bonds (25-27,30), only Jurek et al. discuss the stability issue (29). We feel that it is best to avoid designing block copolymer resists containing $\equiv$Si-OR bonds. However, when block copolymers resists with $\equiv$Si-OR bonds are used the shelf-life and stability during lithographic processing should be monitored. If using a photo-acid generator (PAG) for chemical amplification, or aqueous base as a developer, the chance of the $\equiv$Si-OR bond being cleaved is magnified. This could be especially deleterious if scum-free aqueous base developability is desired, since PDMS homopolymer is not aqueous base soluble.

Photosensitive Silicon-Containing Graft Copolymers. In 1990 Tachibana et al. published a paper on negative tone graft copolymers which were imaged with either 254 or 365 nm exposure tools (33). The polymers were prepared by the random copolymerization of the macromonomer, 3-methacryloxypropyl-oligodimethylsiloxane with methyl methacrylate and glycidyl methacrylate. The glycidyl group was converted to either a cinnamoyl, a phenyl azide or an acrylic group in a subsequent polymer modification reaction (see Table I). The cinnamoyl and phenyl azide modified polymers were sensitive to the exposing radiation, while the acrylic modified polymer was photosensitized when a photo-radical initiator was added. The resulting negative tone resists had sensitivities ranging from 5.7 to 22 mJ/cm^2 and contrasts ranging from 0.8 to 1.5, when organic solvent was used as developer. Similar to the observations made by Brewer (23) and Hartney et al. (21) for negative tone e-beam resists, Tachibana et al. observed that the sensitivity of their resists was at least as good as that of homopolymers of the photo-active component. However, it is not clear that the high sensitivity is caused by microphase separation, since no DSC evidence showing two glass transition temperatures was presented and the TEM micrograph shown did not clearly demonstrate that phase separation was occurring. The molecular weight of the oligodimethylsiloxane was less than 1000 g/mol, which does not give the grafts a large driving force for phase separation.

E-beam Sensitive Silicon-Containing Graft Copolymers. In 1991, DeSimone et al. published a detailed paper on the synthesis, characterization and lithographic properties of poly(1-butene sulfone-*g*-dimethylsiloxane) [PBS-*g*-PDMS] (see Table I) (34). They prepared 5-hexenyl functionalized PDMS macromonomers of 1x10^3, 5x10^3, 10x10^3 and 20x10^3 g/mol by terminating the anionic polymerization of hexamethylcyclotrisiloxane with 5-hexenyldimethylchlorosilane. All the macromonomers were of low polydispersity. They then made a family of PBS-*g*-PDMS by copolymerizing the different macromonomers with 1-butene and sulfur dioxide. The graft copolymers ranged from 2 to 9 wt % silicon and were roughly 50,000 g/mol. All the graft copolymers microphase separated, except for the ones containing the short 1000 g/mol PDMS graft. The microphase separation was confirmed by DSC which showed a Tm corresponding to the PDMS phase and a Tg corresponding to the PBS phase. Also, TEM micrographs convincingly showed a microphase separated structure. Variable angle XPS showed that the surfaces of films of the PBS-*g*-PDMS were enriched with the PDMS component.

Dry etch resistance could expand the use of PBS from its current application of being the e-beam resist used to image chrome plated photo-masks (35). With this in mind, DeSimone et al. designed PBS-*g*-PDMS. The oxygen RIE resistance for a PBS-*g*-PDMS with 6.4 wt % Si was significant, etching 29 times slower than hard-baked novolak. The polymers with 2 wt % Si etched quickly, making them unsuitable for the imageable layer of a bilevel resist. PBS is a positive tone e-beam resist while PDMS is a negative tone e-beam resist. However, PBS has a higher sensitivity than PDMS and minimal crosslinking of the PDMS phase occurred at the dose range used for imaging. These positive tone resists have a sensitivity of 4.5 μC/cm^2 and a contrast less than 2 using 5-methyl-2-hexanone as the developer. 0.2 μm lines with 0.4 μm spaces were successfully imaged using e-beam lithography, and transferred into an organic planarizing layer using oxygen RIE.

E-beam Sensitive Silicon-Containing Block Copolymers. Styrene-diene block copolymers are known e-beam resists and were the first block copolymers used as a resist material (23). In 1992 we first described our work on negative tone e-beam resists (36-38). For these e-beam resists, styrene-diene block copolymers (the most common industrially used block copolymers) were chosen to be a template. A method for attaching hydrosiloxanes to the pendant vinyl groups of the diene block via hydrosilylation was developed (39). Using a Pt based catalyst the 1,2-units of poly(butadiene) and the 1,2- and 3,4-units of poly(isoprene) were hydrosilylated with hydrosiloxanes such as pentamethyldisiloxane (see Table I). DSC showed two distinct Tg's for these polymers. The hydrosilylated poly(styrene-*b*-isoprene) resists were stable with no evidence of degradation during 3 years. In contrast, the hydrosilylated poly(styrene-*b*-butadiene) resists were not stable and crosslinked with time (40). Along with the obvious difference in monomeric composition, the difference in stability may be related to molecular weight. The stability issues deserve further attention, especially given the excellent properties these negative tone e-beam resists exhibit. These silicon-containing block copolymer resists did not contain any ≡Si-OR bonds, and were the first to use a silicon-containing block other than PDMS. The resists containing 12.1 wt % silicon had high resistance to oxygen

RIE, etching 42 times slower than poly(imide). Poly(styrene-*b*-isoprene) hydrosilylated with pentamethyldisiloxane had a sensitivity of 30 $\mu C/cm^2$ and a contrast of 2.8 using a mixture of toluene and acetone (99:1) as the developer. Line-space (0.2 μm) patterns and high density circuit patterns with features as small as 0.1 μm, as shown in Figures 4 and 5, were imaged (41). Using oxygen RIE, features with aspect ratios as high as 4.5 were transferred through a poly(imide) planarizing layer (40). The literature reviewed above was the basis for our most recent work on silicon-containing block copolymers for 193 nm lithography. The remainder of this paper will go into detail about our work in this new area.

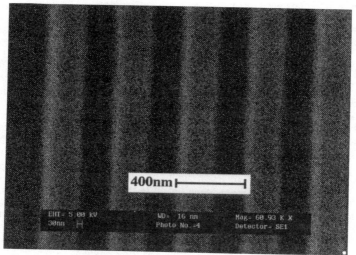

Figure 4. 0.2 μm line space pattern imaged with a poly(styrene) penta-methyldisiloxane modified poly(isoprene) block copolymer (reprinted with permission from reference 41).

Figure 5. Circuit pattern with 0.1 μm features imaged with a poly(styrene) pentamethyldisiloxane modified poly(isoprene) block copolymer (reprinted with permission from reference 41).

Section III: Silicon-Containing Block Copolymers for 193 nm Lithography.

The study of *t*-BMA and 3-methacryloxypropylpentamethyldisiloxane (SiMA) based block copolymer resists for 193 nm lithography began in 1993 (42). These polymers were designed to address the problem of low dry etch resistance of 193 nm resists developed by IBM and MIT Lincoln Laboratory (43,44). The polymers were prepared using the living polymerization method known as group-transfer polymerization (GTP). GTP has many attributes which make it excellent for preparing acrylic based resists, including the ability to control the molecular weight of the chain simply through the ratio of monomer to initiator. GTP requires a catalyst of which oxyanion and bifluorides are the most common. While oxyanions catalyze the GTP of *t*-BMA nicely, the GTP of SiMA using oxyanions has to date only occurred at unacceptably slow rates. The rate of polymerization of SiMA was increased by using the more aggressive tris(dimethylamino)sulfonium bifluoride (TASHF2) catalyst. Also, we report on the free radical polymerization of random copolymers that have the same chemical composition and molecular weight of corresponding block copolymers.

The block copolymers [P(*t*-BMA-*b*-SiMA)] (see Table I) have a low normalized optical absorption at 193 nm of less than $0.1/\mu m$. After exposure P(*t*-BMA-*b*-SiMA) has better development behavior in aqueous base than that of the corresponding random copolymer (41). Also, the copolymers with higher silicon-concentrations are developable as negative tone resists using supercritical CO_2.

Experimental. Methyl trimethylsilyl dimethylketene acetal (MTSDA) and carbon tetrabromide were obtained from Aldrich, *t*-BMA from Scientific Polymer Products, and SiMA from Huls America. Tris(dimethylamino)sulfonium difluorotrimethylsiliconate (TASF) was obtained from PCR and used as received. The THF, used in the GTP, was reagent grade, and was distilled from Na and benzophenone. The THF used in the free radical polymerizations was used as received in SureSeal™ bottles from Aldrich. 2,2'-Azobisisobutyronitrile (AIBN) was obtained from Eastman Chemicals and recrystallized prior to use. The acetonitrile was reagent grade and was distilled from phosphorus pentoxide. The *t*-BMA and SiMA were purified by drying over CaH_2 followed by distillation at reduced pressure. MTSDA was purified by distillation at reduced pressure.

As GTP and free radical polymerizations are sensitive to water and oxygen respectively, the polymerizations were done under argon in dry glassware using cannula and syringe techniques and a double manifold inert atmosphere system (45). Unless otherwise indicated all polymerizations were done at RT. Spectra were obtained using a Varian XL-200 NMR spectrometer. We have measured the T_1 value of protons similar to those of the methyl groups attached to the Si atoms of SiMA and found that they are greater than 2 seconds (40). When taking the NMR spectrum of polymers containing SiMA, a delay time of 15 seconds was used between pulses to obtain proper integration. Number average molecular weight (Mn) and polydispersity (Mw/Mn) were found by GPC (THF) using PMMA standards.

Preparation of TASHF2. Using the purchased TASF, TASHF2 was prepared and recrystallized as described by Sogah et al. (46) ^{1}H NMR (CD_3CN, 200 MHz, -30 °C) δ 2.81 (s, 18 H, CH_3N), δ 16.34 (t, J_{HF} = 121.3 Hz, 1 H, FHF). (Note: The δ value of +16.34 ppm for the proton of the bifluoride differs in sign from the value that Sogah et al. (46) reported due to a typing error in their paper. The literature is confusing as Fujiwara and Martin have measured shieldings of -15.8 to -16.8 ppm for the proton of bifluoride, but shieldings increase in sign opposite to the δ scale (47).) A 0.4 M solution of the TASHF2 was prepared in acetonitrile.

Homopolymerization of SiMA with TASHF2. A 100 mL, three necked round bottom flask, with a stir bar, attached to a THF still was fitted with a thermometer and septum. After distilling 15 mL of THF into the reactor, 0.122 mL (0.6 mmol) of

MTSDA was added. Two min later 0.010 mL (4 μmol) of the TASHF2 solution was added to the reactor. After 10 min, 6.6 mL (21.9 mmol) of SiMA was added, at a rate of 0.5 mL/min using a syringe. The temperature increased from 24.2 °C to 31.4 °C during the addition of the first 4 mL of SiMA but then cooled to 30.1 °C during the addition of the final 2.6 mL. Two hours after the start of the SiMA addition, 0.5 mL of the solution was removed and precipitated into 5 mL of methanol. Two days after the addition of the SiMA, 1 mL (24.8 mmol) of methanol was added to quench the polymerization. After stirring overnight the polymer was precipitated by pouring the solution into 200 mL of methanol. The precipitates were allowed to settled for 30 min and were collected after decanting off the methanol based solution. The sticky white polymer was then washed with two 100 mL aliquots of methanol. The polymer was dried in vacuo at RT to constant weight. The yield, ignoring the effect of sampling, (4.73 g) was 79 %. GPC: Mn(2h) 8910 g/mol, Mw/Mn 1.07, Mn(48h) 8960 g/mol, Mw/Mn 1.07. The theoretical Mn, for the P(SiMA) is 10120 g/mol.

Block Copolymerization of SiMA and *t*-BMA with TASHF2 (Figure 6). A 100 mL, three necked round bottom flask, with a stir bar was fitted with a thermometer, septum and an addition funnel. The top of the addition funnel had a septum port and joint used to attached it to a THF still. After distilling 15 mL of THF into the reactor, 0.122 mL (0.6 mmol) of MTSDA was added. Five min later 0.0075 mL (3 μmol) of the TASHF2 solution was added to the reactor. After 15 min, 6.6 mL (21.9 mmol) of SiMA was added to the addition funnel and then dripped into the round bottom flask at a rate of 0.5 mL/min. The temperature increased from 23.8 °C to 26.3 °C during the addition of the 6.6 mL of SiMA. Four hours after the start of the SiMA addition, 0.5 mL of the solution was removed and precipitated into methanol. At 4.5 hours after the start of the SiMA addition, an additional 0.0075 mL (3 μmol) of the TASHF2 solution was added to the reactor and the temperature increased 2 °C in the next 10 minutes. No increase in temperature was observed after an additional 0.0075 mL (3 μmol) of the TASHF2 solution was added 7.5 hours after the start of the SiMA addition. A 0.5 mL aliquot of the solution was removed and precipitated into methanol. The reaction vessel was the cooled to -25 °C (48). Using a syringe, 6.9 mL (42.4 mmol) of *t*-BMA was added to the addition funnel followed by the distillation into the addition funnel of 14 mL of THF. The solution of *t*-BMA and THF was then dripped into the reaction vessel at a rate of 1.5 mL/min. After stirring overnight, the polymerization was quenched with 1 mL (24.8 mmol) of methanol. Stirring was continued for three days and then the polymer was precipitated by pouring the solution into 400 mL methanol. The precipitates were allowed to settled for one hour and were collected after decanting off the methanol based solution. The white polymer was dissolved in 40 mL of THF, reprecipitated in 400 mL methanol and then collected and dried in vacuo at RT. The yield, ignoring the effect of sampling, (5.2 g) was 43 %. GPC: P(SiMA) samples: Mn(4h) 6196 g/mol, Mw/Mn 1.06, Mn(7.5h) 8010 g/mol, Mw/Mn 1.06. The theoretical Mn, for the P(SiMA) block is 10120 g/mol. P(SiMA-*b*-*t*-BMA): Mn 14860 g/mol, Mw/Mn 1.09. The theoretical Mn, for the block copolymer, ignoring the effect of sampling, is 20,155 g/mol. Analysis of the NMR spectra of the block copolymer revealed that the weight ratio of the SiMA block to the *t*-BMA block was 1.33 (theory 1.01).

Random Copolymerization of SiMA and *t*-BMA. A 300 mL, three necked round bottom flask, with a stir bar, was fitted with a condenser and septum. Carbon tetrabromide (0.093 g, 0.28 mmol) and AIBN (0.135 g, 0.82 mmol) were added along with 100 mL of THF. After adding 3.7 mL (12.3 mmol) of SiMA and 7.6 mL (46.7 mmol) of *t*-BMA, two freeze-thaw cycles were performed. The reaction vessel was heated to 65 °C for 44 hours. The polymer was precipitated by pouring the solution into a mixture of 100 mL water and 400 mL methanol. The polymer was recovered by filtration, dissolved in 40 mL of THF and reprecipitated in a mixture of 100 mL water and 400 mL methanol. The polymer was dried in vacuo at RT. The yield of polymer was 8.77 g (87.7%). GPC: Mn 13,020 g/mol, Mw/Mn 2.24. Analysis of the

NMR spectra of the random copolymer revealed that the weight ratio of the SiMA to the *t*-BMA was 0.52 (theory 0.51).

Results and Discussion. Previously we reported on the preparation of P(*t*-BMA-*b*-SiMA) using oxyanion catalyzed GTP (42). Using oxyanion catalysis resulted in slow rates of polymerization of SiMA. Over the period of 120 hours, the polymer continued to increase in molecular weight until reaching the theoretical molecular weight of 10,000 g/mol. During this period the polydispersity of the polymer stayed less than 1.08 (42). The slow rate of polymerization using oxyanion catalysis was in contrast to the fast polymerization using bifluoride catalysis. For example, the homopolymerization of SiMA to 8910 g/mol in two hours was at least 25 times faster than that obtained using tetrabutylammonium 3-chlorobenzoate (42). This increase in the rate of polymerization is roughly what is expected as it was reported that the bifluoride catalysts are about 100 times more active than the tetrabutylammonium carboxylates (49). Several GTP reactions of SiMA were done using the TASHF2 and the molecular weights were almost always near the theoretical value, and the polydispersities were always lower than 1.10. The few times that the molecular weight of P(SiMA) was not near its theoretical value it was low, indicating incomplete conversion of the SiMA added to the reactor. This only occurred when using TASHF2 which was exposed to the atmosphere and then "restored" by pumping on it at 0.001 Torr overnight as described by Sogah et al (46). When the initial conversion of the SiMA was incomplete it could be increased and the molecular weight of the P(SiMA) edged closer to theoretical by adding additional catalyst as described above in the block copolymer synthesis. With *t*-BMA the GTP always gets deactivated using TASHF2 regardless of how the TASHF2 was handled previous to its addition to the reactor. The low yield of the block copolymer as described in the experimental section, and the weight ratio of SiMA to *t*-BMA skewed from theoretical, both indicate that there is incomplete conversion of the *t*-BMA added to the reactor. We are currently investigating prevention of this deactivation but it is clear that the polymer chains are not dead because the GTP can be reactivated by injecting additional catalyst to the reactor.

A family of block copolymers with different block ratios was prepared using GTP. When the GTP of *t*-BMA was deactivated during the synthesis of any of the polymers belonging to the family, additional catalyst was added to reactivate the polymerization. In this way we were able to polymerize the block copolymers to near their expected molecular weights (based on the ratio of monomer to initiator) and to high yield. Along with making a family of block copolymers with different compositions, a family of corresponding random copolymers was also prepared. The random copolymers were synthesized either by free radical polymerization with a chain transfer agent, as described above, or by GTP through the simultaneous addition of *t*-BMA and SiMA.

With the families of block and random copolymers we are comparing the lithographic properties and obtaining information on the specific benefits that block copolymers offer. Obtaining random and block copolymers of similar molecular weights is necessary since it is known that the dissolution behavior of a resist is affected by molecular weight. For example, the random copolymerization described in this paper leads to a polymer that corresponds very well to a P(*t*-BMA-*b*-SiMA) that was synthesized previously (42). So far, evidence suggests that the block copolymer architecture might allow the incorporation of larger amounts of hydrophobic siloxane-units into a polymer, compared to a corresponding random copolymer, while still maintaining the ability to develop in aqueous base. When mixed with a PAG such as triphenylsulfonium hexafluoroantimonate the block copolymers are aqueous base developable, positive tone resists with sensitivities between 4 and 40 mJ/cm^2. The corresponding random copolymers do not develop in aqueous base except at low silicon-concentrations. We are currently trying to understand the causes of the difference in aqueous base developability between the corresponding random and block copolymers.

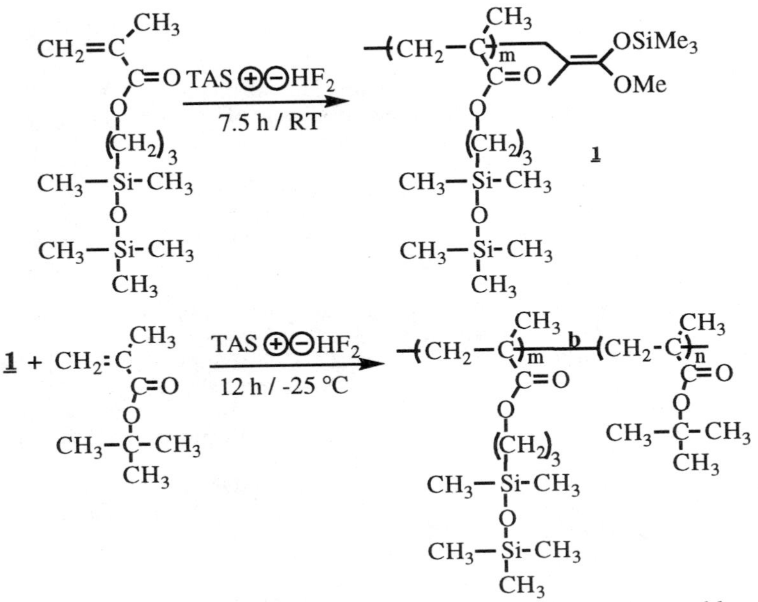

Figure 6. Synthetic scheme used to prepare a 193 nm imageable
block copolymer resist based on SiMA and t-BMA.

Environmentally Friendly Supercritical CO_2 Resist Development. IBM Almaden
and Phasex have recently introduced the concept of using supercritical CO_2 as an
environmentally friendly developer (50). Very few polymers are soluble in
supercritical CO_2, and those which are soluble usually are either fluorinated or
siloxane based (51). However, resists which are soluble in supercritical CO_2 either
before or after exposure, offer the possibility of environmentally friendly
development. In addition, supercritical CO_2 is a very selective solvent allowing
subtle chemical changes within a resist to lead to very high contrast (50). We have
recently demonstrated, in collaboration with Dr. Robert Allen of IBM and Paula
Gallagher-Wetmore of Phasex that our block copolymers with higher silicon-
concentrations are developable as negative tone resists using supercritical CO_2.
These same resists are developable as positive tone resists using aqueous base. In the
future we will continue to both develop dual tone resists, and to specifically tailor
resists for supercritical CO_2 developability.

Conclusions

Silicon-containing resists of controlled architecture have excellent properties many of
which corresponding random copolymers cannot offer. For example, Hartney et al.
demonstrated that a negative tone resist could have sensitivity higher than the
sensitivity of a homopolymer of the more sensitive block. Allen et al. demonstrated
that excellent oxygen RIE resistance could be achieved by properly designing the
block copolymer resist system. Block copolymers have been able to resolve features
feature as small as 0.1 μm using e-beam lithography. The qualities mentioned above,
clearly indicate that block copolymer resists have great potential to not only meet but
also exceed the performance requirements of future resist materials.

We have developed a synthetic approach for preparing block copolymers based
on t-BMA using GTP. With this chemistry the next generation of (193 nm) resists
can take advantage of the improved properties of controlled architecture polymers.

The block copolymers prepared with a large component of hydrophobic siloxane monomeric-units are still developable in aqueous base as positive tone resists. Negative tone development of the block copolymers with higher siloxane-content is also possible using supercritical CO_2.

Acknowledgments

Funding by the SRC Microscience and Technology Program and SRC Lithography Program is gratefully acknowledged. We thank Professor Dotsevi Sogah, Professor Jean Fréchet, Dr. Roderick Kunz and Dr. Robert Allen for their helpful discussions and suggestions. We also wish to thank past and current members of the Ober research group and Lisa Papernik for their help, support and insight. This work was performed in part at the National Nanofabrication Facility which is supported by the National Science Foundation under Grant ECS-8619049, Cornell University and industrial affiliates. This work also made use of MRL Central Facilities supported by the National Science Foundation under Award No. DMR-9121654.

Literature Cited

(1)   Scott, R. L. *J. Chem. Phys.* **1949**, *17*, 279.
(2)   Flory, P. J. *J. Chem. Phys.* **1942**, *10*, 51.
(3)   Bates, F. S.; Wiltzius, P. *J. Chem. Phys* **1989**, *91(5)*, 3258.
(4)   McCrum, N. G.; Buckley, C. P.; Bucknall, C. B. *Principles of Polymer Engineering*; Oxford University Press: New York, 1988.
(5)   Noshay, A.; McGrath, J. E. *Block Copolymers-Overview and Critical Survey*; Academic Press: New York, 1977.
(6)   Kramer, E. J.; Norton, L. J.; Dai, C.; Sha, Y.; Hui, C. *To be Published in Faraday Discuss.* **1994**, *98*.
(7)   Meier, D. J. *J. Polym. Sci.* **1969**, *26C*, 81.
(8)   Helfand, E.; Wasserman, Z. R. In *Developments in Block Copolymers-1* Goodman, I., Ed. Applied Science: New York, 1982; p 99-125.
(9)   Bates, F. S.; Fredrickson, G. H. *Annu. Rev. Phys. Chem.* **1990**, *41*, 525.
(10) Bates, F. S. *Science* **1991**, *251*, 898.
(11) Bates, F. S.; Schulz, M. F.; Khandpur, A. K.; Förster, S.; Rosedale, J. H.; Almdal, K.; Mortensen, K. *To be Published in Faraday Discuss.* **1994**, *98*.
(12) Hatzakis, M.; Paraszczak, J.; Shaw, J. *Proc. Microcircuit Engrg.* **1981**, 396.
(13) Lin, B. J. In *Introduction to Microlithography* Thompson, L. F., Willson, C. G. and Bowden, M. J., Eds.; ACS Symposium Series 219; ACS: Washington DC, 1983; p 288.
(14) Reichmanis, E.; Smolinsky, G.; Wilkins, C. W. *J. Solid State Technology* **1985**, 130.
(15) Reichmanis, E.; Smolinsky, G. *Proc. SPIE* **1984**, *469*, 38.
(16) Chou, N. J.; Tang, C. H.; Paraszczak, J.; Babich, E. *Appl. Phys. Lett.* **1985**, *46*, 31.
(17) Hartney, M. A.; Hess, D. W.; Soane, D. S. *J. Vac. Sci. Technol. B* **1989**, *7(1)*, 1.
(18) McDonnell Bushnell, L. P.; Gregor, L. V.; Lyons, C. F. *Solid State Technology* **1986**, 133.
(19) Gabor, A. H.; Ober, C. K. *Proc. of the A.C.S. Div. of PMSE* **1995**, *72(1)*, 104.
(20) Hartney, M. A.; Novembre, A. E. *Proc. SPIE* **1985**, *539*, 90.
(21) Hartney, M. A.; Novembre, A. E.; Bates, F. S. *J. Vac. Sci. Technol. B* **1985**, *3(5)*, 1346.
(22) Bates, F. S.; Hartney, M. A.; Novembre, A. E. U.S. Patent 4,892,617, 1990.
(23) Brewer, T. L. U.S. Patent 4 061 799, 1977.
(24) Bowden, M. J.; Gozdz, A. S.; Klauser, C.; McGrath, J. E.; Smith, S. In *Polymers for High Technology: Electronics and Photonics* Bowden, M. and Turned, S. R., Eds.; ACS Symposium Series 346; ACS: Washington DC, 1987; p 122.
(25) Crivello, J. V. U.S. Patent 4,689,289, 1987.

(26) Allen, R. D.; MacDonald, S. A. In *3rd Intern. SAMPE Electronics Conf.*; 1989; pp 919-928.

(27) Allen, R. D.; Hedrick, J. L. *Polymer Bulletin* **1988**, *19*, 103.

(28) MacKnight, W. J.; Karasz, F. E.; Fried, J. R. In *Polymer Blends* Paul, D. R. and Newman, S., Eds.; Academic Press: New York, 1978; Vol. 1; p 185.

(29) Jurek, M. J.; Tarascon, R. G.; Reichmanis, E. *Chemistry of Materials* **1989**, *1*, 319-324.

(30) Jurek, M. J.; Reichmanis, E. In *Polymers in Microlithography* Reichmanis, E., MacDonald, S. A. and Iwayanagi, T., Eds.; ACS Symposium Series 412; ACS: Washington, DC, 1989; p 158.

(31) Iler, R. K. *The Chemistry of Silica*; John Wiley & Sons: New York, 1979.

(32) Schmidt, H.; Scholze, H.; Kaiser, A. *J. Non-Crystalline Solids* **1984**, *63*, 1.

(33) Tachibana, Y.; Yasuda, Y.; Jitsumatsu, T.; Koseki, K.; Yamaoka, T. *Polymer* **1990**, *31*, 1553.

(34) DeSimone, J. M.; York, G. A.; McGrath, J. E.; Gozdz, A. S.; Bowden, M. J. *Macromolecules* **1991**, *24*, 5330.

(35) Bowden, M. J.; Gozdz, A. S.; DeSimone, J. M.; McGrath, J. E.; Ito, S.; Matsuda, M. *Makromol. Chem., Macromol. Symp.* **1992**, *53*, 125.

(36) Gabor, A. H.; Lehner, E. A.; Mao, G.; Schneggenburger, L. A.; Ober, C. K. *Polym. Prep.* **1992**, *33(2)*, 136.

(37) Gabor, A. H.; Lehner, E. A.; Long, T. E.; Mao, G.; Rauch, E. C.; Schell, B. A.; Ober, C. K. *Polym. Prep.* **1993**, *34(1)*, 284.

(38) Gabor, A. H.; Lehner, E. A.; Long, T. E.; Mao, G.; Schell, B. A.; Tiberio, R. C.; Ober, C. K. *Proc. SPIE* **1993**, *1925*, 499.

(39) Ober, C. K.; Gabor, A. H.; Lehner, E. A.; Mao, G.; Schneggenburger, L. A. U.S. Patent 5,290,397 and 5,318,877, 1994.

(40) Gabor, A. H.; Lehner, E. A.; Mao, G.; Schneggenburger, L. A.; Ober, C. K. *Chemistry of Materials* **1994**, *6(7)*, 927.

(41) Gabor, A. H.; Chan, M. Y.; Ober, C. K. In *Proc. of the 10th Inter. Conf. on Photopolym*; SPE: Ellenville NY, 1994; p 339.

(42) Gabor, A. H.; Ober, C. K. *Polym. Prep.* **1993**, *34(2)*, 576.

(43) Wallraff, G. M.; Allen, R. D.; Hinsberg, W. D.; Simpson, L. L.; Kunz, R. R. *CHEMTECH* **1993**, *23(4)*, 22.

(44) Kunz, R. R.; Allen, R. D.; Hinsberg, W. D.; Wallraff, G. M. *SPIE* **1993**, *1925*, 167.

(45) Burlitch, J. M. "How to Use Ace No-Air Glassware; Bulletin No. 3841," Ace Glass Co.; Vineland, NJ.

(46) Sogah, D. Y.; Hertler, W. R.; Webster, O. W.; Cohen, G. M. *Macromolecules* **1987**, *20*, 1473.

(47) Fujiwara, F. Y.; Martin, J. S. *Journal of the American Chemical Society* **1974**, *96(25)*, 7625.

(48) Doherty, M. A.; Müller, A. H. E. *Makromol. Chem.* **1989**, *190*, 527.

(49) Webster, O. W.; Anderson, B. C. In *New Methods for Polymer Synthesis* Mijs, W. J., Ed. Plenum Press: New York, 1992.

(50) Gallagher-Wetmore, P.; Wallraff, G. M.; Allen, R. D. *Proc. SPIE* **1995**,

(51) Krukonis, V. *Polymer News* **1985**, *11*, 7.

RECEIVED August 14, 1995

Chapter 20

A Top-Surface Imaging Approach Based on the Light-Induced Formation of Dry-Etch Barriers

U. Schaedeli[1], M. Hofmann[1], E. Tinguely[1], and N. Münzel[2]

[1]Materials Research, Ciba-Geigy Inc., 1723 Marly/Fribourg, Switzerland
[2]OCG Microelectronic Materials AG, Klybeckstrasse 141, Basel, CH–4002, Switzerland

It is believed that optical lithography will continue to play an important role for the manufacturing of future generation IC devices. Surface imaging processes might be required to overcome the small depth of focus associated with new high numerical aperture exposure tools. Our new concept is based on the light induced formation of sites for the fixing of reactive monomers in the top zones of a highly absorbing resist film. These sites, which typically consist of photolytically generated Bronsted acid or radicals, are able to oligomerize silicon-containing, reactive monomers during a baking step. The net result is the formation of a silicon gradient in the top zone of the exposed resist film, which subsequently acts as a dry etch barrier. Negative tone images were obtained by exposing resist films, mainly composed from a linear matrix polymer, a sulfonium salt type photoacid generator, and silicon-containing epoxy monomers, with light of 193 nm or 254 nm wavelength.

Historically, the production of integrated circuits has relied exclusively on photolithography as the method for transferring mask patterns onto silicon chips. It is believed that optical lithography will continue to play an important role for the manufacturing of future generation microdevices, where resolution well below 0.2 micrometer will be needed. Surface imaging processes, where the actual interaction between light and matter is limited to the near surface of a resist film, might be required to overcome the small depth of focus, associated with the new high numerical aperture exposure tools. There has been a continuous trend towards shorter wavelengths to meet the need for smaller feature size. At the forefront of the technology roadmap stands deep-UV lithography (DUV), and 193 nm lithography as the most likely candidate for becoming the technology of choice for future advanced lithography.

0097–6156/95/0614–0299$12.00/0
© 1995 American Chemical Society

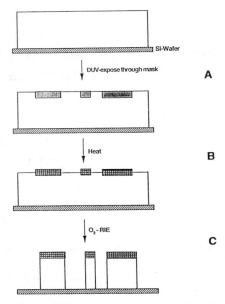

Scheme 1. Process sequence involved in the image forming process.

Figure 1. SEM micrograph of the latent image as formed after post exposure bake on a resist film composed from 9 (25 parts), PS (75 parts), TPST (5 parts), and MA (1 part).

Top surface imaging concepts can mainly be separated in two distinct categories: (I) two layer approaches, where a relatively thin image forming layer, which acts as an etch barrier, is located above of a relatively thick planarizing layer (1-6) and (II) single layer approaches, where the etch barrier is formed in the top zones of a resist film (7-14). In both cases, the image of the mask, which has been defined in the top zones of the film, is subsequently transferred down to the semiconductor substrate by means of dry etching. In approach (I), the etch barrier forming substance is typically an organosilicon compound, and can be present in the original resist formulation. This is contrasted to approach (II), in which a silylation process is needed after the image defining step to allow for the formation of the dry etch barrier. The major drawback of materials based on approach (I) is their need for additional process steps. On the other hand, the silylation process in approach (II) needs to be very strictly controlled in order to obtain reproducible lithographic results.

The inherent advantages of approaches (I) and (II) are combined to design a new resist approach which allows for a process consisting of a minimum number of steps (single layer) and avoids the need for a silylation process. In addition, the concept should have enough flexibility to allow imaging at both 248 nm and 193 nm wavelengths, without the need for introducing changes in the chemistry involved in the image forming process.

Results and Discussion

Imaging Concept. The process steps involved in the image forming sequence of our top surface imaging approach are outlined in Scheme 1. In a first step (A), a resist film which consists of a matrix polymer, a photoacid or photo radical generator, a silicon-containing reactive monomer and an optional light absorber is imaged through a mask. The net result is the generation of sites for fixing the polymerizable monomers in the uppermost zones of the highly absorbing resist film. These sites are composed of photolytically generated Bronsted acid or radicals.

In a second step (B), the wafer containing the illuminated resist film is heated. Thereby the reactive, silicon-containing monomers, which can be considered as a highly boiling solvent, start migrating in the resist matrix. Eventually, these monomers reach the top layers of the film. In areas which have been irradiated before, and which do therefore contain the fixing sites, immobilization of the reactive monomer takes place due to rapid oligomerization. If sufficiently high doses are applied, the reactive monomers cannot leave the film in these irradiated areas. The result is the formation of a silicon gradient in the illuminated zones of the film. However, in the unirradiated zones, where no initiators had been introduced, the reactive monomers can leave the resist film. Whereas the original film thickness is retained in the irradiated zones of the film, a volume loss is observed in the unirradiated zones, leading to the formation of a clearly visible topographic image of the mask pattern (Figure 1).

In the third step (C), the mask pattern, which was generated in the near surface region of the resist film, is transferred down to the semiconductor substrate by means

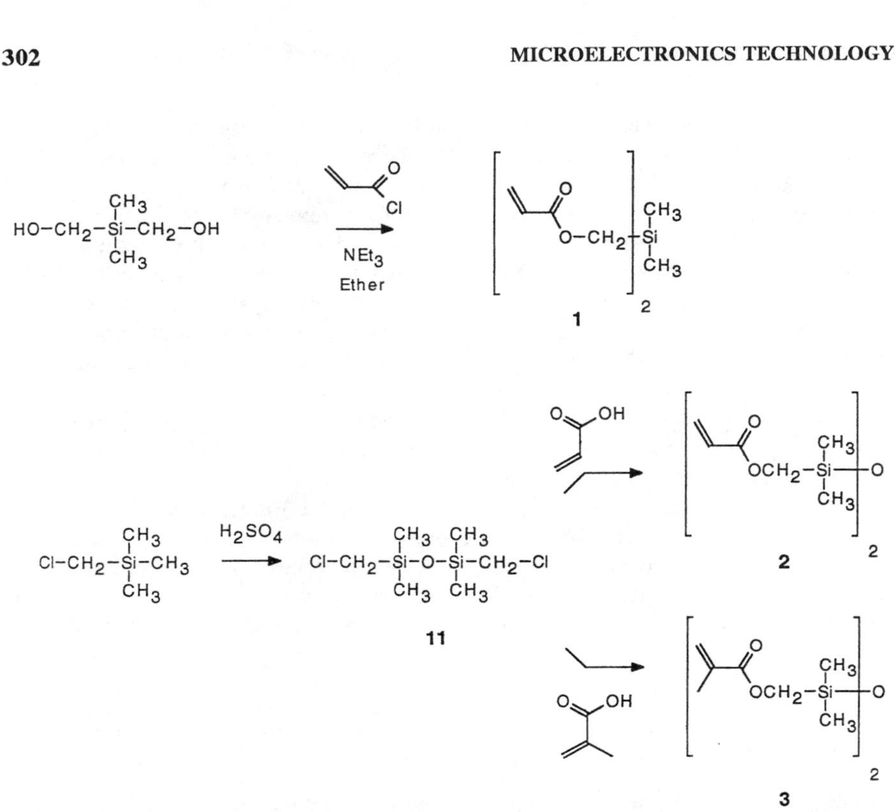

Scheme 2.   Synthetic sequences involved in the preparation of Si-(meth)acrylates.

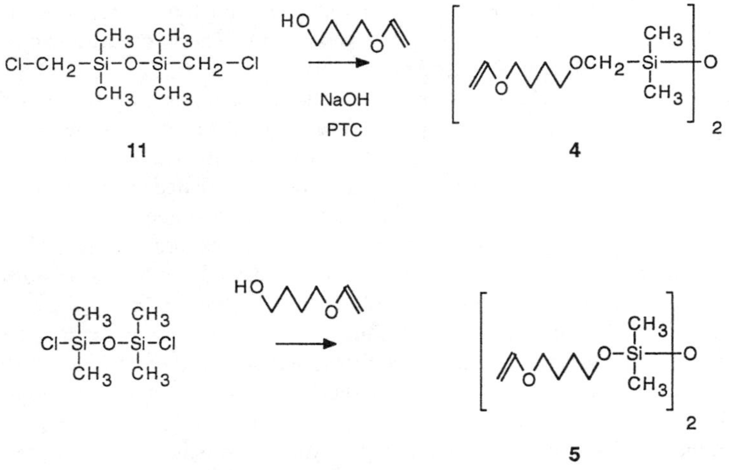

Scheme 3.   Synthetic sequences involved in the preparation of Si-vinyl ethers.

of oxygen plasma etching, since the silicon enriched zones of the film act as effective etch barriers.

Reactive Monomers. A variety of functional groups are known to effectively undergo polymerization in the presence of catalytic amounts of initiating radicals or acid. Since the contact time of the reactive monomers with the initiator species is short, only monomers with highly reactive functional groups were considered in this study. It has been shown before that reactive monomers can be locked in resist films that were imaged using X-ray and photolithography (*15-18*). Whereas (meth)acrylates are most efficiently locked by radical induced oligomerization, vinyl ethers and some epoxides are known to undergo rapid oligomerization in the presence of strong Bronsted acids.

(Meth)acrylates: A series of silicon-containing (meth)acrylates was prepared as outlined in Scheme 2.

Vinyl Ethers: Vinyl ethers are, for acid catalyzed induction, among the fastest oligomerizing functions. However, they are highly susceptible towards trace amounts of impurities. Therefore, actions need to be taken to carefully purify all of the resist components with respect to elimination of acid and metal impurities. Without taking these measures, there is a high risk of obtaining irreproducible lithographic results. The two bifunctional vinyl ethers, which were synthesized according to Scheme 3, were used in this study.

Epoxides: In general, the acid induced polymerization of epoxides proceeds slower compared to vinyl ethers. However, there are considerable differences in reactivity for different types of epoxides: if the epoxy functionality is part of an aliphatic ring system, a much higher reactivity towards acid induced reaction is found compared to epoxides of type 6 or 7. Nevertheless, these highly reactive epoxides are much less susceptible towards trace impurities than vinyl ethers. Besides 6 and 7, which are commercially available, the synthesis of a series of reactive epoxides, starting from the corresponding cyclohexenols or cyclohexene chlorosilanes, was successfully performed (Scheme 4).

The physical properties of the reactive silicon-containing monomers used in this study are summarized in Table I.

Matrix Polymers. It is obvious that the physical properties of the matrix polymer play an important role in our imaging system. First of all, the reactive monomer needs to be sufficiently soluble in a given matrix in order to obtain homogeneous resist films. Incompatibilities between the reactive monomer and the polymeric matrix were determined by observing resist films with the naked eye, with turbid films indicating nonsufficiant miscibility between monomer and matrix. In addition, no undesired side reactions between the reactive monomer and the matrix should occur to avoid an uncontrolled formation of a silicon gradient in the film.

It was found that reactive monomers having low silicon contents (typically <20%) are well miscible with a variety of matrix polymers like polystyrene, polystyrene copolymers, poly-p-hydroxystyrene, poly(methyl methacrylate) or novolaks, yielding

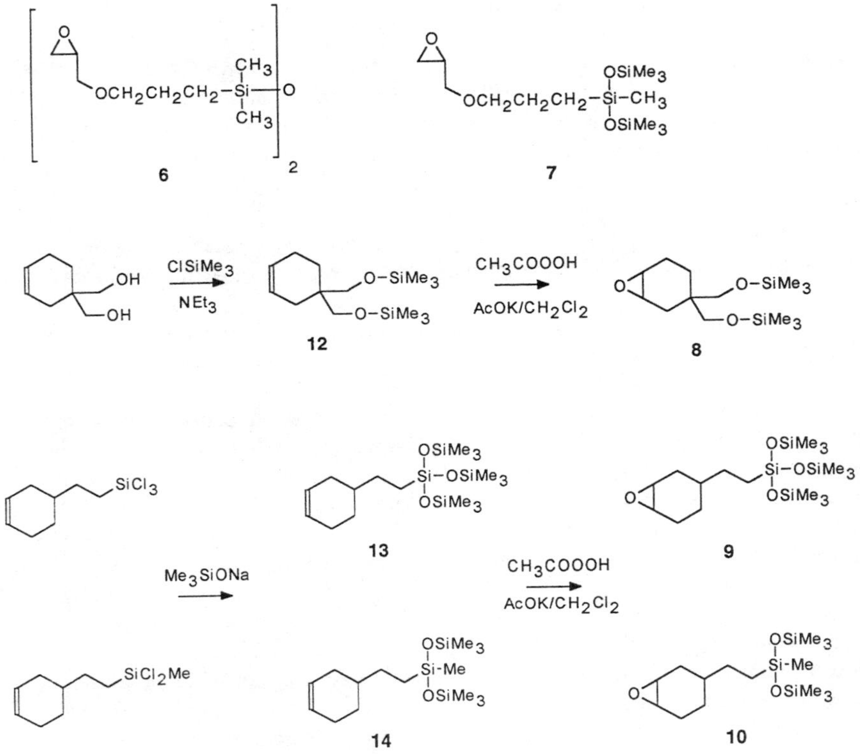

Scheme 4. Synthetic sequences involved in the preparation of Si-epoxides.

Table I. Physical properties of reactive Si-containing monomers.

Monomer	Weight-% Silicon	Boiling Point [°C] @	Vapor Pressure [mbar]
1	12	73	0.1
2	19	70	0.05
3	17	100	0.18
4	14	-*	-*
5	16	150	0.05
6	16	184	2
7	25	96	0.3
8	19	72	0.1
9	27	94	0.04
10	24	95	0.1

* Substance decomposed upon attempted distillation

homogeneous resist films. However, miscibility becomes more critical for reactive monomers with silicon contents >20%, where in some cases the polymer selection might be limited to nonpolar candidates like polystyrene.

In addition, it turned out that the selection of polymer/reactive monomer combinations is further reduced by the fact that some epoxides can react with phenolic groups of the matrix polymer during the baking steps. This undesired side reaction, which leads to the formation of "grass" in the nonexposed zones of the film after plasma etch, is more severe for epoxides 6 or 7, as compared to reactive monomers where the epoxy function is part of aliphatic ring system like in 8.

Another point of concern comes from the fact that photolytically generated radicals or Bronsted acids might migrate during the post exposure baking step, which is typically performed above the glass transition temperature of the resist. As a result, a significantly reduced resolution potential would have to be expected. Therefore the design of a system leading, after illumination, to immobilized active species, seemed desirable. It is well known that the migration sphere of the proton in an acid catalyzed resist system strongly depends on the processing conditions (*19-21*). In a system with the anion of the Bronsted acid covalently linked to the backbone of the matrix polymer, acid migration should be significantly reduced. Therefore the lithographic results should be much less dependent on processing conditions. We were able to successfully synthesize a polymer which incorporated a photoacid generating moiety as outlined in Scheme 5.

Photoactive Compounds. Triphenylsulfonium hexafluoroarsenate was used as the photoacid generator. 1-Hydroxy cyclohexyl phenyl ketone was used as the photo radical generator. To ensure rapid oligomerization of the reactive monomers when passing through the irradiated zones, only photoacid generators yielding extremely strong Bronsted acid were considered. Furthermore, the use of a photo radical generator seemed to be less favorable due to possible quenching of the generated radicals by oxygen.

Additives. Besides the polymeric matrix, the reactive monomer and the photoinitiator, the use of additives was considered to enhance the properties of a resist formulation. In order to obtain resist films with a sufficiently high UV absorbance of >1 for localizing the initiator in the top zones of the film, 9-methylanthracene was added to the formulations, if needed. In addition, trace amounts of polymerization inhibitors were added in some cases in order to prevent the formation of "grass", which might be the result of an uncontrolled degradation of the reactive monomer in the resist film or formulation

Lithographic Evaluation. A large variety of formulations, consisting of the different reactive monomers and matrix polymers, were prepared. The bake and the dry etch conditions needed to be tuned for each individual formulation. It was found that images can be obtained with the (meth)acrylate monomers in combination with the photo radical generator, and with the epoxy monomers in combination with the sulfonium salt photoacid generator. Less favorable results, on the other hand, were obtained with the vinyl ether monomers, probably caused by the well known susceptibility of vinyl ethers towards trace amounts of impurities, which might have been present in the resist films.

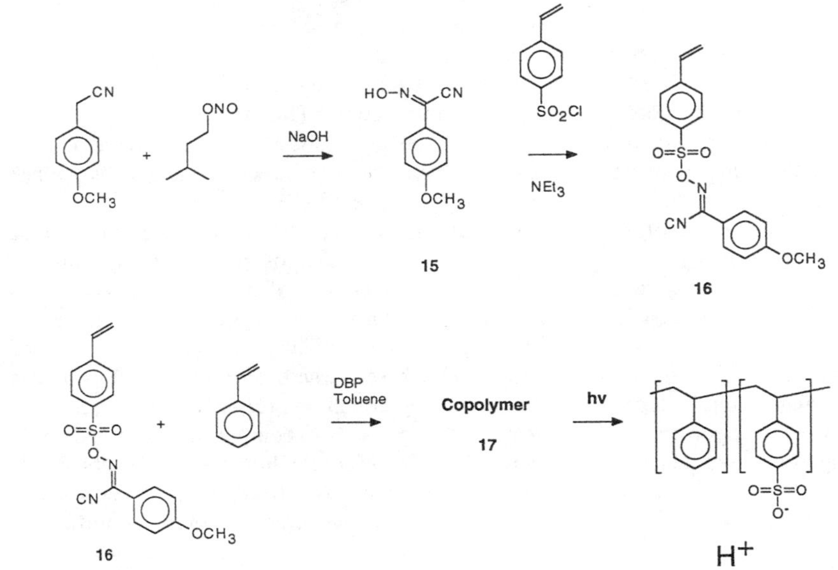

Scheme 5. Synthetic sequence involved in the preparation of copolymer 17.

Table II. Etch rates of resist films after various postbake conditions.

Post Exposure Bake	Etch Rate [nm/min]
Novolak	180
140°C / 30 sec	170
140°C / 1 min	175
140°C / 5 min	**180**
120°C / 1 min	95
120°C / 2 min	135
120°C / 5 min	165
120°C / 15 min	**180**
115°C / 15 min	165
115°C / 30 min	180
110°C / 30 min	150
100°C / 30 min	turbid film

Monomer Selection. As outlined before, there are several criteria a promising candidate monomer needs to fulfill. Among the most important ones are: reactivity of the polymerizable function, silicon content per molecule, and the physical properties, especially the boiling point. Therefore, monomer 9 was chosen from the selection of candidates available in this study. Polystyrene has been found so far to be the only matrix polymer which could be formulated along with 9 to yield homogeneous, nonturbid resist films. TGA data of pure 9 (Figure 2a) and from films consisting of 75 wt. % polystyrene and 25 wt. % 9 (Figure 2b) indicate that a film softbake at 90°C for 1 minute is feasible to prevent significant monomer evaporation. However, complete monomer evaporation from the film seems possible at elevated temperature (~300°C).

Process Evaluation. The UV spectra of 1.5 micron thick films consisting of 75 parts polystyrene (PS), 25 parts 9, 5 parts triphenylsulfonium triflate (TPST), and various amounts of 9-methyl anthracene (MA) were recorded (Figure 3), and from the absorbance values found at 254 nm the corresponding light intensity profiles through the films were calculated (Figure 4). As a result, the formulation containing 1 wt. % MA, showing a UV-absorbance of 1.2 at 254 nm, was chosen for further evaluations, since 55% of the incoming light is absorbed in the top 0.5 micrometer of the resist film.
Post bake conditions for resist films consisting of 25 parts 9, 75 parts PS, 5 parts TPST, and 1 part MA, were optimized by baking 3 inch wafers, coated with a 1.5 micron resist film, at different temperatures for various times (Figure 5). As expected, the monomer evaporation from the resist film is strongly dependent on the baking temperature: Whereas at 140°C heating times below 5 minutes are sufficient, more than 30 minutes are required at 120°C for quantitative monomer evaporation. The contact time of monomers with the photolytically generated fixing sites is longer at low baking temperatures, and as a result more silicon can be retarded in the film. Therefore a more efficient etch barrier is achieved at relatively low post exposure bake temperatures.
A series of etching experiments was performed. The following standard etch conditions (100 Watt, 20 sccm, 1 Pa) were selected, since they turned out to be flexible enough to allow imaging using a variety of different resist formulations (different monomers, polymers, monomer loadings in film). Etch rates of films which were baked at different temperatures were determined (Table II). Reduced etch rates, as compared to novolak, indicate that these films still contain significant amounts of monomer 9. This is in accordance with the conclusions drawn from Figure 5, indicating that an almost quantitative monomer evaporation from the film seems possible at bake conditions of 120°C/15min and 140°C/5min.

Lithographic Results. Wafers coated with 1.5 micron resist films consisting of 25 parts 9, 75 parts PS, 5 parts TPST, and 1 part MA, were irradiated at 254 nm at a dose of 10 mJ/cm^2 in contact mode. Two different post exposure bake conditions were applied: one wafer was baked at 125°C for 10 minutes, the other at 140°C for 1 minute. Subsequently both wafers were dry etched in O_2 (100 Watt, 20 sccm, 1 Pascal). It turned out that a more precise mask image is formed on the wafer which has been

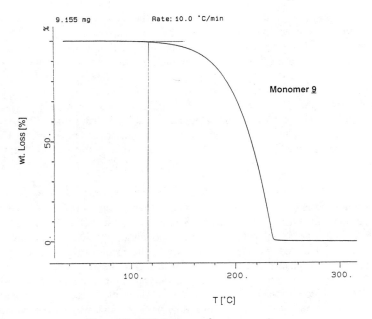

Figure 2a. TGA trace of monomer 9.

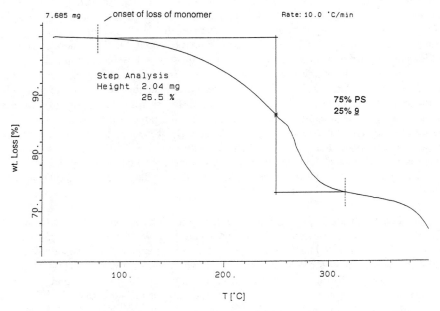

Figure 2b. TGA trace of a resist film composed of polystyrene (75%) and 9 (25%).

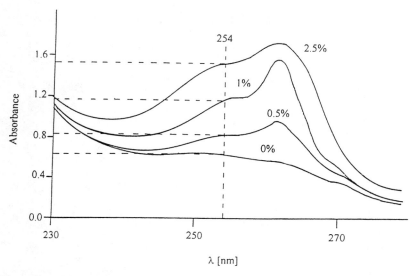

Figure 3.   UV absorption curves of resist films containing different MA concentrations [wt. %]:   0%, Abs.$_{254}$ = 0.65;   0.5%, Abs.$_{254}$ = 0.8;   1%, Abs.$_{254}$ = 1.2;   2.5%, Abs.$_{254}$ = 1.6.

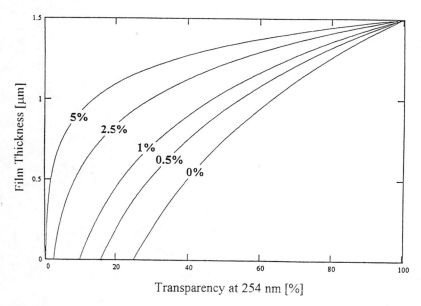

Figure 4.   Light distribution profiles in resist films containing different MA concentrations [wt. % MA in resist film].

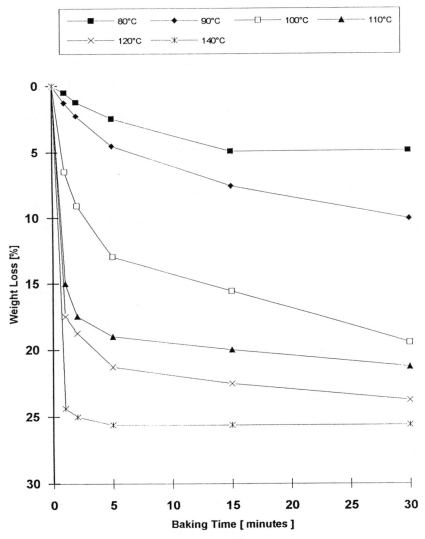

Figure 5. Evaporation of 9 from resist films at various postbake conditions.

baked at the lower temperature (Figure 6a), indicating image detoriation due to acid or excess monomer diffusion into nonexposed zones of the film during post exposure bake at 140°C (Figure 6b).

We found that the lithographic results were strongly dependent on the etching conditions. Wafers coated with a 1.5 micron thick film consisting of 25 parts 9, 75 parts PS, 5 parts TPST, and 1 part MA, irradiated at 254 nm (10 mJ/cm²) and baked at 125°C for 20 minutes, etched differently. Better quality images with respect to linewidth control and profile slopes were obtained at an oxygen pressure of 2 Pascal (Figure 7a), as compared to images obtained at 0.6 Pascal, where sloped profiles, but nevertheless less grass, were formed (Figure 7b). Figure 8 shows equidistant line/space patterns of various sizes.

It is obvious that our surface imaging system at its current state shows poor process latitude. This might be caused by an insufficiently high silicon gradient between exposed and nonexposed zones of the resist film, leading to poor etch selectivity. The grass which is formed at the milder etch conditions (2 Pascal, Figure 7a) indicates that the unirradiated zones of the resist film still contain significant amounts of silicon. This might be caused by trace amounts of acidic impurities present in the resist film, causing a nondesired oligomerization of the highly reactive epoxy monomer 9. This problem is currently being addressed by vigorously cleaning all of the resist components and by the addition of trace amounts of stabilizers to the resist formulation.

When the PAG containing copolymer 17 was formulated with 9, only turbid resist films could be obtained, probably caused by a partial incompatibility of 9 in this particular matrix. No miscibility problems were encountered in a formulation where 7 was used instead, and homogeneous films were obtained. However, no sufficiently active etch barrier was formed in this system, resulting in a significantly reduced etch stability of the irradiated areas of the resist film. We believe that this is caused by the reduced reactivity towards acid-induced oligomerization of 7 as compared to 9, and by the reduced acid strength of the photogenerated sulfonic acid as compared to triflic acid at the fixing sites. Work is currently in progress to overcome the problems mentioned above by chemically modifying monomer 9 in order to improve the miscibility in copolymer 17.

Experimental

General.   Silicon contents were determined by elemental analysis. The visible absorbance spectra were recorded on a Varian Cary 1E UV/Visible spectrometer, film thickness measurements were performed with an Tencor Alpha Step 100 instrument. Deep-UV exposures were performed on an Oriel tool using a 254 nm band pass filter. Pattern transfer was performed in an Alcatel CF 2P plasma etcher. Molecular weights were determined by gel permeation chromatography (GPC, polystyrene calibration), and solution ^{1}H-NMR spectra (CDCl$_3$) were recorded on a 300 MHz Bruker AC 300 instrument.

Resist formulations were prepared by dissolving the matrix polymer, the reactive monomer, the photoacid or photoradical generator, and optionally a light absorbing agent, in cyclopentanone. Resist films of 1-2 μm thickness were obtained by spin casting the resist solution onto 3 inch silicon wafers, and subsequent drying at

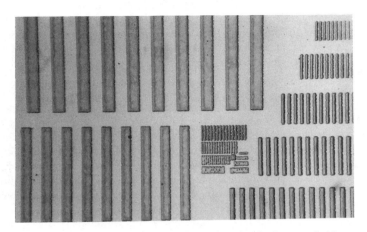

Figure 6a. Optical micrograph of resist images obtained after postbaking at 125°C for 10 min.

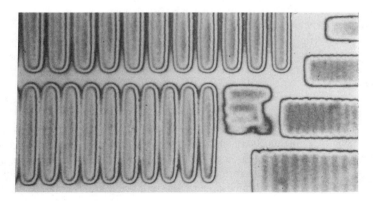

Figure 6b. Optical micrograph of resist images obtained after postbaking at 140°C for 1 min.

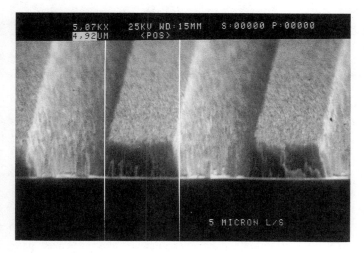

Figure 7a. SEM micrograph depicting 5 μm resist features obtained by applying 100 W, 20 sccm, 2 Pa dry etch conditions with O_2.

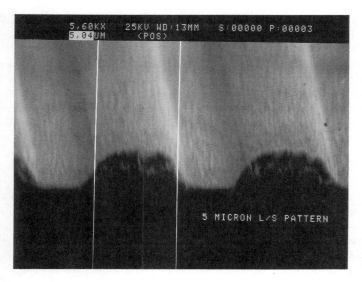

Figure 7b. SEM micrograph depicting 5 μm resist features obtained by applying 100 W, 20 sccm, 0.6 Pa dry etch conditions with O_2.

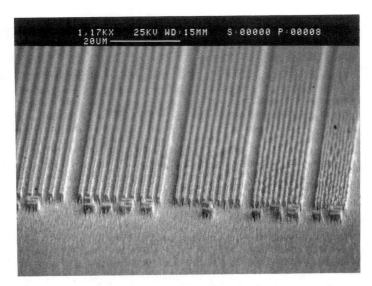

Figure 8. SEM micrograph depicting equidistant line/space mask patterns.

90°C for 60 seconds. Films having UV absorbance of 1 or higher were obtained by addition of a substance which strongly absorbs light at the irradiation wavelength, typically 9-methyl anthracene, to the resist formulation. Resist films were imaged in contact mode, using a mask with 0.5 μm smallest feature size. Typically, doses between 10-30 mJ/cm^2 were applied. Post exposure bake was done at temperatures between 120°C and 140°C for five to sixty minutes, depending on the physical properties of the reactive monomer and the matrix polymer. Typically the following etch conditions were applied: O$_2$ pressure 0.6-6 Pascal, flow 20 sccm, 100-150 Watt, etch time 5-20 min.

Materials. Polystyrene and poly(methyl methacrylate) were purchased from Aldrich, poly(4-hydroxy styrene, PHM-C) was obtained from Maruzen Petrochem. Co. Monomers 6 and 7 were purchased from Petrarch Systems, 4-hydroxybutyl vinylether was obtained from GAF (RAPI-CURE HBVE). 1-Hydroxy cyclohexyl phenyl ketone is a commercial product from Ciba-Geigy Inc. (IRGACURE 184).

Some of the materials have been described before and were prepared according to the literature. Their CAS-numbers are: 2: 4515-14-4, 3: 18415-68-4, 9: 90492-24-3, 10: 65842-29-7, 11: 2362-10-9, 13: 90492-23-2, 14: 101667-50-9, 15: 76104-14-8. The synthesis procedures and characterization of the so far unknown materials are recorded below.

Bis(acryloxymethyl)dimethyl silane 1: A mixture of 51.0 g (425 mmol) bis(hydroxymethyl)dimethyl silane, 76.9 g (850 mmol) acrylic acid, and 86.0 g (850 mmol) triethylamine, dissolved in 600 ml diethyl ether, was reacted for 2h at 10°C. After distillation at 73°C/0.1 mbar, 30.5 g (13.4 mmol, 31%) of 1 was obtained as a clear oil. ^{1}H-NMR: δ 6.38 (dd, 2H); 6.12 (dd, 2H); 5.82 (dd, 2H); 3.96 (s, 4H); 0.17 (s, 6H).

Bis(2,7-dioxaoct-8-enyl)tetramethyl disiloxane 4: A mixture of 11.6 g (50 mmol) bis(chloromethyl) tetramethyl disiloxane 11, 11.6 g (100 mmol) 4-hydroxy-butyl vinylether, and 1.0 g benzyltrimethyl ammoniumchloride, dissolved in 100 ml 1N sodium hydroxyde solution, was refluxed for 3h. The reaction yielded 14.2 g of 4 as a colorless oil, which was used without further purification. ^{1}H-NMR: δ 6.38 (dd, 2H); 4.10 (dd, 2H); 3.91 (dd, 2H); 3.61 (m, 8H); 2.33 (s, 4H); 1.62 (m, 8H); 0.0 (s, 12H).

Bis(1,6-dioxaoct-7-enyl)tetramethyl disiloxane 5: The reaction of 5.0 g (43 mmol) 4-hydroxybutyl vinylether, 6.9 g (43 mmol) hexamethyldisilan, and 4.4 g (21.5 mmol) 1,3-dichlorotetramethyl disiloxane, dissolved in 50 ml diethyl ether, yielded, after stirring for 12h at ambient temperature and distillation at 150°C/0.05 mbar, 1.8 g (3.5 mmol, 16%) of 5 as a colorless liquid. ^{1}H-NMR: δ 6.35 (dd, 2H); 4.05 (dd, 2H); 3.87 (dd, 2H); 3.57 (m, 8H); 1.59 (m, 8H); 0.0 (s, 12H).

Epoxide of 4,4'-bis(trimethylsiloxymethyl)-1-cyclohexene 8: The oxydation of 20 g (70 mmol) 4,4'-bis(trimethylsiloxymethyl)-1-cyclohexene 12 with 26.8g (140 mmol) 40% acetic peracid solution in 200 ml dichloromethane yielded after distillation at 72°C/0.1 mbar 8.6 g (29 mmol, 41%) of 8 as colorless oil. ^{1}H-NMR: δ 3.4-2.9 (m, 6H); 2.0-1.5 (m, 4H); 1.2 (m, 2H); 0.0 (s, 18H).

4,4'-bis(trimethylsiloxymethyl)-1-cyclohexene 12: A mixture of 35.5 g (250 mmol) 4,4'-bis(hydroxymethyl)-1-cyclohexene, 67.9 g (625 mmol) trimethyl chlorosilane, and 64.5 g (637 mmol) triethylamine, dissolved in 550 ml dichloromethane, was stirred for 4h at ambient temperature. Distillation at 60°C/0.07 mbar yielded 55.5 g (194 mmol, 78%) of 12 as a colorless oil. ^{1}H-NMR: δ 5.54 (m, 2H); 3.29 (dd, 4H); 1.89 (m, 2H); 1.71 (m, 2H); 1.37 (m, 2H); 0.0 (s, 18H).

Styryl-4-sulfonatester of hydroxyimino-4-methoxybenzene acetonitrile 16: By reaction of 10.1 g (57 mmol) 4-ethenylbenzene sulfonylchloride, 11.6 g (57 mmol) hydroxyimino-4-methoxybenzene acetonitrile 15, and 6.0 g (59 mmol) triethylamine, dissolved in 200 ml ethyl acetate, for 3h at ambient temperature. After purification by recrystallization from methanol, 13.6 g (40 mmol, 70%) of 16 with a melting point of 115°C was obtained as a yellow powder. ^{1}H-NMR: δ 8.02 (d, 2H); 7.72 (d, 2H); 7.59 (d, 2H); 6.94 (d, 2H); 6.76 (dd, 1H); 5.93 (dd, 1H); 5.49 (dd, 2H); 3.86 (s, 3H). Element. Anal. Calcd for $C_{17}H_{14}N_2SO_4$: C, 59.64; H, 4.12; N, 8.18; S, 9.36. Found C, 59.06; H, 4.23; N, 8.00; S, 9.37.

Copolymer of styrene and 16: 7.0 g (67 mmol) styrene, 3.0 g (9 mmol) of compound 16, and 0.1 g dibenzoyl peroxide were dissolved in 100 g toluene. The mixture was stirred for 24h at 75°C. Precipitation from 1500 ml n-hexane yielded 5.5 g (55%) of copolymer 17 as a yellow powder. Element. Anal. Found: C, 76.24; H, 6.27; N, 3.40; S, 4.21.

Conclusions

A top surface imaging approach, which yields negative tone images of the mask, is presented and discussed. It is based on the light induced formation of reactive sites in the top zones of a resist film. These sites contain polymerization initiators, which are able to graft silicon containing monomers by rapid oligomerization. The resulting dry etch barriers allow for a pattern transfer down to the substrate by means of oxygen plasma etching. Best lithographic results are obtained with a silicon-containing epoxy monomer, in combination with polystyrene and triphenylsulfonium triflate. Submicrometer resolution potential has been demonstrated. However, additional synthetic and process evaluation work is needed to improve image quality.

Acknowledgments

The authors wish to acknowledge Niklaus Bühler of Ciba-Geigy Materials Research, and Reinhard Schulz and Michelle Robeson, both of OCG Microelectronic Materials, for promoting the work. Special thanks to Nadia Reichlin and Béatrice Zurkinden for their help in materials synthesis and screening, and Arnold Grubenmann for helpful discussions.

References

1.   Ito, H.; Ueda, M.; and Renaldo, A. F., *J. Electrochem. Soc.*, **1989**, *136*, 245.

2.   Ito, H.; Ueda, M.; and Schwalm, R., *J. Vac. Sci. Technol.*, **1988**, *B 6*, 2259.

3.   Steinmann, A., *Proc. SPIE*, **1988**, *920*, 13.

4.   Miller, R. D.; Wallraff, G.; Clecak, N.; Sooriyakumaran, R.; Michl, J.; Karatsu, T.; McKinley, A. J.; Klingensmith, K. A.; and Downing, J., *ACS: Polym. Mater. Sci. Eng.*, **1989**, *60*, 49.

5.   Novembre, A. E.; Jurek, M. J.; Kornblit, A.; and Reichmanis, E., *Polymer Eng. Sci.*, **1989**, *29*, 920.

6.   Saotome, Y.; Gokan, H.; Saigo, K.; Suzuki, M.; and Ohnisihi, Y., *J. Electrochem. Soc.*, **1985**, *132*, 909.

7.   MacDonald, S. A.; Schlosser, H.; Ito, H.; Clecak, N. J.; and Willson, C. G., *Chem. Mater*, **1991**, *3*, 435.

8.   Spence, C. A.; MacDonald, S. A.; and Schlosser, H., *Proc. SPIE*, **1990**, *1262*, 344.

9.   Coopmans, F.; and Roland, B., *Proc. SPIE*, **1986**, *633*, 126.

10.   Garza, C. M.; Misium, G.; Doering, R. R.; Roland, B.; and Lombaerts, R., *Proc. SPIE*, **1989**, *1086*, 229.

11.   Nichols, D. N.; Goethals, A. M.; De Geyter, P.; and Van den hove, L., *Microelectr. Eng.*, **1990**, *11*, 515.

12.   Johnson, D. W.; Kunz, R. R.; and Horn, M., *J. Photopolymer Sci. Technol.*, **1993**, *6*, 593.

13.   Hartney, M. A.; Johnson, D. W.; and Spencer, A. C., *Proc. SPIE*, **1991**, *1466*, 238.

14. Sezi, R.; Sebald, M.; Leuschner, R.; Ahne, H.; Birkle, S.; and Borndörfer, H., *Proc. SPIE*, **1990**, *1262*, 84.
15. Taylor, G. N.; Wolf, T. M.; and Moran, J. M., *J. Vac. Sci. Technol.*, **1981**, *19*, 872.
16. Taylor, G. N.; Hellman, M. Y.; Feather, M. D.; and Willenbrock, W. E., *Polymer Eng. Sci.*, **1983**, *23*, 1029.
17. Taylor, G. N.; Wolf, T. M.; and Goldrick, M. R., *J. Electrochem. Soc.*, **1981**, *128*, 361.
18. Taylor, G. N.; and Wolf, T. M., *J. Electrochem. Soc.*, **1980**, *127*, 2665
19. McKean, D. R.; Schaedeli, U.; MacDonald, S. A., *Polymers in Microlithography*, ACS Symp. Ser. **1989**, *412*, 27.
20. Schlegel, L.; Ueno, T.; Hayashi, N.; Iwayanagi, T.; *J. Vac. Sci. Technol.*, **1991**, *B9*, 278.
21. Asakawa, K., *J. Photopol. Sci. Technol.*, **1993**, *6*, 505.

RECEIVED July 7, 1995

Chapter 21

Plasma-Developable Photoresist System Based on Polysiloxane Formation at the Irradiated Surface

A Liquid-Phase Deposition Method

Masamitsu Shirai[1], Norihiko Nogi[1], Masahiro Tsunooka[1], and Takahiro Matsuo[2]

[1]Department of Applied Chemistry, College of Engineering, University of Osaka Prefecture, Sakai, Osaka 593, Japan
[2]Semiconductor Research Center, Matsushita Electric Industrial Company, Ltd., 3–15 Yagumo-Nakamachi, Moriguchi, Osaka 570, Japan

Polysiloxane formation at the near surface of the UV irradiated polymer films using a liquid-phase deposition method was studied. Poly(methacrylates) bearing 1,2,3,4-tetrahydro-1-naphthylideneamino p-styrenesulfonate (NIS) units or 9-fluorenilideneamino p-styrenesulfonate (FIS) units was prepared. When the irradiated polymer films were dipped in a solution containing alkoxysilanes, polysiloxane networks were formed at unirradiated areas was observed. Factors affecting the polysiloxane formation rate were studied. The polymer films irradiated at 254 nm and subsequently modified by the alkoxysilane solution showed good etching resistance to an oxygen plasma. The liquid-phase modification method was compared with the vapor-phase modification method.

A number of photolithographic processes that use surface modification techniques have been studied (1). The predominant approach involves post-exposure silylation of organic polymer films. Coopmans and Roland (2) described the gas-phase silylation of a diazonaphthoquinone/phenolic matrix resin, yielding a negative tone imaging by oxygen reactive ion etching (O_2 RIE). MacDonald and co-workers (3) reported the surface modification of polymer films bearing photochemically formed phenolic -OH groups using gaseous hexamethyldisilazane. Baik and co-workers (4,5) reported the liquid-phase silylation of phenolic polymers (Plasmask 200-G for i-line resist), which gave images by O_2 RIE. It was pointed out that the liquid-phase silylation exhibits several advantages such as an improved silylation contrast, an increased Si concentration, the use of room temperature silylation, and a simple process without post exposure baking steps.

0097–6156/95/0614–0318$12.00/0

In contrast to the silylation process, the selective formation of polysiloxanes, metal oxides or metals at the polymer surface has been reported. Follett and co-workers (6) have reported the plasma-developable electron-beam resists. The essential feature involved selective diffusion of $(CH_3)_2 SiCl_2$ into the irradiated areas of poly(methyl methacrylate), followed by hydrolysis of the chlorosilane by exposure to water vapor, resulting in the formation of an interpenetrating network of polysiloxanes. O_2 RIE yielded a negative tone image. Taylor and co-workers (7-10) reported a photoresist based on the surface oxidation of chlorine-containing polystyrenes using 248.4- or 193-nm light and subsequent treatment with gaseous $TiCl_4$. A surface imaging process using electroless deposition to form a thin metal layer only on the unexposed areas was described (11-13). The metallized film surface provides an effective reactive ion etch barrier.

Recently, we reported oxygen plasma-developable photoresists using photoinduced acid-catalyzed polysiloxane formation at the irradiated polymer surface by a chemical vapor deposition (CVD) method at room temperature (14-16). We have now extended surface modification by the CVD method to a liquid-phase deposition method. Our method is shown in Figure 1. Polymer films containing photoacid generating units appended to their chains are irradiated. When the irradiated films are dipped in a solution containing alkoxysilanes and water, polysiloxane network is formed at the near surface of the irradiated film. No polysiloxane network is formed at unirradiated regions because the photochemically formed acids are necessary for polysiloxane formation by hydrolysis and subsequent condensation of the alkoxysilanes. In this paper we report the polysiloxane formation at the irradiated surface of polymer films by this liquid-phase deposition method and the etching rate of the films modified with polysiloxane networks using oxygen plasma. The present system gives a negative tone image by O_2 RIE. We also compare the liquid-phase deposition method to the vapor-phase method that we previously employed (14-16).

EXPERIMENTAL

Preparation of monomers

1,2,3,4-Tetrahydro-1-naphthylideneamino p-styrenesulfonate (NIS) was synthesized as follows: To a cold (<10°C) solution of 1-tetralone oxime (21.8 g, 0.14 mol) in yridine (50.5 mL) was slowly added p-styrenesulfonyl chloride (30.0 g, 0.15 mol), which was freshly prepared from sodium p-styrenesulfonate and $SOCl_2$ according to the method reported by Kamogawa and co-workers (17). After the addition was complete, the reaction was continued with vigorous stirring at <15°C for 3 h. Then it was poured into 573 mL of ice-cold 5% HCl and extracted with chloroform. After washing the chloroform layer thoroughly with water, the chloroform layer was dried over $K_2 CO_3$ and evaporated under reduced pressure. The oily residue was continuously extracted with hot n-heptane, which yielded white crystals upon cooling. After recrystallization from n-heptane, 38.9 g (88%) of pure product was obtained: mp 80-82°C. IR (KBr): 1370 and 1180 cm^{-1} (O-S-O). Anal. Calcd for $C_{18} H_{17} NO_3 S$; C, 66.06; H, 5.19; N, 4.28. Found C, 66.27; H, 5.39; N, 3.85.

9-Fluorenilideneamino p-styrenesulfonate (FIS) was prepared according to the same method described for NIS. The product was recrystallized from a benzene-heptane

(2:3, v/v) mixture: yield 68%; mp 135-137°C; IR (KBr): 1370 and 1180 cm^{-1} (O-S-O). Anal calcd for $C_{21}H_{15}NO_3S$: C, 69.79; H, 4.18; N, 3.88. Found C, 70.12; H, 4.01; N, 3.98.

Preparation of copolymers

Copolymers (1a-c, 2, and 3) of NIS and methacrylates were prepared by the photochemically-initiated copolymerization of corresponding monomers with 2,2'-azobis(isobutyronitrile) (AIBN) as an initiator at 29°C by irradiation with light at wavelengths above 350 nm. The flux was 4 mW/cm^2. The photodecomposition of NIS did not occur by the light at wavelengths above 350 nm. The concentrations of total monomer and AIBN in benzene were usually 4.5 and 1.6×10^{-2} mol/L, respectively. The sample solution was degassed under vacuum by repeating freeze-thaw cycles before polymerization. The content of NIS units in polymers was determined by measuring the absorbance at 254 nm in CH_2Cl_2. The molar extinction coefficient of the NIS units was estimated to be equal to that of the model compound 1,2,3,4-tetrahydro-1-naphthylideneamino p-toluenesulfonate, ε being 1.53×10^4 L/mol-cm at 254 nm in CH_2Cl_2 at room temperature. Although the copolymers could be obtained by the conventional thermally initiated copolymerization of the corresponding monomers with AIBN, they showed broad molecular weight distributions ($M_w/M_n > 3.5$).

The copolymer of FIS and methyl methacrylate (4) was prepared by the conventional thermally initiated copolymerization with AIBN. The concentrations of total monomer and AIBN in N-N-dimethylformamide (DMF) were 4.5 and 1.6×10^{-2} mol/L, respectively. The content of the FIS units in the polymer was determined by measuring the absorbance at 254 nm in CH_2Cl_2, ε being 6.1×10^4 L/mol-cm at 254 nm in CH_2Cl_2 at room temperature. Detailed polymerization conditions and polymer properties are shown in Table 1. The structure of the polymers used is shown in Figure 2.

Reagents

Tetramethyl orthosilicate (TMOS), tetraethyl orthosilicate (TEOS), tetra-n-propyl orthosilicate (TPOS), methyltrimethoxysilane (MTMOS), methyltriethoxysilane (ETEOS), ethyltriethoxysilane (ETEOS), and benzyltriethoxysilane (BTESO) were of reagent grade and used without further purification. Methyl methacrylate (MMA), i-propyl methacrylate (IMPA), and benzyl methacrylate (BMA) were used after distillation.

Deposition of polysiloxanes

The polymer films (8.8×22 mm) were prepared on glass plates (8.8×50 mm) by casting from chloroform solution and drying under vacuum at room temperature. The prebake of the films at 90°C for 1 min did not affect the polysiloxane formation rate. The polymer weight on the glass plate was usually 2×10^{-4} g. If the polymer is assumed to have a density of 1 g/cm^3, the thickness of the film is ≈ 1 μm. After irradiation with 254-nm light, the glass plate coated with polymer film was transferred to a glass vessel containing 0.2 mol/L alkoxysilane solution (15 mL). A mixed solvent of n-hexane and acetone (9:1, v/v) was used. Alkoxysilanes were stable in the mixed solvent without acids. The glass vessel was held either at room

Table 1   Polymerization Conditions and Polymer Properties a)

Polymer b)	Monomer in Feed					Polymerization				Photoacid Generating Units in Polymer (mol %)	T_g c) (°C)
	NIS (g)	FIS (g)	MMA (mL)	IPMA (mL)	BMA (mL)	Time (h)	Conversion (%)	$\overline{M}_n$ x10^{-4}	$\overline{M}_w/\overline{M}_n$		
1a	1.0		3.5			20	33	10.0	2.5	15	-d)
1b		0.41	3.0			15	47	11.5	2.0	7.4	114
1c		0.15	2.5			6	26	8.1	2.2	4.7	110
2		0.41		1.8		15	45	13.1	1.9	15	105
3		0.39			2.5	8	42	15.5	2.3	18	83
4	0.75		2.4			12	38	4.3	4.2	13	135

a) [Total monomer]=4.5 mol/L, [AIBN]=1.6x10^{-2} mol/L.   Benzene was used as a solvent for the preparation of 1a-c, 2, and 3.   N,N-Dimethylformamide was used as a solvent for the preparation of **4**.   b) Structure was shown in Figure 2.   c) Glass transition temperature.   d) Distinct T_g was not observed.

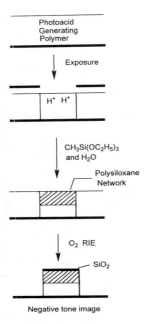

Figure 1.   Imaging method based on photoinduced acid-catalyzed formation of polysiloxanes on polymer surfaces.

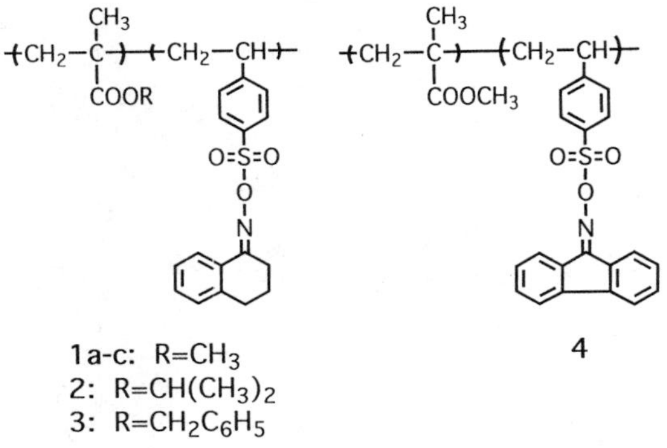

$$\text{-}(CH_2\text{-}\underset{\underset{COOR}{|}}{\overset{\overset{CH_3}{|}}{C}}\text{-})\text{-}(CH_2\text{-}CH)\text{-} \qquad \text{-}(CH_2\text{-}\underset{\underset{COOCH_3}{|}}{\overset{\overset{CH_3}{|}}{C}}\text{-})\text{-}(CH_2\text{-}CH)\text{-}$$

1a-c: R=CH₃
2: R=CH(CH₃)₂
3: R=CH₂C₆H₅

4

Figure 2.   Structure of copolymers containing acid forming iminoxysulfonate groups bonded to styrene units in the main chain: tetralone oxime based system (1-3) and fluorenone oxime based system (4).

temperature or 35°C. The sample plate was taken out from the alkoxysilane solution and it was thoroughly rinsed with the solvent. After drying in vacuo the amounts of polysiloxanes formed at the near surface of the irradiated films were determined from the difference between the weight of the sample plate before and after dipping in the alkoxysilane solutions.

Hydrolysis of alkoxysilanes

The acid-catalyzed hydrolysis of alkoxysilanes was carried out in acetone at 25°C. Concentrations of alkoxysilane, water, and p-toluenesufonic acid as a catalyst were 9.3×10^{-2}, 1.04, and 1.14×10^{-4} mol/L, respectively. The rate of hydrolysis was measured by following the decrease of alkoxysilanes or the increase of alcohols as products. Gas chromatography (column: Apiezon-L 10%) was used to check the substrates in the reaction mixture. Pseudo-first-order rate constants (k) were obtained from the slope of the linear plots of log [alkoxysilane] versus time.

Etching with an oxygen plasma

Oxygen plasma etching was carried out at room temperature using a laboratory-constructed apparatus where the oxygen plasma was generated using two parallel electrodes and RF power supplies. The typical etching conditions were as follows: 20W power (13.56 MHz), power density of 1.0 W/cm^2, 125 mTorr, and oxygen flow of 1 sccm.

RESULTS AND DISCUSSION

Photochemistry of polymers

It has been reported that upon UV irradiation the cleavage of -O-N= bonds in the photoacid generating units and the subsequent abstraction of hydrogen atoms from residual solvent in the polymer film or from polymer molecules leads to the formation of sulfonic acid, azines and ketones (18). The reaction mechanism for the photolysis of the photoacid generating units is shown in Scheme 1. For the photolysis of the photoacid generating units on polymer chains in air, the reaction of imino radicals with oxygen is dominant compared to their dimerization and hydrogen abstraction reactions (19).

Polymers 1a-c, 2, 3, and 4 showed an absorption peak at 254 nm. Figure 3 shows the spectral changes of the 1b film upon irradiation with 254-nm light. The absorbance at 254 nm decreased with irradiation time and an isosbestic point was observed at 305 nm. The slight increase in absorbance above 300 nm is due to the formation of tetralone azine. The quantum yields (φ) for the photolysis of NIS units in the polymer films at 254 nm were ≈ 0.3. They were not strongly dependent either on the structure of the methacrylate units or the NIS content of the polymer. A similar spectral change was observed for the 4 film on irradiation at 254 nm. The φ value for the photolysis of FIS units was almost the same as that for NIS units.

Deposition of polysiloxane

In the presence of water and strong acids, the hydrolysis and subsequent polycondensation reactions of MTEOS and its analogues lead to the formation of polysiloxane networks, which are well known as the sol-gel

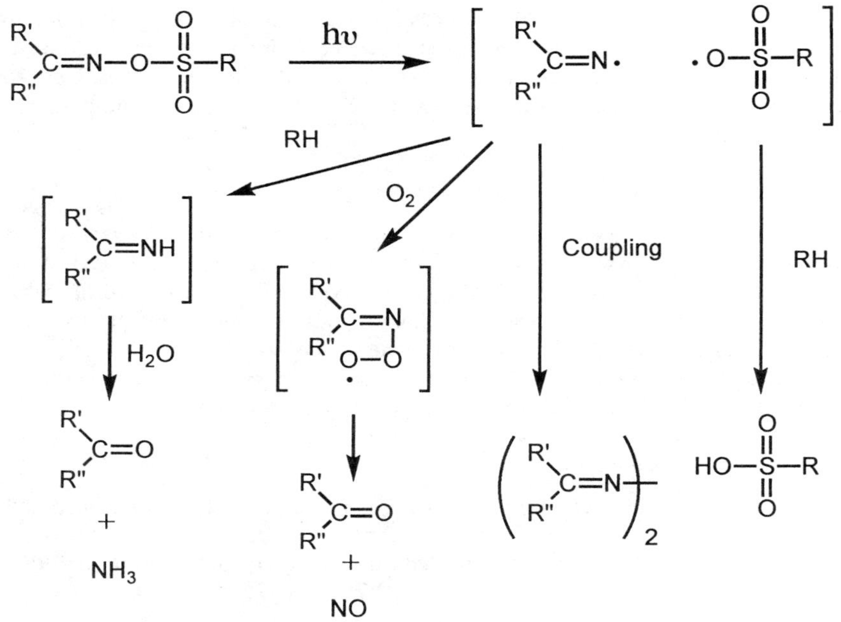

Scheme 1   Reaction mechanism for the photolysis of the photoacid
            generators

$$n\ CH_3\ Si(OC_2H_5)_3 + 1.5n\ H_2O \xrightarrow{H^+} Polysiloxanes + 3n\ C_2H_5OH$$

process for the silica glass formation (20,21). When the irradiated polymer films bearing the NIS units were dipped in a liquid-phase modification bath (n-hexane-acetone mixture (9:1, v/v) containing MTEOS and small amounts of water), polysiloxane networks were formed in the near surface region of the films, which was confirmed by IR analysis (see Figure 4). The irradiated films treated with the solution of alkoxysilanes showed new peaks at 3500 (Si-OH), 1272 (Si-CH$_3$), 1000-1200 (Si-O-Si), and 790 cm^{-1} (Si-CH$_3$). The presence of the peak due to Si-OH suggests the incomplete polycondensation of the silanol. The hydrolysis and subsequent polycondensation of alkoxysilanes occur at the film-solution interface and beneath the film surface. When the unirradiated polymer films bearing the NIS units were dipped in the liquid-phase modification bath, no polysiloxane networks were formed. This was confirmed by IR spectroscopy. Although MTEOS molecules may diffuse into both irradiated areas are hydrolyzed to form polysiloxane networks. MTEOS molecules diffused into the unirradiated region can be removed by the rinse and drying after the treatment of the liquid-phase modification. Thus the p-styrenesulfonic acid units formed photochemically in the film were essential for the formation of polysiloxane networks by the hydrolysis and subsequent polycondensation of MTEOS and its analogues.

In the present system the concentration of water in the modification bath (a mixture of n-hexane and acetone (9:1, v/v) containing alkoxysilanes) affected the rate of polysiloxane formation and the properties of the film surface modified with polysiloxanes. The polysiloxane formation rate increased with the increase of the water concentration in the bath. The water content in the bath was determined by the Karl Fischer method. The mixture of n-hexane and acetone (9:1, v/v) which were reagent grade contained 1.9×10^{-2} mol/L of water. The n-hexane - acetone mixture which was saturated with water contained 6.5×10^{-2} mol/L of water. The alkoxysilane solution with a low water content gave a transparent surface modified with polysiloxane networks, while the alkoxysilane solution having higher water content gave an opaque surface.

The fraction of acetone in the liquid-phase modification bath significantly affected the rate of polysiloxane network formation in the near surface of the irradiated films. n-Hexane and acetone are poor and good solvents, respectively, for the polymers. The polysiloxane formation rates on the irradiated 1b film using MTEOS were 1.50×10^{-7}, 7.0×10^{-7}, and 4.95×10^{-6} g/cm^2 min for the liquid-phase modification bath with acetone content in the bath enhanced the polysiloxane formation rate, the fraction of acetone must be kept below 10 vol % because the polymer films partially dissolved in solutions with higher acetone concentration during the surface modification treatment. The thickness loss of the film was not observed during the surface modification treatment if the modification bath with acetone content below 10 vol % was used. If n-hexane containing MTEOS without acetone was used, polysiloxane formation at the irradiated polymer surface was negligibly slow. Acetone was necessary as a diffusion promoter of alkoxysilane molecules. Similar results were reported for the liquid-phase silylation of phenolic resists, where xylene and N-methyl-2-pyrrolidinone were used as a solvent and a diffusion promoter, respectively (4,5).

Figure 5 shows the effect of MTEOS concentration on the polysiloxane formation rate at the irradiated surface of 1b film. The rate of polysiloxane network formation increased with the concentration of MTEOS in the liquid-phase modification bath, passed through a maximum value and decreased at even higher concentrations of MTEOS. The decreased rate of polysiloxane formation at higher concentrations of MTEOS may be due to the fact that the rapid formation of polysiloxane networks at film surface prevents further diffusion of MTEOS molecules into the film. A similar phenomenon was reported for the liquid-phase silylation of phenolic resins (4,5).

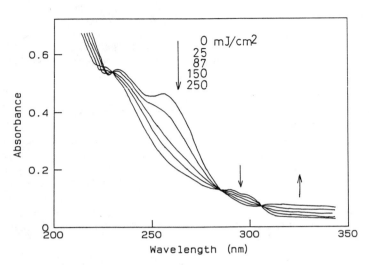

Figure 3.   Spectral change of the 1b flim upon irradiation with 254 nm light. Film
             thickness: 0.4 μm.

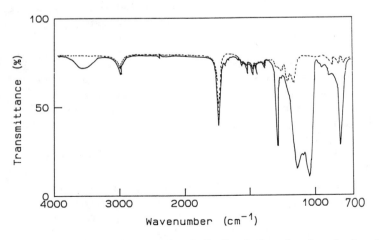

Figure 4.   FT-IR spectra of the irradiated 1b film before (-----) and after (——)
             dipping into the MTEOS solution for 30 min. Exposure dose: 379
             mJ/cm².

Table 2 shows the effect of alkoxysilane structure on the polysiloxane formation rate at the surface of the irradiated 1c film. The rate of polysiloxane formation decreased in the order MTMOS > MTEOS > TMOS >> ETEOS ~ BTEOS ~ TEOS ~ TPOS. Pseudo-first-order rate constants (k) for the acid-catalyzed hydrolysis of ETEOS and BTEOS were 4.36×10^{-4} and 6.08×10^{-4} sec^{-1} at 25°C in aqueous acetone, respectively. Although these values were close to the k values for MTMOS and MTEOS, the rates of polysiloxane formation using ETEOS and BTEOS were negligibly low. This finding suggests that the polysiloxane formation rate was dependent both on the hydrolytic reactivity of alkoxysilanes and the diffusion rate of alkoxysilanes into the polymer films. The diffusion rates of alkoxysilanes with large alkyl groups on Si into the irradiated film is slow. For the vapor-phase modification method, the polysiloxane formation rate at the irradiated surface was dependent on the hydrolytic reactivity and the vapor pressure of alkoxysilanes at room temperature.

Figure 6 shows the relationship between irradiation time and polysiloxane formation at the irradiated 1b film using MTEOS. The polysiloxane formation increased with irradiation time. This means that the polysiloxane formation rate is proportional to the number of p-styrenesulfonic acid units formed photochemically in the polymer matrix. In the present system the photolysis degree of NIS units in 1b film was 100, 80, and 40 mol %, respectively, for the irradiation with 379, 126, and 63 mJ/cm². No significant differences were observed for the polysiloxane formation rates at the irradiated surface of the 1a and 1b films, suggesting that for the polymers with high content of NIS units, the diffusion of alkoxysilanes into the irradiated film is the rate-determining step for the polysiloxane network formation. It is worthwhile to note that very small amounts of the acid units (≈ 3.0 mol %) effectively worked as a catalyst for the hydrolysis of alkoxysilane to form polysiloxane networks. In the case of the vapor-phase modification at 30°C using MTEOS, no formation of polysiloxanes was observed for poly(methyl methacrylate) bearing NIS units less than 10 mol % (15).

Figure 7 shows the polysiloxane formation on the irradiated films of 1a, 2, 3, and 4. The fraction of the photoacid generating units of the polymers was 13-18 mol %. The polysiloxane formation rate was almost independent on the alkyl group of the methacrylate units, though the hydrophobic nature of the methacrylate ester groups strongly reduced the polysiloxane formation at the irradiated surface by a vapor-phase modification using alkoxysilane vapor (14). Although the φ value for FIS was almost the same as that for NIS, the polysiloxane formation rate for the irradiated 4 film was slower than that for 1a. The photochemically formed 9-fluorenone and fluorenone azine which are more hydrophobic than tetralone and its azine may prevent the water sorption at the irradiated areas, resulting in the slow rate of the polysiloxane formation.

Oxygen plasma etching

Figure 8 shows the effect of surface modification of 1b film on the oxygen plasma etching. The etching rate of 1b flm was 0.1 μm/min in the present etching conditions. The sample film was irradiated (126 mJ/cm²) and subsequently dipped in the liquid-phase modification bath (n-hexane-acetone (9:1, v/v) mixture containing 0.2 mol/L of MTEOS). The initial rapid-rate region corresponds to the process that polymers are

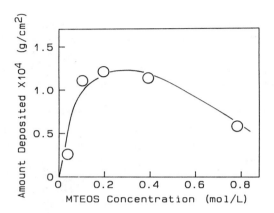

Figure 5.   Effect of MTEOS concentration on polysiloxane formation at the irradiated surface of a 1b film. Exposure dose: 372 mJ/cm^2.

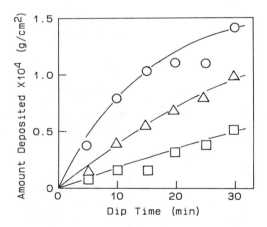

Figure 6.   Polysiloxane formation at the irradiated surface of a 1b film using MTEOS solution. Exposure dose: (○) 379, (△) 126, (□) 63 mJ/cm^2.

Table 2 Effect of alkoxysilane structure on the rate of polysiloxane formation at the irradiated surface of 1c film[a]

Alkoxysilane		$k \times 10^4$ (sec^{-1})[b]	Polysiloxane formed ($g/cm^2/min$)	
structure	abbreviation		$[H_2O]=1.7 \times 10^{-4}$ mol/L[c]	$[H_2O]=2.8 \times 10^{-2}$ mol/L[c]
$CH_3Si(OCH_3)_3$	MTMOS	5.28	4.67×10^{-6}	1.56×10^{-5}
$CH_3Si(OC_2H_5)_3$	MTEOS	6.55	3.69×10^{-6}	7.14×10^{-6}
$Si(OCH_3)_4$	TMOS	2.80	$-$[d]	5.97×10^{-6}
$Si(OC_2H_5)_4$	TEOS	0.25	$-$[d]	$-$[d]
$Si(OC_3H_7)_4$	TPOS	0.23	$-$[d]	$-$[d]
$C_6H_5CH_2Si(OC_2H_5)_3$	BTEOS	6.08	$-$[d]	$-$[d]
$C_2H_5Si(OC_2H_5)_3$	ETEOS	4.36	$-$[d]	$-$[d]

a) 1c film irradiated with 254-nm light at 379 mJ/cm2 was dipped in the alkoxysilane solution for 20 min at 35 °C. b) Pseudo-first-order rate constant. c) Water content in the alkoxysilane solution. d) Very small.

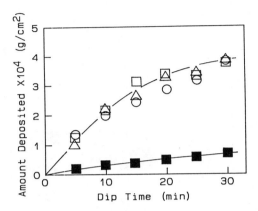

Figure 7.   Effect of polymer structure on the polysiloxane formation rate at the irradiated surface of the films. Polymer: (○) 1a, (△) 2, (□) 3, (■) 4.

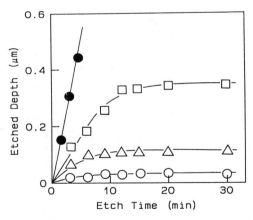

Figure 8.   Oxygen plasma etched depth of 1b films modified by polysiloxane networks.
Dip time of the irradiated 1b film in the 0.2 mol/L of MTEOS solution: (●) 0, (□) 5, (△) 15, (○) 25 min. Exposed dose: 126 mJ/cm^2.

etched to make SiO_2 layer acting as etch barrier to oxygen plasma. The initial thickness loss was dependent on the period treated with MTEOS solution. In the very slow-rate region the SiO_2 layer acted as a good etch barrier, where the etching rate was 13-40 times slower than that of 1b film. The etch rate could be controlled by exposure dose, the treatment time of the film with alkoxysilane solution, and the type of alkoxysilanes.

CONCLUSIONS

Polymers having 1,2,3,4-tetrahydro-1-naphthylideneamino p-styrenesulfonate (NIS) units of 9-fluorenilideneamino p-styrenesulfonate (FIS), which formed p-styrenesulfonic acid units upon irradiation with 254-nm light, were prepared. When the UV-irradiated polymer films were dipped in a liquid-phase modification bath (n-hexane-acetone (9:1, v/v) mixture containing alkoxysilanes) at 35°C, the polysiloxane networks were formed at the near surface of the film. No polysiloxane was formed at the unirradiated areas of the film. The polysiloxane formation rate was dependent on the number of the photochemically formed acid and the structure of alkoxysilane. The film surface modified with polysiloxane networks showed a good resistance to an oxygen plasma etching.

ACKNOWLEDGEMENT

This work was partly supported by a Grant-in-Aid from the Ministry of Education, Science, and Culture, Japan (No. 06805084).

REFERENCES

1. Thompson, L. F.; Willson, C. G.; Bowden, M. J. *Introduction to Microlithography Second Ed.*, American Chemical Society, Washington, DC 1994; chapter 3.

2. Coopmans, F.; Roland, B. *Proc. SPIE* 1986, 631, 34.

3. MacDonald, S. A.; Ito, H.; Hiraoka, H.; Willson, C. G. *Proceedings of the Regional Technical Conference on Photopolymers*, New York; Mid-Hudson Section, SPE: Ellenville, New York, 1985; p 177.

4. Baik, K.; Ronse, K.; Van den hove, L.; Roland, B. *Proc. SPIE* 1992, 1672, 362.

5. Baik, K.; Ronse, K.; Van den hove, L.; Roland, B. *Proc. SPIE* 1993, 1925, 302.

6. Follett, D.; Weiss, K.; Moore, J. A.; Steckl, A. J.; Liu, W. T. *The Electrochemical Society Extended Abstracts*; The Electrochemical Society: Pennington, NJ, 1982; Vol. 82-2, Abstract 201, p 321.

7. Stillwagen, L. E.; Vasile, M. J.; Baiocchi, F. A.; Silverman, P. J.; Taylor, G. N. *Microelectron. Eng.* 1987, 6, 381.

8. Taylor, G. N.; Nalamasu, O.; Stillwagon, L. E. *Microelectron. Eng.* 1989, 9, 513.

9. Nalamasu, O.; Taylor, G. N. *Proc. SPIE* 1989, 1086, 186.

10. Taylor, G. N.; Nalamasu, O.; Hutton, R. S. *Polym. News* 1990, 15, 268.

11. Calvert, J. M.; Dulcey, C. S.; Peckerar, M. C.; Schnur, J. M.; Georger Jr., J. H.; Calabrese, G. S.; Sricharoenchaikit, P. *Solid State Technol.* 1991, 34, 77.

12. Abali, L. N.; Bobbio, S. M.; Bohland, S. M.; Calabrese, G. S.; Gulla, M.; Pavelchek, E. K.; Sricharoenchaikit, P. *Microelectron. Eng.* 1991, 13, 93.

13. Calabrese, G. S.; Abali, L. N.; Bohland, J. F.; Pavelchek, E. K.; Sricharoenchaikit, P.; Vizvary, G.; Bobbio, S. M.; Smith, P. *Proc. SPIE* 1991, 1466, 528.

14. Shirai, M.; Hayashi, M.; Tsunooka, M. *Macromolecules* 1992, 25, 195.

15. Shirai, M.; Kinoshita, H.; Sumino, T.; Miwa, T.; Tsunooka, M. *Chem. Mater.* 1993, 5, 98.

16. Shirai, M.; Miwa, T.; Sumino, T.; Tsunooka, M. *J. Mater. Chem.* 1993, 3, 133.

17. Kamogawa, H.; Kanzawa, A.; Kadoya, M.; Naito, F.; Nanasawa, M. *Bull. Chem. Soc. Jpn.* 1983, 56, 762.

18. Shirai, M.; Masuda, T.; Ishida, H.; Tsunooka, M.; Tanaka, M. *Eur. Polym. J.* 1985, 21, 781.

19. Shirai, M.; Kinoshita, H.; Tsunooka, M. *Eur. Polym. J.* 1992, 28, 379.

20. Bradley, D. C. *Chem. Rev.* 1989, 89, 1317.

21. Hench, L. L.; West, J. K. *Chem. Rev.* 1990, 90, 33.

RECEIVED July 17, 1995

Chapter 22

New Polysiloxanes for Chemically Amplified Resist Applications

J. C. van de Grampel[1], R. Puyenbroek[1], A. Meetsma[1],
B. A. C. Rousseeuw[2], E. W. J. M. van der Drift[2]

[1]Department of Polymer Chemistry, University of Groningen,
Nijenborgh 4, 9747 AG Groningen, Netherlands
[2]Delft Institute of Microelectronics and Submircron Technology, Delft
University of Technology, P.O. Box 5053,
2600 GB Delft, Netherlands

New polymers have been developed for application in bilayer resist systems. The products are sensitive for e-beam and DUV exposure via chemical amplification and contain both silicon and acid-labile groups. Incorporation of silicon, necessary for a high resistance to reactive ion etching (RIE) in oxygen plasmas, was achieved by polysiloxanes present either as graft or as main chain units. Graft copolymers were prepared by radical polymerization involving styrene and methacrylate derivatives and silicon-containing macromers. In polymers where silicon is incorporated in the main chain, acid-labile groups were coupled to the siloxane chain by means of hydrosilylation starting from a poly(methylhydrosiloxane) and compounds possessing an unsaturated organic moiety. t-Butoxycarbonyloxy groups and epoxy groups were introduced as acid-labile moieties.
To raise the glass transition temperatures of the systems studied bulky side-groups were attached to the polymer chain. In another approach acid-labile groups are attached to a rigid poly(phenylsilsesquioxane) framework.
Depending on the developer used the polymers showed positive or negative images. In particular graft copolymers with t-butoxycarbonyloxy groups appeared to be very suitable as resists in DUV and e-beam exposure experiments.

In designing polymers suitable as resists in microlithography, several aspects have to be considered (1-2). For instance, the polymer should possess reactive groups to facilitate pattern distinction after irradiation. The polymer should be soluble in solvents that allow the formation of uniform thin films during spin coating. The thermal stability should be such that the polymer is stable under the microlithographic conditions employed. In general materials which are thermally stable above 150 °C and exhibit a glass transition temperature above 90 °C are desired.

0097–6156/95/0614–0333$12.50/0
© 1995 American Chemical Society

Important parameters in the application of a resist are: sensitivity, contrast, resolution and dry-etching resistance. Sensitivity determines the measure of how efficiently a resist responds to a given amount of radiation and depends to a large extent on reactive groups in the polymer and their rate of chemical change upon exposure. An approach to improve the sensitivity involves the concept of chemical amplification (3-11). Chemical amplification is characterized by photogeneration of protons which catalyze reactions resulting in differences in solubility between the exposed and unexposed regions. In these reactions the proton is being regenerated and thus continuously available for subsequent reactions, hence the "amplification nature" of the system. An example is given below. The cation of the photo acid generator triphenylsulfonium hexafluoroantimonate liberates a proton, that, upon subsequent baking at about 90 °C, catalyzes the cleavage of the t-butoxycarbonyloxy group to give amongst others poly(hydroxystyrene) and a proton (Figure 1).

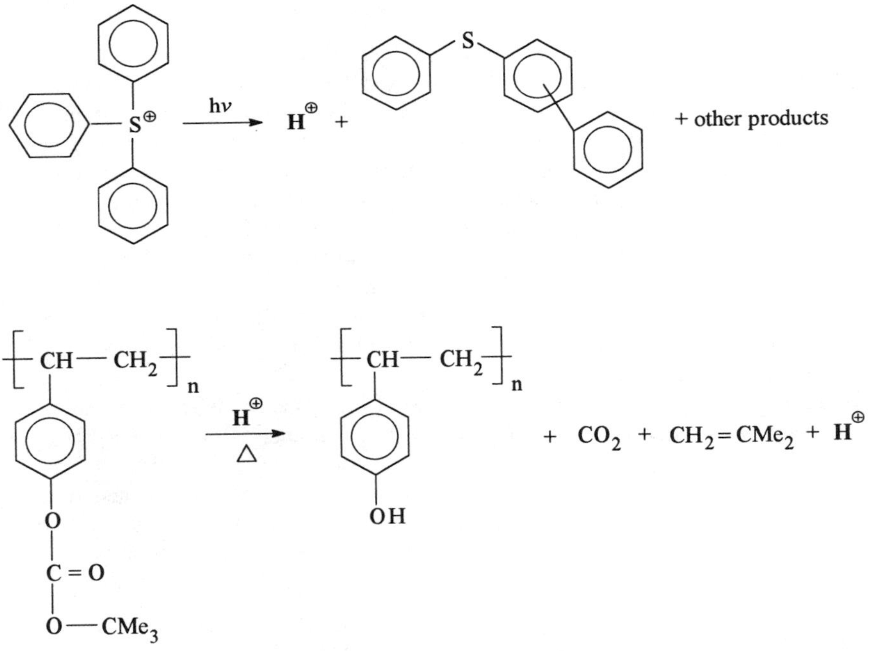

Figure 1. A scheme for chemical amplification with Ph_3S^+ as acid generator.

In the absence of an acid generator, cleavage of the t-butoxycarbonyloxy group only takes place at temperatures above 190 °C. The deprotection reaction given above results in a large polarity change in the exposed region of the resist. This change offers the possibility of either positive or negative images depending on the developer used subsequently. Non-polar solvents remove the unexposed areas giving rise to negative tone images, whereas solvents like tetramethylammoniumhydroxide (TMAH) or alcohols remove the exposed areas generating positive tone images. Resists based on chemical amplification processes have shown a considerable increase in sensitivity although unwanted interference between catalyst and polymer have been reported (12). Moreover, H^+ ions tend to diffuse to unexposed regions during postbake, which will affect the resolution in a negative sense.

Two other examples of chemical amplification, which will be discussed in more detail, are acid catalyzed crosslink reactions and polycondensation reactions. Irradiation of epoxy functionalized polymer films containing small amounts of a photo acid generator results in cationic polymerization of epoxy moieties (13-14). With A representing a polymer system (Figure 2a) chains are coupled together by crosslinking rendering the exposed area less soluble.

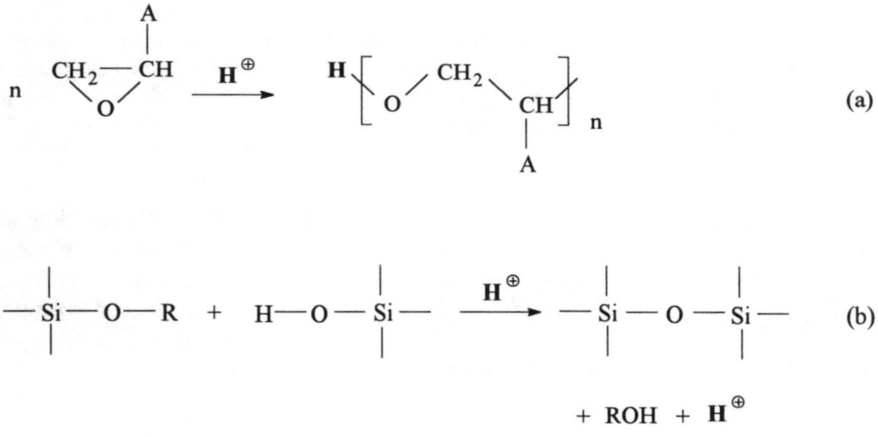

(a)

(b)

Figure 2. Chemical amplification involving (a) ring-opening polymerization and (b) polycondensation.

A similar result can be obtained by applying condensation chemistry to low-molecular weight silicon-containing polymers with alcohol and ether functionalities (Figure 2b) Photo acid generation starts condensation reactions by which Si-O-Si crosslinks are formed, resulting in a polymer with reduced solubility (15-16).

Resists for the top-layer in bilayer masks should display sufficient dry-etching resistance to reactive ion etching (RIE) in oxygen plasmas. Incorporation of silicon appears to be very effective to improve this etching resistance by the formation of a SiO_2 network after the first few seconds of oxygen plasma processing (17-20). This network acts as a mask for anisotropic etching in the same plasma step. Although polysiloxanes are very good candidates in this respect their low glass transition temperatures may hamper their application. The glass transition temperature of polysiloxanes can be raised by incorporation of bulky side-groups at the polymer chain. Another method that can be used is the application of ladder polymers, *e.g.* polysilsesquioxanes (21).

This paper describes the synthesis of silicon-containing polymers and their microlithographic evaluation in bilayer resist systems. Silicon is incorporated by using polysiloxanes, either in the main chain or in graft copolymers. Also poly(phenylsilsesquioxanes) are considered in this study. Functional groups such as t-butoxycarbonyloxy, t-butoxy or epoxy, all capable of chemical amplification, are present in all systems. The influence of halo-styrenes or chloro-substituted cyclophosphazenes as side-groups in combination with epoxy groups has been discussed in terms of microlithographic properties.

Experimental Part: Synthesis

General. All reactions were carried out in a dry oxygen-free nitrogen atmosphere using inert gas-vacuum techniques and Schlenk type glassware. Solvents were purified and dried according to conventional methods. Commercial products were used as received or purified by conventional methods.

Measurements. NMR spectra (^{1}H, ^{13}C, ^{29}Si) were recorded from $CDCl_3$ solutions on a Varian VXR 300 spectrometer. Molecular weights were obtained by means of vapor pressure osmometry (Knauer osmometer), and by membrane osmometry (Knauer osmometer equipped with a Knauer y-1244 membrane). Glass transition temperatures were recorded on a Perkin Elmer DSC7 unit. X-ray measurements were carried out on an Enraf-Nonius CAD-4F diffractometer.

Acid generators. The sulfonium acid generator (GE Silicones, UVE1014) was obtained as a solution of triphenylsulfonium hexafluoroantimonate, $[Ph_3S]^+[SbF_6]^-$, (30-60 wt%) and propylene carbonate (30-60 wt%) and used as received. The iodonium acid generator (GE Silicones, UV9310C) was obtained as a viscous solution of bis(4-dodecylphenyl)iodonium hexafluoroantimonate, $[p\text{-}Me(CH_2)_{11})Ph]_2I^+SbF_6^-$, (30-60 wt%) and 2-ethyl-1,3-hexanediol (30-60 wt%) and used as received.
Tri(methylsulfonyloxy)benzene (MESB) was prepared according to the literature (22). The yellow crystalline material obtained was recrystallized twice from acetone to yield 47 % of colorless crystals of tri(methylsulfonyloxy)benzene, mp 154-155 °C [lit. (23) 159 °C]. IR: 794, 842, 969, 1008, 1266, 1336, 1359, 1375, 1485, 1592, 2940, 3033 cm^{-1}. Elemental analysis: Calc. C 30.0, S 26.69, H 3.36. Found C 30.36, S 26.19, H 3.42.

The structure of MESB was confirmed by a X-ray structure analysis (Figure 3). Crystallographic data have been deposited with the Cambridge Crystallographic Data Centre.

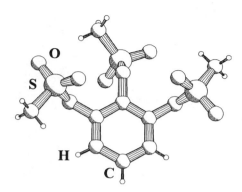

Figure 3. Molecular structure of MESB.

Synthesis of polymers. The synthetic procedures for the graft copolymers of the CPS and CPM series (Figure 4) and graft copolymers and terpolymers of the EPOXY series (Figure 5) have been described before (24-26). For the preparation of polymers to be applied in microlithographic experiments macromers (Figure 6) with molecular weights of 2.8×10^3 (CPS and CPM) and 17×10^3 or 4.5×10^3 (EPOXY) were used. The polysiloxanes (JASIC, Figure 7) with acid-labile side-groups have been prepared by hydrosilylation, starting from poly(methylhydrosiloxanes) and the appropriate alkenes (25). Only the synthesis of poly[methyl(5-norbornane-2-*t*-butoxy-carbonyloxy)siloxane] (NORSOX, Figure 7) will be given here. The synthesis of polysilsesquioxanes (SILSESQUI, Figure 8) has been given in ref. 21.

Synthesis of 5-norbornene-2-t-butoxycarbonate. A mixture of 5.01 g (45 mmol) of 5-(norbornene)-2-ol and 5.10 g (45 mmol) of potassium-*t*-butoxide in 100 ml of tetrahydrofuran (THF) was stirred for 1 hour at room temperature. After the addition of 10.30 g (47 mmol) of di(*t*-butyl)dicarbonate and 150 ml of THF the light-orange suspension was stirred for an additional period of 72 hours at room temperature and subsequently poured into 200 ml of ice-water. The aqueous mixture was extracted with 3 portions of 50 ml of diethyl ether. After drying over $MgSO_4$ the combined extracts were filtered and evaporated to dryness to yield a white solid (9.00 g, 43 mmol, 90 %), mp 41-43 °C. The compound was characterized by ^1H and ^{13}C NMR.

Synthesis of poly[methyl(5-norbornane-2-t-butoxycarbonyloxy)siloxane], NORSOX. A solution of 0.40 g (0.18 mmol) of $Me_3SiO[(Si(MeH)O]_{35}SiMe_3$ and 2.5 g (11.89 mmol) of 5-norbornene-2-*t*-butoxycarbonate in 15 ml of toluene and 0.09 ml of Karstedt-catalyst (Hüls-Petrarch) was stirred at room temperature. After 72 hours the polymer was precipitated into methanol. The white powder obtained after decantation was washed with methanol (2x25 ml) and dried at 50 °C in vacuo.

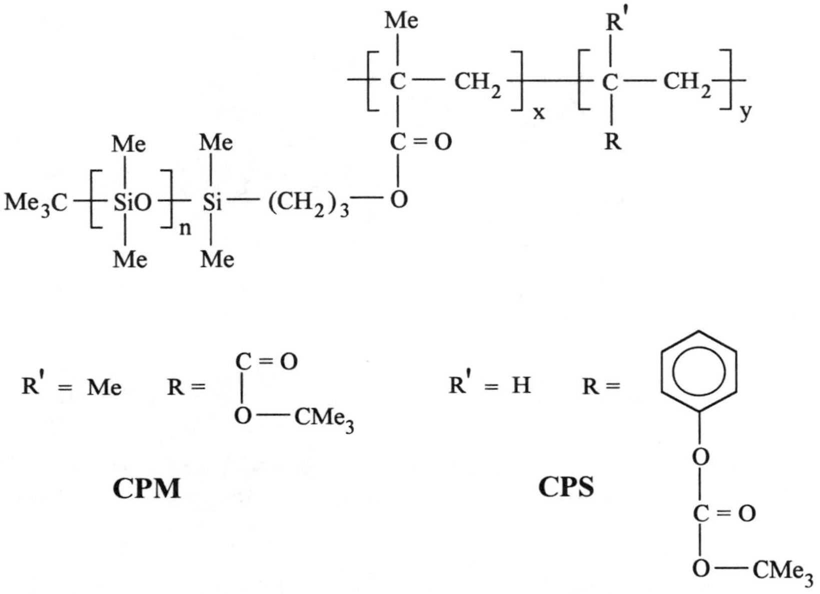

Figure 4. CPM and CPS series.

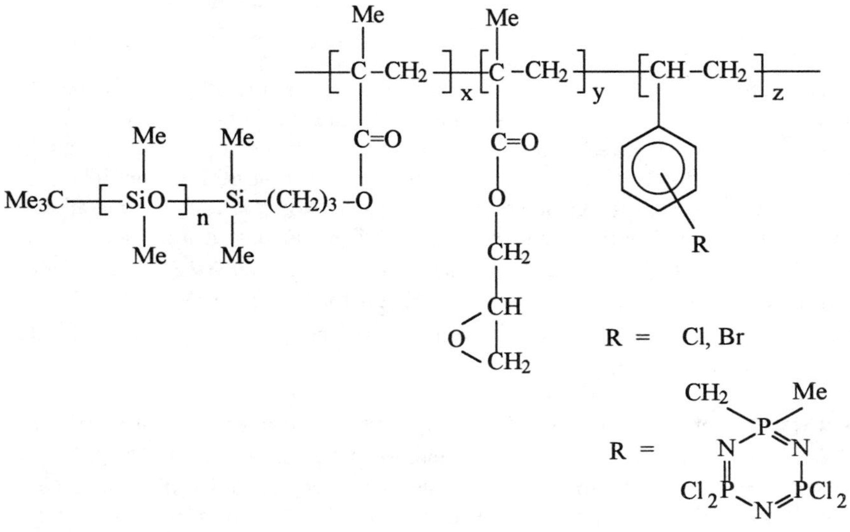

Figure 5. EPOXY series, $z = 0$ (graft copolymer), $z \neq 0$ (graft terpolymer).

Yield 0.84 g (0.09 mmol, 50%) of NORSOX. M_n (calc) 9.6×10^3, M_n (found by GPC) 9×10^3, D 1.5, T_g 60 °C. ^{29}Si NMR δ 7.2 (SiMe$_3$)$_3$, -23 to -25 ppm (SiMeR).

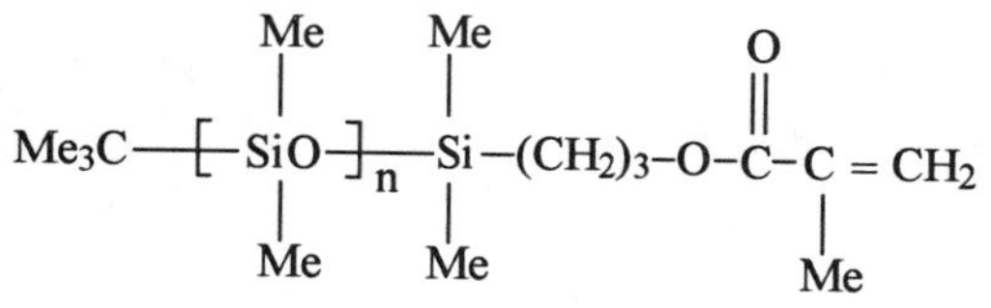

$$Me_3C \longrightarrow \left[\begin{array}{c} Me \\ | \\ SiO \\ | \\ Me \end{array} \right]_n \begin{array}{c} Me \\ | \\ Si \\ | \\ Me \end{array} -(CH_2)_3-O-\overset{\overset{\displaystyle O}{\|}}{C}-\underset{\underset{\displaystyle Me}{|}}{C}=CH_2$$

Figure 6. Structure of a macromer with a methacrylate end group.

Experimental; Microlithography

Solution, preparation and bake conditions. Prebake and postbake were carried out on hot plates in air. Prebake was performed at temperatures above the glass transition temperatures of the polymers to ensure complete removal of the solvent. Postbake was carried out at temperatures at which the chemical amplification reactions proceed most efficiently. By preference postbake temperatures were taken below the T_g of the polymer to avoid pattern deformation and to reduce acid diffusion.

CPS/CPM, JASIC: Standard resist solutions were 7.5 wt% in propyleneglycol monomethyl ether acetate (PGMEA) with 1 wt% bis(4-dodecylphenyl)iodonium hexafluoroantimonate as acid generator. Polymers were spin-coated from solution on a silicon wafer and subsequently baked for 3 minutes at 90 °C. Film thickness of the polymer layer was about 275 nm.

NORSOX: A solution of 8.5 wt% of NORSOX in bis(2-methoxyethyl)ether (DME) with 1 wt% of tri(methylsulfonyloxy)benzene (MESB), was spin-coated on a silicon wafer and prebaked at 90 °C for 3 minutes to form a 200 nm thick polymer layer.

EPOXY: A solution of 9.5 wt% co- or terpolymer in PGMEA with 3 wt% triphenylsulfonium hexafluoroantimonate was spun on a silicon wafer, and prebaked at 100 °C for 3 minutes to form a 300 nm thick polymer layer.

SILSESQUI: Standard resist solutions were 10 wt% in PGMEA with 3 wt% bis(4-dodecylphenyl)iodonium hexafluoroantimonate or 3 wt% triphenylsulfonium hexafluoroantimonate as acid generator. Solutions were spin-coated on Si wafers. To improve the adhesion of the polymer to the surface the wafer was primed with hexamethyldisilazane (HMDS). Prebake took place for 3 minutes at 90 °C. Film thickness was about 300 nm.

The formulations of polymer solutions given above, essential for the lithographic performance of the resists, were obtained after extensive optimization studies.

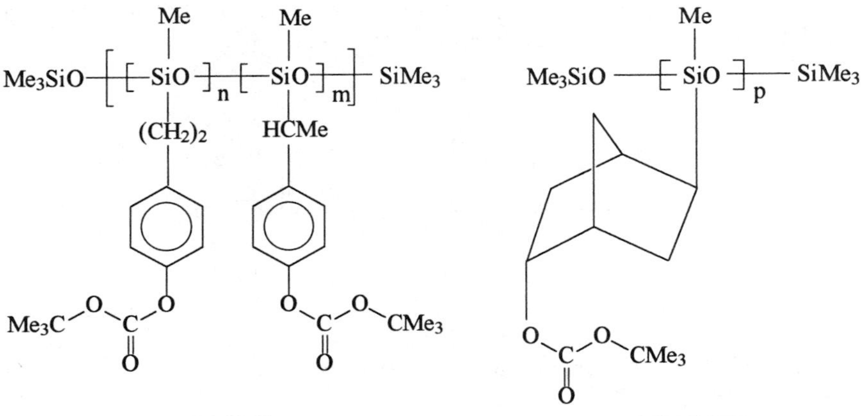

Figure 7. Structure of JASIC (n+m=35) and NORSOX (p=35).

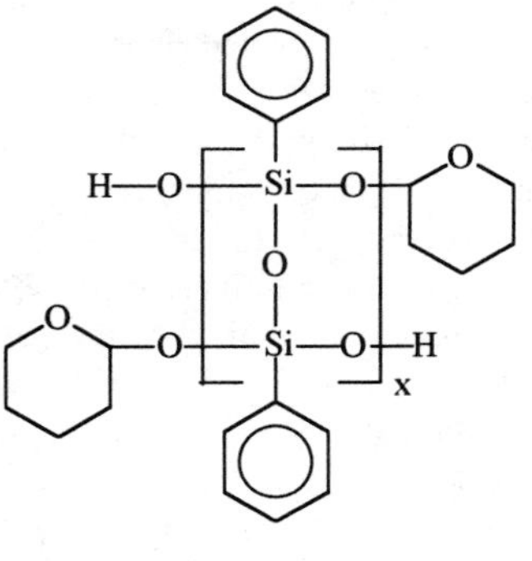

Figure 8. Polysilsesquioxane derivative.

Exposure methods. E-beam exposures were carried out with a Philips EBPG/03 operating at 50 keV. DUV exposures (quartz chromium mask plate) were performed in a home-built system operating at 248 nm [standard mercury lamp (Osram)].

Pattern fabrication. CPS/CPM, JASIC: Immediately after exposure resist layers were postbaked for 3 minutes at 90 °C and patterns were developed in either 5 % tetramethylammoniumhydroxide (TMAH) (positive tone) for 2.5 minutes or in a (3:1) mixture of methylisobutylketone and hexane (negative tone) for 5 seconds.

NORSOX: After exposure the resist layer was postbaked at 80 °C for 3 min. The patterns were developed in either DME (negative tone) for 5 seconds or 5% (TMAH) (positive tone) for 2.5 minutes.

EPOXY: After exposure the resist layer was postbaked at 80 °C for 3 minutes and subsequently developed with PGMEA or mixtures of PGMEA and hexane. In the case of the copolymers the postbake temperature was varied between 50 and 80 °C.

SILSESQUI: Immediately after exposure the resist layer was postbaked for 3 minutes at 110 °C and patterns were developed in 3:1 mixture of methylisobutylketone (MIBK) and hexane (negative tone) for 5 seconds.

Etch conditions. AZ photoresist (Shipley Co.) was baked during 1 hour at 200 °C. A 750 nm thick hard-baked AZ resist layer was used as the thick bottom layer in bilevel experiments. Thickness of CPS, JASIC and EPOXY layers are given before.

Dry-etching (O_2-RIE) of CPS/AZ, JASIC/AZ and EPOXY/AZ double layers was performed in a parallel plate reactive ion etcher Z401 (Leybold) in a low pressure oxygen plasma (RIE) at 3 μBar, 0.07 W/cm^2 rf power at 13.56 MHz with dc bias voltage of about -180 V and 20 sccm gas flow. Dry-etching of NORSOX and SILSESQUI resists was performed as single layers under the same conditions as described above.

Analysis of microlithographic methods. Development results were measured with a stylus apparatus (Tencor Alphastep 200) on 25 μm wide isolated lines. Sensitivity was defined as thickness normalized to resist thickness after postbake (PEB) (positive tone) or after development (negative tone). Pattern quality was checked by scanning electron microscopy (JEOL JSM 840). XPS measurements were carried out on a PHI 5400 ESCA apparatus. IR spectra were recorded on a Mattson 4020 Galaxy FT-IR spectrometer.

Results and Discussion

Graft Copolymers. Two classes of graft polymers have been synthesized. The first class contains polymers prepared by polymerization of t-butoxycarbonyloxystyrene with silicon-containing macromers (CPS series) while the second class concerns polymers made by polymerization of the same macromers with t-butylmethacrylate (CPM series). Acid-labile moieties are represented by t-butoxycarbonyloxy and t-butylmethacrylate groups. Upon acid-catalyzed decarboxylation, hydroxyl groups are formed with CPS while methacrylic acid groups are formed when CPM is used as resist. Micro phase separation between the "organic" and "silicon" phase has been observed for incorporation of macromers for the CPS and CPM series, in particular for those polymers with molecular weights higher than about 3.5×10^3 (24). This phenomenon makes these graft copolymers less suitable for microlithography, as O_2-dry-etching leads to non-uniform SiO_2 networks. As shown by TEM experiments graft copolymers with grafts of lower molecular weight show a reduced phase separation (24). For the CPS and CPM series described here copolymers are used with grafts having a molecular weight of 2.8×10^3.

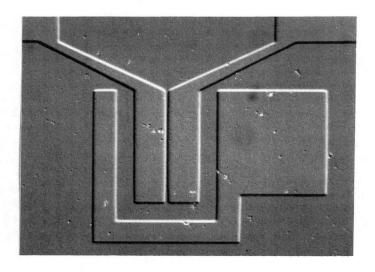

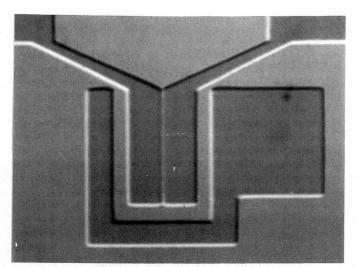

Figure 9. Optical micrographs of negative (top) and positive (bottom) tone patterns in CPS resist after DUV exposure. Dose 10 mJ/cm^2, postbake 90 ^{o}C/3 min, developers: negative tone, MIBK-hexane mixture (3:1) for 5 seconds, positive tone 5% TMAH for 2.5 minutes. Small bar in the center is 2 μm wide. Reproduced with the permission of reference 30.

Microlithographic experiments with CPM type polymers showed these polymers not to be suitable as resists under e-beam exposure. In the negative tone, a resolution below 1 micron could not be obtained, and the quality of the patterns was poor. In the positive mode, side reactions of the methacrylate acid groups resulted in crosslinked products which could not be developed in either aqueous base solutions or alcohols. These side reactions were also observed by Ito et al. (27), Shiraishi et al. (28) and Allen et al. (29), who studied poly(t-butylmethacrylate) as positive resist.

The CPS series showed more promising results. Detailed investigations were carried out on a graft copolymer, having a molecular weight of 82×10^3 and containing 33 wt% of a macromer with a molecular weight of 2.8×10^3. The upper glass transition temperature of this copolymer, abbreviated as CPS, is 106 °C (24). As mentioned before, CPS can exhibit a positive tone as well as a negative tone, depending on the developer used. An example of DUV exposure results is given in Figure 9.

Exposure curves for CPS in the presence of acid generator in the positive, and negative tone and for plain CPS in the negative tone are shown in Figure 10. In Table I the lithographic characteristics of CPS are compiled including the dry etch selectivity ratios.

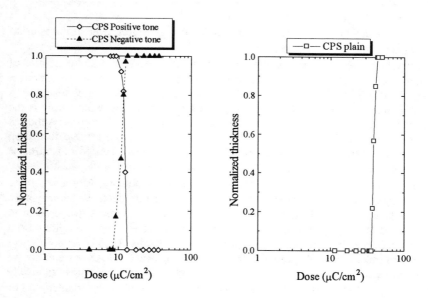

Figure 10. E-beam exposure characteristics of CPS with acid generator in positive and negative mode, and plain CPS in negative mode at 50 keV e-beam exposure.

Surprisingly, resist formulations of CPS without acid generator also show images after e-beam exposure. However, images could only be obtained at high doses (about

40 μC/cm^2) and in the negative tone only. The underlying mechanism can be attributed to crosslinking initiated by high energy electrons reacting with the methyl groups of the silicon-containing branches (31). In contrast, DUV experiments show

Table I. Lithographic characteristics for CPS resist (negative and positive tone) and plain CPS resist (negative tone)

	Negative tone		Positive tone
	no acid	acid	acid
sensitivity (50 keV e-beam, μC/cm^2)	42	13	13
contrast	13.7	5.7	-12.6
sensitivity (248 nm DUV, mJ/cm^2)	>150	7.5	7.5
etch selectivity (relative to AZ)	6.3	7.0	3.5

no imaging features in resist formulations without acid generator, even at doses up to 150 mJ.cm^{-1}.

In the negative tone, the acid generator lowers the clearance dose by a factor of 3, and the exposures for 'acid' and 'no acid' formulations are well separated. Contrast values (defined as slope of the linear parts of sensitivity curves) are comparable to results reported for t-butyloxycarbonyloxy (t-BOC) systems (32). Noteworthy is the extremely high contrast for CPS with acid generator in the positive tone. Although positive images can be obtained most of the images suffer from residue formation which can grow to about 20 nm thickness. XPS measurements carried out on these residues suggested the presence of siloxane groups. Also FT-IR spectra show absorptions around 1050 cm^{-1}, characteristic for siloxane moieties. Moreover, these spectra suggest the presence of insoluble phenyl-containing products, which may arise from irradiated bis(4-dodecylphenyl)iodonium hexafluoroantimonate (used as acid generator). Earlier publications also point to negative tone effects resulting from insoluble decomposition products of irradiated arylonium salts (12, 33).

Pattern transfer in acid-containing formulation of bilayer CPS-AZ systems shows that when pattern dimensions shrink to micron region the clearance dose gradually increases from 13 to 35 μC/cm^2, i.e. comparable to the level of plain formulation. For negative tone patterns it is observed that at high dose the t-BOC reaction easily extends into the unexposed areas, too far to be accounted for by proximity exposure effects. The observations can be ascribed to a relative increase of diffusion of H$^+$ into the unexposed regions with decreasing pattern dimensions. From line-broadening a diffusion constant of 1.1×10^{-4} μm^2/sec may be derived, which is comparable to values obtained by Schlegel et al. (34). As an example, they found a diffusion coefficient of 1.3×10^{-4} μm^2/sec at 80 °C for a [Ph$_2$I]$^+$[PF$_6$]$^-$ acid generator in combination with tetrahydropyranyl protected polyvinylphenol.

Figure 11 shows SEM micrographs of negative tone images obtained for CPS-AZ systems after e-beam exposure, development and O_2-RIE. In order to obtain 1.0 and 0.5 μm imaging features a dose of 30 $\mu C/cm^2$ is required, while a dose of 50 $\mu C/cm^2$ is necessary in getting 0.25 μm negative tone images. For CPS resists the etch rate selectivity is moderate when compared with the polysiloxane based resist SNR ($[SiPh(C_6H_4Cl)O]_n$). SNR processed under identical plasma circumstances shows an almost infinite dry etch resistance (35). For CPS, however, a value of about 7 was observed. The lower value can partly be explained by thinning of the resist layer (about 25%) due to the formation of volatile products (CO_2 and isobutene) during postbake. O_2 RIE experiments showed that 33 wt% macromer incorporation, which corresponds with 12.5 % silicon, is necessary to ensure a moderate etch rate selectivity. As CPS is a graft copolymer of organic segments and inorganic branches, a porous network of SiO_2 will be left after imaging and plasma treatment. Moderate etch rate selectivity of the CPS series is also caused by microphase separation of organic backbone chains and silicon-containing branches, as is observed in methacrylic-siloxane graft copolymers (36).

When increasing the molecular weight of the macromers an increase in sensitivity is observed. This increase is, however, acompagnied by a decrease of the etch resistance. Graft copolymers with silicon-containing branches with molecular weights of 2.5×10^3 exhibit sensitivities around 13 $\mu C/cm^2$ while sensitivity is increased to 4 $\mu C/cm^2$ for macromers with a molecular weight of 17×10^3. Microphase separation in case of high molecular weight macromers results in a 'clustering' of t-BOC groups and hence an increase in sensitivity. However, this 'clustering' results in a non-uniform etch resistance of the film leading to a decrease of the overall etch resistance of the film.

Systems containing at least 33 wt% silicon with branches having a molecular weight of about 1×10^3 are currently under investigation.

Polysiloxane Derivatives, JASIC and NORSOX. In contrast to the CPS graft copolymers, in the case of the polysiloxanes silicon is incorporated in the main chain. Not only a higher silicon content is now achieved, but also the negative influence of microphase separation is excluded. Similar to the CPS graft copolymers, acid-labile moieties are represented by t-butoxycarbonyloxy groups which are converted to hydroxyl groups upon acid-catalyzed decarboxylation.

A disadvantage of polysiloxanes is their relatively low glass transition temperatures. For instance, JASIC (Figure 6) has a T_g of 37 °C. Higher glass transition temperatures can be obtained by a direct attachment of bulky groups to the polymer chain, which is illustrated by the polymer NORSOX (Figure 6; T_g = 60 °C). The norbornane group acts as a bulky spacer between the functional group and the siloxane chain. JASIC and NORSOX show positive and negative tone images after DUV exposure, comparable to those obtained for CPS. E-beam exposure curves for negative tone images in presence of acid generator is given in Figure 12. Lithographic data of JASIC and NORSOX are compiled in Table II. As observed for CPS,

(a)

(b)

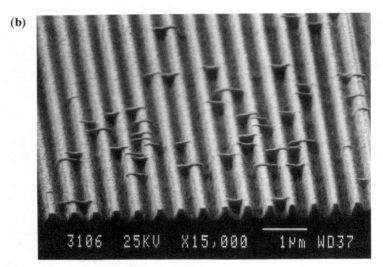

Figure 11. SEM micrographs of negative tone images obtained in a CPS/AZ bilevel
resist system after e-beam exposure (50 keV, development with a MIBK-
hexane mixture (3:1) for 5 seconds and O_2-RIE).
a) 1.0 μm, (dose 30 μC/cm^2) and b) 0.25 μm (dose 50 μC/cm^2) lines and
spaces.

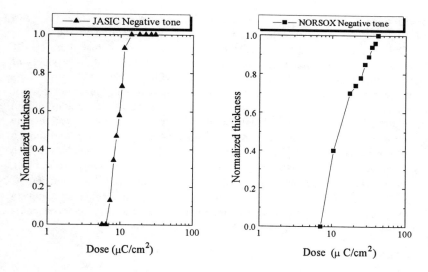

Figure 12. E-beam exposure characteristics of JASIC and NORSOX with acid generator at 50 keV.

negative tone images are formed for JASIC and NORSOX in absence of any acid generator, when a dose of about 40 $\mu C/cm^2$ is applied. The negative tone images may be ascribed to crosslinking initiated by electrons via the methyl groups linked to silicon (30). As may be seen from Table II the use of an acid generator lowers the clearance dose by about a factor of three. Exposure curves in the positive mode are

Table II. Lithographic characteristics for JASIC (negative tone) and NORSOX (negative tone)

	JASIC	NORSOX
sensitivity (50 keV e-beam, $\mu C/cm^2$)	13	11
contrast	3.8	2.3
sensitivity (248 nm DUV, mJ/cm^2)	7.5	7.5
etch selectivity (relative to AZ)	10	6

seriously hampered by residue formation. This phenomenon , which is more pronounced (up to 50% of the initial layer thickness) for the polysiloxane derivatives than for CPS, can be ascribed for the major part to reactions involving the polysiloxane framework, or more precisely to:

a) acid catalyzed crosslink reactions involving Me_3Si end-groups (16).

b) side reactions at the methyl groups initiated by high-energy electrons (31).

In both systems residues exhibit a pronounced IR absorption in the Si-O-Si region. Only for JASIC phenyl containing residues could be detected by FT-IR measurements. In this context it has to emphasized that decomposition products of irradiated acid generators can also contribute to residue formation.

Acid generator concentrations above 3 wt% lead to poor film structures for JASIC. The main cause is low solubility of the generator in the JASIC medium, which lead to non-homogeneous distributions of the acid. Figure 13 shows negative tone images obtained for JASIC with 1 wt% acid generator. At that concentration a homogeneous distribution of the acid generator is obtained accompanied by an improvement of the negative tone images. However, resolution is limited to 0.5 μm. This limitation is mainly caused by the postbake temperature at 90 °C (about 60 °C above the Tg of JASIC) during 3 minutes which is necessary to accomplish complete t-BOC conversion. Pattern deformation at that temperature leads to a line-broadening of the negative tone images, and a limited resolution is obtained.

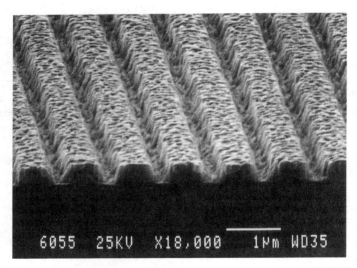

Figure 13. SEM micrograph of negative tone images obtained in a JASIC/AZ double layer resist system with 1 wt% of the iodonium containing acid generator. Images are obtained after e-beam exposure (50 keV, 25μC/cm^2), postbake (90 °C for 3 minutes), development (MIBK-hexane mixture (3:1) for 5 seconds) and O$_2$ RIE.

The etch selectivity for JASIC is moderate when compared with the polysiloxane based resist SNR (35). Under exposure to O$_2$ plasma rapid formation of a dense SiO$_2$

network may be expected as an efficient mask for the photoresist under-layer. However, the acid induced decarboxylation reaction in JASIC yields highly volatile products like CO_2 and isobutene, and a rather porous film structure is left. The rough structure of the JASIC containing upper-layer in Figure 13 illustrates this effect. In addition, thinning of the resist layer after postbake (about 25%), as was found for CPS, decreases the etch resistance. The etch ratio is sufficient to withstand the oxygen RIE process necessary to pattern a 750 nm thick underlying AZ resist layer in a bilevel resist system using a 200 nm thick JASIC layer as top-layer.

The lithographic evaluation of NORSOX as the upper-layer in a bilevel resist system leads to analogous characteristics as obtained for JASIC. In presence of MESB as acid generator the e-beam resolution of negative tone images is limited to 0.5 µm, when applying a postbake temperature of 80 °C (about 15 °C above the glass transition of the polymer) for three minutes. Postbake at 50 °C for 5 minutes shows a decreased pattern quality as a result of incomplete decarboxylation.

Epoxy series. As shown in Figure 5 two classes of polymers have been investigated, those with $z = 0$ (epoxy graft copolymers) and those with $z \neq 0$ (epoxy graft terpolymers). Copolymers of glycidylmethacrylate (GMA) and silicon-containing macromers (molecular weight 15000 or 4500) with Tg values ranging from 72 °C to 76 °C show poor negative tone images in DUV as well as in e-beam exposure experiments. Independent of the amount of incorporation or size of the macromers, e-beam resolution is limited to approximately 1-2 µm at doses ranging from 1 to 10 µC/cm^2. Only by decreasing the amount of the acid generator ($[Ph_3S^+][SbF_6^-]$) from 3 to 0.5 wt%, a slight improvement of the features was observed. Line-broadening of patterns after development with PGMEA during 5 seconds and PEB temperature of 80 °C for 3 minutes causes substantial pattern distortions. A line-broadening of 400 nm was observed which corresponds to a diffusion coefficient of about 4.4×10^{-4} µm^2/sec. Probably a highly effective polymerization of the epoxy moieties results in cross-linking extending into the unexposed areas. Moreover, the bake temperature is higher than the T_g value, which may give rise to pattern deformation. Reducing the postbake period from 3 minutes to 30 seconds did not affect the poor lithographic performance of the copolymers. Postbake temperatures below 80 °C showed even worse negative tone images as the result of incomplete cross-linking of the exposed areas. The influence of the postbake temperature is also reported by Hatzakis et al. (37), who showed a postbake temperature of 90 °C during 3 minutes to be necessary to ensure complete crosslinking of the exposed regions.

As reported by Taylor and Wolf (38) incorporation of halogens influences the kinetics of cationic polymerization of epoxy-containing polymers. Styrene-based polymers with halogen groups enables radiation-induced crosslinking with high crosslink efficiency. The localized nature of the crosslinking, as opposed to the chain propagation mechanism in the epoxy-containing resists, depresses the line-broadening effects of epoxy materials (39-40). For this reason bromo- or chlorostyrene were incorporated to suppress the highly efficient acid-catalyzed reaction. An additional advantage of incorporating styrene containing monomers is the increase of the T_g by about 20 °C. The higher T_g allows postbake, which is most efficient at temperatures of 80 °C, below the glass transition temperature. Higher resolution can therefore be expected.

Indeed a terpolymer containing 73 wt% GMA, 13 wt% macromer (molecular weight 4500), and 14 wt% bromostyrene showed an improvement of the resolution after e-beam exposure to approximately 0.5 µm. Line-broadening after development and postbake is reduced but still the resist patterns showed a distortion of the line-space (L/S) ratio. Normally the ratio should be 1, but appeared to be 1.2 : 0.8 in case of 1 µm lines and spaces. Proton diffusion into the unexposed regions may easily explain this observation. Increasing the molecular weight of the silicon-containing macromer from 4500 to 17000 or increasing the amount of incorporation of the macromers from 6 to 20 wt% did not have any significant effect on the resolution. The terpolymers containing chlorostyrene show similar results.

Figure 14 shows the sensitivity curve for e-beam exposure (50 keV) of the above mentioned terpolymer containing bromostyrene (BRTERP). The shape of the sensitivity curve is remarkable in that it contains two inflection points. Probably two

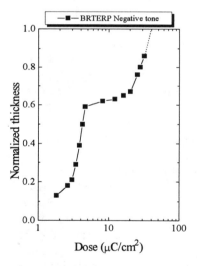

Figure 14. E-beam exposure characteristic of BRTERP with acid generator in negative tone at 50 keV.

mechanisms are responsible for the performance of this polymer as negative resist. In the range of 1-7 µC/cm^2 acid-catalyzed cationic polymerization of epoxy moieties and cleavage of the carbon-halogen bonds are the main processes in creating negative tone patterns. Above 15 µC/cm^2 a second process contributes to the cross-linking of exposed regions. As was found in the case of the CPS graft copolymers, electron induced side-reactions with the silicon-containing branches are likely to occur. This second process reduces the sensitivity of the resist system because less energy is available for the acid-catalyzed and radical induced crosslink reactions. As an example, Hatzakis et al. reported sensitivities of about 0.5 $\mu C/cm^2$ at 50 keV using an epoxy-containing novolac resist (37). These sensitivities are considerably higher than those found for the corresponding epoxy conversion in the terpolymer (about 3-4 $\mu C/cm^2$ at 50 keV).

The copolymers of GMA and macromers possess etch ratios of about 9 compared with the AZ-type resists. As was found for CPS, an increase of the amount of silicon does not result in higher etch ratios probably as a result of a non-uniform distribution of silicon-containing branches.

In addition two phosphazene-containing terpolymers were synthesized (26); one (PHTERP1, T_g value = 90 °C, containing 34 wt% gem-methylvinylbenzyl-tetrachlorocyclotriphosphazene (STP), 33 wt% GMA and 33 wt% of a macromer (molecular weight 4.5×10^3) and another, (PHTERP2, T_g = 95 °C), containing 47 wt% STP, 20 wt% GMA and 33 wt% of the same macromer . E-beam characteristics are summarized in Table III.

Table III. Lithographic characteristics for PHTERP1 and PHTERP2 (negative tone)

	PHTERP1	PHTERP2
sensitivity (50 keV e-beam, $\mu C/cm^2$)	20	20
contrast	1.4	1.1
etch selectivity (relative to AZ)	10	11

Resolution of PHTERP1 for e-beam in negative tone is limited to 1.0 μm. For PHTERP2 a higher resolution could be accomplished, viz. 0.5 μm. The increase of the phosphazene content apparently suppresses crosslink reactions into the unexposed regions. As shown in Table III the terpolymers discussed are capable of forming a homogeneous SiO_2 top layer, providing a good etch resistance towards oxygen plasma reactions. The terpolymers containing halo-styrene moieties exhibit the same etch ratios.

Polysilsesquioxanes. Preliminary DUV lithographic experiments were carried out on polysilsesquioxanes partially protected by tetrahydropyranyl groups (21). As described before lithographic performance of protected poly(phenylsilsesquioxanes) is based on polycondensation reactions (Figure 2b). Exposed resist films result in less soluble materials, leading to possible application as negative resists. Watanabe et al. (15) and Sakata et al. (16) reported polysilsesquioxanes with ethoxy end-groups. Using these systems it was possible to achieve 0.4 μm negative tone images with an e-beam exposure dose of 0.2 $\mu C/cm^2$ at 25 keV. An advantage of the polysilsesquioxanes is the high silicon content of the polymers (about 21 wt%), giving an excellent O_2 RIE resistance. In the system discussed here tetrahydropyranyl groups are incorporated instead of ethoxy groups, providing a higher sensitivity towards acid as compared with the ethoxy derivatives. The polymer, which is stable up to 150 °C, exhibits no distinct glass transition temperature below that temperature (21).

Using the iodonium containing acid generator good films were obtained, particularly when silicon wafers were pre-coated with HMDS, which ensures improved adhesion of the resist layer to the wafer. At a dose of 10 mJ/cm^2, an acid generator concentration of 3 wt% and a postbake temperature of 110 °C negative tone patterns were obtained. Concentrations of the acid generator below 3 wt% gave poor negative tone patterns caused by incomplete polycondensation reactions. Below 110 °C pattern distinction was absent. Postbake above 150 °C resulted in completely insoluble films due to polymer degradation reactions. Line-broadening was not observed which means acid diffusion occurs in a much smaller range than in the resists described before, viz. CPS, or JASIC.

Due to the high silicon content and the high network-like structure, the etch selectivity of tetrahydropyranyl protected poly(phenylsilsesquioxanes) is excellent and amounts to 19 when compared to commercial organic AZ-resists.

Conclusions

Evaluating the preliminary lithographic experiments carried out on the four classes of silicon-containing polymers, CPS graft copolymers with t-BOC moieties and silicon-containing branches are the most promising resists. These graft copolymers are suitable in DUV and e-beam lithography.

Optimum lithographic results were obtained with a CPS material having a molecular weight of 82×10^3 with 12.4 % Si incorporated and a glass transition temperature of about 106 °C . This resist has a resolution of 0.25 μm in the negative mode (Figure 11b) with a sensitivity of 13 $\mu C/cm^2$, a contrast value of 5.7 and an etch selectivity ratio of about 7. Although positive images could be obtained most of these patterns suffer from residues in particular when e-beam radiation is used. These residues are caused by side reactions of the decomposition products of the irradiated photo acid generator or electron induced side reactions with the polysiloxane branches and can be avoided by using DUV exposure techniques.

Epoxy containing polymers are extremely sensitive negative resists in DUV as well as in e-beam lithography. However copolymers of GMA and a silicon-containing macromer are not suitable as high resolution resists as a result of chain propagation of the epoxy groups. Halogen incorporation decreases the catalytic chain efficiency of the acid generator. The resulting decrease of acid diffusion leads to negative tone patterns of about 0.5 μm for chloro- and bromostyrene incorporation.

Phosphazene containing terpolymers seem to be the most suitable halogen-containing resists because of their high halogen content and excellent film qualities.

Chemical amplification based on t-BOC chemistry provides negative as well as positive tone images depending on the developer used. A major disadvantage of resists based on chemical amplification is the acid diffusion which limits the resolution.

Acknowledgements

This work was fully supported by the Technology Foundation of The Netherlands (STW). GE Plastics (Bergen op Zoom, The Netherlands) is acknowledged for providing acid generators UVE1014 and UV9310C. The authors are indebted to Dr. H.T. van de Grampel for valuable discussions.

Literature cited

1. Reichmanis, E.; Thompson, L.F. In *Polymers in Microlithography*; Reichmanis, E.; MacDonald, S.A.; Iwayanagi, T., Eds., ACS Symp. Ser. No. 412, ACS Washington, DC, 1989, 1.
2. Reichmanis, E.; Novembre, A.E. *Annual Rev. Mater. Sci.* **1993**, 23, 11.
3. Ito, H.; Willson C.G; Fréchet, J.M.J.; Farrall, M.J.; Eichler, E. *Macromolecules* **1983**, 16, 510.
4. Ito, H.; Willson C.G. In *Polymers in Electronics*; Davidson, T. Ed. ACS Symp. Ser. No. 242, ACS, Washington, DC, 1984, 11.
5. Willson, C.G.; Ito, H.; Fréchet, J.M.J.; Tessier, T.G.; Houlihan, F.M. *J. Electrochem. Soc.* **1986**, 133, 181.
6. Crivello, J.V.; Lam, J.H.W. *Macromolecules* **1977**, 10, 1307.
7. Crivello, J.V. In *Polymers in Electronics*; Davidson, T. Ed. ACS Symp. Ser. No. 242, ACS, Washington, DC, 1984, 3.
8. Crivello, J.V.; Lee, J.L.; Conlon, D.A. *Makromol. Chem. Macromol. Symp.*, **1988**, 13/14, 145.
9. Conlon, D.A.; Crivello, J.V.; Lee, J.L.; O'Brien, M.J., *Macromolecules*, **1989**, 22, 509.
10. Houlihan, F.M.; Reichmanis, E.; Thompson, L.F.; Tarascon, R.G. In *Polymers in Microlithography*. Reichmanis, E.; MacDonald, S.A.; Iwayanagi, T., Eds., ACS Symp. Ser. No. 412, ACS Washington, DC, 1989, 39.
11. Reichmanis, E.; Houlihan, F.M.; Nalamasu, O.; Neenan, T.X. *Chem. Mater.* **1991**, 3, 394.
12. Uhrich, K.E.; Reichmanis, E.; Hefner, S.A.; Kometani, J.M.; Nalamasu, O. *Chem. Mater.* **1994**, 6, 287.
13. Schlesinger, S.I. *Polym. Eng. Sci.* **1974**, 14, 513.
14. Stewart, K.J.; Hatzakis, M.; Shaw, J.M.; Seeger, D.E.; Neumann, E. *J. Vac. Sci. Technol.* **1989**, B7, 1734.
15. Watanabe, H.; Todokoro, Y.; Inoue, M. *Microelectron. Eng.* **1991**, 13, 69.
16. Sakata, M.; Ito, T.; Yamashita, Y. *Jpn. J. Appl. Phys.* **1991**, 30, 3116.

17. Taylor, G.N.;Wolf, T.M. *Polym. Eng. Sci.* **1980**, 20, 1087.
18. Taylor, G.N.;Wolf, T.M. *Solid State Technology* **1980**, 23(5), 73.
19. Reichmanis, E.; Novembre, A.E.; Tarascon, R.G.; Shugard, A.; Thompson, L.F. In *Silicon-Based Polymer Science*; Zeigler, J.M.; Gordon Fearon, F.M., Eds.; Advances in Chemistry. Ser. No. 224, ACS, Washington, DC, **1990**, 265 and references cited therein.
20. Hashimoto, K.; Katsuyama, A.; Endo, M.; Sasago, M. *J. Vac. Sci. Technol.* **1994**, B12, 37.
21. Puyenbroek, R.; Van de Grampel, J.C.; Rousseeuw, B.A.C.; Van der Drift, E.W. J.M. *Polymer Commun.* **1994**, 35, 3131.
22. Ueno, T.; Shiraishi, H.; Schlegel, L.; Hayashi, N.; Iwayanagi, T. In *Polymers for Microelectronics-Science and Technology*, Kodansha, Tokyo, **1990**, 413.
23. Helferich, B.;Papalambrou, P. *Liebigs Ann. Chem.* **1942**, 551, 235.
24. Puyenbroek, R.; Werkman, P.J.; Rousseeuw, B.A.C.; Van der Drift, E.W.J.M.; Van de Grampel, J.C. *J. Inorg. Organometall. Polymers* **1994**, 4, 289.
25. Van de Grampel, J.C.; Puyenbroek, R.; Rousseeuw, B.A. C.; Van der Drift, E.W.J.M. In *Inorganic and Organometallic Polymers II*; Wisian-Neilson, P., Allcock, H.R. and Wynne, K.J., Eds., ACS Symp. Ser. No. 572, ACS, Washington, DC, **1994**, 81.
26. Puyenbroek, R.; Bosscher, G.; Van de Grampel, J.C.; Rousseeuw, B.A. C.; Van der Drift, E.W.J.M. *Phosphorus Sulfur Silicon Relat. Elem.* **1994**, 93-94, 277.
27. Ito, H.; Ueda, M. *Macromolecules* **1988**, 21, 1475.
28. Shiraishi, H.; Hayashi, N.; Ueno, T.; Sakamizu, T.; Murai, F. *J. Vac. Sci. Technol.* **1991**, B9, 3343
29. Allen, R.D.; Wallraff, G.M.; Hinsberg, W.D.; Simpson, L.L. *J. Vac. Sci. Technol.* **1991**, B9, 3357.
30. Rousseeuw, B.A.C.; Puyenbroek, R.; Van der Drift, E.; Van de Grampel, J.C. *Microelectr. Eng.* **1995**, 27, 375.
31. Babich, E.; Paraszczak, J.; Hatzakis, M.; Shaw, J.; Grenon, B.J. *Microelectr. Eng.* **1985**, 3, 279.
32. Gozdz, A.S.; Shelburne, J. *SPIE* **1992**, 1672, 184.
33. Crivello, J.V.; Shim, S.-Y.; Smith, B.W. *Chem. Mater.* **1994**, 6, 2167.
34. Schlegel, L.; Ueno, T.; Hayashi, N.; Iwayanagi, T. *J. Vac. Sci. Technol.* **1991**, B9, 278; *Jpn. J. of Appl. Phys.* **1991**, 30, 3132.
35. Tamamura, T.; Tanaka, A. In *Polymers for High Technology: Electronics and Photonics*; Bowden, M.J., Turner, S.R., Eds., ACS Symp.Ser. No. 346, ACS, Washington, DC, **1987**, 67.
36. Smith, S.D.; DeSimone, J.M.; Huang, H.; York, G.; Dwight, D.W.; Wilkes, G.L.; MacGrath, J.E. *Macromolecules* **1992**, 25, 2575.
37. Hatzakis, M.; Stewart, K.J.; Shaw, J.M.; Rishton, S. *Microelectr. Eng.* **1990**, 11, 487.
38. Taylor, G.N.; Wolf, T.M. *J. Electrochem. Soc.* **1980**, 127, 2665.
39. Thompson, L.F.; Doerries, E.M. *J. Electrochem. Soc.* **1979**, 126, 1699.
40. Thompson, L.F.; Yau, L.; Doerries, E.M. *J. Electrochem. Soc.* **1979**, 126, 1703.

RECEIVED August 21, 1995

Chapter 23

Environmentally Friendly Polysilane Photoresists

James V. Beach[1], Douglas A. Loy[1], Yu-Ling Hsiao[2],
and Robert M. Waymouth[2]

[1]Properties of Organic Materials Department, Sandia National
Laboratories, Albuquerque, NM 87185–1407
[2]Department of Chemistry, Stanford University, Stanford, CA 94305

Several novel polysilanes synthesized by the free-radical hydrosilation of oligomeric polyphenylsilane or poly(*p-tert*-butylphenylsilane) were examined for lithographic behavior. This recently developed route into substituted polysilanes has allowed for the rational design of a variety of polysilanes with atypical chemical properties such as alcohol and aqueous base solubility. Many of the polysilane resists made could be developed in aqueous sodium carbonate and sodium bicarbonate solutions. These materials represent environmentally friendly polysilane resists in both their synthesis and processing.

For over a decade, polysilanes have received a great deal of attention as unusual materials with many potential applications. Polysilanes owe much of their unusual properties to the delocalization of the Si-Si sigma bonds along the polymer backbone (*1*). This delocalization allows the silicon catenates to absorb strongly in the UV region. Polysilanes have potential applications as photoconductors (*2,3,4*), semiconductors (*5*), nonlinear optical materials (*6*), LEDs (*7*) and photoresists (*8*).

Polysilanes can perform as positive photoresists because exposure to UV light can lead to depolymerization via homolytic cleavage of the Si-Si bond (*9*). With some polysilanes, depolymerization leads cleanly to volatile products. Miller and co-workers (*10,11*) and Zeigler (*12*) demonstrated early on that some polysilanes behave as self-developing resists (a resist that does not require solvents to remove the exposed areas). Resists that self-develop would have a large competitive advantage over traditional resists because they require less processing to create an image. More importantly, dry developing does not generate the large volumes of hazardous waste associated with wet developing (increasingly restrictive EPA regulations on waste treatment have recently put a large financial burden on the electronics industry). Polysilanes also show submicron resolution and excellent oxygen-etch resistance in bilayer microlithography (*10-13*).

Despite their potential, polysilanes have yet to see industrial implementation in the United States (polysilanes are produced on an industrial scale in Japan as precursors to silicon carbide). A major factor hindering the use of polysilanes in industry is the method of their synthesis. Currently, the best synthesis of high molecular weight polysilanes is the Wurtz coupling of dichlorosilanes (Equation 1).

0097–6156/95/0614–0355$12.00/0
© 1995 American Chemical Society

This reaction has several drawbacks. The Wurtz couple gives low yields (5-60%) and variable molecular weights. As large quantities of molten sodium metal are used, the reaction is inherently dangerous even on a small scale. The harsh conditions exclude the incorporation of polar side groups on the monomer so only a limited number of monomers can be used. Consequently, most polysilanes synthesized by this method are soluble only in non-polar solvents. The reaction also generates large volumes of metal and solvent waste.

$$
\begin{array}{c}
R_{\text{nonpolar}} \\
\diagdown \\
\quad SiCl_2 \\
\diagup \\
R_{\text{nonpolar}}
\end{array}
\xrightarrow[\text{toluene}]{\text{Na}}
\left(\begin{array}{c} R_{\text{nonpolar}} \\ | \\ Si \\ | \\ R_{\text{nonpolar}} \end{array} \right)_n
\qquad (1)
$$

An alternative and less hazardous method for the preparation of polysilanes was discovered by Harrod in the mid 1980's (14,15,16). Titanium and zirconium metallocene complexes were found to perform dehydrogenative coupling on phenylsilane to form oligomeric polyphenylsilane. Although this reaction gives only oligomeric silane catenates, it is high-yielding, runs under very mild conditions and produces very little metal waste (14-16,17,18). Recently, Waymouth and Hsiao demonstrated that the Si-H bonds of polyphenylsilane undergo mild free-radical hydrosilations across a wide variety of primary olefins and carbonyl compounds (19). This synthesis opens the door to hundreds of novel polysilanes inaccessible by the Wurtz coupling method (Equation 2).

$$
SiH_3 \text{—} \langle \text{Ph} \rangle
\xrightarrow{Cp_2ZrHCl}
H \left(\begin{array}{c} H \\ | \\ Si \\ | \\ \langle Ph \rangle \end{array} \right)_n H
\xrightarrow[\text{AIBN}]{\underset{X}{\overset{R}{\diagup}}}
H \left(\begin{array}{c} \overset{R}{\underset{\diagup}{X}} \\ | \\ Si \\ | \\ \langle Ph \rangle \end{array} \right)_n H
\qquad (2)
$$

$$
n = 30\text{-}40 \qquad X = CH_2 \text{ , O}
$$

The above synthesis represents a commercially attractive method into polysilanes. Not only is the synthesis mild and high yielding, the versatility of the reaction allows for fine-tuning of the physical and chemical properties of the polymer to meet specific applications. Because polar side groups can be attached directly to the polymer, polysilanes can be rendered soluble in environmentally benign solvents such as water or alcohols.

We have begun examining a variety of these new polysilanes for their potential as self-developing photoresists. The printed wiring board (PWB) industry is currently under governmental pressure to reduce the amount of waste water it produces. A typical shop creates about 125 gallons of waste water for every one-square-foot of board made. Use of a self-developing photoresist could reduce the amount of water needed by up to 80%. Other benefits of a self-developing resist include the reduction in the number of processing steps needed to produce a board and a significant reduction in the amount of equipment required. Any resist used in the manufacture of PWBs must show resistance to acidic plating baths and acidic/basic etching solutions. Optimally, the resist films should cast from environmentally benign solvents such as alcohols rather than nonpolar solvents like toluene or methylethyl ketone. While the resolution requirements of the resist need be no finer than 50 microns, thick films (25-50 microns) are required for plating applications. This means that, for a self developing resist, a relatively large amount of material must photovolatilize in order to create a useful image. Moreover, the resist must self develop under exposure from a common and inexpensive UV light source (Most reported self-developing resists require eximer lasers. Eximer lasers would be too expensive for use in PWB shops).

Given these demanding requirements on a self developing resist, our primary focus was on the design of a polysilane that gave clean and rapid imaging. Solubility in alcohols, however, was also desirable. The synthesis shown in Equation 2 is limited in the type of precursor polymers that can be made (low molecular weight polyarylsilanes) but virtually unlimited in the type of side group that can be attached. We explored several different types of side group (Figure 1) in an attempt to use attached functionality to engineer the desired chemical and photochemical behavior of the resist.

Experimental

Substituted Polyphenylsilanes

Polyphenylsilane was synthesized as described by Banovetz and Waymouth using zirconocene chloride hydride (*20*). A typical synthesis of substituted polymer involved dissolving polyphenylsilane (1.0 g) in dry THF (3.0 mL). AIBN (0.16 g, 0.98 mmole) and a slight excess of olefin (10.4 mmole) was added. The solution was degassed by three freeze pump thaw cycles and sealed under vacuum. The resulting solution was heated to 65 °C for 4 hours. The solution was concentrated and the polymer purified by precipitation from THF/MeOH. Polymer **1** was made by bubbling ethylene gas through a toluene solution of polyphenylsilane and AIBN at 65 °C. Polymer **2** was made by charging a pressure tube containing a toluene solution of polyphenylsilane and AIBN with 25 psi of isobutylene. The reaction was allowed to run at 90 °C for 4 hours. Poly(cyclohexyloxyphenylsilane) (**3**) could be made by heating a solution of polyphenylsilane and AIBN in cyclohexanone to 80 °C for 4 hours. Copolymers **7** and **8** were made from polyphenylsilane in a single reaction by adding 50 mole% vinyl acetic acid and a slight excess of 2-methoxypropene or vinyltrimethylsilane to the THF/polymer solution. The polymers were purified by precipitation from THF/benzene.

Substituted Poly(*p-tert*-butylphenylsilanes)

The *p-tert*-butylphenylsilane monomer was made in a two-step reaction. To dry diethyl ether (40 mL) was added silicon tetrachloride (16.2 g, 95.3 mmole). A 2.0 M solution (in ether) of *p-tert*-butylphenyl magnesium bromide was added dropwise over a period of 12 hours (48 mL, 96 mmole). The reaction was allowed to run for a total of 24 hours. The salts were filtered from the solution and the filtrate was added dropwise to a mixture of dry diethyl ether (50 mL) and LiAlH4 (95% dispersion, 3.0 g, 79.0 mmole) at 0 °C. The reaction was allowed to warm to room temperature. After 2 hours, the mixture was cooled to 0 °C and carefully quenched with 2.5 M HCl. Caution! Phenylsilanes in the presence of AlClx can redistribute to form silane gas (*21*). It is important that the ether is not lost during the quench to prevent redistribution. Additional diethyl ether may need to be added during the quench to replace any that boils off . The mixture was placed in a separatory funnel and the ether layer collected. The ether solution was washed with 3x50 mL of 2.5 M HCl, dried over MgSO4 and concentrated. The residue was purified by distillation to yield purified product (3.2 g, 20% overall yield, unoptimized).

Poly(*p-tert*-butylphenylsilane) was made with zirconocene chloride hydride following the same procedure used to make polyphenylsilane. The polymerization reaction was allowed to run for 3 weeks and the resulting polymer could not be fractionated with toluene/pentane. No other attempt was made to separate out the low molecular weight fraction. The resulting polymer was a colorless solid with a weight-average molecular weight of 2300 (unoptimized). The free radical substitution reactions were performed as described above.

Lithography

A typical process involved dissolving 0.1 g of polymer in 2.0 mL of solvent (toluene or cyclohexanone). The solution was filtered through a 0.2 μm Nylon filter and concentrated to a volume of ~0.2 mL. The solution was spin-coated onto a silicon wafer at 2000 rpm and allowed to dry under vacuum for 1-24 hours. Exposure was performed in air using an unfiltered, uncollimated 200 watt Hg/Xe lamp through either a shadow mask or a chrome-on-glass mask. Aqueous development, if any, was performed in a beaker with 3% (by mass) aqueous $NaHCO_3$ or 15% aqueous Na_2CO_3.

Results and Discussion

Eight polysilanes were prepared with different functional groups designed to aid in imaging or alcohol solubility. Of the polymers made for this study, only polymers **1** and **2** can be made directly from a Wurtz couple. The rest of the polysilanes are novel materials. Polymer **3** is particularly interesting in that oxygen is attached directly to the silicon backbone. Full characterization of these polymers will be presented elsewhere. Though these new polysilanes are interesting materials in their own right, our immediate concern was screening these oligomers for lithographic behavior.

Table 1.   Characterization of Novel Polysilane Oligomers

	% substitution[a]	UVmax	MW[b]	polydisp.
1	54%	314 nm	4322	1.88
2	51%	319 nm	---	---
3	80%	340 nm	3339	1.51
4	88 %	322 nm	3557	1.33
5	77%	315 nm	5896	1.46
6	76%	317 nm	---	---
7	~90%*	306 nm	3514	1.33
8	~90%*	313 nm	7978	1.51

[a]Determined by ^{1}H NMR. [b]Determined by GPC, polystyrene standards. *Broad overlapping peaks do not allow for accurate determination of the percent of substitution by ^{1}H NMR.

As shown in Table 1, the free radical hydrosilation of polyphenylsilane gives 50-90% substitution. The rest of the polymer units contain a Si-H linkage. Polymers **1** and **2** showed the lowest values, probably resulting from the low concentration of the olefin present during the reaction. Polymers **3-8** show substitution of 76-90% which are comparable to those values reported by Waymouth and Hsiao. With the exception of polymer **3**, the UV absorption maximum of the polysilane oligomers all range from 306-322 nm. High molecular weight aryl polysilanes generally show a λ_{max} ranging from 335-347 nm. The absorption wavelength maximum of polysilanes and their extinction coefficients both increase sharply with increasing degrees of polymerization (DP) but quickly approach limiting values at a DP of 40-50 (*9*). The DP of the polymers listed in Table 1 range from 30-35 so one would expect the values to be blue shifted slightly. High molecular weight alkyl polysilanes absorb at 303-325 nm. Polymer **3** shows a significant red shift in comparison to the other oligomers, presumably from the electronic effects of the oxygen substituent (*20*).

The small blue shift of our polymers relative to high molecular weight material is of little concern because all polysilane resists bleach significantly during exposure. Bleaching is an important property to photoresists since it allows the UV light to penetrate through to the bottom of the film for an even exposure. All of the oligomers listed in Table 1 show photobleaching characteristic to polysilanes. Figure 2 shows the photobleaching of a thin film of **3** spin-coated onto a quartz plate and given a blanket exposure with a 200 watt Hg/Xe lamp. A quantitative comparison of bleaching rates has not yet been done. One can see in Figure 2 that the polymer bleaches over the entire range from 250 nm to 400 nm. This bleaching, though desirable for an even exposure through a film, makes it increasingly difficult to cleave the smaller polymer chains (since the shorter chains are less photosensitive). Complete depolymerization is necessary for clean self developing. It is therefore desirable to use a UV lamp that contains a substantial emission near 250 nm so that dimers and trimers can absorb light and be cleaved (*12,22*).

Polysilanes **1-8** were examined as both self-developed and as solvent-developed systems. Thin films (5-10 μm) of the novel polysilanes were spin-coated from toluene or cyclohexanone. All polysilanes with the exception of **6** gave good quality films. Polymer **6** always gave films with a mottled surface. If any of the polymer solutions were allowed to stand in air for a period of several hours to several weeks before spin coating, films began to streak. It is assumed that the polymer solutions are slowly crosslinking by the oxidation of unsubstituted silicon units. Evidence for crosslinking has been obtained through GPC (both the molecular weight and the polydispersity of polymer solutions are observed to increase upon standing in air for several days).

High molecular weight poly(methylphenylsilane) has been reported to self-develop under eximer lasers (*23*). Polymer **1** is similar in structure to poly(methylphenylsilane) though it has a significant number of Si-H groups along the backbone. Exposure of a thin film of **1** (~1 μm) with the 200 watt lamp gave a visible image after 5 minutes but SEMs show that little material actually ablated from the film. The fluence for the 5 minute exposure was calculated to be 2700 mJ/cm^2 (measured from 300-400 nm, centered at 365 nm). Since **1** is transparent at 365 nm, this number overestimates the amount of exposure energy interacting with the film. The lack of significant ablation under these conditions is not surprising as polyarylsilanes are known to often crosslink upon exposure, presumably from attack of the silyl radical on the aromatic ring (*9,24,25,26*). Since only arylsilane monomers can be polymerized to a reasonable extent with metallocene catalysts, crosslinking is a concern with any polymer made by this method.

Polymer **2** is similar to **1** but contains a branched substituent. Zeigler has shown that branched substituents improve the photosensitivity of polysilanes, presumably by forcing the polymer chains into an all trans conformation (12). Though the substituent of **2** is slightly heavier than that of **1** (and, therefore, the fragmentation products might be less volatile), it was thought that abstraction of a hydrogen atom from the side chain could lead to the loss of isobutylene (Figure 3). Polymer **2**, however, did not give an improved image over **1**.

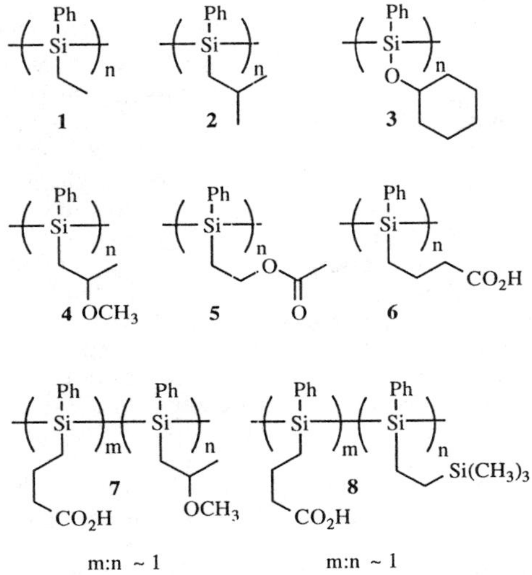

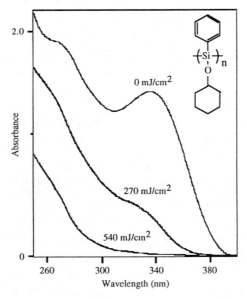

Figure 1. Structures of the polysilanes examined in this study for lithographic behavior. The polymers were all synthesized by the free radical hydrosilation of polyphenylsilane.

Figure 2. Photobleaching of a thin film of poly(cyclohexyloxyphenylsilane) on quartz. The UV spectra were taken after an accumulated fluence of 0, 270 and 540 mJ/cm^2.

Polymer **3** shows the longest λ_{max} of those listed in Table 1. Since the absorbance of this polymer is red shifted at least 20 nm from the other polymers in Table 1, it was thought that this polymer might show superior lithographic performance. Unfortunately, exposure of **3** gave no discernible image even after 16.2 j/cm^2 exposure. Upon developing the exposed film with pentane, an image quickly appeared. However, as shown in Figure 4, developing left considerable residue in the exposed regions. The porous residue left behind is assumed to be crosslinked silicon oxides.

Polymer **4** has a similar branched structure to **2** but contains an ether linkage. This polymer behaves similarly to polymers **1** and **2** even though it has a greater degree a substitution. The ether linkage does not impart enough polarity to the polymer to render it soluble in alcohol. Polymer **5** incorporates a carbonyl chromophore in the side chain. The presence of a carbonyl in the side chain could have two possible interactions with the polysilane. It can aid in the absorption of UV light and it could interact with the silyl radical to help prevent crosslinking (it should be mentioned that even though polyphenylsilane can hydrosilate aldehydes and ketones under free radical conditions, it was found to be unreactive towards esters). Qualitatively, polymer **5** gave a better image than polymers **1-4**. Use of other chromophores may improve performance but any chromophore attached would have to be small in size as to not impede volatility.

While these polysilane oligomers were not showing immediate promise as self-developing resists, the opportunity to create an aqueous developed polysilane was apparent. It was reasoned that the attachment of carboxylic acid groups to the polysilane backbone would create a positive resist that could be developed in aqueous carbonate or bicarbonate solutions. Polymer **6** was found to be too soluble in aqueous base to give a good developed image. The number of carboxylic acid groups, therefore, was reduced in polymer **7**. Surprisingly, polymer **7** gave a dramatically improved self-developed image over any of the other polymers shown in Figure 1. Figure 5 shows that a substantial amount of material was lost upon exposure. It is not yet known why **7** performed so much better than the other polymers but it is hoped that further investigations of this polymer and the conditions of exposure will give complete self-developing. Both polymers show good solubility in isopropanol.

In addition to an improved self-developed image, Polymer **7** could be developed with a 3% solution of sodium bicarbonate to yield quality features with out residue. Figure 6 shows 35 μm lines produced after aqueous development of a thin film of **7** exposed for 5 minutes through a chrome-on-glass mask (35 μm resolution is more than adequate for current printed-wiring-board manufacture). Polymer **8**, with its less polar side groups, would only partially develop in saturated sodium bicarbonate. The polymer, however, could be developed completely in 15% sodium carbonate solution. Polymers **6**, **7** and **8** demonstrate how the chemical behavior of the resist can be easily adjusted to give desired properties.

This control is not limited to adjusting the solubility of a resist in a specific solvent. One can theoretically adjust oxygen etch behaviors of the polysilanes in bilayer microlithography. It is known that a high silicon content improves a polymer's oxygen etch resistance (27). One would expect, therefore, polymer **8** to have improved oxygen etch resistance over polymer **7**. By adjusting the number of silicon-containing side groups, one should see a corresponding response in etch resistance (28).

In an effort to further improve the lithographic performance of our materials, a poly(*p-tert*-butylphenylsilane) precursor was synthesized. It is known that poly(*p-tert*-butylphenylethylsilane) has a higher photosensitivity than poly(phenylethylsilane) (29,30). The increased sensitivity is presumably due to the *tert*-butyl group blocking

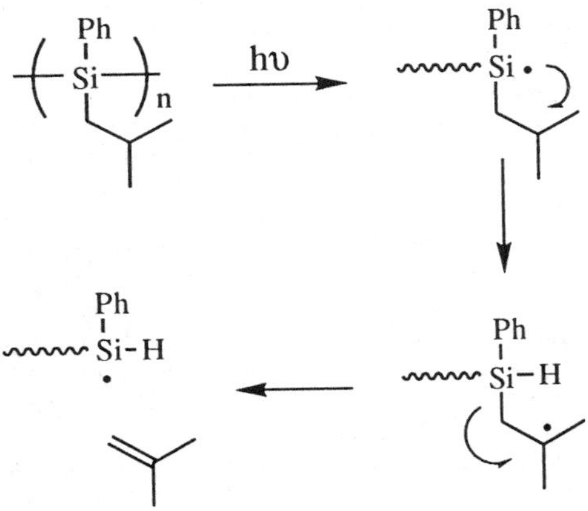

Figure 3. A mechanism for the proposed loss of isobutylene during the photodepolymerization of polymer **2**.

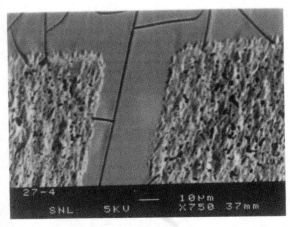

Figure 4. SEM of solvent-developed **3** after a 5 minute exposure through a chrome-on-glass mask. The unexposed areas cracked upon development. The exposed areas left behind a porous residue.

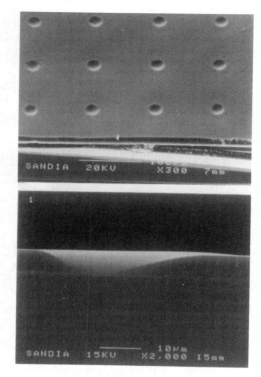

Figure 5. Self-developed image of **5**. Exposure time was 5 min. through a shadow mask.

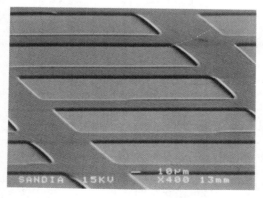

Figure 6. Aqueous-developed image of **5**. Exposure time was 5 min. through a chrome-on-glass mask.

the aryl ring from attack by the silyl radicals making it more difficult to form crosslinks. Poly(*p-tert*-butylphenylsilane) (**9**) was made in three steps from *p-tert*-butylmagnesium bromide and tetrachlorosilane (Equation 3). The polymerization reaction was noticeably slower than the polymerization of phenylsilane but conditions have not yet been optimized.

(3)

The resulting polymer had a weight-average molecular weight of 2300 (corresponding to a DP of about 14, unoptimized) and a polydispersity of 1.18. The polymer could be easily substituted in the same manner as polyphenylsilane with a terminal olefin. Free radical hydrosilation of vinyl acetic acid gave a Polymer **10** with > 95% substitution (measured by ^{1}H NMR) and a UV absorbance maximum of 317 nm. Exposure of a 1.5 μm thick film of polymer **10** (540 mJ/cm^2) gave a poor self-developed image prior to aqueous development. However, wet developing with 15% aqueous sodium carbonate gave well defined features with 15 μm resolution (Figure 7). The resolution capabilities of this polymer will be studied later under optimal exposure and developing conditions.

Conclusions

We have synthesized several novel polysilane oligomers that show complete development in aqueous sodium bicarbonate and sodium carbonate solutions. These polymers represent a significant advancement in polysilane lithography in that the polymers were synthesized in a high yielding two-step reaction from commercially available starting materials can produce a clean image using inexpensive exposure tools (Hg/Xe lamp and a glass mask) and environmentally friendly developing solvents (aqueous NaHCO$_3$ or Na$_2$CO$_3$). Past work on polysilanes had always involved a low-yielding and dangerous synthesis, expensive eximer laser exposure tools and/or organic developing solvents. This work also demonstrates that a great deal of flexibility and control can be exercised over the chemical properties of polysilanes. We are currently investigating how different side groups influence self development of the phenylsilane polymers. It is hoped that judicial choice of the side

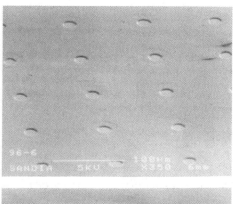

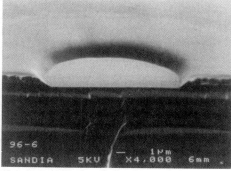

Figure 7. SEM of a 1.5 μm thick film of polymer **10** exposed for 1 minute and developed in aqueous Na₂CO₃.

groups can hinder crosslinking and permit clean photovolatilization. We see potential in the *p-tert*-butylphenylsilane polymers as aqueous resists for deep UV microlithography. This work is represents a quick survey of the many novel polysilanes that can be synthesized by this new method for potential lithographic application. Optimization and a detailed comparison of polymer properties is needed for better understanding of the influence complex organic side groups can have on polysilanes. This knowledge should lead to the design of new and better materials for lithography as well as other electronic applications.

Acknowledgments

This research was supported by the United States Department of Energy under Contract No. DE-A C04-94AL85000.

Literature Cited

1. For a reveiw on polysilanes and their properties, see Miller, R. D.; Michl, J. *Chem. Rev.* **1989**,*89*, 1359.

2. Kepler, R. G.; Zeigler, J. M.; Harrah, L. A.; Kurtz, S. R. *Bull. Am. Phys. Soc.* **1983**, *28*, 362.
3. Kepler, R. G.; Zeigler, J. M.; Harrah, L. A.; Kurtz, S. R. *Phys. Rev . B* **1987**, *35*, 2818.
4. Fujino, M. *Chem. Phys. Lett.* **1987**, *136*, 451.
5. West, R.; David, L. D.; Djurovich, P. I.; Stearley, K. L.; Srinivasan, K. S. V.; Yu, H. *J. Am. Chem. Soc.* **1981**, *103*, 7352.
6. Kajzar, F.; Messier, J.; Rosilio, C. *J. Appl. Phys.* **1986**, *60*, 3040.
7. Suzuki, H.; Meyer, H.; Simmerer, J.; Yang, J.; Haarer, D. *Adv. Mater.* **1993**, *5*, 743.
8. Miller, R. D.; MacDonald, S. A. *J. Imaging Sci.* **1987**, *31*, 43.
9. Trefonas, P.; West, R.; Miller, R. D.; Hofer, D. C. *J. Polym. Sci, Polym. Lett.* **1983**, *21*, 823.
10. Miller, R. D.; Hofer, D.; McKean, D. R.; Willson, C. G.; West, R.; Trefonas, P. T. *Materials for Microlithography ACS Symposium Series*, 266, Thompson, L. F.; Willson, C. G.; Frechet, J. M. J. Eds., American Chemical Society, Washington, D. C., 1984, Chapter 14.
11. Miller, R. D.; Hofer, D.; Fickes, G. N.; Wilson, C. G.; Marinero, E.; Trefonas, P.; West, R. *Photopolymers: Principles, Processes, and Materials*, Proc. of Mid-Hudson Section of Society of Plastics Engineers, Ellenville, N. T., October 1985.
12. Zeigler, J. M.; Harrah, L. A.; Johnson, A. W. *SPIE Proc.* **1985**, *539*, 166.
13. Hofer, D. C.; Miller, R. D.; Willson, C. G. *SPIE Proc.* **1984**, *469*, 16.
14. Aitken, C. T.; Harrod, J. F.; Samuel, E. *J. Organomet. Chem .* **1985**, *279*, C11.
15. Aitken, C. T.; Harrod, J. F.; Samuel, E. *J. Am. Chem. Soc.* **1986**, *108*, 4059.
16. Aitken, C. T.; Harrod, J. F.; Samuel, E. *Can. J. Chem.* **1986**, *64*, 1677.
17. Woo, H. G.; Walzer, J. F.; Tilley, T. D. *J. Am. Chem. Soc.* **1992**, *114*, 5698.
18. Woo, H. G.; Walzer, J. F.; Tilley, T. D. *J. Am. Chem. Soc.* **1992**, *114*, 7047.
19. Hsiao, Y.; Waymouth, R. M. *J. Am. Chem. Soc.* **1994**,*116*, 9779.
20. Banovetz, J. P.; Stein, K. M.; Waymouth, R. M. *Organometallics* **1991**, *10*, 3430.
21. Speier, J. L.; Zimmerman, R. E. *J. Am. Chem. Soc.* **1955**, *77*, 6395.
22. Trefonas, P.; West, R.; Miller, R. D. *J. Am. Chem Soc.* **1985**, *107*, 2737.
23. Hofer, D. C.; Jain, K.; Miller, R. D. *IBM Tech. Disclos. Bull.* **1984**, *26*, 5683.
24. Hofer, D. C.; Miller, R. D.; Willson, C. G.; Neureuther, A. R. *SPIE Proc.* **1984**, *465*, 108.
25. Miller, R. D.; Hofer, D. C.; Fickes, G. N.; Willson, C. G.; Marinero, E.; Trefonas, P.; West, R. *Polym. Eng. and Sci.* **1986**, *25*, 1129.
26. West, R.; Zhang, X-H.; Djurovich, R. I.; Stüger, H. *Science of Ceramic Chemical Processing*, Hench, L. L.; Ulrich, D. R., Eds.; Wiley, New York, N. Y., 1986; Chapter 36, pp 337-344.
27. Taylor, G. N.; Wolf, T. M. *Polym. Eng. and Sci.* **1980**, *20*, 1087.
28. Reichmanis, E.; Smolinsky, G. *SPIE* **1984**, *469*, 38.
29. Wallfaff, G. M.; Miller, R. D.; Clecak, N.; Baier M. *SPIE Proc.* **1991**, *1466*, 211.
30. Miller, R. D.; Hofer, D. C.; Fickes, G. N.; Willson, C. G.; Marinero, E. *Poly. Eng. and Sci.* **1986**, *26*, 1129.

RECEIVED August 14, 1995

POLYMER DIELECTRICS FOR MICROELECTRONIC APPLICATIONS

Elsa Reichmanis

Polymers are used in microelectronics for a variety of applications, but perhaps their most important function in the final electronic device is as a low dielectric material. In this role, polymers serve as adhesives, encapsulants, substrates and as thin film dielectric layers. As demands increase even in low-end desk top computers, challenges for improvements in such materials include a reduction in dielectric constant, higher thermal stability, photoimageability and low dielectric loss. Coefficient of thermal expansion (CTE) and Young's modulus are also of concern. New forms of processing are becoming increasingly important and include organic analogs to CVD processes largely based on plasma polymerization of thin films.

The focus for work on dielectrics falls into two basic applications: unfilled interlayer dielectrics where the polymer acts to separate small metal lines and permits high density chip-to-carrier connection, and filled systems where the polymer acts either as a printed circuit board (PCB), substrate or may be used as an encapsulant. In the former situation, it is important to realize that processing temperatures may be quite high (>350°C) if solder reflow is needed and the polymer must be able to withstand these conditions for extended periods of time. Circuit density is also much higher, so that a low dielectric constant is more critical. In the latter case, thermal stability and dielectric constant are less critical, but as circuit densities become higher in the case of printed circuit boards even here high performance criteria are being applied. This has resulted in the recent use of *fluorinated polymers* in such applications rather than the more traditional and less expensive thermosets. There is also an interest in developing polymers where parallel assembly of chip substrates is possible using polymers that are not thermosets, but could be extruded in sheets and bonded together. Issues such as adhesion are important particularly if polymers like *polyimides* are considered.

0097–6156/95/0614–0367$12.00/0

Several chapters in this section deal with methods for reducing the dielectric constant (ε). This goal is related to the ability to reduce circuit dimensions if a lower dielectric material can be prepared. Increased circuit density will of course enable smaller chip design and greater circuit speed, but cross-talk must be avoided. Polymers generally have dielectric constants ranging from 2.3 for highly fluorinated materials up to 3.8 for conjugated structures. Several strategies are being followed, therefore to reduce ε. One is to use fluorinated polymers to reduce the dielectric constant of the base material. A recent innovation is instead to introduce microscopic bubbles into the polymer matrix and thereby decrease the overall dielectric constant of the nanofoam. Both approaches are effective and are described in some of the following chapters. If moisture uptake is of concern, of the two approaches, fluorination most directly addresses this.

Imageability is an aspect of polymer processing that is essential to lithographic processing of semiconductors, but which is becoming increasingly important in low dielectric polymers. Structures with the resolution of 50 μm or more are currently possible, but as the demands of second level electronic packaging approaches that of the chip related structures, the ability to image smaller and smaller feature sizes will be needed in low dielectric polymers. The methods of imaging polyimide, for example, including crosslinking of methacrylate groups (which are subsequently pyrolyzed off) will need to give way to cleaner, more precise photochemistry. Other photoimageable, robust, low ε polymers will need to be prepared.

Low CTE polymers are also of importance due to their ability to minimize the stresses that can build up in thin film and bulk structures when CTE mismatch occurs. Most polymers expand when heated 60 or more ppm°C while metals or semiconductors do so only 10 or less ppm°C. This leads to stresses in electronic structures during thermal cycling and can greatly reduce the fatigue life of an electronic device. If the CTE behavior of polymers or polymer composites can be brought in line with that of the metals, then a number of problems will be reduced. Polymers such as polyimides or thermoplastic LC polymers can display very low CTE values in their oriented direction. Thermosets are also offering possible solutions to this problem due to their crosslinked structure. All of these materials offer high modulus which can be important for properties ranging from mechanical robustness to reduced electron migration in metal lines.

Finally, *plasma (or CVD) processing* of polymers provides yet another alternative to the formation of low ε polymers. These materials are applied as a reactive vapor of monomer which will polymerize on the substrate. This process is capable of producing highly crosslinked, thermally stable polymer films of reasonable dielectric constant. Processing steps familiar to those processing silicon are involved. Much study needs to be carried out in this area to develop completely reproducible structures, but it does hold great promise.

RECEIVED August 21, 1995

Chapter 24

Fluoropolymers with Low Dielectric Constants: Triallyl Ether–Hydrosiloxane Resins

Henry S.-W. Hu[1], James R. Griffith[2], Leonard J. Buckley[2], and Arthur W. Snow[2]

[1]Geo-Center, Inc., 10903 Indian Head Highway, Fort Washington, MD 20744
[2]Naval Research Laboratory, 4555 Overlook Avenue, S.W., Building 207, Washington, DC 20375

The preparation of a class of processable heavily fluorinated triallyl ether homo- and co-polymers is carried out to elucidate the structure-property relationships in comparison with the acrylic analogs. The triallyl ether monomer was prepared in good yield through the condensation of the triol, 1,3,5-tris(2-hydroxy-hexafluoro-2-propyl)benzene, in dry acetone with allyl bromide in the presence of potassium carbonate. Homopolymers were obtained and cured through a slow free radical polymerization, while copolymers were obtained through a fast hydrosilylation with catalyst. The dielectric constant of the copolymer of the triallyl ether and an equivalence of polymethylhydrosiloxane (PMHS) catalyzed by a trace of dicyclopentadienylplatinum(II) chloride is 2.33 at 13.2 GHZ with a dissipation factor of 0.004. The factors which affect the dielectric constant and thermal stability are the fluorine content, the polymer type and the molecular architecture.

With recent trends toward microminiaturization of electronic systems and utilization of very thin conductor lines, close spacings, and very thin insulation, greater demands are being placed on the insulating layer. Reductions in such parasitic capacitance can be achieved in a number of ways through the proper selection of materials and the design of circuit geometry. In 1988 St. Clair et al.[1a] reported a reduction of dielectric constant to 2.39 by modifying a polyimide to reduce the interchain interactions and by the incorporation of carbon-fluorine bonds. In 1991 Kane et al.[1b] reported a reduction of dielectric constant to 2.32 for the hexafluoroisopropylidene-containing polyacrylates and copolyacrylates. In 1992 Snow et al.[1c] reported that the thermally induced trimerization of a perfluorohexamethylene linked aromatic cyanate resin to a cyanurate linked network gave a dielectric constant between 2.3 and 2.4. In 1993 Babb et al.[1d] stated that the

0097–6156/95/0614–0369$12.00/0
© 1995 American Chemical Society

thermally induced cyclodimerization of a trifluorovinyl aryl ether to a perfluorocyclobutane aromatic ether polymer gave a dielectric constant of 2.40.

Recently we reported the preparation of a series of processable heavily fluorinated acrylic and methacrylic homo- and co-polymers which exhibit dielectric constants as low as 2.06,[2] very close to the minimum known values of 2.0-2.08 for Teflon® and 1.89-1.93 for Teflon AF®.[3] However, these measurements were made using a coaxial transmission line method with samples in cylindrical donut molds that are not optimized for high accuracy. The factors which affect the reduction of dielectric constant from structure-property relationships have been elucidated from our experimental findings.

The polar nature of the C-F bond has been used to provide some high-performance characteristics in comparison with their hydrogen-containing or other halogen-containing analogues. Fluorine-containing epoxies or acrylics generally exhibit resistance to water penetration, to chemical reaction, and to environmental degradation; they also show unusual values for surface tension, friction coefficient, optical clarity, refractive index, vapor transmission rate, and electromagnetic radiation resistance. Fluorinated polyethers are the key intermediates for new types of practical organic coatings and plastics which have fluorocarbon properties including high thermal stability in some cases.

Polysiloxanes offer other unusual properties such as very low glass transition temperature, high permeability to gas, oxidative stability and the ability to be fabricated into useful products. It is this versatility that has established the reputation of siloxane containing polymers in many applications.[4] Polysiloxane elastomers capped with functional groups have been used to increase flexibility, to improve thermal stability, and to lower internal stress of cured epoxy resins.[5]

The hydrosilylation, an addition reaction of Si-H compounds to unsaturated organic molecules with the aid of a platinum or rhodium complex catalyst, has been used widely for Si-C bond formation. Siloxane-containing polymers obtained from hydrosilation normally have several advantages in properties and processing.[6] The curing proceeds in the presence of oxygen, with little shrinking, and with good dimensional stability. In addition, the hydrosilation occurred with exclusive anti-Markovnikov addition.[7]

In this paper we report a class of processable heavily fluorinated triallylic ether homo- and co-polymers which are readily prepared in rectangular block molds to elucidate the structure-property relationships in comparison with the acrylic analogs.

EXPERIMENTAL

Only the general preparations are described here, the detailed procedures will be reported else where.

Materials

Allyl bromide was obtained from Aldrich Chemical Co. and was distilled to collect the fraction with bp 71-72°C/1 atm as a clear liquid (the fraction before that is cloudy). Acetone was dried over potassium carbonate, decanted, and distilled from potassium carbonate before use. Dicyclopentadienylplatium(II) chloride was prepared according to the procedure of Apfel et al.[8] Alumina (neutral, Brockman activity 1, 80-200 mesh) was obtained from Fisher Scientific Co. Polymethylhydrosiloxane (PMHS) was obtained from Aldrich (#17620-6) and a Si-H equivalent formula weight of 63.13 g/mole is used for calculation. All other reagents were used as received or purified by standard procedures.

Preparation of triallyl ether 2 from triol 1. 1,3,5-Tris(2-hydroxy-hexafluoro-2-propyl)benzene 1 was prepared by a multistep route according to the procedure of Soulen and Griffith.[9a] This compound was hygroscopic as observed by Griffith and O'Rear.[9b]

Over the course of 30 min, allyl bromide (31.6 g, 261 mmol) was added dropwise to a solution of triol 1 (40.0 g, 69.4 mmol) in dry acetone (500 mL) in an ice-water bath under nitroge. After 10 min, potassium carbonate (32.0 g, 231 mmol) was added in portions in 3 min and stired 1/2 hr at 0 °C followed by 1 hr at room temperature. The mixture was then slowly heated to reflux over 1 hr and kept refluxing for 12hr. The reaction was worked up by filtering through Celite, and evaporated at reduced pressure and in vacuo at 30 °C for 3 hr to yield a liquid (48.4 g).

The liquid was dissolved in hexanes (200 mL), percolated through a column of neutral alumina (80 g) twice and each time washed with hexanes (150 ml). It was evaporated at aspirator pressure and then in vacuo at room temperature for 4 hr to give a colorless liquid (40.9 g) of triallyl ether 2, yield 85%. R_f 0.81, Hexane/CH_2Cl_2 (2:1) (v/v), 0.41 (Hexane); IR(neat film) 3104, 2952, 2891, 1652, 1609, 1457, 1429, 1414, 1373, 1350-1100 (C-F), 1051, 1012, 979, 933, 889, 734, 709 cm[-1]; ^{1}H NMR($CDCl_3$/TMS) δ 7.99 (s, 3H, Ar-H), 5.92 (d,d,d, J=17.2, 10.5, 5.1 Hz, 3H, CH_2=CH-), 5.43 (d, d, J=17.2, 1.3 Hz, 3H, cisoid CH_2=CH-), 5.32 (d,d, J=10.5, 1.3 Hz, 3H, transoid CH_2=CH-), 4.09 (d, J=5.1 Hz, 6H, -CH_2O-); ^{19}F NMR($CDCL_3$/$CFCL_3$) -71.51.

Preparation of Molded Samples

Semitransparent homopolymers and 50/50 equimolar copolymers were obtained as follows. Hydrosilylation of triallyl ether 2 with PMHS: Triallyl ether 2 (1.99 g, 2.86 mmol) and PMHS 4 (0.55 g, 8.71 mmol) were mixed with [(Cp$_2$)Pt]Cl$_2$ (0.8 mg) at room temperature and transferred into a rectangular block mold (16.02x8.15x8.94mm, lxwxh) made from General Electric RTV 11 silicone molding compound. To effect cure, the temperature was slowly raised to 150 °C over 2 days and kept at 150 °C for 1 hr. Homopolymerization of triallyl ethers 2: Triallyl ether 2 was mixed with a trace amount of solid azobisisobutyronitrile (AIBN) or liquid

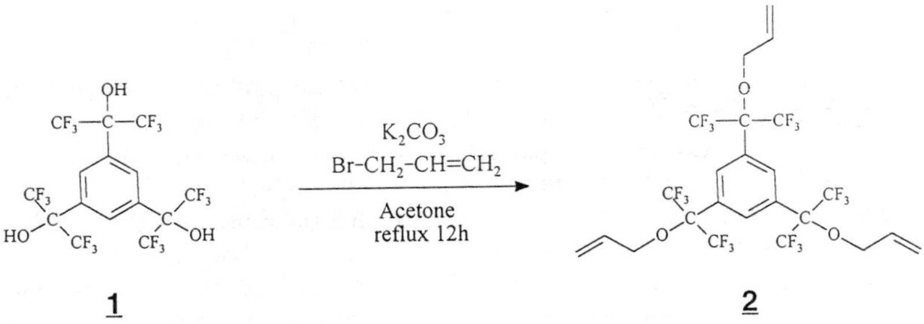

Scheme 1

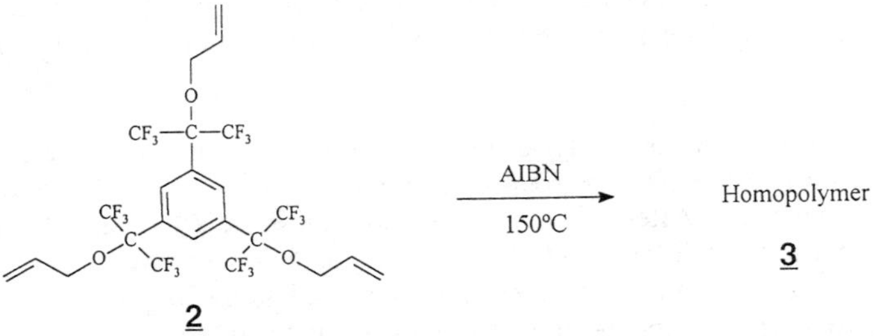

Scheme 2

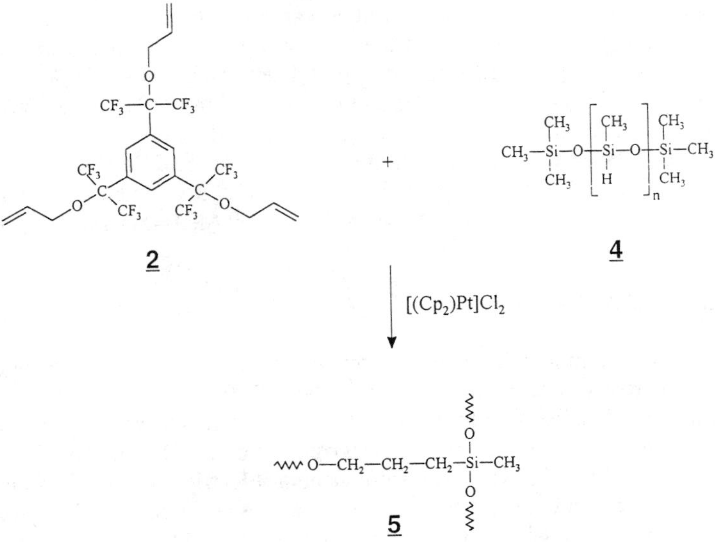

Scheme 3

methylethylketone peroxide (MEKP) in a sealed tube and was slowly heated to 150 °C over 4 days and kept at 150 °C for 3 days.

Dielectric constant measurements. Dielectric constant values are reported as "complex permittivity" with the symbols ϵ' and ϵ''. The measurements were performed on a Hewlett Packard 8722C Automated Network Analyzer using a transmission line method with rectangular waveguides after a full two-port internal waveguide calibration. The complex permittivity was calculated from the measured scattering parameters using a Nicolson and Ross algorithm.[10a] Reflection measurements using this configuration were pioneered by Von Hippel and are still commonly used today.[10b]

RESULTS AND DISCUSSION

Monomers and polymers. The multistep preparation of fluorinated triol **1** was carried out according to the reported procedure.[9] The triallyl ether **2** was synthesized by the SN$_2$ reaction of triol **1** with allyl bromide (Scheme 1). Purification of **2** by distillation was not feasible and attempts resulted impurities from polymerization because of the temperature required even under high vacuum, but purification by percolation of a solution through neutral alumina resulted in products of good purity characterized by TLC, FT-IR, and ^{1}H, ^{13}C and ^{19}F FT-NMR spectroscopy.

The ^{1}H NMR spectra shown in Figure 1 reveals that the triallyl ether **2** has a characteristic allylic long-range coupling pattern in the region of δ 6.1-4.0, while the acrylates showed a characteristic ABX pattern in the region of δ 6.8-6.0 with a pair of doublet couplings for each vinyl proton. The methacrylates showed a characteristic AB pattern in the region of δ 6.5-5.8 with an equivalent singlet peak for each vinyl proton. All the monomers freshly purified by percolation over alumina contained no detectable hydrate water or polymerized impurities. The stability of triallyl ether **2** is better than the triacrylate analog, and it is found that no change can be detected by IR and NMR after storage in air for three months.

Liquid monomers cure as thermosets without phase separation. For the slow radical homopolymerization as shown in Scheme 2, the use of solid azobisisobutyronitrile (AIBN) or liquid methylethylketone peroxide (MEKP) is convenient although the induced allyl ether radicals are stable. The reaction mechanism of this radical induced polymerization is complicated and it is difficult to determine the exact nature of the product. During the course of slow curing up to 150 °C for 7 days, the problems of surface inhibition of free radicals by oxygen of the air and the evaporization of monomers can be avoided by inert gas blanketing in a sealed tube. It is known that due to the degradative chain transfer of allyl radicals, the allylic monomers such as allyl acetate polymerize at abnormally low rates, and the degree of polymerization is very low and indepedent of the polymerization rate.[11]

As shown in Scheme 3, the use of polymethylhydrosiloxanes **4** in the catalytic presence of [(Cp$_2$)Pt]Cl$_2$ is practical for the copolymerization through hydrosilylation. The curing proceeds in the air and with little evaporation.

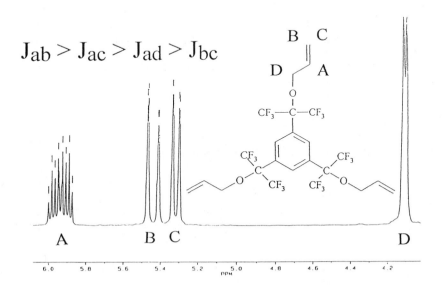

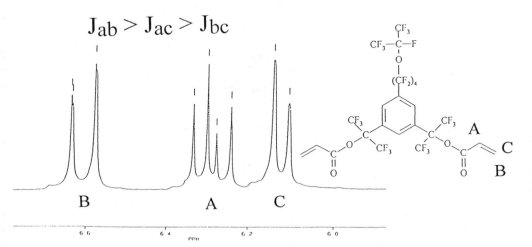

Figure 1.   ¹H NMR spectra comparison of allyl ether and acrylate.

The homopolymers 3 and the 50/50 equimolar copolymers 5 are semitransparent, hard solids and some shrinkage in volume is observed during curing. In an undercured state they are frangible, but when totally cured they acquire a more resilient character. The degree of hydrosilation or conversion of monomers can be easily monitored by FT-IR, by examining the intensity of the absorption frequency at 2171 cm^{-1} which is assigned to the Si-H functional groups.[12]

Since each allyl group is difunctional, the triallyl ether is hexafunctional. For radical polymerization of the polyfunctional monomers at sufficiently high degrees of conversion, the branching must result in the formation of cross-links to give a three dimensional network.

A DSC thermogram of the triallyl ether hydrosiloxane copolymer 5 cured at 150 °C displays an exotherm at 272 °C in nitrogen. An isothermal TGA analysis for 2 hr in air showed weight loss of 5% at 250 °C, 16% at 300 °C, and 47% at 350 °C. A TMA thermogram of the copolymer 5 cured at 215 °C displays a glass transition temperature (T_g) at 85 °C with coefficients of thermal expansion (CTE) of 120 x 10^{-6}/°C below T_g and 194 x 10^{-6}/°C above T_g in helium.

Upon being heated from room temperature up to 500 °C, the triallyl ether homopolymer 3 showed a higher thermal stability (400 °C in nitrogen, 325 °C in air) in comparison with that of triacrylate homopolymer (370 °C in nitrogen).[13a] This result demonstrates that the network cured from the allyl ether functional group is more thermally stable than that from the acrylate. Since the C=C double bonds are consumed during the radical polymerization, it is speculated that the thermal stability difference is derived from the ether linkage being more stable than the ester linkage. The decomposition of ester by heat or light will generate a stable acyl free radical from homolytic cracking.[13b] The fluoroallylic ether polymers are tough, rugged materials not easily damaged by impact or mechanical abuse, and can be handled as free-standing thin samples.

Dielectric constant (ϵ) and structure-property relationships. Rectangular blocks of polymer were used for the measurement of ϵ on a Hewlett Packard 8722C Automated Network Analyzer. As shown in Figure 2, the dielectric constant of the copolymer of the triallyl ether 2 and an equivalence of polymethylhydrosiloxane (PMHS) 4 catalyzed by a trace of dicyclopentadienylplatinum(II) chloride is 2.33 at 13.2 GHZ with a dissipation factor of 0.004. The dielectric constants measured at freuency region of 12.4-18.0 GHz showed a small frequency dependence. This result is fairly consistent with our recent findings of fluorine content from acrylic samples for dielectric constant measurements using a coaxial transmission line method.[2] Based on the assumption of pure starting materials and complete reactions, the fluorine content for triallyl ether homopolymer 3 is 49.11%, while the 50/50 equimolar copolymer 5 of triallyl ether/PMHS/catalyst (78.40/21.57/0.03) by weight has a 38.50% fluorine content.

It is known that glass is an amorphous material composed principally of silicone dioxide, and that clear silica glass has a DE of 3.81.[14] Polymethylhydrosiloxane 4

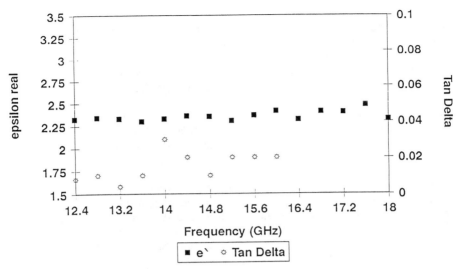

Figure 2.   Dielectric constant of cured equimolar triallyl ether-polymethyl-
hydrosiloxane copolymer <u>5</u>.

contains siloxane linkages in the structure, and the substituted alkyl groups are
nonpolar hydrocarbon which lower the DE below that of glass. This effect was
reported in the case of polydimethylsiloxane (PDMS) by Bass and Kauppi,[15] the type-
500 fluid with viscosity 3 cS has a DE of 2.41-2.39, while the type-200 fluid with
higher viscosity 250 cS has a DE of 2.74-2.70.

Lower dielectric constants are obtained as fluorine content on the polymer backbone
or sidechain increases, when acrylate is replaced by methacrylate, when ether
linkages are present in the fluorocarbon and when aromatic structure is symmetrically
meta-substituted.[2] The wide range of frequency independence of these low ϵ polymers
and the processability of these monomers suggest many potential applications.

Lower dielectric constants. Dielectric constants of these materials can be further
lowered by known means such as by incorporating voids in the materials.   A
difference of a couple of hundredths in the dielectric constant value may be important
when one is at the low extremes thereof. Singh et al. calculated the dielectric
constants of polyimide films from the measured free volume fraction and found that
the calculated values for the dielectric constants are close to the experimental
results.[16]

In 1991 Groh and Zimmermann[17] estimated the theoretical lower limit of the refractive index of amorphous organic polymers by using the Lorenz-Lorentz equation and reported the lower limit to be very close to 1.29. Using the Maxwell Relation[18] of $DE = n^2$ for the case of nonmagnetic materials without free charges this refractive index translates to a dielectric constant of 1.664. They reported that functional groups with a high fluorine content, like CF_3 and CF_2, have the lowest refractive index contribution. The value for the ether group is also remarkably low, while the values for the carbonyl and carboxyl groups are high.[17] Using this theoretical guide, synthetic modifications are continuing with hopes of achieving lower dielectric constants.

CONCLUSIONS

In this work we have demonstrated that a new class of heavily fluorinated triallyl ether resins can be efficiently synthesized and then cured to solid form with a catalyst at elevated temperatures. Homopolymers were obtained and cured through a slow free radical polymerization, while copolymers were obtained through a fast hydrosilylation. The factors which affect the dielectric constant and thermal stability are the fluorine content, the polymer type and the molecular architecture.

ACKNOWLEDGEMENT

Partial funding support from the Office of Naval Research is gratefully acknowledged.

REFERENCES AND NOTES

(1) (a) St. Clair, A. K.; St. Clair, T. L.; Winfree, W. P. Polym. Mater. Sci. Eng. 1988, 59, 28. (b) Kane, K. M.; Wells, L. A.; Cassidy, P. E. High Perform. Polym. 1991, 3(3), 191. (c) Snow, A. W.; Griffith, J. R.; Soulen, R. L.; Greathouse, J. A.; Lodge, J. K. Polym. Mater. Sci. Eng. 1992, 66, 466. (d) Babb, D. A.; Ezzell, B. R.; Clement, K. S.; Richey, W. R.; Kennedy, A. P. Polym. Prepr. 1993, 34(1), 413.; J. Polym. Sci. Part A: Polym. Chem. Ed. 1993, 31, 3465.

(2) (a) Hu, H. S.-W.; Griffith, J. R. Polym. Mater. Sci. Eng. 1992, 66, 261. (b) Hu, H. S.-W.; Griffith, J. R. In Polymers for Microelectronics, Willson, G.; Thompson, L. F.; Tagawa, S. Eds., ACS Symposium Series 1994, 537, 507. (c) Hu, H. S.-W.; Griffith, J. R. Polym. Prepr. 1993, 34(1), 401.(d) Griffith, J. R.; Hu, H. S.-W. U. S. Patent 5,292,927. March 8, 1994. (e) Griffith, J.R.; Hu. H. S.-W. U. S. Patent 5,405,677. April 11, 1995.

(3) (a) Licari, J. J.; Hughes, L. A. Handbook of Polymer Coating for Electronics; Noyes Publications: Park Ridge, NJ, 1990. see p. 378, Table A-13: Dielectric Constants for Polymer Coatings (at 25°C). (b) Resnick, P. R. Polym. Prepr. 1990, 31(1), 312.

(4) Iwahara, Y.; Kusakabe, M.; Chiba, M.; Yonezawa, K. J. Appl. Polym. Sci. 1993, 50, 825.

(5) Tong, J.; Bai, R; Zou, Y.; Pan, C.; Ichimura, S. J. Appl. Polym. Sci. 1994, 52, 1373.

(6) Crivello, J. V.; Bi, D. J. Polym. Sci. Part A: Polym. Chem. Ed. 1993, 31, 2729.

(7) Mathias, L. J.; Lewis, C. M. Macromolecules 1993, 26, 4070.

(8) Apfel, M. A.;Finklemann, H.; Janini, G. M.; Laub, R. J.; Lühmann, B. -H.; Price, A.; Roberts, W. L.; Shaw, T. J.; Smith, C. A. Anal. Chem. 1985, 57, 651.

(9) (a) Soulen, R. L.; Griffith, J. R. J. Fluorine Chem. 1989, 44, 210. (b) Griffith, J. R.; O'Rear, J. G. Polym. Mater. Sci. Eng. 1985, 53, 766.

(10) (a) Nicolson, A.; Ross, G. IEEE Trans. Instrumentation and Measurement 1970, IM-19, 377. (b) Von Hippel, A. R. Dielectric Materials and Applications; M.I.T. Press: Cambridge, MA, 1954.

(11) Odian, G. Principles of Polymerization; 3rd ed., Wiley: New York, NY, 1991. see p. 266.

(12) Pouchert, C. J. The Aldrich Library of FT-IR Spectra; Aldrich Chemical: Milwaukee, WI, 1985.

(13) (a) Hu, H. S.-W. et al. unpublished results. (b) Carey, F. A.; Sundberg, R. J. Advanced Organic Chemistry; Plenum: New York, NY, 1984. see p. 635 and 653.

(14) In Handbook of Chemistry and Physics; Chemical Rubber Co.: Cleveland, OH. see p. E-66.

(15) Clarson, S. J.; Semlyen, J. A. Siloxane Polymers; PTR Prentice Hall: Englewood Cliffs, NJ, 1993. see p. 420 for ref. 26. (16) (a) Singh, J. J.; Eftekhari, A.; St. Clare, T. L. NASA Memorandum 102625, 1990. (b) Eftekhari, A.; St. Clare, A. K.; Stoakley, D. M.; Kuppa, S.; Singh, J. J. Polym. Mater. Sci. Eng. 1992, 66, 279.

(17) Groh, W.; Zimmermann, A. Macromolecules 1991, 24, 6660.

(18) Lin, L.; Bidstrup, S. A. J. Appl. Polym. Sci., 1994, 54, 553.

RECEIVED July 17, 1995

Chapter 25

Photophysics, Photochemistry, and Intramolecular Charge Transfer of Polyimides

Masatoshi Hasegawa, Yoichi Shindo, and Tokuko Sugimura

Department of Chemistry, Faculty of Science, Toho University,
2–2–1 Miyama, Funabashi, Chiba 274, Japan

Photophysical properties of a biphenyl-type polyimide (PI) were modeled by measuring the ultraviolet-visible absorption, emission spectrum and lifetime of its model compounds in dilute solution at room temperature and 77 K. On the basis of a photophysical model, the rate constant of the intramolecular charge-transfer (CT) process in the excited state of the biphenyldiimide model compound was estimated to be more than 5×10^{11} s^{-1}. The relationship between the intramolecular CT and photoreaction, i.e., the photo-induced hydrogen abstraction for a benzophenone-containing PI, was discussed on the basis of the photochemical kinetic parameters for its model compounds, determined by the Stern-Volmer analysis. A relation between the intramolecular CT character of the model compounds and the kinetic parameters led to the postulation of a photophysical mechanism in which intersystem crossing followed by the hydrogen abstraction competes with the intramolecular CT process followed by effective deactivation.

Polyimide precursor, polyamic acids (PAA) are generally polymerized from aromatic dianhydride and aromatic diamine. The reaction proceeds through CT interaction between dianhydride as an electron acceptor and diamine as a donor.(*1*) It has been believed that yellow color of model compound crystals(*2*) and films(*3*) of aromatic PIs such as KAPTON results from the CT interaction. However, these systems include necessarily contribution of both intra- and intermolecular interactions. The aim of this work is to elucidate the presence of CT interactions in aromatic PIs and to examine relationship between the intramolecular CT and photophysical(*4*) and photochemical(*5-8*) properties of PIs. We focus on the model compound systems in dilute solution which precludes intermolecular interactions.

A biphenyl-type PI was selected for photophysical studies because the films of polypyromellitimides are relatively nonfluorescent.(*9,10*) For photoreaction studies, a benzophenone-type PI was chosen. This PI is known to be useful as a thermally stable photoresist material. Upon uv-irradiation the triplet state of the benzophenonediimide unit in the PI main chain, formed by intersystem crossing, abstracts intermolecularly hydrogen atom from the adjacent alkyl groups, followed

0097–6156/95/0614–0379$12.00/0
© 1995 American Chemical Society

by subsequent coupling between the radicals formed to form a crosslinking point.(*11*) The present study also describes comparison of the photoreactivity for various benzophenonediimides and benzophenone (BP).

EXPERIMENTAL SECTION

Materials. The model compounds of biphenyl- and benzophenone-type polyimides were prepared from 3,4,3',4',-biphenyltetracarboxylic dianhydride (BPDA) or benzophenonetetracarboxylic dianhydride (BTDA) with the stoichiometric amount of monoamines; dianhydride recrystallized from acetic anhydride was added to the dried N,N'-dimethylacetamide solution of the monoamine with continuous stirring at room temperature for 2 h, and then the solution was refluxed for 1 h. The acylation of aromatic and aliphatic monoamines with dianhydride occurred quantitatively as well as polycondensation of polyamic acid. The precipitate was recrystallized twice from a suitable solvent.

The abbreviations of amines used are as follows: p-phenylenediamine (PDA), cyclohexylamine (CHA), 2-methylcyclohexylamine (MCHA), 2-methylaniline (2-MA), 2-isopropylaniline (2-iPrA), 2,6-diethylaniline (2,6-DEA), 2,4-dimethylaniline (2,4-DMA), 3-ethylaniline (3-EA). The symbol of model compounds was expressed using these abbreviations of dianhydrides and monoamines.

Measurements. The uv-vis absorption spectra of the model compounds were recorded on Jasco Ubest-30 spectrophotometer. The corrected luminescence spectra of PI films and the model compounds were measured at room temperature and 77 K. Fluorescence lifetimes were measured using a single photon counting system equipped with a hydrogen flash lamp (fwhm: 2 ns, frequency: 30000 Hz, excitation: 300 nm) and sharp-cut filter in front of detector. Phosphorescence lifetimes τ_P of the model compounds and BP in a transparent rigid glass (MTHF/ethanol=9/1) were measured at 77 K using Hitachi 850 fluorescence spectrometer equipped with a phosphorescence lifetime measurement apparatus based on the sampling method. The emission quantum yields for the model compounds in solution were determined by comparing with the integral fluorescence intensity of quinine sulfate in 1N H_2SO_4 ($\Phi_f = 0.55$). Transient triplet-triplet (T-T) absorption and emission spectra of the model compounds in CH_3CN were measured by excitation at 355 nm (third harmonics of Nd-YAG laser, Continuum SL I-10: 6 ns fwhm, 20 mJ per pulse) with a detection system (Tokyo Instrument) composed of a multichannel diode array (Princeton IRY-512G: 18 ns gate) with a SPEX 270M monochromator (resolution: 0.3 nm/channel).

Various photochemical kinetic parameters were determined on the basis of the Stern-Volmer plots with respect to the quantum yield for disappearance of the model compounds, Φ_M, vs the concentration of hydrogen donor or triplet quencher (Q = naphthalene). Due to the limited solubility of the model compounds, CH_2Cl_2 was chosen as a hydrogen donor. The degassed CH_2Cl_2/CH_3CN solutions of the model compounds (ca. 10^{-3} mol dm^{-3}) were irradiated for 0.5-2 h with continuous stirring through a bandpass filter (Toshiba, UV-D36A) by using a high-pressure mercury lamp at 25°C on a merry-go-round together with the degassed cyclohexane solution of BP as a actinometer.

The conversion for the photoreduction of the model compounds was determined from the change in the integral absorption intensity ratio of 1290 cm^{-1} band characteristic of benzophenone-carbonyl group to the benzene ring stretching band at 1462 cm^{-1} as an internal standard or by HPLC. Both methods provided the same results.

RESULTS AND DISCUSSION

Photophysics of Biphenyl-type Polyimide Model Compounds. It is known that the pyromellitimide units in PI backbone have strong electron-accepting character similar to pyromellitic dianhydride (PMDA) as a monomer component of PAA.(12) There are large number of papers on PMDA/aromatic compounds CT complex systems. Figure 1 shows a linear relationship between the wavenumber of CT absorption band reported in the literature and the ionization potential I_P of aromatic compounds, fitting to the Mulliken's CT theory, i.e., $h\nu=I_P+C$ (C: constant).(13) The plot indicates that CT complex could be formed in nonpolar solvents when $I_P \leq$ 9.24 eV (benzene). Since the N-aryl group in PI chains is a stronger electron donor than benzene, aromatic PI chains consist of alternative donor-acceptor sequence, i.e., -D-A-D-A-. To eliminate the contribution of intermolecular interaction, use of low molecular weight model compounds isolated in dilute solution is useful.

Biphenyl-type polyimide PI(BPDA/PDA) was selected for photophysical study instead of relatively nonfluorescent KAPTON-PI. Figure 2a shows uv-vis absorption, fluorescence, and excitation spectra of a PI(BPDA/PDA) thin film. The broad fluorescence peaking around 550 nm was assigned to an intermolecular CT fluorescence from the relation between the fluorescence intensity and film density.($14,15$) Correlation of the excitation spectra and the absorption spectrum means that the CT fluorescence results from energy transfer from excited structural units to emissive CT sites.

The uv-vis absorption spectra were measured to examine the structural unit contributing to the lowest energy absorption as shown in Figures 2b-d. The N-aryl group (Figure 2b) has no absorption above 280 nm. On the other hand, the biphenyldiimide unit (Figures 2c and d) shows strong absorption at longer wavelengths (over 300 to 400 nm). The magnitude of the molar extinction coefficient ($\varepsilon = 8000$ M^{-1} cm^{-1} at 325 nm) suggests that the strong absorption band corresponds to the (π,π^*) transition.

Figure 3 shows the fluorescence polarization of a model compound M(BPDA/CHA) dispersed molecularly in PMMA film (10^{-4} M). The fluorescence, upon excitation within the 325 nm band, is not practically subjected to depolarization. This indicates that the 325 nm band corresponds the $S_0{\rightarrow}S_1$ transition because the fluorescence would be depolarized if the 325 nm band was due to the $S_0{\rightarrow}S_2$ transition. The mirror image also supports this assignment. M(BPDA/CHA) emitted a phosphorescence around 550 nm (τ_P = several sec) in toluene/ethanol at 77 K in addition to the fluorescence peaking at 385 nm.

Figure 4 illustrates absorption spectra of four kinds of biphenyldiimides possessing different amine components. The model compounds possessing ortho-alkyl phenyl group, i.e., M(BPDA/2-MA) and M(BPDA/2-iPrA) show spectra similar to the non-CT compound, M(BPDA/CHA). This is probably due to steric inhibition of conjugation caused by a larger dihedral angle between phthalimide and N-aryl molecular planes. The ab initio calculation (GAUSSIAN 92)(16) of the dihedral angles (73° for N-2,6-diethylphenylphthalimide and 33° for N-3-ethylphenylphthalimide) and the rotational barriers (essentially forbidden for the former and free for the latter) supports the interpretation of the absorption spectra for biphenyldiimides. The result in Figure 4 predicts that M(BPDA/2-MA) and M(BPDA/2-iPrA) should show a normal fluorescence similar to the non-CT model M(BPDA/CHA). However as shown in Figure 5, M(BPDA/2-MA) and M(BPDA/2-iPrA) emitted broad and longer wavelength fluorescences quite different from that of the non-CT model. These fluorescences red-shifted with an increase in solvent polarity. These results lead us to conclude that these broad fluorescence bands arise from a photo-induced intramolecular CT state.

From these results, a photophysical diagram is depicted in Figure 6. The quantum yield of the 385 nm fluorescence for the non-CT compound, M(BPDA/CHA), is given by

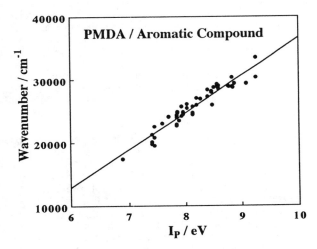

Figure 1. Relationship between the peak wavenumber of CT absorption band and the ionization potential of donors for PMDA/aromatic compounds systems in nonpolar solvents.

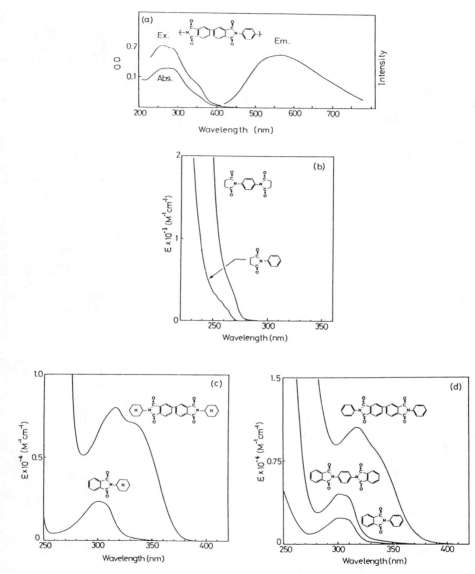

Figure 2. Uv-vis absorption and fluorescence spectra of (a) PI(BPDA/PDA) thin film and (b-c) absorption spectra of its model compounds corresponding to each structural unit. (Reproduced with permission from ref. 4. Copyright 1994 John Wiley & Sons, Inc.)

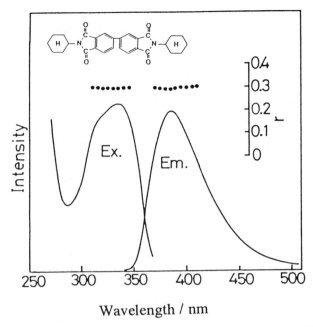

Figure 3 Fluorescence, excitation, and polarization spectra of M(BPDA/CHA) in PMMA. (Reproduced with permission from ref. 4. Copyright 1994 John Wiley & Sons, Inc.)

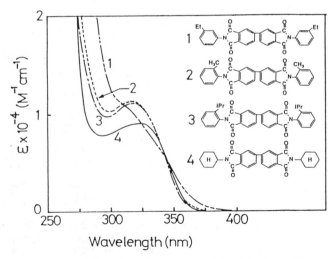

Figure 4. Uv-vis absorption spectra of biphenyldiimides in dichloromethane. (Reproduced with permission from ref. 4. Copyright 1994 John Wiley & Sons, Inc.)

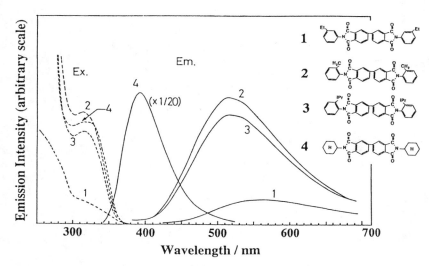

Figure 5. Fluorescence and excitation spectra of biphenyldiimides in CH_2Cl_2. (Reproduced with permission from ref. 4. Copyright 1994 John Wiley & Sons, Inc.)

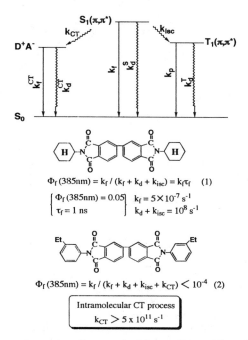

$$\Phi_f(385nm) = k_f / (k_f + k_d + k_{isc}) = k_f\tau_f \quad (1)$$

$$\begin{cases} \Phi_f(385nm) = 0.05 \\ \tau_f = 1 \text{ ns} \end{cases} \quad \begin{cases} k_f = 5 \times 10^{-7} \text{ s}^{-1} \\ k_d + k_{isc} = 10^8 \text{ s}^{-1} \end{cases}$$

$$\Phi_f(385nm) = k_f / (k_f + k_d + k_{isc} + k_{CT}) < 10^{-4} \quad (2)$$

Intramolecular CT process
$$k_{CT} > 5 \times 10^{11} \text{ s}^{-1}$$

Figure 6. Schematic energy diagram for biphenyl-type PI model compounds.

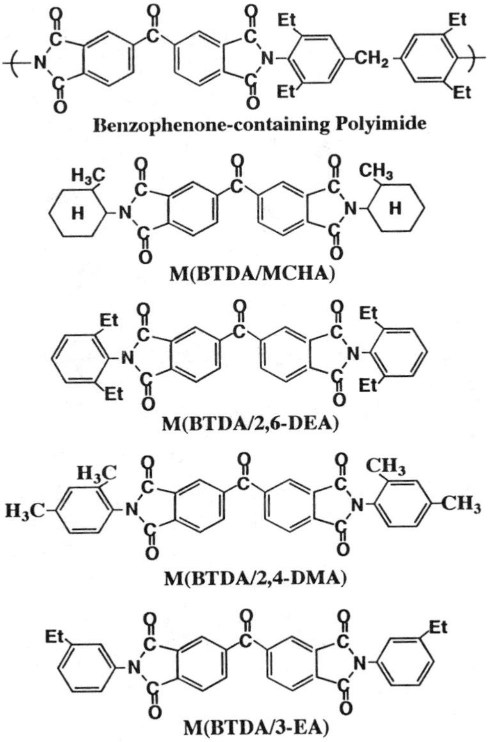

Benzophenone-containing Polyimide

M(BTDA/MCHA)

M(BTDA/2,6-DEA)

M(BTDA/2,4-DMA)

M(BTDA/3-EA)

Figure 7. Molecular structures and symbols of benzophenone-type PI model compounds.

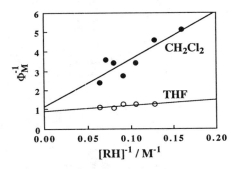

Figure 8. Stern-Volmer plots for M(BTDA/2,6-DEA) in CH_2Cl_2/CH_3CN and THF/CH_3CN in the absence of triplet quencher (naphthalene).

$$\Phi_f (385nm) = k_f /(k_f + k_d + k_{isc}) = k_f \tau_f \tag{1}$$

where k_f, k_d, k_{isc}, and τ_f are rate constants of the 385 nm fluorescence emission, deactivation, intersystem crossing, and lifetime. Using the experimental values of Φ_f (385nm) =0.05 and τ_f = 1 ns for M(BPDA/CHA) in PMMA film, and we obtained k_f = 5x10^7 s^{-1} and $k_d + k_{isc} = 10^8$ s^{-1}. For the CT-compound M(BPDA/3-EA), no appreciable 385 nm fluorescence was observed. Therefore,

$$\Phi_f (385nm) = k_f /(k_f + k_d + k_{isc} + k_{CT}) \quad < 10^{-4} \tag{2}$$

where k_{CT} is the rate constant for intramolecular CT formation from the singlet (π,π^*) state. Eq.(2) indicates that $k_{CT} > 5 \times 10^{11}$ s^{-1}. We propose that such an extremely fast intramolecular CT process, followed by effective deactivation of aromatic PIs, is responsible for their high UV radiation resistance of aromatic PIs.

Photochemistry of Benzophenone-type Polyimide Model Compounds.
Molecular structures and symbols of the model compounds of the benzophenone-type PI are shown in Figure 7. The extremely fast intramolecular CT process for PI model compounds expressed as D-A-D is comparable to the intersystem crossing for benzophenone (BP) (k_{isc}=2x10^{11} s^{-1}).(*17*) For benzophenonediimides possessing both of benzophenone- and imide-carbonyl groups, we observed no appreciable decrease in the absorbance of the imide carbonyl stretching band (1774 cm^{-1}) and complete disappearance of the benzophenone-carbonyl bands (1670, 1290 cm^{-1}) after photoirradiation of the models in CH$_2$Cl$_2$ at 365 nm for 4 h. The result indicates that the photoreduction of benzophenonediimides occurs preferentially at the benzophenone-carbonyl group.

Various photochemical kinetic parameters of the model compounds were determined on the basis of the Stern-Volmer plot (*18*):

$$\Phi_M^{-1} = \Phi_{isc}^{-1} + (k_d + k_q[Q]) / (k_r[RH]\Phi_{isc}) \tag{3}$$

where Φ_M and Φ_{isc} are the quantum yields for photo-induced hydrogen abstraction and intersystem crossing. k_d, k_r, and k_q represent the rate constants for the triplet deactivation, hydrogen abstraction, and triplet quenching by naphthalene (nearly diffusion-controlled), respectively. [RH] and [Q] (in mol dm^{-3}) are the concentration of hydrogen donor and triplet quencher, respectively. Naphthalene has little absorption at 365 nm.

Figure 8 shows the Stern-Volmer plots for M(BTDA/2,6-DEA) in CH$_2$Cl$_2$/CH$_3$CN or THF/CH$_3$CN in the absence of triplet quencher. The linearity of the plots suggests that no physical quenching of the triplet M(BTDA/2,6-DEA) occurs by the added hydrogen donor molecules. It is noted that the extrapolated reciprocal intersystem crossing yields (Φ_{isc}^{-1}) are nearly unity regardless of hydrogen donor solvent. The difference between the slopes (k_d/k_r) of the linear plots for both systems is attributable to the difference of the hydrogen donor ability of the solvents; the hydrogen abstraction from THF occurs about ten times rapidly compared with that from dichloromethane.

We estimated k_q to be 2.0x10^9 M^{-1}s^{-1} for M(BTDA/2,6-DEA)/naphthalene system and 5.7x10^9 M^{-1}s^{-1} for BP/naphthalene system by the transient T-T absorption decay measured in CH$_3$CN at 20°C on the basis of the relation $\tau_0 / \tau = 1+k_q\tau_0[Q]$. For several model compounds, the Stern-Volmer analysis were

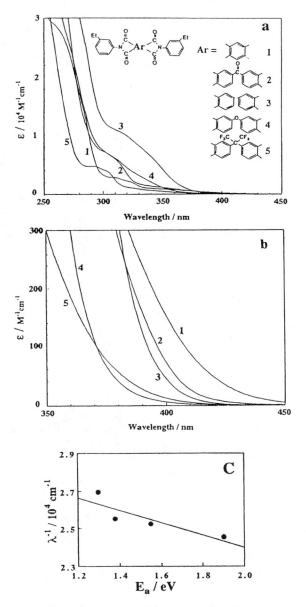

Figure 9. Uv-vis absorption spectra of a series of diimide compounds derived from various dianhydrides and 3-ethylaniline in CH_2Cl_2. (a) 5×10^{-5} M, (b) 5×10^{-3} M, (c) Relation between the reciprocal wavelength at $\varepsilon = 100$ $M^{-1}cm^{-1}$ and the electron affinity E_A of the anhydride components.

conducted in the absence and the presence of triplet quencher (naphthalene) to obtain various photochemical kinetic parameters as summarized in Table I.

As described above for biphenyldiimide compounds, the intramolecular CT character decreases in the following order depending on the position of the alkyl-substituents on the N-aryl group and the aromaticity of the amine components: meta- > ortho-alkyl substituted phenyl group > aliphatic group. This is based on the fluorescence peak position and magnitude of conjugation associated with the dihedral angle. Accordingly, it is most likely expected that in benzophenone-type PI model compounds the CT character decreases in the following order: M(BTDA/3-EA) > M(BTDA/2,4-DMA) > M(BTDA/2,6-DEA) > M(BTDA/MCHA).

Figures 9a and b show the uv-vis absorption spectra of a series of bisimides derived from various aromatic dianhydrides with 3-EA and their longer wavelength absorption tails, respectively. The broad absorption bands extending up to 450 nm are observed in Figure 9b. By contrast, non-CT compounds composed of various dianhydrides with MCHA had little absorption over 400-450 nm. Comparing the wavelength (λ) at a constant ε (100 M^{-1}cm^{-1}) because there is no absorption peak in the longer wavelength region, a linear relation between λ^{-1} and the electron affinity E_A of the dianhydride components according to the Mulliken's CT theory was observed for the former system as demonstrated in Figure 9c. On the other hand, no linear plot was obtained for the non-CT compound systems. Consequently, the existence of the intramolecular CT transition was evidenced for benzophenonediimides.

Figure 10a and b give that the uv-vis absorption spectra and their absorption tails of benzophenonediimides with various amine components. The spectral shift of the absorption tails corresponds to the order of CT ability predicted above for the model compounds. In addition, the order coincides with the result expected from the transient CT fluorescence peak position for benzophenonediimides in Figure 11.

Table I suggests that the values of Φ_M are closely related with the intramolecular CT character; with an increase in the CT character, Φ_M decreases parallel to Φ_{isc}. The change in Φ_{isc} determined by the Stern-Volmer analysis at 20°C corresponds well to the change in Φ_P measured spectroscopically at 77 K (Figure 12a). On the other hand, no correlation of Φ_M with k_r and with k_d is observed (Figure 12b). Assuming that a process competing with the intersystem crossing from S_1 to T_1 is the intramolecular CT, the results summarized in Table I can be satisfactorily explained with a schematic energy diagram depicted in Figure 13. In the mechanism, Φ_{isc} should decrease with an increase in the CT character (increase in k_{CT}) so that an decrease in Φ_M. In fact, M(BTDA/3-EA) having the strongest CT character provides the lowest value of Φ_{isc} and Φ_M as shown in Table I. Thus, the photoreactivity of benzophenonediimides is strongly affected by the intramolecular CT character associated closely with the conformation around the N-aryl linkage.

Phosphorescence spectra of non-CT model M(BTDA/MCHA) at 77 K were quite similar to those of all other models having CT character (Figure 14), and the lifetimes of all the models were less than 20 ms. In addition, the values of k_r at 20°C are independent of the CT character. Therefore, the lowest triplet state T_1 of the model compounds is assigned as pure (n,π*) (not a mixed state of (n,π*) with (π,π*) or with CT state) regardless of CT character. This means that the lower photoreactivity of M(BTDA/3-EA) is attributable not to the mixing of the ^{3}(n,π*) and ^{3}CT states but most likely to the decrease in Φ_{isc} resulting from the increase in k_{CT}.

Figure 15 shows two kinds of transient emission spectra of M(BTDA/2,6-DEA) in CH$_3$CN at 20°C. We believe that the broad emission observed at 150 ns after excitation is the sum of the intramolecular CT fluorescence and the phosphorescence. At 500 ns after excitation, the CT emission disappeared,

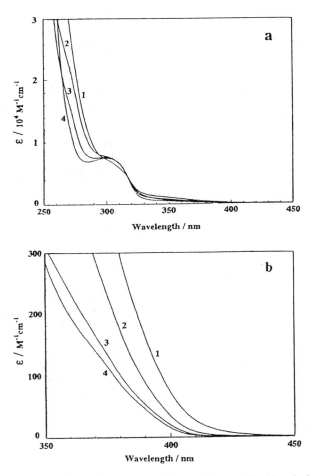

Figure 10. Uv-vis absorption spectra of benzophenonediimides derived from BTDA and various monoamines in CH_2Cl_2. (a) 5×10^{-5} M, (b) 5×10^{-3} M. (1) M(BTDA/3-EA), (2) M(BTDA/2,4-DMA), (3) M(BTDA/2,6-DEA), (4) M(BTDA/MCHA).

Table I. Photochemical kinetic parameters of the models and BP. Dichloromethane was used as hydrogen-donor solvent

MODEL	20°C				-196°C	
	Φ_M	Φ_{ISC}	$k_r(M^{-1}s^{-1})$	$k_d(s^{-1})$	$\tau_P(ms)$	Φ_P
M(BTDA/MCHA)	0.38	1.0	1.3×10^5	3.3×10^6	15.6	0.71
M(BTDA/2,6-DEA)	0.41	0.87	3.4×10^5	7.6×10^6	7.8	0.50
M(BTDA/2,4-DMA)	0.22	0.35	4.6×10^5	9.2×10^6	9.9	0.12
M(BTDA/3-EA)	0.096	0.19	2.3×10^5	2.9×10^6	6.6	0.051
BP	0.40	1.0	3.0×10^5	6.6×10^6	2.6	0.82

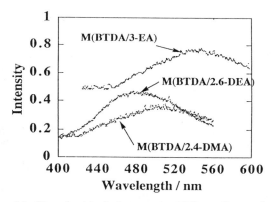

Figure 11. Transient emission spectra (100 ns after excitation) of benzophenonediimides in CH_3CN.

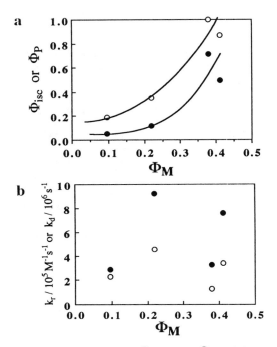

Figure 12. Relation of (a) Φ_M vs Φ_{isc} ($\bigcirc$) and Φ_P ($\bullet$) and (b) Φ_M vs k_r ($\bigcirc$) and k_d ($\bullet$).

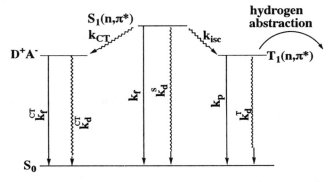

Figure 13. Schematic energy diagram for the model compounds of benzophenone-type PI.

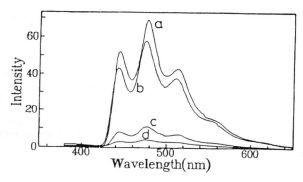

Figure 14. Phosphorescence spectra of various benzophenonediimides at 77 K.
(a) M(BTDA/MCHA), (b) M(BTDA/2,6-DEA), (c) M(BTDA/2,4-DMA), and (d) M(BTDA/3-EA).

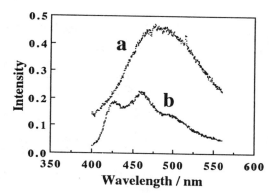

Figure 15. Transient emission spectra of M(BTDA/2,6-DEA) in CH_3CN at 20°C at (a) 150 ns and (b) 500 ns after excitation.

consequently, the phosphorescence spectrum similar to that observed at 77 K remained. The result supports the photophysical processes proposed in Figure 13.

Table I also gives the comparison of the photoreactivity of the model compounds with BP. Benzophenonediimides provide k_r no less than BP unless there is specific interaction with solvents. This is probably due to pure (n,π^*) state of T_1 for both the model compounds and BP.

CONCLUSIONS

Photophysical properties of biphenyl-type PI were modeled using appropriate model compounds in dilute solution. The model compounds possessing N-aryl group emitted broad intramolecular CT fluorescences. On the basis of a photophysical model, the rate constant for formation of the intramolecular CT state was estimated to be more than 5×10^{11} s^{-1}.

The photophysical kinetic parameters for the model compounds of benzophenone-containing PI were determined to discuss the photophysical processes associated closely with the photo-induced hydrogen abstraction of the model compounds from hydrogen donor solvents. The relationship between the kinetic parameters obtained by the Stern-Volmer analysis and the intramolecular CT character elucidated the photophysical mechanism in which the intramolecular CT process competes with the intersystem crossing followed by photoreduction.

Literature Cited
(1) *Polyimides;* Bessonov, M.I.; Koton, M.M.; Kudryavtsev, V.V.; Laius, L.A., Eds.; Plenum Press: New York, 1987; pp 14.
(2) Dine-Hart, R.A.; Wright, W.W. *Makromol. Chem.* **1971**, *143*, 189.
(3) Kotov, B.V.; Gordina, T.A.; Voishchev, V.S.; Kolninov, O.V.;Paravednikov, A.N. *Vysokomol. Soyed.* **1977**, *A19*, 614. (*Polym. Sci. U.S.S.R.*, *19*, 711.)
(4) Hasegawa, M.; Shindo, Y.; Sugimura, T.; Ohshima, S.; Horie, K.; Kochi, M.; Yokota, R.; Mita, I. *J. Polym. Sci. Part B* **1993**, *31*, 1617.
(5) Hasegawa, M.; Sonobe, Y.; Shindo, Y.; Sugimura, T.; Karatsu, T.; Kitamura, A. *J. Phys. Chem.* **1994**, 42, 10771.
(6) Scaiano, J.C.; Becknell, A.F.; Small, R.D. *J. Photochem. Photobiol. A* **1988**, *44*, 99.
(7) Higuchi, H.; Yamashita, T.; Horie, K.; Mita, I. *Chem. Materials* **1991**, *3*, 188.
(8) Hoyle, C.E.; Anzures, E.T.; Subramanian, P.; Nagarajyan, R.; Creed, D. *Macromolecules* **1992**, *25*, 6651.
(9) Hasegawa, M.; Kochi, M.; Mita, I.; Yokota, R. *J. Polym. Sci.: Part C* **1989**, *27*, 263.
(10) Arjavalingam, G.; Hougham, G.; LaFemina, J.P. *Polymer,* **1990**, *31*, 840.
(11) Lin, A.A.; Sastri, V.R.; Tesoro, G.; Reiser, A.; Eachus, R. *Macromolecules* **1988**, *21*, 1165.
(12) Sep, W.J.; Verhoeven, J.W.; de Boer, T.J. *Tetrahedron* **1975**, *31*, 1065.
(13) Mulliken, R.S. *J. Am. Chem. Soc.* **1952**, *74*, 811; *J. Phys. Chem.* **1952**, *56*, 801.
(14) Hasegawa, M.; Kochi, M.; Mita, I.; Yokota, R. *Eur. Polym. J.* **1989**, *25*, 349.
(15) Hasegawa, M.; Arai, H.; Mita, I.; Yokota, R. *Polym. J.* **1990**, *22*, 875.
(16) Tokita, Y.; Ino, Y.; Okamoto, A.; Hasegawa, M.; Shindo, Y.; Sugimura, T. *Kobunshi Ronbunshu* (*Jpn. J. Polym. Sci. Technol.*) **1994**, *51*, 245.
(17) Rentzepis, P.M. Science **1970**, *169*, 239.
(18) Beckett, A.; Porter, G. *Trans. Faraday Soc.* **1963**, *59*, 2038.

RECEIVED July 7, 1995

Chapter 26

Structure, Properties, and Intermolecular Charge Transfer of Polyimides

Masatoshi Hasegawa[1], Junichi Ishii[1], Takahumi Matano[1], Yoichi Shindo[1],
Tokuko Sugimura[1], Takao Miwa[2], Mina Ishida[2], Yoshiaki Okabe[2],
and Akio Takahashi[2]

[1]Department of Chemistry, Faculty of Science, Toho University,
2–2–1 Miyama, Funabashi, Chiba 274, Japan
[2]Hitachi Research Laboratory, Hitachi Ltd., 7–7–1, Ohmika-cho,
Ibaraki 319–12, Japan

The presence of intermolecular donor-acceptor interaction in solid
state aromatic polyimides (PIs) was confirmed by spectroscopic
means. It was found that the thermal expansion coefficient is closely
associated not only with the degree of in-plane orientation, but also
with the degree of molecular stacking. These two factors influence
each other, and are not completely independent. The intermolecular
CT fluorescence spectra were a sensitive reflection of the molecular
stacking process. Dynamic mechanical measurements of PIs
suggested that the CT complexes probably function as physical
crosslinks.

Aromatic polyimides (PIs), especially those possessing a rigid (linear) polymer
backbone, e.g., PI(BPDA/PDA) (BPDA: 3,4,3',4',-biphenyltetracarboxylic
dianhydride, PDA: p-phenylenediamine), show properties depending strongly on
the cure method (chemical and thermal), thermal imidization conditions (final cure
temperature and heating rate), presence of substrate, and film thickness. This is
attributed to the difference of the degree of molecular orientation and molecular
stacking which are closely related with molecular mobility during thermal cure of
polyamic acid (PAA). The aim of the work is to elucidate the presence and
contribution of intermolecular charge-transfer (CT) interaction on various physical
properties and to examine the relationship between physical properties and the
structures of molecular aggregates. The intermolecular charge-transfer (CT)
fluorescence is useful as an indicator of the degree of PI chain packing(1-3) and the
miscibility of PI/PI binary blends.(4)

Recently, Okabe et al. (5) reported that a PI derived from a polyamic acid
ester (PAE) has a thermal expansion coefficient (TEC) higher than the PI made
from the corresponding PAA. The present paper also describes the difference of
molecular aggregation structure between PIs from PAA and PAE, characterized by
the intrinsic intermolecular CT fluorescence(1,3) and the dichroic absorption spectra
measured at a tilt angle for a rigid-rod dye dispersed in PI film.(6)

EXPERIMENTAL SECTION
Materials. PAAs were polymerized by adding equimolar amounts of dianhydride
powder into a N,N-dimethylacetamide (DMAc) solution of diamine at room

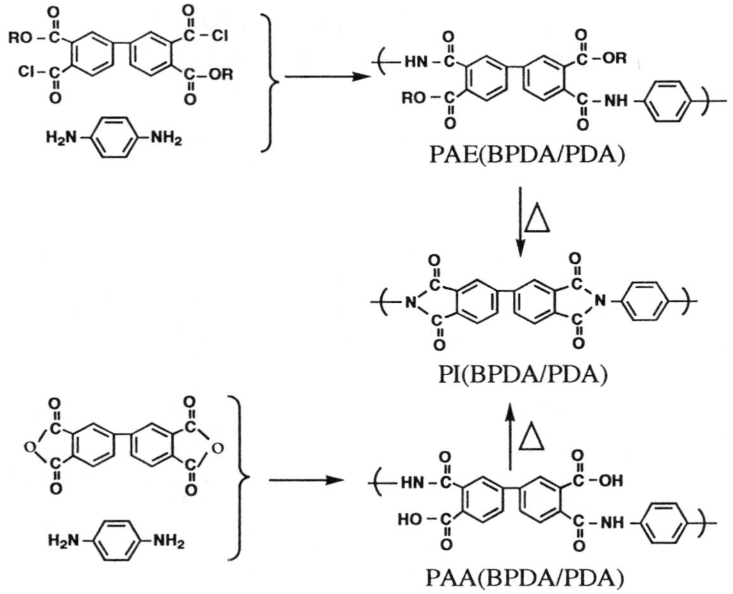

Figure 1. Reaction scheme for PAA and PAE synthesis and imidization.

Table I Comparison of donor-acceptor ability of monomers with that of respective structural units of PI backbone

	Monomer	Ability	Structural Unit in PI Chain
Acceptor	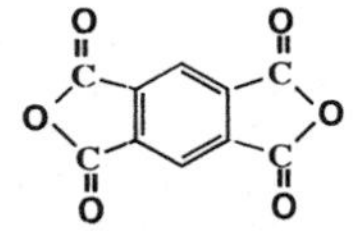	≈	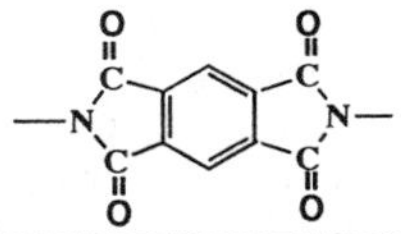
Donor	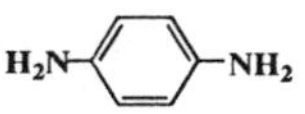 (I_P = 6.87 eV)	>	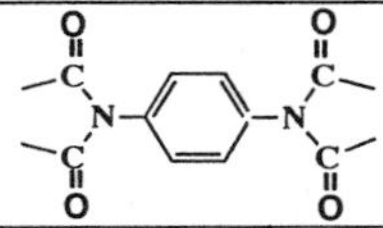

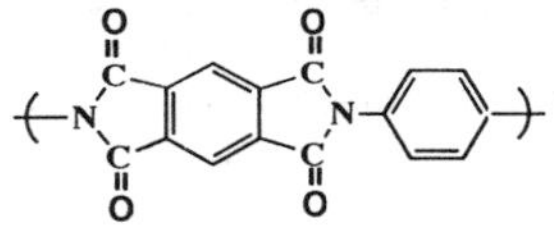 **PI(PMDA/PDA)**

temperature. PAE(BPDA/PDA) was synthesized by polycondensation of diacid chloride with PDA in the presence of pyridine in *N*-methyl-2-pyrrolidone (NMP) at 25°C as shown in Figure 1.(5) A typical electron acceptor, *p*-chloranil (CA), was recrystallized from benzene and vacuum-dried at 80°C. The DMAc solution of PAA was cast after mixing with CA, and then the CA-containing PAA films were cured at 200°C for 2 hr in vacuum.

Measurements. The corrected fluorescence spectra of PI films were measured at room temperature after thermal annealing at an established temperature in the front face arrangement to minimize reabsorption (excitation: 350 nm, sharp-cut filter: L-39) using a Hitachi F-2000 Fluorescence Spectrometer. The uv-vis absorption spectra of PI films were measured at room temperature using a Jasco Ubest-30 spectrometer. Polarized infrared absorption spectra of the thin PI films on silicon wafer were measured at a constant tilt angle of 60° using a Jasco FT-IR 5300 equipped with KRS-5 polarizer. The degree of in-plane orientation of the PI chains was estimated from the dichroic ratio, measured at an incidence angle, of *N,N'*-bis(2-methylcyclohexyl)perylenediimide (PEDI) dispersed in the film. The detailed procedure is described in ref.*6-8*. Film density was measured using a density gradient column composed of xylene-CCl$_4$ mixtures. The dynamic mechanical measurements (heating rate: 5°C/min, frequency: 0.1 Hz, load: 10 g) were conducted using a thermal mechanical analyzer (Mac Science TMA-4010)

RESULTS AND DISCUSSION
Donor-Acceptor Alternative Sequence in Aromatic PI Chains.
The structural units in the PI backbone were compared with the corresponding of monomers for the PAA polymerization as listed in Table I. As described previously(9), a typical dianhydride monomer, pyromellitic dianhydride (PMDA) acts as a strong electron acceptor and forms CT complexes in nonpolar solvents with a large number of aromatic compounds. Similarly, *p*-phenylenediamine (PDA) monomer is a very strong electron donor (Ionization Potential I_P = 6.87 eV). Unlike the fact that the acceptor ability of pyromellitimide is equivalent to that of PMDA monomer(10), it is predicted that the *N*-aryl group structural unit in the PI backbone would show donor ability weaker than PDA monomer since the *N,N'*-tetra carbonyl groups are electron-withdrawing. Gordina *et al.*(11) observed that a new visible absorption band appeared by adding strong electron acceptors into PI films derived from oxydiphthalic anhydride (ODPA) and 4,4'-diamino-*N*-methyl diphenylamine. Accordingly, we obtained the difference absorption spectrum of a *p*-chloranil-containing PI(ODPA/PDA) film and the film without the additive to estimate the donor ability of the *N*-aryl unit in the PI chains as shown in Figure 2. The new absorption band peaking at 410 nm in the difference spectrum suggests intermolecular CT complex formation between *N*-aryl units and *p*-chloranil. Figure 3 illustrates that the relationship between the I_P of donors and the peak wavenumber of the CT absorption band for *p*-chloranil/aromatic compound systems in nonpolar solvents reported in numerous papers is linear as predicted by Mulliken's CT theory: $hv = I_P + C$ (*C*: constant).(12) From the peak position of the difference spectrum in Figure 2, the I_P (donor ability) of the *N*-aryl unit in the PI backbone was estimated to be 8.50 eV. This indicates that the *N*-aryl unit has a donor ability similar to that of *p*-xylene (I_P = 8.48 eV). Therefore, it is most likely expected that both intra- and intermolecular CT interactions exist in the solid state of aromatic PIs expressed as -D-A-D-A-.

Aromatic PI films such as KAPTON usually do not show an individual CT absorption band but only an absorption tail in the visible range.(13,14) We found that a new absorption band appeared in the longer wavelength region upon blending of PI(BPDA/CHDA) (CHDA: trans 1,4-cyclohexyldiamine) with PI(DMCDA/PDA) (DMCDA: 5-(2,5-dioxotetrahydrofuryl)-3-methyl-3-cyclohexene-1,2-dicarboxylic

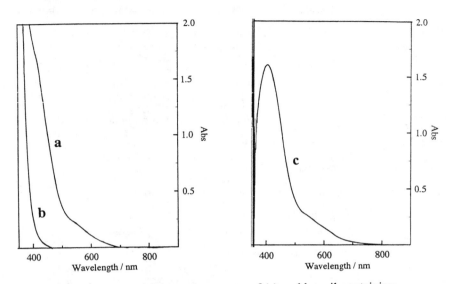

Figure 2. Uv-vis absorption spectra of (a) p-chloranil-containing
PI(ODPA/PDA) and (b) pure PI, and (c) the difference spectrum.

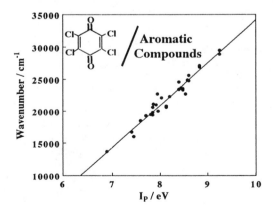

Figure 3. Relationship between I_P of donors and the peak wavenumber of CT
absorption band for p-chloranil / aromatic compounds systems.

anhydride) which are both PIs possessing no CT character as shown in Figure 4. Thus, for the blended PI system an intermolecular donor-acceptor interaction was demonstrated from the results in Figure 4.

Molecular Aggregation and Intermolecular CT Fluorescence.
Kotov *et al.* (*15*) illustrated the presence of CT interaction from the linear plot of $h\nu$ (absorption edge of PI films) vs I_P of the N-aryl unit in PI backbone. But they pointed out that to separate the contribution of intra- and intermolecular CT is difficult. In the present work, we will show CT fluorescence behavior associated with molecular aggregation of PI chains.

Figure 5a shows the fluorescence spectra of PIs derived from several dianhydrides and a fixed diamine (PDA). PI(PMDA/PDA) was almost nonfluorescent compared with PI(BPDA/PDA). The plot of λ^{-1} vs the electron affinity E_A of the dianhydride components [taken from ref.(*16*)] showed a linear relationship, according to CT theory (Figure 5b). The effect of thermal annealing on the CT fluorescence intensity leads us to assign whether it is from intra- or intermolecular CT complex. For semi-rigid PI(BPDA/PDA), the film density depends strongly on the cure program, probably owing to low molecular mobility during imidization.

Figure 6 indicates that the CT fluorescence is nearly proportional to the film density. This is interpreted to result from an increase in the intermolecular CT complex population by enhancement of molecular stacking. The intermolecular CT fluorescence is useful as a probe for PI chain stacking. In fact, the CT fluorescence was applied to studies of morphological changes of PIs during thermal cure(*2,3*) and miscibility of PI binary blends.(*4*)

Physical Properties Affected by In-plane Molecular Orientation.
Numata *et al* .(*17*) reported that PI films cured on a substrate show TECs lower than for those cured in free-standing state, and they demonstrated that this tendency is marked for PIs possessing rigid chains. They speculated that the in-plane orientation enhanced by thermal curing on a substrate is responsible for the lower TEC. The degree of in-plane orientation for PI(PMDA/ODA) (ODA: oxydianiline) was estimated by Russell *et al.* (*18*) by means of birefringence and X-ray diffraction measurements.

We used a spectroscopic technique to estimate quantitatively the degree of in-plane orientation of both PAA and PI chains. The visible dichroic absorption spectra of N,N'-bis(2-methylcyclohehxyl)perylenediimide (PEDI) dispersed in the film were measured at a tilt angle as depicted schematically in Figure 7. The refraction angle was kept constant ($\alpha = 28°$) for all samples. The incidence angle was adjusted on the basis of the refractive indices.(*6-8*) We defined $f = (1-D)/(1-D_0)$ as an in-plane orientation parameter: where the dichroic ratio D is the absorbance ratio [at peak wavelength (ca. 535 nm) of PEDI] of the linearly polarized lights parallel (P-polarized light) and perpendicular (S-polarized light) to the incidence plane, i.e., (A_P /A_S), and $D_0 (= \cos^2\alpha)$ is the dichroic ratio for complete in-plane orientation (two-dimensionally random). The value of f ranges from 0 (three dimensionally random) to unity (complete in-plane orientation).

Figure 8 shows the structural changes for a BPDA/PDA system during the course of a stepwise thermal treatment (maintained for 10 min at the established temperatures). It is noted that the increase in the degree of in-plane orientation proceeds parallel to the progress of imidization as illustrated in Figure 8a and 8b. The in-plane orientation of PI chain segments (f value) did not change appreciably upon stepwise thermal annealing above 250°C where imidization is almost completed. This means that the increase in the degree of in-plane orientation must have some degree of molecular mobility available before complete imidization. On

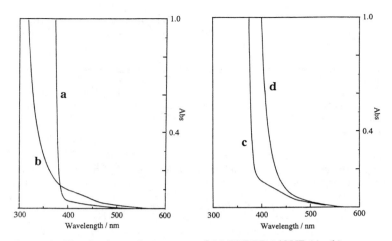

Figure 4. Uv-vis absorption spectra of (a) PI(BPDA/CHDA), (b) PI(DMCDA/PDA), (c) simple sum of (a) and (b), and (c) the blend (blend ratio: 5/5).

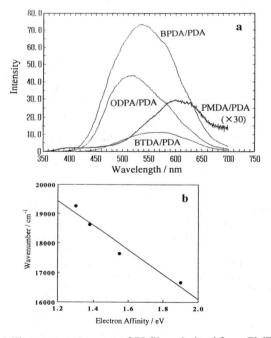

Figure 5. (a) Fluorescence spectra of PI films derived from PMDA, BTDA, BPDA, and ODPA with PDA. (b) Relationship between E_A of dianhydride components and the peak wavenumber of CT fluorescence band taken from Figure 5a.

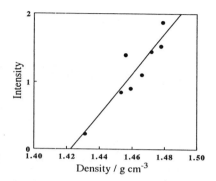

Figure 6. Change in the intermolecular CT fluorescence intensity with film density for PI(BPDA/PDA) system.

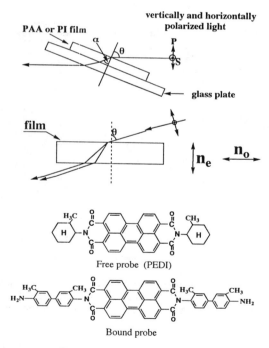

Figure 7. Schematic diagram for measurement of the visible dichroic absorption spectra of PAA and PI films containing PEDI molecule.

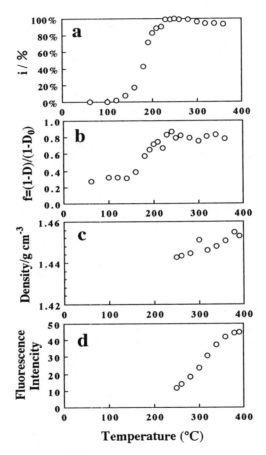

Figure 8. Structural changes for BPDA/PDA system in the course of stepwise thermal annealing. (a) extent of imidization, (b) the degree of in-plane orientation, (c) film density, and (d) intermolecular CT fluorescence intensity.

the other hand, film density (Figure 8c) and the intermolecular CT fluorescence intensity (Figure 8d), reflecting the CT complex population (1-3), increased gradually with increasing annealing temperature above 250°C. In addition, for PI(BPDA/PDA) the orientation of the phthalimide molecular plane to the film plane, which is estimated from the infrared dichroic absorption spectra measured at a constant tilt angle of 60° for the bands characteristic of the imide ring (1774, 1720, and 738 cm^{-1}), did not noticeably change upon stepwise thermal annealing above 250°C. The results indicate that molecular stacking of rigid PI chains (changes in density and CT fluorescence intensity) could proceed even upon only local molecular motions, in contrast to the orientation of PI chain segments and phthalimide molecular planes.

TEC of PI(BPDA/PDA) films cured on substrate decreased somewhat upon thermal annealing at higher temperature. This is probably due to an enhanced intermolecular interaction, *e.g.*, CT complex formation or crystallization which cause a decrease in the free volume fraction), not due to the change in the degree of in-plane orientation of PI chain segments. Interestingly, the CT fluorescence spectra red-shifted gradually with increasing annealing temperature as shown in Figure 9, probably reflecting a structural change such as the donor-acceptor separation distance in the CT complexes. Figure 10 demonstrates that TEC of PI(BPDA/PDA) films (8-50 μm thick) obtained by cure of the NMP-cast PAA film is closely associated with the degree of in-plane orientation. Thus, TEC is a function of both the degree of in-plane orientation and molecular stacking.

Figure 11 shows that PI(BPDA/PDA) thin films cured on substrates provide extremely high in-plane orientation while the *f* values of free-standing cured thick PI films were not very high. For comparison, the intermolecular CT fluorescence (excited at 350 nm) and excitation spectra (monitored at 600 nm) for thin (cured on quartz plate, ca. 1 μm thick) and thick films (cured in free-standing state, 54 μm thick) were measured as shown in Figure 12. The fluorescence position and especially the shape of excitation spectra of the thin film differs significantly from that of the thick film. Although we have no clear explanation yet, obviously, the in-plane orientation influences local structure such as CT complex.

Comparison of the Properties of PIs from PAA and PAE.

Miwa *et al.* (19) reported that PI(BPDA/PDA) from PAE has a higher TEC than for PI(BPDA/PDA) produced from PAA. Figure 13 illustrates the values of TEC for PIs from PAA and PAE (cured at 350°C) as a function of casting temperature. The higher values of TEC of the PI from PAE are attributed to both loose molecular stacking and slightly lower degrees of in-plane orientation compared with the PI from PAA, as illustrated in Figure 14 and 15, respectively. The CT fluorescence position of the PI from PAE was different from that of the PI from PAA, reflecting the difference in local structure. Interestingly, the degree of molecular stacking estimated by the film density and the CT fluorescence intensity for the PI produced from PAE became nearly equivalent upon annealing at 400°C. Why the differences of the orientation and the stacking of the resulting PIs generated from PAA and PAE ? Possible reasons are the differences in hydrogen-bonding ability with the solvent, imidization rate, and/or the size of the volatile group outgassed during imidization.(5,19)

Comparison of properties of PI(BPDA/PDA) and PI(BPDA/CHDA).

Dynamic mechanical properties of PI(BPDA/PDA) and PI(BPDA/CHDA) (CHDA: trans 1,4-cyclohexyldiamine) (ca. 60 μm thick and cured in free-standing state for both samples) were compared. PI(BPDA/CHDA) has high chain linearity as well as PI(BPDA/PDA). Figure 16 shows that the loss energy peak and the decrease in E' at Tg lowered upon annealing at higher temperature for the former (aromatic PI). The

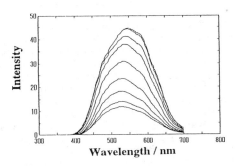

Figure 9. Change in the CT fluorescence spectra accompanied with stepwise thermal annealing. (from top to bottom: 390, 380, 360, 340, 320, 300, 280, 260, 250°C)

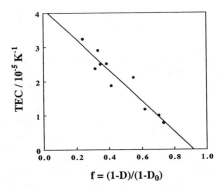

Figure 10. Relationship between TEC and the degree of in-plane orientation for PI(BPDA/PDA) films annealed at 330°C in free-standing state after cured on a substrate at 250°C.

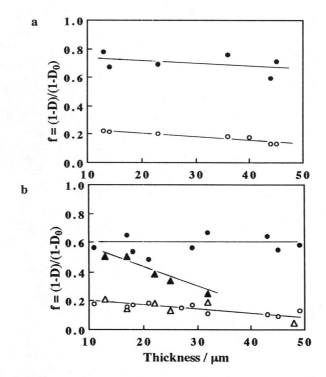

Figure 11. The degree of in-plane orientation for BPDA/PDA system. (a) cured on substrate and (b) cured in frame (●) and in free-standing state (▲). Open and closed marks mean PAA and PI.

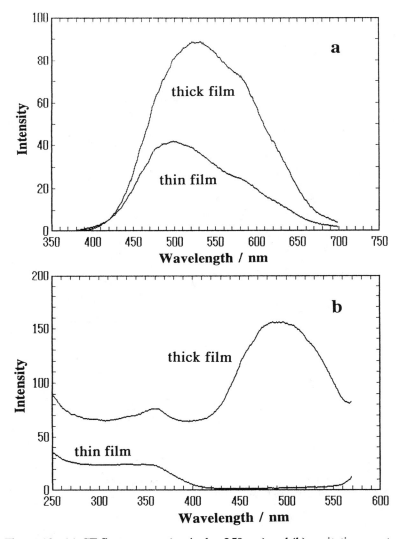

Figure 12. (a) CT fluorescence (excited at 350 nm) and (b) excitation spectra (monitored at 600 nm) of thin (ca. 1 μm, spin-coated, cured on substrate) and thick (54 μm, cured in free-standing state) PI(BPDA/PDA) films cured at 250 °C; 1h + 300°C; 0.5 h.

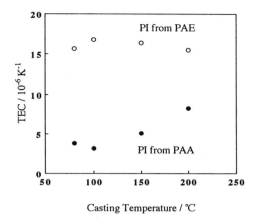

Figure 13. TEC of PI(BPDA/PDA) films cured at 350°C from PAA and PAE as a function of casting temperature.

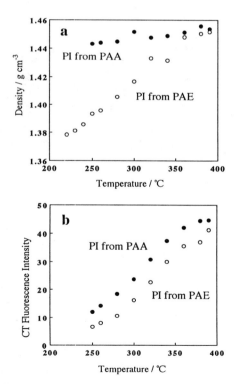

Figure 14. Change in film density and CT fluorescence intensity for PI (BPDA/PDA) films obtained from PAA and PAE with increasing annealing temperature.

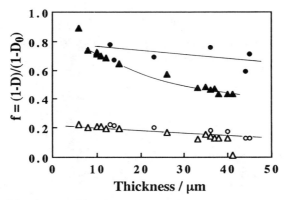

Figure 15. The degree of in-plane orientation of PI(BPDA/PDA) films (●,▲) prepared from PAA (○) and PAE (△) by curing at 250°C on substrate.

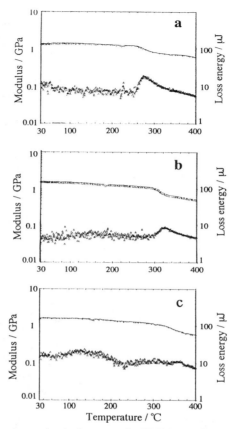

Figure 16. Dynamic mechanical spectra for PI(BPDA/PDA) cured in free-standing state. Annealing conditions: (a) 250°C; 1hr, (b) (a)+ 300°C; 1hr, (c) (b)+ 330°C; 1hr.

decrease in E' for the latter (aliphatic PI) remained clear after annealing compared with the former system (Figure 17). The film densities of the latter (< 1.39 g cm^{-3}) were much lower than those of the former (1.4332 g cm^{-3} for (a) 250°C; 1hr, 1.4510 g cm^{-3} for (b) (a) + 300°C; 1hr, 1.4594 g cm^{-3} for (c) (b) + 330°C; 1hr). On the other hand, there was little difference of TEC (2.6x10^{-5} for the former and 3.0x10^{-5} K^{-1} for the latter). We believe that the lower E' decrease at Tg for PI(BPDA/PDA) results from the intermolecular CT interaction as physical crosslink. Then, how much is the concentration of the CT complex in solid state PI and how does it influence some physical properties ?   These important problems remain less understood.

PI(BPDA/CHDA) showed the values of f lower than PI(BPDA/PDA) as plotted in Figure 18. The difference between them became marked with the increase in film thickness. Dense molecular stacking of the PI chains may be required to attain considerably high degree of the in-plane orientation. Considering the results in Figures 11 and 12, the in-plane orientation and the local molecular stacking influence each other, and are not completely independent.

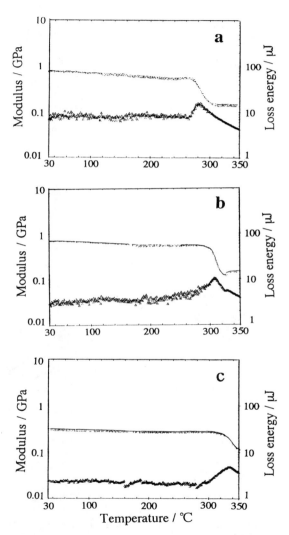

Figure 17. Dynamic mechanical spectra for PI(BPDA/CHDA) cured in free-standing state. Annealing conditions (a-c) are as in Fig. 16.

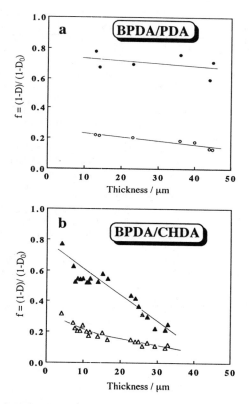

Figure 18. The degree of in-plane orientation for (a) PI(BPDA/PDA) and (b) PI(BPDA/CHDA) cured at 250°C on substrate as a function of PAA film thickness. Open and closed marks mean PAA and PI.

Acknowledgment   We wish to thank Dr. L.Vladimirov of Russian Academy of Science for translation of ref.*11* into English.

Literature Cited
(1) Hasegawa, M.; Kochi, M.; Mita, I.; Yokota, R. *Eur. Polym. J.* **1989**, *25*, 349.
(2) Wachsman, E.D. and Frank, C.W. *Polymer* **1988**, *29*, 1191.
(3) Hasegawa, M.; Arai, H.; Mita, I.; Yokota, R. *Polym. J.* **1990**, *22*, 875.
(4) Hasegawa, M.; Mita, I.; Kochi, M.; Yokota, R. *Polymer* **1991**, *32*, 3225.
(5) Okabe, Y.; Miwa, T.; Takahashi, A.; Numata, S. *Kobunshi Ronbunshu* (*Jpn. J. Polym. Sci. Technol.*), **1993**, *50*, 947.
(6) Hasegawa, M.; Matano, T.; Shindo, Y.; Sugimura, T. *J. Photopolym. Sci. Technol.* **1994**, *7*, 275.
(7) Hasegawa, M.; Matano, T.; Shindo, Y.; Sugimura, T. In*Polymeric Materials for Microelectronic Applications: Science and Technology;* Ito, H.; Tagawa, S.; Horie, K., Eds.; ACS Symposium Series 579. 1994, pp 234.
(8) Hasegawa, M.; Matano, T.; Shindo, Y.; Sugimura, T. Proceedins of 5th International Conference on *Polyimides*, 1994, New York, Ellenville in press.
(9) Hasegawa, M.; Shindo, Y.; Sugimura, T. the preceding paper in this symposium.
(10) Sep, W.J.; Verhoeven, J.W.; de Boer, T.J. *Tetrahedron* **1975**, *31*, 1065.
(11) Gordina, T.A.; Kotov, B.V.; Kolninov, O.V.; Pravednikov, A.N. *Vysokomol. soyed.* **1973**, *B15*, 378.
(12) Mulliken, R.S. *J. Am. Chem. Soc.* **1952**, *74*, 811; *J. Phys. Chem.* **1952**, *56*, 801.
(13) Ishida, H; Wellinghoff, S.T.; Baer, E.; Koenig, J.L. *Macromolecules* **1980**, *13*, 826.
(14) Hasegawa, M.; Shindo, Y.; Sugimura, T.; Ohshima, S.; Horie, K.; Kochi, M.; Yokota, R.; Mita, I. *J. Polym. Sci.: B* **1993**, *31*, 1617.
(15) Kotov, B.V.; Gordina, T.A.; Voishchev, V.S.; Kolninov, O.V.; Pravednikov, A.N.*Vysokomol. Soyed.* **1977**, *A19*, 614. (*Polym. Sci. USSR., 19,* 711)
(16) *Polyimides*; Bessonov, M.I.; Koton, M.M.; Kudryavtsev, V.V.; Laius, L.A. Plenum Press: New York, 1987; pp 16.
(17) Numata, S.; Fujisaki, K.; Kinjo, N. *Polymer* **1987**, *28*, 2282.
(18) Russell, T.P.; Gugger, H.; Swalen, J.D. *J. Polym. Sci., Phys. Ed.* **1983**, *21*, 1745.
(19) Miwa, T.; Ishida, M.; Okabe, Y.; Takahashi, A,; Hasegawa, M.; Matano, T.; Shindo, Y.; Sugimura, T. Proceedings of 5th International Conference on *Polyimides,* 1994, New York, Ellenville, in press.

RECEIVED July 7, 1995

Chapter 27

Application of Polyisoimide as a Polyimide Precursor to Polymer Adhesives and Photosensitive Polymers

Amane Mochizuki and Mitsuru Ueda

Department of Materials Science and Engineering, Faculty of Engineering, Yamagata University, Yonezawa 992, Japan

Recent developments and application of polyisoimide (PII) to high temperature adhesive and photosensitive polymers are described. First, new high temperature adhesives based on PII have been developed. PIIs showed stronger adhesions to copper foils because of favorable anchoring between PII and copper foil due to a good flow of PII. Next, a positive working photosensitive polyimide precursor based on PII and 2,3,4-tris[1-oxo-2-diazonaphthoquinone-4-sulfonyloxy] benzophenone (5) as a photoreactive compound has been developed. The photosensitive polyimide-precursor containing 20 wt% of 5 showed a sensitivity of 250 mJ/cm^2 and a contrast of 2.4 with 435 nm-light when it was postbaked at 150 °C for 10 min followed by developing with 5 % aqueous tetramethylammonium hydroxide solution at 45 °C. Further, a new amine photo-generator{[(4,5-dimethoxy-2-nitrobenzyl)oxy] carbonyl}-2,6-dimethyl piperidine (6) was prepared from 2,6-dimethyl piperidine and 4,5-dimethoxy-2-nitrobenzyl-p-nitrophenylcarbonate. The PII containing 10 wt% of 6 functioned as photosensitive resist having a sensitivity of 900 mJ/cm^2 and a contrast of 3.4 with 365 nm-light when it was postbaked at 150 °C for 5 min followed by development with cyclohexanone at 45 °C.

High-performance plastics are currently receiving considerable attention for their potential uses in aerospace, automotive, electronic and related industries. In particular, polyimides (PIs) have been widely used as insulation materials for microelectronic devices because of their excellent properties, such as thermal and chemical stability, and low dielectric constants (1). However, aromatic PIs are usually intractable because they are insoluble and high melting temperatures. Therefore, Polyamic acids (PAAs) are prepared first, processed into shaped objects and then converted to PIs by thermal cyclization.

On the other hand, polyisoimides (PIIs) are the processable isomeric form of the corresponding PIs and a commercial acetylene-terminated isoimide oligomer (IP-600) is an example of a high temperature thermo-setting resin (2). In a recent paper (3), we reported the preparation and properties of polyisoimide (PII) as a PI-precursor, and found that PII has a lower glass transition temperature (Tg) than that of the corresponding PI, and is easily converted to PI without elimination of volatile compounds.

0097–6156/95/0614–0413$12.00/0

This paper describes the recent developments and application of PIIs to   high temperature adhesive and photosensitive polymers.

Experimental

Materials. N-Methyl-2-pyrrolidone (NMP), triethylamine (TEA), and pyridine (Py) were   purified   by   distillation.   4,4´-[Hexafluoroisopropylidenebis(p-phenyleneoxy)]dianiline   (BAPF)   (**2a**)   and   4,4´-[isopropylidenebis(p-phenyleneoxy)]dianiline (BAPP) (**2b**) were purified by recrystallization from cyclohexane and 2-propanol, respectively. 4,4´-Hexafluoroisopropylidenebis(phthalic anhydride) (6FDA) (**1a**) was obtained from American Hoechst Co.   Oxydiphthalic anhydride (ODPA) (**1b**) was obtained from Occidental Chem Corp. Rolled copper foil (35 μm thickness) was used as an adherent. Other reagents and solvents were obtained commercially and used as received.

Polymer Synthesis. A typical example is as follows.
Polyisoimide (**3a**) from **1a** and **2a**. A solution of   **2a** (2.22 g, 5.0 mmol)   in NMP (43.2 ml), was cooled with an ice-water bath.   While stirring, **1a** (2.59 g, 5.0 mmol) was added to this solution. The resulting mixture was stirred at room temperature for 4 h.   The viscous solution was diluted with NMP (48.2 ml) and TEA (1.4 ml, 10.0 mmol) was added dropwise with continued stirring.   The reaction mixture was then cooled with an ice-water bath, and trifluoroacetic anhydride (1.54 ml, 11.0 mmol) was added dropwise with stirring . The mixture was stirred at room temperature for 4 h and then poured into 2-propanol (1000 ml) to precipitate the polymers.   The precipitated polymer was filtered off and dried *in vacuo* at 40 °C.   The yield was 4.54 g (98 %). The inherent viscosity of the polymer in DMAc was 0.40 dl/g at a concentration of 0.5 g/dl at 30 °C. IR (KBr), ν (cm^{-1}) 1810, (C=O), 920 (C-O) Analysis: Calculated for $C_{46}H_{22}N_2O_6F_{12} \cdot H_2O$: C, 58.48; H, 2.56; N, 2.96. Found: C, 58.56; H, 2.78; N, 2.90.

Evaluation of Adhesion Strength. A typical procedure is as follows.

Film Casting. A solution of polymer **4a** was made by dissolving the polymer in chloroform to afford a 15 wt% solution. Films cast on glass plates were prebaked at 80 °C for 60 min and dried *in vacuo* at 150 °C for 12 h. The thickness of polymer film was 25 μm.

Hot Press.   Test press No. 1541 (Gonno Hydraulic Press Manufacturing Co.) was used for the adhesion between a polymer film and a copper foil. The polymer film was placed between 50 mm x 50 mm pieces of copper foil. The copper foil was washed with acetone prior to use. The assembly was placed in the hot press and melt processed at 200-260 °C under 14.7 Mpa. After holding 15 min, the platens were cooled under pressure to room temperature. At least five samples were tested and the average values are reported.   The peel strength was measured by Tensilon UTM-III-500 (Toyo Baldwin) at a peel rate of 50 mm/min at room temperature (θ^0-peel method) *(19)*.

Photosensitivity. 3- 5 μm- thick PII films on a silicone wafer were exposed to 365 nm or 435 nm radiation using a filtered super high pressure mercury lamp.   Exposed films were postbaked at 150 °C for 5 min, developed in 5 % aqueous tetramethylammonium hydroxide (TMAH) solution or cyclohexanone at 45 °C, and subsequently rinsed with 2-propanol. The characteristic sensitivity curve was obtained by plotting a normalized film thickness against logarithmic exposure energy.

Measurement. The infrared spectra were recorded on a Hitachi I-5020 FT-IR spectrophotometer. Viscosity measurements were carried out by using an Ostwald viscometer at 30 °C. Thermal analyses were performed on a Seiko SSS 5000-TG/DTA 200 instrument at a heating rate of 10 °C/min for TG and a Seiko SSS 5000-DSC220 at a heating rate of 10 °C/min for differential scanning calorimetry (DSC) under nitrogen. Molecular weights were determined by a gel permeation chromatography (GPC) with polystyrene calibration using a JASCO HPLC system equipped with a Shodex KD-80M column at 40 °C in DMF. The film thickness was measured using a Dektak 3030 system (Veeco Instruments Inc.). Dynamic mechanical analysis (DMA) was performed with a DVE-V4 FT Rheospectra (Rheology Co., LTD) in tensile mode at a frequency of 100 Hz and at a heating rate of 2 °C min^{-1} with film specimens 5 mm mode and 25 μm thick.

Result and Discussion

Polymer Synthesis. We prepared the polyisoimides (PIIs)(3) and polyimides (PIs) (4) by chemical cyclization methods to investigate their properties as melt processable adhesives and photosensitive polyimide precursors. Although many dehydrating agents had been proposed for the preparation of isoimides from amic acids (4), it had not been clear which reagents would be the best for PII synthesis. In a previous investigation (3), we found that trifluoroacetic anhydride (TFAA)-triethylamine (TEA) was the best dehydrating agent for the formation of isoimide. Thus, ring-opening polyadditions of tetracarboxylic dianhydrides and diamines were carried out in NMP for 4h at room temperature equation 1, yielding poly(amic acids), which were subsequently converted to PIIs and PIs by using TFAA-TEA and acetic anhydride-pyridine (Py), respectively. Table I indicates that polymers 3 and 4 were produced in excellent yields with inherent viscosities of up to 0.75 dl/g . Polymers 3 and 4 were characterized by infrared spectroscopy and elemental analysis.

Ar₁: **1a** [structure with CF₃ groups]

Ar₂: **2a** [structure with CF₃ groups]

1b [structure with O linkage]

2b [structure with CH₃ groups]

2c [structure with SO₂ linkage]

Table I. Synthesis of Polyisoimides (3) and Polyimides (4) [a]

Ar₁	Ar₂	polymer	Yield(%)	$\eta_{inh}(dl/g)$[b]	Tg (°C)
1a	2a	3a	96	0.40	215
1a	2a	4a	99	0.45	250
1a	2b	3b	94	0.75	190
1a	2b	4b	95	0.71	240
1a	2c	3c	96	0.40	203
1b	2c	3d	98	0.36	190

a **Ring-opening polyaddition was carried out with 2mmol of each monomer in DMAc at room temperature.**

b **Measured at a concentration of 0.5g/dl in DMAc at 30°C.**

Evaluation of PII as High Temperature Adhesives *(5).* Thermoplastic PI adhesives *(6- 8)* such as Lark-TPI are prepared by the   condensation reaction of aromatic tetracarboxylic dianhydrides and aromatic diamines.   These PIs are linear thermoplastic polymers that generally have a high melt viscosity and must be processed at relatively high temperatures (> 300 °C).   On the other hand, a polyisoimide (PII) has a lower glass transition temperature (Tg) than that of corresponding PI and can be converted easily to PI without elimination of volatile compounds. Therefore, PII is of considerable interest as a candidate for high temperature adhesives.

Adhesion Properties of Polymer 3 and 4. Films of polymer **3a** and **4a** were prepared by casting cyclohexanone solutions onto glass plates which were then heated on a hot plate at 60 °C, and subsequently dried at 150 °C for 12h *in vacuo.* Dynamic mechanical analysis (DMA) was performed in tensile mode at a frequency of 100 Hz with film specimens 5 mm and 25 μm thick. Figure 1 shows the dynamic storage modulus of polymer **3a** and **4a**. The storage modulus for polymer **3a** declined sharply around 200 °C, which is agreed with the glass transition temperature of polymer **3a** dertermined by DSC.   Polymer **4a** maintained mechanical integrity   at this temperature and the storage modulus did not drop until 250 °C. These results indicate that a large difference in the flow behavior between polymers **3a** and **4a** can be expected above 200 °C.

The adhesion test was evaluated by the peel strength of copper foil at 180 degrees from the polymer film as shown in Figure 2. The heat compression was conducted employing the process as illustrated in Figure 3. The assembly was placed in the hot press and compressed at 200-260 °C for 15 min under 14.7 Mpa. Subsequently, the platens were cooled to a temperature below 50 °C and the assembly was removed from the heat press tool.

The resulting peel strengths of PIIs (**3a-b**) and PIs (**4a-b**) films on copper foils are shown in Table II. PII (**3a**) exhibited peel strengths of 780 gcm^{-1} and 400 gcm^{-1} at the press temperatures of 260 °C and 230 °C, respectively. On the other hand, the peel strengths of PI (**4a**) film on copper foils were very weak even at a press temperature of 260 °C. In addition, PII (**3a**) film was flexible and free of voids after compression at 260 °C, presumably due to the lower Tg. The isomerization to the corresponding PI was confirmed by IR spectroscopy.

Table II. Peel strength of polymer films on copper foil [a]

Run	Polymer	Press Temp.(°C)	Peel Strength (g/cm)
1	3a	260	780
2	3a	230	415
3	3a	200	_[b]
4	4a	260	77
5	4a	230	_[b]
6	4a	200	_[b]
7	3b	250	590
8	3b	200	180
9	4b	250	110
10	4b	200	_[b]

[a] **bonded at 14.7 MPa ,** [b] **no adhesion**

The greater adhesive nature of PII can be elucidated by considering the lower Tg and the drastic changes in elasticity above it's Tg compared to that of the corresponding PI. In addition, the advantage of using the PII films as a high temperature adhesive has also been shown with the conversion to the corresponding PIs during the heat compression process without the generation of volatile compounds.

Application of PII to the Photosensitive Polyimide (PSPI) System using Diazonaphthoquinone (NQD) *(9).* Positive resists based on novolak resins with o-diazonaphthoquinone (NQD) are standard materials used in semiconductor manufacturing, where NQD acts as a dissolution inhibitor for aqueous base development of the novolac resin. Several groups *(10-11)* have reported the resists consisting of polyamic acids and NQD. However, dissolution rates of polyamic acids

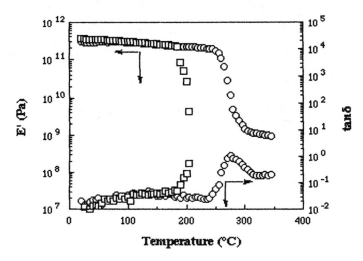

Figure 1. Dynamic mechanical analysis of polymer films: (□) polymer **4a**;
(○) Polymer **5a.**

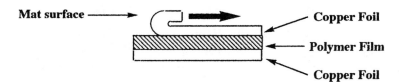

Figure 2. Measurement of peel strength.

are essentially too high to get a sufficient dissolution contrast, and as a result the dissolution rates of polyamic acids have to be reduced by prebaking, or post exposure bake (PEB).

Herein we describe a new photosensitive polyimide (PSPI) system consisting of PII and NQD as the polymer matrix and a photosensitive compound, respectively. We have chosen PII (**3c**) and 2,3,4-tris[1-oxo-2-diazonaphthoquinone-4-sulfonyloxy] benzophenone (**5**) as the candidate polymer matrix and photoreactive compound, respectively. In order to investigate dissolution behavior of the exposed (500 mJ/cm²) and unexposed area, the effect of the loading of **5** on the dissolution rate in a developer after post exposure bake (PEB) at 150 °C for 10 min was studied, and the results are shown in Figure 4. Development was performed at 45 °C using 5 % TMAH solution as a developer. The dissolution rate of the exposed area increased clearly with an increase in **5**, and the exposed film dissolved faster than the unexposed film at all concentrations of **5**. Furthermore, dissolution rates of unexposed films is not effected by **5**. This result indicates that **5** acts as a dissolution promoter in PII (**3c**) film, and that 20 wt% of **5** is necessary to achieve good dissolution contrast.

The most widely used positive resists are generally two-component materials consisting of an alkaline soluble matrix resin that is rendered insoluble in aqueous alkaline solutions through addition of hydrophobic radiation-sensitive materials. In this case, although the PII matrix is insoluble in aqueous alkaline solutions, the PII is susceptible to hydrolysis to give alkaline soluble polyamic acid. Hydrolysis of PII by aqueous alkaline may occur in the exposed area due to the diffusion of the developer into the PII film. Presumably, this diffusion is facilitated by the presence of hydrophilic moieties (e.g., carboxylic acid, sulfonic acid or hydroxyl) in the exposed regions. Therefore, the accelerated dissolution behavior in the exposed regions can be attributed to the polarity change of PII, as a result of PII hydrolysis in the development process and the photochemical reaction of compound **5**.

After a preliminary optimization study involving compound **5** loading, postbaking temperature, and developing temperature, we prepared a photosensitive polyimide-precursor system consisting of PII and 20 wt% of **5**. The sensitivity curve for a 5 μm thick PII film shown in Figure 5 was consistent with the dissolution behavior studied above, indicating that the sensitivity (D^0) and contrast (γ^0) were 300 mJ/cm² and 4.5,

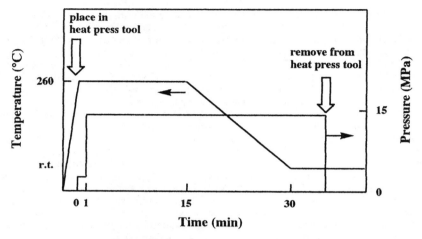

Figure 3. Typical hot press cycle.

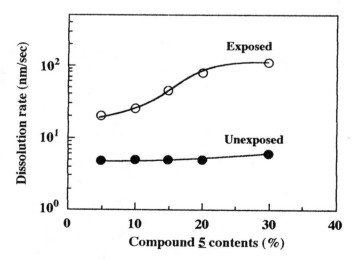

Figure 4. Relationship between Compound **5** Contents and Dissolution rate.

with 365 nm-light, respectively after PEB treatment at 150 °C for 10 min. Furthermore, with 436 nm-light the values of D^0 and γ^0 were 250 mJ/cm^2 and 2.4. Figure 6 illustrates the scanning electron micrograph of the positive image projection printed in PII by postbaking at 150 °C for 10 min after the film exposed to 400 mJ/cm^2. This resist is capable of resolving a 2.5 μm feature when a 5 μm thick films is used.

Application of PII to the PSPI System using Amine Photo-generator *(12).* It is well known that aromatic nitro compounds containing benzylic hydrogens at the *ortho* to the nitro group are light sensitive *(13).* These photosensitive protecting groups are of potential importance in many areas of synthetic chemistry and formulation of resist materials *(14-15).*

Recently, we found that the isomerization reaction of isoimide to imide was accelerated by a catalytic amount of base *(2).* This finding led us to synthesize a new amine photo-generator, {[(4,5-dimethoxy-2-nitrobenzyl)oxy]carbonyl}-2,6-dimethyl piperidine (**6**). 2, 6-dimethylpiperidine (DMP) was chosen as the hindered photo-generated base in order to avoid the nucleophilic addition of amine to isoimide. It is known that isoimide tend to react with nucleophiles such as amines or alcohols *(16-17).*

The new amine photo-generator **6** was prepared by the reactions shown in equation 3. The reduction of 4,5-dimethoxy-2-nitrobenzaldehyde gave 6-nitroveratryl alcohol. This compound was converted to 4,5-dimethoxy-2-nitrobenzyl-p-nitrophenylcarbonate by the treatment with p-nitrophenyl chloroformate in the presence of triethylamine. The reaction of 4,5-dimethoxy-2-nitrobenzyl-p-nitrophenylcarbonate with DMP in the presence of 1-hydroxybenztriazole (HOBt) *(18)* yielded the new amine photo-generator **6**. Recrystallization from a mixture of benzene and n-hexane gave pale yellow needles. The structure of amine photo-generator (**6**) was confirmed to be the corresponding carbamate by ^{13}C-NMR, infrared spectroscopy, and elemental analysis. The IR spectra exhibited characteristic absorptions at 1700, 1525, and 1320 cm^{-1} due to carbamate carbonyl, nitro asymmetric, and nitro symmetric stretching vibrations, respectively.

The base catalyzed isomerization of PII to the corresponding PI was carried out by irradiating PII (**3d**) films containing 10 wt % of **6** (Figure 7). PII (**3d**) is significantly converted to the corresponding PI (**4d**) upon exposure to 1000 mJ/cm^2 of 365 nm irradiation followed by post-exposure bake (PEB) at 150 °C, as evidenced by strong absorptions due to imide C=O at 1780 cm^{-1} and imide C-O at 1380 cm^{-1}. On the other hand, the unexposed polymer (**3d**) film containing 10 wt% of **6** was quite stable under the thermal treatment at 150 °C for 5 min.

6

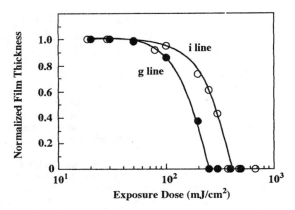

Figure 5. Exposure characteristic curves for the system of PII and **5**.

Figure 6. Scanning electron micrograph of pattern from PII containing **5**. (After development with 5% TMAH solution)

After preliminary optimization studies involving the loading of **6**, postbaking temperature, developer, and developing temperature, we formulated a photosensitive polyimide- precursor system consisting of PII and 10 wt% of **6**. The PII films (ca. 3 μm thick) containing 10 wt % of **6** were exposed to 365 nm UV irradiation, postbaked at 150 °C for 5 min and developed with cyclohexanone at 45 °C.

The sensitivity curve for a 3-μm-thick PII film shown in Figure 8 is consistent with the isomerization study, indicated that the sensitivity ($D^{0.5}$) and contrast ($\gamma^{0.5}$) were 900 mJ/cm^2 and 3.4, respectively. Figure 9 shows the scanning electron micrograph of a negative image projection-printed in PII by postbaking at 150 °C for 5 min after exposure to 1000 mJ/cm^2.

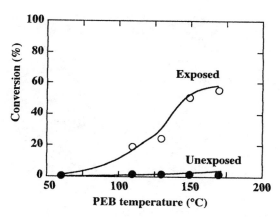

Figure 7. Thermal isomerization of PII in the presence of 10 wt % of **6**.

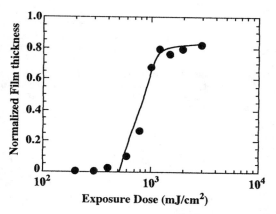

Figure 8. Exposure characteristic curves for the system of PII and 10 wt % of **6**.

Conclusions

PIIs were successfully prepared from the corresponding polyamic acids using trifluoroacetic anhydride-TEA system as a dehydrating agent. PII exhibited better flow properties than the corresponding PI and also functioned as a suitable high-temperature adhesive for bonding copper foil. The PII containing 20 wt% of **5** was found to be a positive-type PSPI-precursor, in which **5** acts as the dissolution controller. Furthermore, the new amine photo-generator **6** was effective for the isomerization of PII to PI. PII containing 10 wt% of **6** functioned as a negative type-PSPI with good contrast due to a large difference in the solubility between PII and PI.

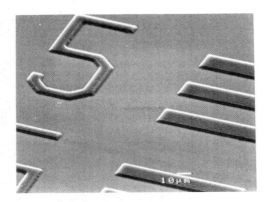

Figure 9. Scanning electron micrograph of pattern from PII containing **6.** (After development with cyclohexanone)

Literature Cited
1. Wood, A.S.; *Modern plastics international.*, **1989**,*June*, 26
2. Landis, A. L.; Naselow, A. B.*"Polyimides: Synthesis, Characterization, and Aplications"*, *vol.1, 39, K. L.Mittal, Ed, Plenum Publishing Corp., New York (1982) , p 39*
3. Mochizuki, A.; Teranishi T.; Ueda, M. *Polym.J.*, **1994**, *26 (3)*, 315.
4. R.J.Cotter, C.K.Sauers, and J.M.Whelan, *J.Org.Chem.*, **1961**, *26*, 10
5. A. Mochizuki, T. Teranishi, and M. Ueda, *Polymer*, **1994**, *35 (18)*, 4022 .
6. St. Clair, A. K., and St. Clair,T. L. *Sci. Adv. Mater. Process Eng. Ser.* **1981**, *26*, 165
7. Hergenrother, P. M., Wakelyn, N. T. and Havens, S. J. *J.Polym.Sci.*, *Polym.Chem.Ed.*, **1987**, *25*, 1093
8. Pratt, J. R., Blackwell, D. A., and St. Clair,T. L. *Polym. Eng. Sci.*, **1989**, *29*, 63
9. A. Mochizuki, T. Teranishi, K. Matsushita, and M. Ueda, Polymer, in press
10. Moss, M.G., Cuzmar, R.M. and Brewer, T. *SPIE Advances in Resist Technology and Processing VI*, **1989**, *1086*, 396
11. Hayase, S., Takano, K., Mikogami, Y. and Nakano, Y. *J. Electrochem. Soc.*, **1991**, *138*, 3625
12. A. Mochizuki, T. Teranishi, and M. Ueda, *Macromolecules*, **1995**, *28,*365(1995).
13. Ciamician, G.; Silber, P. *Ber.*, **1901**, *34*, 2040
14. Cameron, J. F.; Fréchet, J. M. J. *J.Am.Chem.Soc.*, **1991**, *113*, 4303
15. Mckean, D. R.; Briffaud, T.; Volksen, W.; Hacker, N. P.; Labadie, J. W.*Polymer preprints*, **1994**, *35(1)*, 387
16. Hedaya, E.; Hinman, R. L. Theodoropulos, S. *J. Org. Chem.*, **1966**, *31*, 1311
17. Fan, Y. L.; Pollart, D. F. *J. Org. Chem.*, **1968**, *33*, 4372
18. Konig, W.; Geiger, R. *Chem, Ber.*, **1973**, *106*, 3625
19. Wake, W. C. *Polymer*, **1978**, *19*, 291

RECEIVED July 7, 1995

Chapter 28

Polyimide Nanofoams Prepared from Styrenic Block Copolymers

J. L. Hedrick[1], T. P. Russell[1], C. Hawker[1], M. Sanchez[1], K. Carter[1], Richard A. DiPietro[1], and R. Jerome[2]

[1]IBM Almaden Research Center, 650 Harry Road, San Jose, CA 95120–6099
[2]Universite de l'Etat a Liege, Start Tilman, Liege, Belgium

New routes for the synthesis of high T_g thermally stable polymer foams with pore sizes in the nanometer regime have been developed. Foams were prepared from well-defined microphase separated block copolymers comprised of a thermally stable block and a thermally labile material. Upon heating, the thermally unstable block undergoes thermolysis generating pores, the size and shape of which are dictated by the initial copolymer morphology. Several labile blocks were investigated including poly(styrene), poly(α-methylstyrene) and copolymers of α-methylstyrene-styrene and styrene. A common feature of these labile blocks is essentially complete unzipping to monomer upon heating, however, the rate is substantially slower for poly(styrene). The copolymers were synthesized through either the poly(amic-acid) precursor, followed by chemical imidization to the corresponding polyimide, or the poly(amic alkyl ester) precursor followed by thermal imidization. The decomposition of the labile coblock was studied by thermogravimetric and dynamic mechanical analysis. Upon decomposition, the foams showed pore sizes in the nanometer regime along with the expected reduction in mass density.

High performance polymers play an important role as packaging materials in the manufacture of microelectronic devices and components. Material requirements generally include high thermal stability, low thermal expansion coefficient, minimal residual stress from thermal cycling and good mechanical properties. One key advantage of using polymers instead of inorganic materials is the lower dielectric constant which results in shorter machine cycle time and minimizes the distance between signal lines (1). As a class of materials, polyimides have best satisfied the requirements for these applications. To further reduce the dielectric properties of polyimides, perfluoroalkyl groups have been incorporated and, while this approach produces films with dielectric constants just below 3.0, it is often at the expense of both solvent

0097–6156/95/0614–0425$12.00/0
© 1995 American Chemical Society

resistance and mechanical properties. A new approach has been developed to further reduce the dielectric constant while maintaining the desired thermal and mechanical properties. This involves the generation of a nanometer scale foam or porous structure where the reduction in the dielectric constant is realized by replacing the polymer with air which has a dielectric constant of 1. The approach discussed herein uses microphase separated block copolymers where a thermally labile component is dispersed in a matrix (2−4) of a thermally stable matrix. Upon heating, the thermally unstable block undergoes thermolysis leaving pores with sizes and shapes commensurate with the initial copolymer morphology and, as such, are called nanofoams.

Nanofoams have been demonstrated using poly(propylene oxide) as the labile component with both polyimide and poly(phenylquinoxaline) as the matrix polymer. In each case, the volume fraction of voids generated in the foam was significantly less than the initial volume fraction of propylene oxide in the copolymer. The dominant decomposition products of poly(propylene oxide), acetaldehyde and acetone, were shown to swell the imide matrix, thus plasticizing the polymer. The glass transition temperature of the matrix is, therefore, reduced leading to a partial collapse of the foam structure. This results suggests that there is a limiting void size and volume fraction that can be produced using an imide matrix with poly(propylene oxide) as the thermally labile coblock.

To circumvent this limitation, monodispersed oligomers of poly(α-methylstyrene), poly(styrene) and copolymers of styrene and α-methylstyrene having controlled molecular weight and end group functionality were studied. Poly(α-methylstyrene) undergoes rapid thermal decomposition by depolymerization. The average number of monomer units generated per radical formed via initiation or intermolecular transfer defines the zip length (6,7). For poly(α-methylstyrene) the zip length is extremely high, ∼ 1200, with a nearly quantitative monomer yield (6,7). By comparison, poly(styrene) has a zip length of only 60. In each case, the decomposition temperature is sufficiently high to allow solvent removal and imidization of the matrix. In this article, the use of poly(styrene), poly(α-methylstyrene) and styrene/α-methylstyrene copolymers as labile coblocks to generated polyimide nanofoams will be reported.

Experimental

Materials. N-methyl-2-pyrrolidone (NMP), pyridine and acetic anhydride were purchased from Aldrich and used without further purification. The 9,9′-bis(4-aminophenyl) fluorene (FDA) and pyromellitic dianhydride (PMDA) (Chriskev Co.) were sublimed 3× prior to use. The α-methylstyrene and styrene were dried over CaH_2 at 25 °C, and distilled over fluoroenllithium before polymerization. Ethylene oxide was also dried over CaH_2 and distilled over n-butyllithium before use. The 1,1-bis(4-aminophenyl)-1-phenyl-2,2,2-trifluoroethane FDA and diethyl pyromellitate diacyl chloride were prepared according to a literature procedure (8,9). Hydroxyl terminated poly(styrene), poly(α-methylstyrene) and oligomers of styrene and α-methylstyrene were prepared anionically and subsequently derivatized to the amino-end group according to literature procedures. Poly(styrene) oligomers of controlled molecular weight, distribution and amino-end functionality were also prepared via free radical methods.

The copolymers were prepared via the poly(amic-acid) route followed by chemical imidization or through the poly(amic alkyl ester) route followed by chemical imidization.

Foam Formation. The copolymers were dissolved in NMP at a concentration of $9-15\%$ solids. Coatings 10 to 25 μm in thickness were obtained by doctor blading. The removal of the solvent and imidization, if required, was accomplished by heating the polymer films to 300 °C at 5 °C/min and maintaining them at 300 °C for 2 h in a nitrogen atmosphere. The films were then heated over a 4 h period to 340 °C and held at this temperature for 4.5 h to effect the decomposition of the styrenic coblock.

Characterization. Glass transition temperatures were measured on a DuPont 1090 instrument at a heating rate of 10 °C/min. A Polymer Laboratories' Dynamic Mechanical Thermal Analyzer (DMTA) operating at 10 Hz with a heating rate of 10 °C/min in the tension mode was used to measure the dynamic mechanical properties. Isothermal and variable temperature (5 °C/min heating rate) thermal gravimetric analysis (TGA) measurements were performed on a Perkin-Elmer model TGA-7 in a nitrogen atmosphere. Density measurements were obtained with a density gradient column composed of water and calcium nitrate. The column was calibrated against a set of beads of known densities at 25 °C. At least two specimens were used for each density measurement.

Fourier transform infrared, FTIR, measurements were made on an IBM Instruments IR44 with an MCT detector. Transmission microscopy analysis was performed on a Phillips 12 520 Instrument 100 kV. The samples for TEM were prepared by ultramicrotoming the films at room temperature.

Results and Discussion

The polyimide chosen for this study was derived from pyromellitic dianhydride and 1,1-bis(4-aminophenyl)-1-phenyl-2,2,2-trifluoroethane (3FDA/PMDA), since this was the polyimide used in the previous study, thus allowing direct comparisons (5). The 3FDA/PMDA polyimide has a T_g of 440 °C and a decomposition temperature of 500 °C. 3FDA/PMDA polyimide is also soluble in common organic solvents, so that significant synthetic flexibility is obtainable through thermal, chemical or solution imidization. The choice of processing route has previously been shown to have a major influence on the morphology of foams obtained from 3FDA/PMDA imide-α-methylstyrene triblock copolymers (5). Consequently, both chemically imidized copolymers and thermally imidized films cast from the poly(amic alkyl ester) precursor are considered.

Poly(α-methylstyrene) and poly(styrene) are ideally suited for use as thermally labile coblocks, since well-defined functional oligomers can be prepared and both depolymerize to monomer which can readily diffuse through the matrix. The TGA's of the polymers, shown in Figure 1, show rapid and quantitative thermal degradation. The decomposition temperatures are sufficiently high to allow conventional film processing and solvent removal while remaining well below the T_g of the polyimide. This defines a substantial processing window. Furthermore, the decomposition temperature is relatively independent of the atmosphere.

An anionic and a free radical approach were investigated to amino terminate styrene and α-methylstyrene oligomers. In the first case, hydroxyl terminated poly(styrene) and poly(α-methylstyrene) oligomers were prepared by the living anionic polymerization of styrene in THF, at $-78\,°C$ (Scheme 1). The polymerizations were initiated with sec-BuLi and end-capped with ethylene oxide. The monofunctional hydroxy terminated oligomers were converted into mono-amino derivatives (**1a-c**) using the aminophenyl carbonate in a manner similar to that previously reported for the derivatization of poly(propylene oxide) oligomers. An amino-functionalized polystyrene **1d** was also prepared by a novel "living" free-radical polymerization procedure using the appropriately functionalized AIBN initiator and 2,2,6,6-tetramethylpiperidinyloxy (TEMPO) as the "counter" radical (Scheme 2) (*10*). Comparison of molecular weights, determined by GPC and titration, confirmed the introduction of a single amine end group. In a similar fashion, an α-methylstyrene oligomer (**1a**) and three oligomers of styrene and α-methylstyrene oligomers were prepared with styrene to α-methylstyrene compositions of 4:1, 1:1 and 1:4. The characteristics of the samples are also shown in Table I.

Table I. Characteristic of Aromatic Amine Functional Thermally Labile Oligomers

Sample Entry	Thermally Labile Block Type	Polymerization Method	Molecular Weight, g/mol	T_g °C
1a	Poly(α-methylstyrene)	anionic	12,000	155
1b	Poly(styrene)	anionic	14,000	100
1c	Styrene/α-methylstyrene copolymer	(4/1) anionic	14,000	100
1d	Polystyrene	free radical	13,000	100

The 3FDA/PMDA-based copolymers were prepared using a poly(amic-acid) precursor followed by chemical imidization or using a poly(amic alkyl ester) followed by thermal imidization (Schemes 3 and 4, respectively). The synthesis of the poly(amic acid) involved the addition of solid PMDA to a solution of the styrene oligomer and 3FDA to yield the corresponding poly(amic acids). Chemical imidization of the poly(amic-acid) solutions was carried out in situ by reaction with excess acetic anhydride and pyridine for $6-8$ h at $100\,°C$. The synthesis of the poly(amic alkyl ester)-based copolymers involved the incremental addition of PMDA diethyl ester diacyl chloride in methylene chloride to a solution of the oligomer and 3FDA in NMP containing pyridine as the acid acceptor (Scheme 4). The meta isomer of PMDA diethyl ester diacyl chloride was used primarily due to its enhanced solubility, and to facilitate comparison with previous studies. The hydrolytically stable poly(amic alkyl ester) precursors can be isolated, characterized, and washed to remove possible homopolymer contamination prior to imidization. See Table II.

Table II. Characteristics of 3FDA/PMDA Polyimide-based Block Copolymers

Sample Entry	Polyimide Form	Thermally Labile Block Type (Synthetic Method)	Thermally Labile Block Composition, wt.%			Volume Fraction of Labile Block, %
			Charge	Incorporated		
				^{1}H NMR	TGA	
2	Chemically imidized	α-methylstyrene	25	–20	24	27
3	Poly(amic alkyl ester)	α-methylstyrene	25	—	24	27
4	Chemically imidized	Styrene (anionic)	15	—	15	18
5	Chemically imidized	Styrene/α-methylstyrene (anionic)	15	—	14	18
6	Poly(amic alkyl ester)	Styrene (anionic)	20	18	19	22
7	Poly(amic alkyl ester)	Styrene/α-methylstyrene (anionic)	20	15	14	18
8	Chemically imidized	Styrene (free radical)	15	—	9	11
9	Poly(amic alkyl ester)	Styrene (free radical)	20	19	15	18

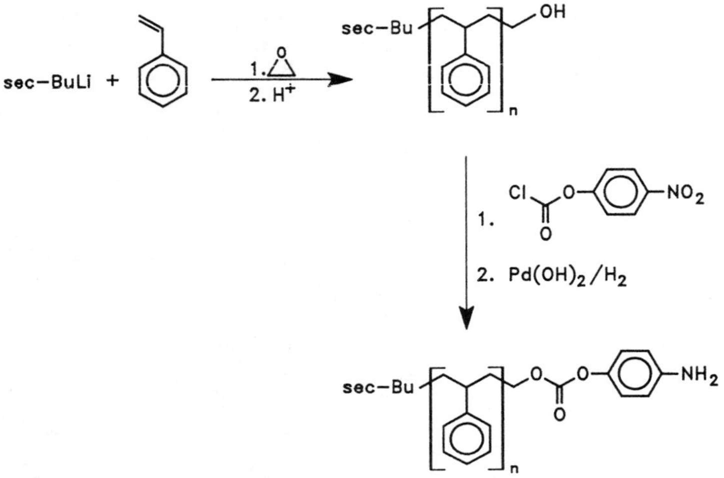

Scheme 1

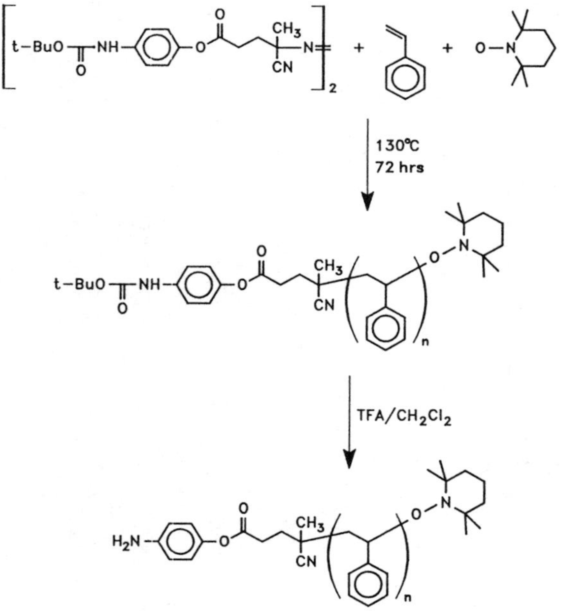

Scheme 2

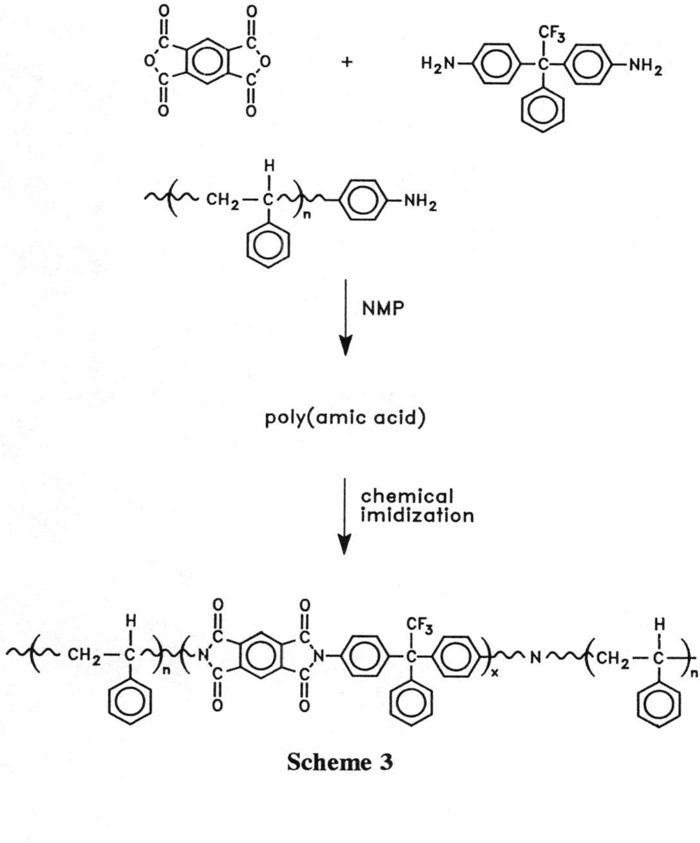

Scheme 3

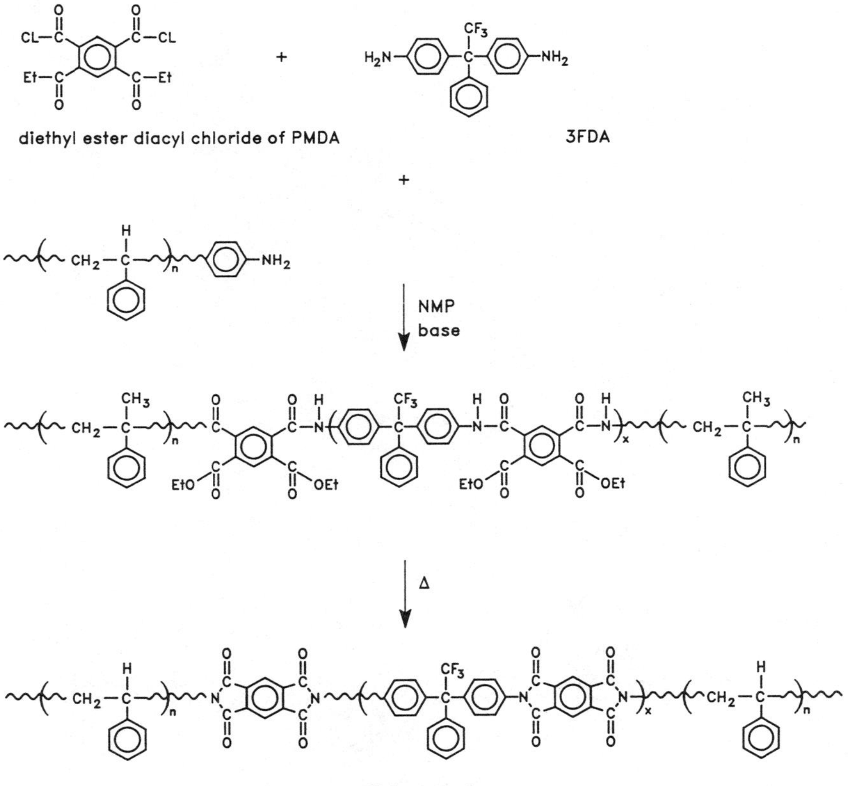

diethyl ester diacyl chloride of PMDA   3FDA

Scheme 4

The 3FDA/PMDA imide-based copolymers prepared are shown in Table III. Owing to the solubility of the 3FDA/PMDA polyimide, copolymers were prepared in both the fully imidized form (copolymers **2**, **4**, **5** and **8**) and as the poly(amic alkyl ester) precursor which was converted to the polyimide (copolymers **3**, **6**, **7** and **9**). The labile coblocks examined included poly(α-methylstyrene), poly(styrene), prepared by both anionic (oligomer **1b**) and free radical (oligomer **1d**) methods, and a (4:1) copolymer of styrene and α-methylstyrene (oligomer **1c**). The weight percentage of the labile blocks in the copolymers was intentionally maintained low (∼ 20 wt%) to produce discrete domains of the labile block dispersed in the polyimide matrix. At higher loadings, phase separation of the block could, in principle, produce cylindrical or more interconnected structures which are undesirable. The loading of the labile block in the copolymer agreed closely with that expected from the feed ratios.

Table III. Characteristic of 3FDA/PMDA Polyimide Foams

Copolymer Entry	Initial Labile Block Composition, Vol%	Density g cm^{-3}	Volume Fraction of Voids (Porosity) %
3FDA/PMDA polyimide (control)	—	1.35	—
2	27	1.13	16
3	27	—	30†
4	18	1.24	8.2
5	18	1.26	7.0
6	22	—	—
7	18	—	—
8	11	1.22	9.0
9	18	1.23	8.2

The cure temperature of the imide/α-methylstyrene copolymers was limited to 265 °C by the decomposition temperature of poly(α-methylstyrene) and, while this treatment is sufficient to remove the casting solvent, the degree of imidization was limited to ∼ 94%, as determined by ^{1}H NMR. The poly(styrene) and α-methylstyrene/styrene copolymers, on the other hand, show higher decomposition temperatures which permits cure temperatures of 300 °C, significantly higher than that of the poly(α-methylstyrene)-based copolymers. The cured films of each copolymer were transparent with no evidence of large scale phase separation, characteristic of homopolymer contamination. The dynamic mechanical spectra for selected copolymers are shown in Figure 2. It is essential that separation occur with high phase purity in

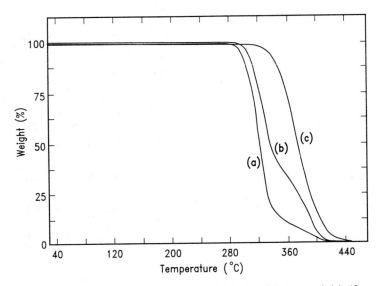

Figure 1. TGA thermograms of (a) **1a**, (b) **1c**, and (c) **1b**.

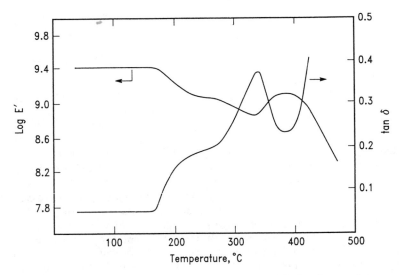

Figure 2. Dynamic mechanical spectra of copolymer **2**.

order to obtain a nanofoam while minimizing collapse. Two transitions indicative of a microphase separated morphology were observed in each case. The first transition occurs in the proximity of the styrenic based coblock, while the polyimide transition shows a strong dependence on the styrenic block type and composition. High labile block compositions (i.e., > 15%) exhibit a T_g in the proximity of the decomposition temperature due to plasticization of the matrix by the degradation products.

Previous studies have shown that the retention of the foam structure depends upon a balance between the rate of decomposition of the labile coblock, the solubility of the degradation products in the imide matrix and the rate at which these products diffuse out of the matrix. If the rate of decomposition is significantly higher than the diffusion coefficient, accumulation of the degradation products in the polyimide matrix occurs. This reduces the T_g of the matrix which leads to the development of a coarse void structure. In the case of the imide/α-methylstyrene copolymers, the α-methylstyrene block thermally degrades by depolymerization at extremely high rates, and, since the degradation products strongly interact with the polyimide matrix, plasticization of the matrix occurs, which, in turn, leads to pores significantly larger than expected. To overcome this blowing effect, attempts were made to decrease the depolymerization rate by the use of poly(styrene) and styrene/α-methylstyrene copolymers as the labile blocks. Furthermore, previous studies have shown that both the heating rate and residence time at final temperature are crucial variables in minimizing plasticization and optimizing the foam content. The optimum cure schedule for the styrene-based copolymers was found to be 300 °C for 2 h to effect imidization and solvent removal followed by a slow heating to 350 °C over a 10 h period and holding at temperature for 10 h. Isothermal gravimetric analysis shows quantitative degradation of the styrenic coblock.

The formation of the foam was assessed by density measurements. These data show that the porosity obtained is considerably less than the volume fraction of labile block incorporated in the copolymer, i.e., a partial collapse of the foam occurred (Table III). However, the interpretation of the porosity values measured by flotation in the density gradient column for the foamed polymers was difficult. In previous studies, the fluid was observed to penetrate the pores, giving a density similar to the homopolymer (5). Infra-red spectroscopy was also used to determine the void content in the foamed copolymers (8). However, due to the large size of the pores, the scattering of the IR beam precluded an accurate determination.

Transmission electron microscopy (TEM) was used to assess the size of the pores. The translucency or, in some cases, opacity suggest a pore size significantly larger than the microphase separated domains of the α-methylstyrene, styrene or styrene/α-methylstyrene labile coblock. Shown in Figure 3 are the TEM micrographs of typical microtomed sections of copolymers 3, 6 and 7, prepared from the anionically synthesized labile coblocks via the poly(amic alkyl ester) route. The dark regions of the micrographs correspond to the polyimide phase, while the light regions correspond to the pores. As can be seen, a highly porous structure was obtained with pores having cross-sections ranging from 20 to 200 nm for copolymer 3 and 40 to 60 nm for copolymers 6 and 7. These pores are several times larger than would be anticipated from the initial block copolymer morphology and, as expected, the films were somewhat translucent. However, it should be

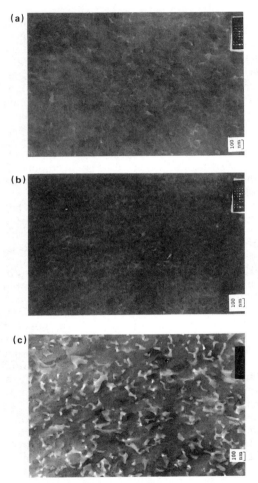

Figure 3. TEM micrographs of **3, 6, 7**.

pointed out that the styrene-based copolymers showed pores significantly smaller than those observed from the polyimide foams where α-methylstyrene was used as the labile coblock. Conversely, the foams prepared from copolymer **8**, synthesized with the free radically polymerized poly(styrene) via the poly(amic alkyl ester) route, showed an even larger distribution in pore size (~ 20 to 250 nm), thus giving rise to the observed opacity in the films (Figure 4).

Figure 4. TEM micrographs of copolymer **8**.

Summary

A route to high T_g thermally stable polyimide foams has been developed which uses microphase separated block copolymers comprising a thermally stable block and a thermally labile block, where thermally labile block is the dispersed phase. Several labile blocks were studied including poly(styrene), poly(α-methylstyrene) and several α-methylstyrene-styrene copolymers. The copolymers were synthesized through either the poly(amic acid) precursor, followed by chemical imidization to the polyimide foam, or the poly(amic alkyl ester) precursor followed by thermal imidization. The foam prepared from the anionically prepared labile coblocks showed pore sizes with diameters in the 100 nm region, along with a 7 to 12% reduction in density. The reduced decomposition rate afforded by the poly(styrene) minimized the plasticization and blowing observed in previous studies. Conversely, the copolymers prepared with the free radically polymerized poly(styrene) blocks showed markedly different behavior. In this case, a large distribution of pore sizes was observed, and the larger size is attributed to the lower decomposition temperature.

Literature Cited

1. Tummala, R. R.; Rymaszewski, E. J. *Microelectronics Packaging Handbook*; Van Nostrand Reinhold: New York, NY, 1989a; Chap. 1.
2. Hedrick, J. L.; Labadie, J. W.; Russell, T.; Wakharkar, V. *Polymer* **1993**, *34*, 122.
3. Charlier, Y.; Hedrick, J. L.; Russell, T. P.; Volksen, W. *Polymer* **1994**, accepted.
4. Jayaraman, S.; Srinivas, S.; Wilkes, G. L.; McGrath, J. E.; Hedrick, J. L.; Volksen, W.; Labadie, J. *Polymer* **1994**, submitted.
5. Charlier, Y.; Hedrick, J. L.; Russell, T. P.; DiPietro, R.; Jerome, R. *Polymer* **1994**, accepted.
6. Bywater, S.; Black, P. E. *J. Phys. Chem.* **1965**, *69*, 2967.
7. Sinaha, R.; Wall, L. A.; Bram, T. *J. Chem. Phys.* **1958**, *29*, 894.
8. Rogers, M. E.; Moy, T. M.; Kim, Y. S.; McGrath, J. E. *Mat. Plas. Soc. Symp.* **1992**, *13*, 264.
9. Volksen, W.; Yoon, D. Y.; Hedrick, J. L.; Hofer, D. *Mat. Res. Soc. Symp. Proc.* **1991**, *227*, 23.
10. Hawker, C. J. *J. Am. Chem. Soc.* **1994**, *116*, xxxx.
11. Hawker, C. J.; Hedrick, J. L. *Macromolecules* **1994**, submitted.
12. McNeil, I. C. *Comprehensive Polymer Chemistry*; Pergamon Press: New York, NY, 1989; Vol. 6, Chap. 15, p. 451.

RECEIVED August 31, 1995

Chapter 29

Internal Acetylene Unit as a Cross-Link Site for Polyimides

Tsutomu Takeichi and Masaaki Tanikawa

Materials Science, Toyohashi University of Technology, Toyohashi 441, Japan

Polyimides having internal acetylene units that are linked *para* to the aromatic rings were prepared using 4, 4'-diaminodiphenylacetylene (*p*-intA) as diamine monomer and pyromellitic dianhydride (PMDA), biphenyltetracarboxylic dianhydride (BPDA), and 2,2'-bis(3,4-dicarboxyphenyl)hexafluoropropane dianhydride (6FDA) as acid dianhydride. Crosslinking behavior of the polyimides having acetylene units were examined using DSC. Exotherm appeared for every polyimide containing *p*-intA units that is thermally imidized at 250°C, confirming that acetylene units that are linked *para* to the aromatic rings thermally crosslink. The onset of the exotherm depends on the structure of the polyimides : at 330°C for 6FDA/*p*-intA, at 379°C for BPDA/*p*-intA, and at 387°C for PMDA/*p*-intA. Crosslinking was performed by thermally treating at 400°C for 30 min. After crosslinking, glass transition temperature of the polyimides increased to above 400°C. Tensile strength of the polyimide films that consist of *p*-intA units was higher than that consisting of *m*-intA units. Especially, the rigid polyimides such as BPDA/*p*-intA and PMDA/*p*-intA had considerably high modulus. Cold-drawing of the films afforded still higher modulus for the rigid polyimides.

Polyimide is one of the most promising thermally stable polymer. Initial polyimide has been known for their excellent thermal and physical properties. The insoluble and infusible nature of the polyimide has been a major defect, however, making processability very difficult. Preparation of soluble or thermoplastic polyimide has been a major research interest. The processable polyimide, however, should have drawbacks in lower solvent resistance and lower thermal properties.

A promising approach to this problem is the use of crosslink site which reacts, on thermal cure at or after the processing, to become insoluble and highly thermally stable structures. The crosslink sites for polyimdes have been studied in relation to using polyimide as matrix resin of FRP. Nadic end-cap and acetylene end-cap are the representative crosslink sites well documented [1]. They have been applied for

0097–6156/95/0614–0439$12.00/0

high performance FRP. Most crosslink sites are end-cap type, however, and little has been studied for the crosslink sites that can be introduced into the polyimide backbone [2]. Versatile design of polymer structure is possible by using internal crosslink sites, which cannot be achieved by using the end-cap type crosslink sites.

We have been studying crosslinking behavior of internal acetylene units that are linked *meta* to the aromatic connecting units utilizing 3,3'-diaminodiphenylacetylene (*m*-intA). The *m*-intA was introduced into polyimide and polyamide backbone, and the effect of crosslinking on thermal and physical properties of the polymers have been studied. For example, *m*-intA was introduced into oligoimide, which gave crosslinked polyimide that have excellent thermal properties, such as high Tg and excellent thermal stability [3,4]. The oligoimide was shown to be suitable as a matrix resin for graphite fiber composite, which maintained physical properties at 316°C for 1200h [5]. The *m*-intA was also introduced into high molecular weight polyimide backbone to give reactive polyimide, in which the acetylene units crosslink to give polyimide having high performance at high temperature [6-8]. It must be emphasized that the control of crosslink density is easily done by changing the ratio of introduced acetylene monomer, thus enabling the control of thermal and physical properties. The crosslinkable polyimide enabled laminate processing of high performance polyimide/polyimide molecular composite films [9,10].

We report here the preparation of polyimides having acetylene units that are linked *para* to the aromatic rings utilizing 4,4'-diaminodiphenylacetylene (*p*-intA), and crosslinking behavior and properties of polyimides in comparison to the above mentioned polyimides having *meta* acetylenes (*m*-intA). The structures of monomers, *m*-intA and *p*-intA, are shown in Fig. 1. Preparation and thermal properties of the polyimide having *p*-intA are reported [11], but little is known so far about the crosslinking behavior.

Experimental.

Preparation of Polyimides.
Polyimides were prepared by the reaction of corresponding acid dianhydride and diamine in N-methyl-2-pyrrolidone (NMP), followed by thermal imidization. As acid dianhydrides, biphenyltetracarboxylic dianhydride (BPDA), pyromellitic dianhydride (PMDA), and 2,2'-bis(3,4-dicarboxyphenyl)hexafluoropropane dianhydride (6FDA) were employed. As diamines, 4,4'-diaminodiphenylacetylene (*p*-intA) [11] and 1,4-phenylenediamine (PDA) were employed.

As a representative example, preparation of BPDA/*p*-intA is shown below. Into a flask containing 2 ml of NMP were added *p*-intA (2.0 mmol, 0.4165 g) and then about 3 ml of NMP to wash the diamine monomer into the flask. Eqiumolar amount of BPDA (2.0 mmol, 0.5884 g) was added to the solution, and about 4 ml of NMP was then added so that total amount of NMP is 9.1 g to wash the dianhydride into the flask. The mixture was stirred at room temperature for an overnight. Polyamide acid film, obtained by casting the solution on glass plates followed by drying at 50°C in vacuo, was peeled off the glass plate and imidized thermally by heating at 100°C for 1h, 200°C for 1h, and then at 250°C for 1h. Some films were further heat treated for crosslinking at 300°C for 1h and then at 400°C for 30 min.

Preparation of Uniaxially Oriented Polyimide Films.
Polyamide acid films, typically 6 cm × 1 cm, were cold-drawn at the drawing rate of 0.1 mm/min, and then thermally imidized as fixed.

Measurements.
Viscosities of polyamide acids were measured at the concentration of 0.5 % in NMP at 30°C. Thermal analyses were performed with a Rigaku Thermal Analysis Station, TAS 100, equipped with a DSC cell base at a heating rate of 10°C/min under nitrogen atmosphere. Thermogravimetric analyses (TGA) were carried out with a SEIKO I TG/DTA 300, at a heating rate of 5°C/min under Ar. Dynamic viscoelastic measurements were performed using ORIENTEC Automatic Viscoelastometer RHEOVIBRON Model DDV-01FP at 35 Hz at the heating rate of 4°C/min in air. Tensile properties were investigated using a IMADA Seisaku-sho Model SV-3 at room temperature.

Results and Discussion.

Preparation of Polyimides Having *p*-IntA.
Three kinds of polyimides, BPDA/*p*-intA, PMDA/*p*-intA, and 6FDA/*p*-intA, were prepared. The structures are shown in Fig. 2. The reaction of *p*-intA with acid dianhydrides gave high molecular weight polyamide acids that have *para*-linked internal acetylene units in the backbone. The viscosities (η_{red}) of polyamide acids were 0.8 - 1.3 dl/g, 0.5 - 0.6 dl/g, and 0.9 - 1.1 dl/g, for BPDA/*p*-intA, PMDA/*p*-intA, and 6FDA/*p*-intA, respectively. Polyamide acids were cast on glass plates and imidized thermally. Polyimides utilizing PDA instead of *p*-intA were also prepared for comparison.

IR spectra of PMDA/*p*-intA thermally imidized at 250°C and 400°C are shown in Fig. 3. It can be seen from the spectra that the imidization is almost complete by the treatment at 250°C. It should be noted from the absorption at 1850 cm^{-1} in Fig. 3a that small amount of anhydride is present after 250°C treatment, suggesting the cleavage of amide acid into anhydride and amine groups [12]. Disappearance of anhydride after 400°C treatment is confirmed from Fig. 3b.

Crosslinking of *p*-IntA.
The polyimides containing *p*-intA units were analyzed by DSC to see the progress of crosslinking. In the case of BPDA/*p*-intA imidized at 250°C, onset of exotherm appeared at 379°C, reaching maximum at 462°C (Fig. 4). The amount of exotherm was calculated to be 11.7 cal/g. With higher heat treatment, the onset temperature of the exotherm became higher and the amount of the exotherm became smaller. The exotherm became unappreciable after 400°C treatment. In the case of rod-like polyimide, PMDA/*p*-intA, onset of the exotherm was as high as 387°C (Fig. 5). In the case of less rigid 6FDA/*p*-intA, onset of the exotherm appeared as low as 330°C (Fig. 6). Thus, the easiness of crosslinking depends on the chemical structure of the polyimides. In these cases also, exotherm became unappreciable after 400°C treatment.

It should be mentioned that the exotherm for the polyimides having *p*-intA units appeared at higher temperature than that for the polyimide having *m*-intA units (Table 1). The onset of the exotherm was 379°C with BPDA/*p*-intA, which is ca. 40°C higher than that of BPDA/*m*-intA. The same was true with PMDA/*p*-intA: Onset of the exotherm was 387°C and ca. 50°C higher than that of PMDA/*m*-intA. In the case of less rigid 6FDA/*p*-intA, however, onset of the exotherm appeared at 330°C, which is only 10°C higher than that of 6FDA/*m*-intA.

This shows that the crosslinking of *p*-intA needs higher energy than that of *m*-intA. The reason for the lower reactivity of *p*-intA than *m*-intA is considered to be primarily steric hindrance, suggested by our previous work on aromatic oligoamides

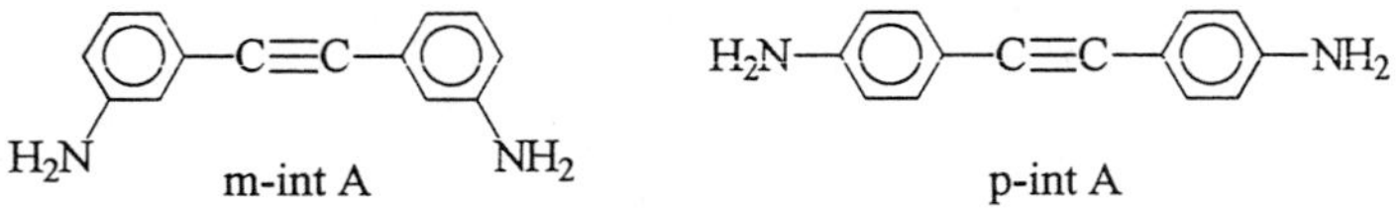

Figure 1. Structure of monomers.

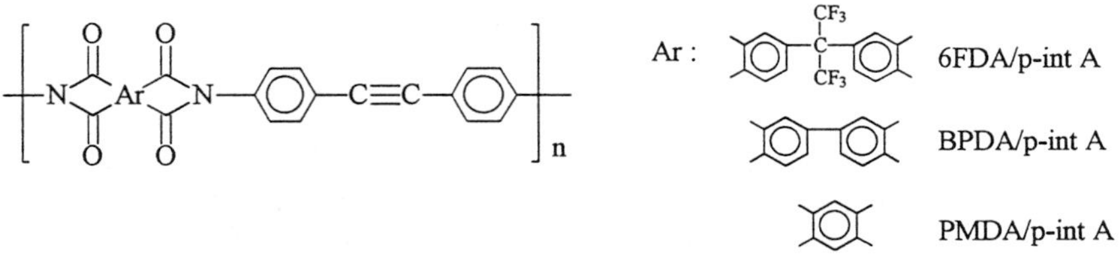

Figure 2. Structure of polymers.

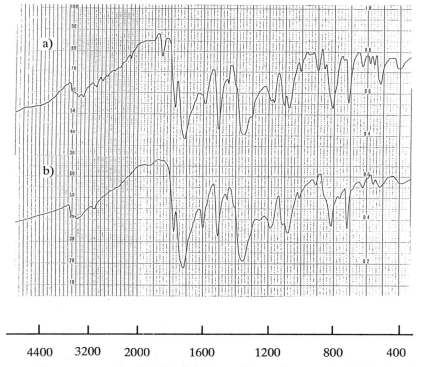

Figure 3. IR spectra of PMDA/p-intA.   a) Thermally treated at 250°C, b) thermally treated at 400°C.

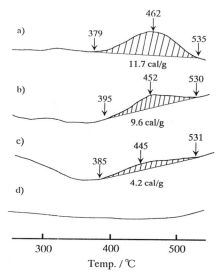

Figure 4. DSC of BPDA/p-intA. a) Thermally treated at 250°C, b) thermally treated at 300°C, c) thermally treated at 350°C, d) thermally treated at 400°C.

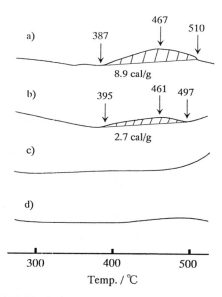

Figure 5. DSC of PMDA/p-intA. a) Thermally treated at 250°C, b) thermally treated at 300°C, c) thermally treated at 350°C, d) thermally treated at 400°C.

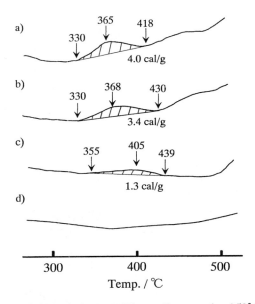

Figure 6. DSC of 6FDA/p-intA. a) Thermally treated at 250°C, b) thermally treated at 300°C, c) thermally treated at 350°C, d) thermally treated at 400°C.

Table 1. Onset of exotherm of polyimides containing reactive acetylene units on DSC.[a]

Exotherm	BPDA		PMDA		6FDA	
Onset	p-intA	m-intA	p-intA	m-intA	p-intA	m-intA
°C	379	340	387	341	330	321

a) Polyimides were thermally imidized at 250°C.

having *m*-intA, in which case exotherm of the oligoamides that have all *meta*-phenylene linkages appeared at lower temperature than that of oligoamides having *para*-phenylene linkages, and that melt-processing gave high quality films from oligoamides having all *meta*-phenylene linkages but only very brittle films from oligoamides having *para*-phenylene linkages [13]. Apparently the rigidity introduced by the *para*-phenylene linkages lowered the crosslinking reactivity of *m*-intA. However, electronic effect also should be considered as seen from the lower crosslinking temperature of oligoimides end-capped with *meta*-acetylene ($T_{exo,max}$ ca. 254°C) than the oligoimides end-capped with *para*-acetylene ($T_{exo,max}$ 310 - 320°C) [14].

Properties of Polyimides Having *p*-IntA.
The viscoelastic analysis showed glass transition temperature (Tg) at 290°C for BPDA/*p*-intA imidized at 250°C. After treating at 400°C, Tg increased to above 400°C, confirming the progress of crosslinking (Fig. 7). 6FDA/*p*-intA also showed Tg at 330°C after 250°C treatment, and above 400°C after 400°C treatment (Fig. 8). In the case of PMDA/*p*-intA, Tg was not observed below 400°C in both cases (Fig. 9).

Tensile measurements of the polyimides that have *para*-linked acetylene were performed. Uncrosslinked polyimides (treated at 250°C) and crosslinked polyimides (treated at 400°C) had almost the same modulus. Elongation became smaller after 400°C treatment, suggesting the progress of crosslinking.

It was observed that the polyimides that have *p*-intA showed higher modulus than the polyimides having *m*-intA (Fig. 10). For example, with BPDA as an acid dianhydride, modulus was 5.8 and 3.1 GPa for the polyimides with *p*-intA and *m*-intA. With PMDA, modulus was 5.2 and 3.8 GPa for the polyimides with *p*-intA and *m*-intA. With 6FDA, modulus was 2.9 and 2.5 GPa for the polyimides with *p*-intA and *m*-intA.

It is well known that the rigid polyimides become to have higher modulus by drawing the polyamide acid films followed by imidization [9,10,15]. Rigid polyimides are prepared from *p*-intA when combined with BPDA and PMDA, and higher modulus is expected. It was confirmed that rigid polyimide films, drawn at the stage of polyamide acid, gave higher modulus depending on the drawing ratio, as shown in Figure 11 in the case of BPDA/*p*-intA.

Conclusions.

1) Polyimides that have internal acetylene units linked *para* to the aromatic rings (*p*-intA) were prepared.
2) It was confirmed from DSC that *p*-intA are thermally crosslinkable, but crosslinking was less easy to occur than *m*-intA. The progress of crosslinking was also confirmed by the viscoelastic analyses in which, with the thermal treatment at 400°C, increase of Tg up to above 400°C was observed.
3) Rigid polyimides such as BPDA/*p*-intA and PMDA/*p*-intA have higher modulus than the *m*-intA counterparts. Further increase of modulus was achieved by cold-drawing the polyamide acid films followed by imidization.

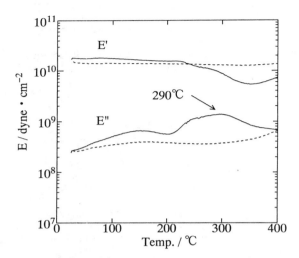

Figure 7. Viscoelastic analyses of BPDA/p-intA. Thermally treated at 250°C
(—————), thermally treated at 400°C (- - - - - -).

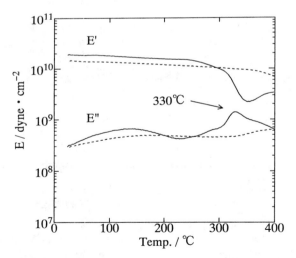

Figure 8. Viscoelastic analyses of 6FDA/p-intA. Thermally treated at 250°C
(—————), thermally treated at 400°C (- - - - - -).

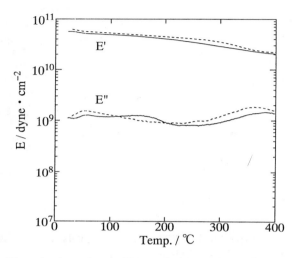

Figure 9. Viscoelastic analyses of PMDA/p-intA. Thermally treated at 250°C
(— — — — —), thermally treated at 400°C (- - - - - -).

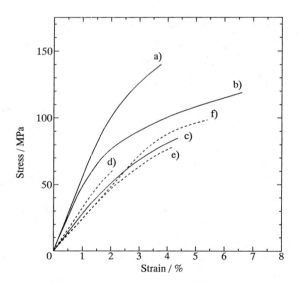

Figure 10. Stress-strain curves of various polyimide films thermally treated at
400°C. a) BPDA/p-intA, b) PMDA/p-intA, c) 6FDA/p-intA, d) BPDA/m-intA,
e) PMDA/m-intA, f) 6FDA/m-intA.

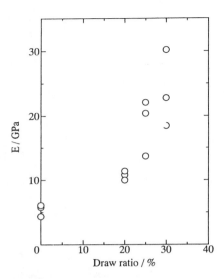

Figure 11. Tensile modulus of BPDA/p-intA thermally treated at 400°C vs. draw ratio of polyamide acid films.

References.

1. *Polyimides*; Wilson, D., Stenzenberger, H. D., Hergenrother, P. M., Eds.; Blackie: Glasgow, UK, 1990.
2. Reinhardt, B. A.; Arnold, F. E. *J. Appl. Polym. Sci.*, **1981**, *26*, 2679.
3. Takeichi, T.; Stille, J. K. *Macromolecules*, **1986**, *19*, 2103.
4. Takeichi, T.; Stille, J. K. *Macromolecules*, **1986**, *19*, 2108.
5. Stoessel, S.: Takeichi, T.: Stille, J. K.: Alston, W. B. *J. Appl. Polym. Sci.*, **1988**, *36*, 1847.
6. Takeichi, T.: Date, H.: Takayama, Y. *J. Polym. Sci.: Part A: Polym. Chem.*, **1990**, *28*, 1989.
7. Takeichi, T.: Date, H.: Takayama, Y. *J. Polym. Sci.: Part A: Polym. Chem.*, **1990**, *28*, 3377.
8. Takeichi, T.: Ogura, S.: Takayama, Y. *J. Polym. Sci.: Part A: Polym. Chem.*, **1994**, *32*, 579.
9. Takeichi, T.: Takahashi, N.: Yokota, R. *J. Polym. Sci.: Part A: Polym. Chem.*, **1994**, *32*, 167.
10. Takeichi, T.: Miyaguchi, N.: Yokota, R. *High Performance Polymers*, in press.
11. Inoue, K: Imai, Y. *J. Polym. Sci., Polym. Chem. Ed.*, **1976**, *14*, 1599 .
12. Young, P. R.; Davis, J. R. J.; Chang, A. C.; Richardson, J. N. *J. Polym. Sci.: Part A: Polym. Chem.*, **1990**, *28*, 3107.
13. Takeichi, T.: Kobayashi, A.: Takayama, Y. *J. Polym. Sci.: Part A: Polym. Chem.*, **1992**, *30*, 2645.
14. Takeichi, T.: Stille, J. K. *Macromolecules*, **1986**, *19*, 2093.
15. Yokota, R.: Horiuchi, R.: Kochi, M.: Soma, H.: Mita, I. *J. Polym. Sci.,: Part C: Polym. Lett.*, **1988**, *26*, 215.

RECEIVED August 8, 1995

Chapter 30

Vapor-Depositable Polymers with Low Dielectric Constants

J. A. Moore[1], Chi-I Lang[1], T.-M. Lu[2], and G.-R. Yang[2]

[1]Department of Chemistry, Polymer Science and Engineering Program and [2]Center for Integrated Electronics and Department of Physics, Rensselaer Polytechnic Institute, Troy, NY 12180–3590

Chemical vapor deposition (CVD) has been one of the preferred processes for generating thin films of inorganic insulators in the fabrication of microelectronic devices. Currently, this methodology is being studied to generate organic thin films for use as dielectrics. Interaction between synthesis of the potential organic precursors for vapor deposition polymerization and design of the deposition apparatus has become a very active and demanding area in our research effort. Several thin polymer films, such as Parylene-N, Parylene-F (fluorinated), Teflon, Teflon AF (amorphous), poly(naphthalene) (PNT-N), poly(fluorinated naphthalene) (PNT-F) and poly(bis-benzocyclobutene-F$_8$) (fluorinated), which have very low dielectric constants(<2.7) have been deposited in different vapor deposition apparati.

There are three main components in the fabrication of computer chips: the integration of devices, the wiring(or the interconnects), and packaging. Research has been very active in all three areas to provide higher performance (speed) and density. Active research areas include the scaling of devices, the search for novel materials, and processing technologies for interconnects and packaging. Because speed is now largely limited by packaging technology, rather than the state-of-the-art chip construction, the bottleneck to computer speed improvement resides in further developments of packaging schemes and materials. The development of new insulators for interconnects and packaging is one approach to increase the speed of pulse propagation. As the operating frequencies of electronic devices enter the gigahertz range and as the dimensions of electronic devices approach the submicron level, dielectric media with low dielectric constants(<3) become increasingly important for the reduction of signal coupling between transmission lines. In addition, because larger silicon wafers (≥ 8" with the advent of 12" wafers looming on the horizon) are being used (currently, the technical difficulty of controlling uniformity for larger wafers has been a challenging area for the spin coating of polymer dielectrics), the ability to deposit polymers with appropriate properties as conformal films from the gas phase would be a useful advance yielding high purity and uniform films on larger wafers.

0097–6156/95/0614–0449$12.50/0

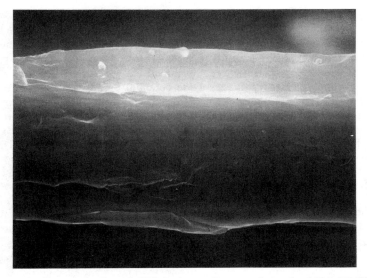

Figure 1. Cross-sectional SEM view of an as-deposited PA-N film.

For a material to be suitable as an interconnect dielectric, it should have a low dielectric constant and be able to withstand temperatures higher than 500°C. Other important considerations are compatibility with other materials, long term thermal, chemical and electro-chemical stability, ease of fabrication and low cost. SiO_2, which has a dielectric constant of 3.5-4.0, has been used as the interlayer dielectric material in the industry. Further breakthroughs in high performance chips hinge critically on the development of new insulators with dielectric constants much lower than that of SiO_2. It is generally believed that to achieve such a low dielectric constant one must consider organic polymeric materials instead of the traditional inorganic materials. Although there are several different types of potential organic polymers which posses low dielectric constants, they are excluded from microelectronic application because many polymers fall short either in their lack of chemical and/or thermal stability, or are difficult to process. As a result, to date, much research has been focused on spin-on materials such as the polyimide family. Polyimides, through a two-step process, solution coating of the polyimide precursor poly(amic acid) and subsequent curing (*1*) can be deposited as thin films formed by spin coating and, therefore, are suitable for microelectronic applications. Most of the members of the polyimide family have a dielectric constant around 3.5-3.6. These values are a function of film thickness and are highly dependent on cure conditions and moisture content (*2*). Polyimides also require toxic solvents, and extra efforts must be implemented to take care of the hazardous waste and impurities.

New dielectric materials must not only possess proper electrical properties but must also minimize or eliminate the use of solvents. Because of this need, we have initiated our research in two different ways, synthesis of potential precursors and design of vapor deposition processes. To prepare pure, more uniform and less water absorptive thin polymer films for use as dielectrics, several new vapor depositable precursors and novel thin film vapor deposition processes as well, have been developed in our laboratory. Polymers such as Parylene-N, Parylene-F (fluorinated), Teflon, Teflon AF 1600 (amorphous fluoropolymer), poly(naphthalene) (PNT-N), poly (fluorinated naphthalene) (PNT-F) and poly(bis-benzocyclobut- ene-F_8) (fluorinated), which have extremely low dielectric constants (below 2.7) in thin film form, have been synthesized and deposited. Because of the extremely low dielectric constants and high temperature stability of these vapor depositable materials, they may become very attractive for future electronic applications as interlayer dielectrics.

Chemical vapor deposition (CVD) is one approach which provides a clean, environmentally safe process to deposit films. Solvent is not involved in these processes and the film obtained as the final product therefore does not have to be soluble. CVD also simplifies the fabrication process. The central theme of this methodology involves thermal activation of a volatile precursor without eliminating any gaseous fragments. These reactive intermediates are condensed onto wafers in a vacuum chamber and little or no chemical waste is produced. Further- more, cold traps can be used to collect and reuse any uncondensed monomers.

We are able to monitor the thickness, deposition rate and uniformity of the films by controlling the reaction parameters of the chemical vapor deposition process. As an example **Figure 1** shown a scanning electron micrograph (SEM) of Parylene N film. The cross-sectional view shows the uniformity of this 10 μm thickness of PA-N as deposited from the vapor.

The chemical nature and chemical reaction mechanisms of the precursors are two very significant factors which control the vapor deposition process and subsequently affect the reactor design. Additionally, the potential precursors should satisfy several requirements: First, they should have adequate vapor pressure so that they are able to sublime slowly if they are solid, or they should evaporate gently without decomposition at the operational pressure if they are liquid. The other major concern is that no gaseous by-products are liberated when pyrolyzing the precursors in the high temperature chamber because this occurrence results in pressure fluctuations, which, in turn, complicate the deposition process.

Design Of The Vapor Deposition Process

To deposit the activated polymer precursors we synthesized or obtained from suppliers, three different types of vapor deposition processes have been adapted in our lab and they have been used successfully to deposit these polymers as films on the substrates:

Hot Wall Vapor Deposition

This process is one in which monomer is activated in a hot chamber and polymerized in another cold chamber. For instance, Parylene F(PA-F) as shown in **Scheme 2**, made from dimer or a liquid precursor, has been synthesized in this type of apparatus. The vapor of the source material was introduced into the pyrolysis chamber where the monomer was activated. The reactive species was transported to cooled substrates (-10°C to -30°C), where the polymer was deposited as a tough, transpar- ent film. The pyrolysis temperature at which quantitative conversion of source material to intermediate occurs depends on the temperature and contact time in the pyrolysis zone. The source material could be activated by heat alone, or with catalysts, and subsequently flow into a low temperature chamber containing a cold substrate to deposit as film. For Parylene F (PA-F), we need a temperature of 720°C to crack the Parylene F dimer to generate the reactive intermediate but the required temperature for the brominated precursor (dibromotetrafluoro-p-xylylene) with zinc is only 350 - 400°C.

Cold Wall Vapor Deposition

This process is one in which the only hot zone is where the polymerization process take place. Poly(naphthalene) (PNT) analogs as shown in **Scheme 4**, have been deposited by the cold wall vapor deposition technique. Instead of bond dissociation, a chemical bond rearrangement occurs at the high temperature of the substrate surface which is the only high temperature zone in this process. In our lab, this design required lower temperature than the hot wall apparatus we have been using. The rest of the reactor was maintained at room temperature or lower to prevent unnecessary polymerization. The unreacted source material can be collected in a cold trap and reintroduced into the system. This advantage can enable the growth of thicker films without interrupting the deposition process. The range of potential precursors which can be used in this approach is limited by the necessity of obtaining high conversion to active intermediate on the substrate surface.

UV-assisted Vapor Deposition (3)

In this approach a monomer is activated in the gas phase only by ultraviolet radiation and subsequently polymerizes on the substrate. As mentioned earlier, the vapor deposition process is very dependent on the chemical properties of the precursors. Bis-benzocyclobutene-F_8 (shown in **Scheme 7**) was designed for use in a hot wall vapor deposition system, but did not polymerize efficiently within a convenient temperature range. However we were able to activate the source material, bis-benzocyclobutene-F_8 by UV irradiation. This discovery led us to investigate the UV-assisted vapor deposition process (vide infra).

Vapor Depositable Dielectric Materials

I. Parylene-N

Poly(p-xylylene) was discovered by Szwarc (4) in 1957 and then commercialized by Gorham (5) (**Scheme 1**). Gorham has reported that di-p-xylylene is quantitatively cleaved by vacuum vapor-phase pyrolysis at 600°C to form two molecules of the reactive intermediate, p-xylylene, which subsequently polymerize on the cold substrate. In a system maintained at less than 1 Torr, p-xylylene spontaneously polymerizes on surfaces below 30 °C to form high molecular weight, linear poly-p-xylylene. The facile cleavage of di-p-xylylenes to p-xylylene is, in all probability, due to the high degree of steric strain in the dimeric species, and to the comparatively stable nature of p-xylylene. Several different poly-p-xylylene derivatives, such as chlorinated, brominated or acetyl poly-p-xylylene have also been investigated by Gorham and Yeh (6).

A schematic drawing of the apparatus used for the preparation of poly(p-xylylene) and poly(tetrafluoro-p-xylylene) in our research is shown in **Figure 2**. The unit consists basically of four sections: a source (dibromotetrafluoro-p-xylene) vessel with a needle valve, xylylene vapor generator, a deposition chamber with substrate holder and a pumping system. Cooling of the activated species can be accomplished with internal or external condensers, cooling coils, tubes or the like immediately after the pyrolysis zone. Another cold trap is placed right before the system outlet to recover unreacted precursors or dimer.

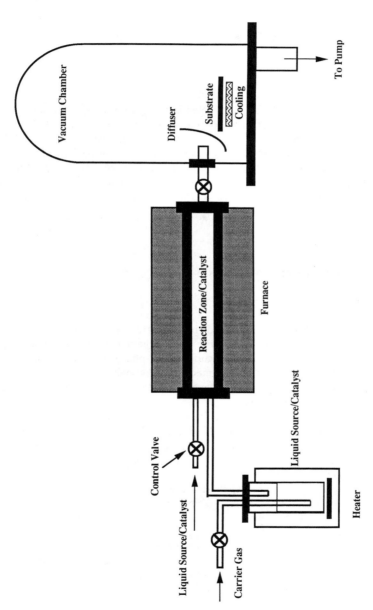

Figure 2.   The current New Method CVD system for PA-F films by deposition from a precursor.

Scheme 1

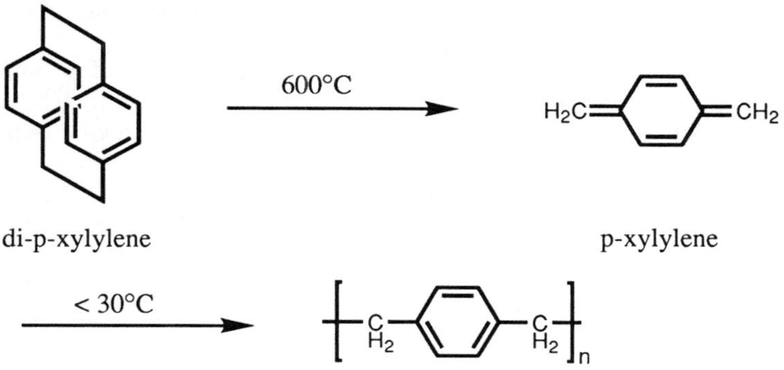

di-p-xylylene

p-xylylene

poly-p-xylylene
(Parylene)

Parylene has a variety of useful properties (**7**) such as low dielectric constant and conformal deposition (planarization). The major drawback in this application exhibited by poly(paraxylylene) is that it reverts to monomer when thin films are heated above ~ 400 °C and it cracks when the films are annealed at 300-350 °C in nitrogen. As an example, shown in **Figure 3** is an optical micrograph(150x) of a Parylene N film showing surface cracking of the PA-N film.

During module assembly the chip joining (soldering) process causes short exposure to temperatures between 300-450 °C and therefore precludes use of this material under these conditions. It is well known that replacing a C-H bond with a C-F bond not only enhances the thermal stability of the resulting polymer, but also reduces the dielectric constant. Joesten (**8**) has reported that the decomposition temperature of poly(tetrafluoroparaxylylene) (PA-F) is ~530 °C. Thus, it seems this fluorinated analog would satisfy many of the exacting requirements for utility as an on-chip dielectric medium.

II. Parylene-F

Hertler (**9**) was the first to report the preparation of poly(tetrafluoro-p-xylylene) by a multi-step synthesis. Pyrolysis (330 °C, 0.025 Torr) of dibromotetrafluoro-p-xylene (Br$_2$F$_4$C$_8$H$_4$) over zinc led to deposition of the polymer film in a cold trap. Fuqua and coworkers (**10**) developed a much shorter route from α,α,α',α'-tetrafluoro-p-xylene but were unsuccessful in generating polymer because they conducted their pyrolyses at 820-925 °C. Chow and co-workers (**11**) developed a unique multi-step synthesis for the commercial production of tetrafluoroparylene which used octafluoro[2.2]paracyclophane (PA-F dimer) as the precursor to polymer. PA-F dimer was cracked at 720-730°C and polymer was deposited on a substrate at -25 to -35°C. Chow (12) also attempted to pyrolyze $Br_2F_4C_8H_4$ at very high temperatures. The film which was deposited was of poor quality compared to that prepared from dimer.

We sought to adapt these approaches to the generation of polymer film directly from the vapor state in a conventional vacuum system in a manner compatible with the deposition of metal on silicon wafers. Cracking dimer does provide a clean and efficient process to deposit film but the main hindrance to produce PA-F thin film in large quantity by this method is the cost of preparation and the short supply of PA-F dimer, which is currently commercially unavailable (recently a new synthetic route to octafluoro[2.2]paracyclophane has been reported by Dolbier (13)). Therefore, this situation provided the impetus to develop an alternative process to make PA-F film.

Precursor Method

When vapors of $Br_2F_4C_8H_4$ were led through a bed of mossy zinc at 350-400°C and then allowed to pass over a silicon wafer (-15°C ; 0.2 - 0.3 Torr), films ranging in thickness from 2000 - 8000 Å were deposited. Thicker film could be deposited by using more source material or a longer reaction path. FT-IR spectroscopy indicates that the films, as deposited, are essentially identical to that prepared by the conventional route. The average dielectric constant of the films deposited by our approach was 2.6 ± 0.1, while the value measured on films generated by cracking the dimer was 2.38 ± 0.5. X-ray Photoelectron Spectroscopy (XPS) measurements of the as-deposited film revealed the presence of O (5.86-9.00 at.-%), Zn (4.03 at.-%) and Br (5.4 at.-%) (14) contaminants. Annealing the films in nitrogen at 530 °C for 30 min. removed the Br and reduced the O and Zn levels to 3.63% and 3.23%, respectively. Further improvements in the design of the apparatus and optimization of the deposition conditions are expected to remove the remaining discrepancies. To simplify the synthetic effort required to deposit such films we attempted to deposit films by pyrolyzing tetrafluoro-p-xylene. Under similar reaction conditions, a polymer film is deposited which is different from poly(tetrafluoro-p-xylylene) because the FT-IR spectrum indicates that it contains more hydrogen and less fluorine. Presumably HF is preferentially eliminated rather than H_2.

New Method

As an extension of this approach, we have found that the commercially available material, $(C_8H_4F_6)$ 1,4-bis(trifluoro- methyl)benzene, in conjunction with a catalyst/initiator, can be used to make PA-F film (15). We examined the co-pyrolysis of 1 wt-% dibromotetrafluoro-p-xylylene with commercially avail- able hexafluoro-p-xylene (from Aldrich) and were gratified to find that it was indeed possible to prepare films which were spectroscopically indistinguishable from those deposited from dimer. The films obtained are of excellent quality with a dielectric constant of 2.2-2.3 at 1 MHz and have a dissociation temperature of 530°C in N_2.

A uniformity of better than 10% can be routinely achieved with a 0.5 μm thick film on a 5" silicon wafer with no measurable impurities as determined by XPS. During a typical deposition run, the precursor source was maintained at 50°C, the reaction zone (a ceramic tube packed with Cu or Ni) was kept at 375-550°C and the substrate was cooled to -10°C to -20°C. The film deposited had an atomic composition, C : F : O = 66 : 33: 1 ± 3 as determined by XPS. Except for O, no other impurities were detected. Within instrumental error, the film is stoichiometric Parylene-F which has a theoretical composition of C : F = 2 : 1. The major concern here is the catalysts we used in the system. They became deactivated after the surface of the metal had been in contact with the precursor and the film growing process was interrupted. The synthetic route is shown in **Scheme 2**.

Scheme 2

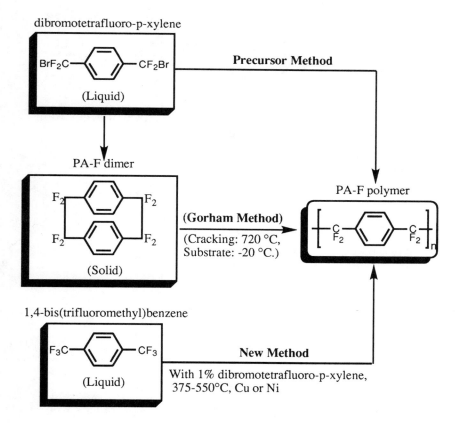

dibromotetrafluoro-p-xylene

BrF_2C—⟨benzene⟩—CF_2Br

(Liquid)

Precursor Method

PA-F dimer

F_2 ... F_2 / F_2 ... F_2

(Solid)

(Gorham Method)
(Cracking: 720 °C,
Substrate: -20 °C.)

PA-F polymer

$\left[C_{F_2} - \text{⟨benzene⟩} - C_{F_2} \right]_n$

1,4-bis(trifluoromethyl)benzene

F_3C—⟨benzene⟩—CF_3

(Liquid)

New Method
With 1% dibromotetrafluoro-p-xylene,
375-550°C, Cu or Ni

An interesting result was observed during the annealing process of Parylene-F films from cracking dimer and from the New Method. Some (30%) of the films deposited from dimer develop cracks when the films are annealed at 510 °C in

Figure 3. Optical micrograph of the cracked surface of PA-N film after annealing in Nitrogen.

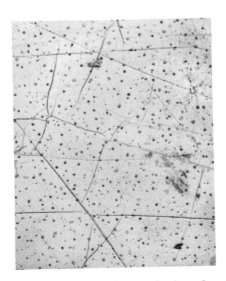

Figure 4. Optical micrograph of the cracked surface of PA-F film deposited from dimer after annealing in Nitrogen.

nitrogen but we do not observe any cracks in the films deposited by the New Method. **Figure 4** and **Figure 5** show optical micrographs (150x) of this observation.

III. Teflon and Teflon AF (1600)

We also initiated a study of the deposition of simpler fluoropolymers (16). It has been reported (17) that Teflon can be pyrolyzed to gaseous fragments and redeposited as a coating. Teflon has for many years been the polymer recognized as having the lowest dielectric constant, high volume resistivity and good mechanical strength. However problems with adhesion, creep, processibility and thermal stability have hindered its use in microelectronics applications. Recently, the DuPont corporation has produced a family of materials called Teflon AF (as shown in **Scheme 3**), which are amorphous fluoro-copolymers composed of varying amount of tetrafluoroethylene and 2,2-bis(trifluoromethyl)-4,5-difluoro-1,3-dioxole (18). These new materials possess some structural and physical properties which are superior to regular Teflon (19). They have an even lower dielectric constant and refractive index, are chemically inert and are amorphous (20). The properties for which improvement is claimed include transparency, no creep and it is also reasonable to expect adhesion to other materials would also be enhanced. The dielectric constant of Teflon AF (1600) (containing 33% tetrafluoroethylene) is reported to be 1.93 which is less than Teflon, and it is the lowest of any known material with reasonable mechanical strength.

Scheme 3

TEFLON

TEFLON AF

Thin film Teflon AF is usually prepared by spin coating, a technique which uses expensive and potentially toxic perfluorinated solvent, such as Fluorinert(FC-75). The use of organic solvents in the preparation of the film tends to incorporate them. as well as water and other impurities. Although the deposition process for Teflon AF (1600) is similar to the hot wall vapor deposition process, the substrate temperatures can be varied from -70°C to 250°C, a temperature close to the application temperature of a cold wall reactor. In this discussion , we still consider this process to be hot wall deposition.

A two step vapor deposition process has been developed to produce smooth, uniform, and pin-hole free Teflon AF (1600) films from 0.1 μm to 10 μm in

Figure 5. Optical micrograph of the surface of PA-F film deposited
by the New Method after annealing in Nitrogen.

thickness. First, heating Teflon AF (1600) in a graphite crucible with a nozzle of 2-3 mm diameter at 300 °C for 5 minutes enables the source material to melt completely and build up a quasi-equilibrium state. In the same crucible, the source material is then continuously heated to about 400 °C with a base pressure of 3×10^{-6} Torr. Silicon wafers are used as the substrates and their temperatures are controlled from -70 °C to 250 °C. Deposition takes place at a rate of 0.5-3 Å/sec. The thickness of these films is 0.1 μm to 10 μm. However, pinholes were observed at deposition rates > 10 Å/s. Spectroscopic analysis (XPS) indicates the material suffered little change in overall composition upon evaporation and redeposition and there was no change observed in the infrared spectrum between Teflon AF (1600) before and after deposition.

In comparison to the conventional one step deposition process, two step deposition has been shown to give an even smoother surface, and more uniform films. It has also been shown that Teflon AF (1600) thin films deposited by the two step deposition process are much smoother than laser-ablated thin films (21). The key difference between the conventional one step deposition and the two step deposition processes is the annealing period at 300°C, which melts the source material completely with no large solid particles remaining during the deposition process. We believe the rough surface we observed from the one step deposition process is caused by the presence large solid particles. The powder-free and completely pyrolyzed Teflon AF (1600) gives better and smoother films. The vapor-deposited Teflon AF (1600) thin films possess a reasonably high breakdown field strength (6×10^5 V/cm). However, this value is lower than that of spin-coated thin films (1.5×10^6 V/cm). We believe that the lower breakdown field strengths of vapor deposited thin films may be primarily attributed to the lack of complete polymerization. In the vapor deposition process, Teflon AF (1600) is first pyrolyzed into a vapor, and subsequently repolymerized on the substrate (22). As is well-known, the repolymerization process strongly depends on the deposition conditions and the resulting polymer usually has a shorter chain length than the bulk material. The fact that the vapor deposited Teflon AF (1600) thin films show a lower mechanical strength than spin-coated thin films supports this argument.

Although Teflon and Teflon AF (1600) possess an outstanding combination of physical, electrical, and chemical properties, the thermal stability of these polymers is still a major concern for microelectronic application. The reported thermal stability of Teflon AF (1600) is only 360 °C in air.

IV. Poly(naphthalene) and Poly(fluorinated naphthalene)

Fundamental considerations of the parameters which must be dealt with in designing a thermally stable, low dielectric constant polymer naturally lead one to consider, aromatic rigid rod polymers containing no "lossy" functional groups. A structure such as poly(naphthalene) (**Scheme 4**) is a likely candidate. The formation of C-C bonds between aromatics rings is an important step in many organic syntheses and may be accomplished by chemical, photochemical, or electrochemical means. Poly(naphthalene) is chemically similar to poly(p-phenylene), which is insoluble, infusible, resistant to oxidation and radiation degradation. The successful synthesis of high molecular weight samples of these rigid-rod macromolecules has been an ongoing problem. Via conventional synthetic methods the poor solubility of these aromatic polymers result- ed in low molecular weight and poor processability, attributes which excluded them from thin film applications.

Bergman's study of the thermal cycloaromatization of enediynes led to the suggestion of a benzene 1,4-radical intermediate (23,24). Tour and coworkers theorized that 1,4-naphthalene diradicals generated in solution might couple to eventually form a polymer (25). Even though large excesses of radical terminators were employed in their work, polymeriza- tion was still the predominant process .

The solubility and processability problems for these types of rigid-rod polymers should be surmountable using chemical vapor deposition from appropriate precursors which can be activated to form intermediates which polymerize readily.

The requisite monomers, 1,2-diethynylbenzene or 1,2-di-ethynyltetrafluorobenzene, are thermodynamically labile, espe- cially when the polymerization is carried out in a sealed tube. They are, however, easily and safely deposited in the vapor deposition apparatus. The polymerization process is outlined in **Scheme 4**. The vapor deposition polymerization was followed by following the pressure changes accompanying the beginning of the reaction. The vapors of the monomers caused an increase in pressure which returned to the initial base vacuum of the system when the reaction reached completion.

Scheme 4

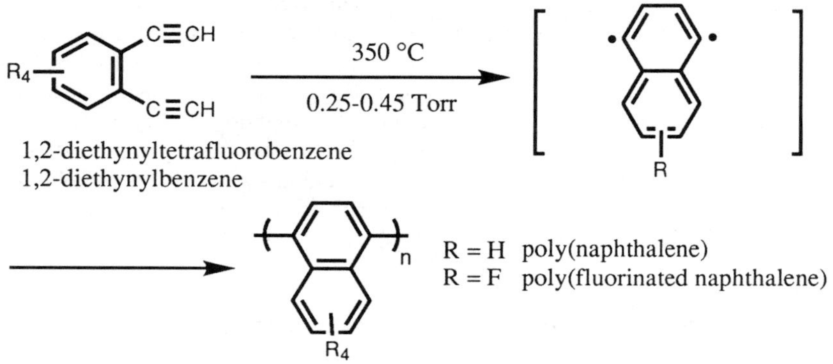

The general syntheses of these monomers are described in a current publication by Professor Tour (24). Polynaphthalene (PNT-N) and poly(fluorinated naphthalene) (PNT-F) were synthesized by vaporizing 1,2-diethynylbenzene or 1,2-diethynyl tetrafluorobenzene in vacuo and condensing the vapor on a hot surface at 350 °C. A schematic of the apparatus used for the preparation of PNT-N and PNT-F in this work is shown in **Figure 6**. The unit consists, basically, of four sections: a source (1,2-diethynylbenzene) vessel with a needle valve, a vapor introduc- tion channel, a high temperature area with substrate holder, a recycling trap and a pumping system. No catalyst or solvent was involved in this process and unreacted monomer could be recovered from the cold trap. The best film was obtained when oxygen was rigorously excluded from the system. Film which was deposited in an oxygen rich ambient exhibited lower thermal stability and poor adhesion. These films have low dielectric constants of 2.2 to 2.5. In comparison to PA-F films made from dimer, PNT-N and PNT-F films have higher dissociation temperatures (>570 °C), better thermal stability (>530 °C), and no film cracking until annealed at 600 °C in nitrogen. The presence of an inert vaporous diluent in the pyrolysis process is not preferred because the films deposited with argon as a carrier gas do not adhere well to silicon surfaces. This puzzling result is different from the vapor deposition of PA-F film, in which inert, vaporous diluents such as nitrogen, argon, carbon dioxide, steam and the like can be employed to change the total effective pressure in the system (26).

The films, generated by vapor deposition on hot surfaces such as glass or silicon, are not soluble in common laboratory solvents. A difference in reactivity between 1,2-diethynyl- benzene and 1,2-diethynyltetrafluorobenzene was observed in the polymerization process. These polymers (PNT-N and PNT-F) have been synthesized by both chemical vapor deposition and solution polymerization and the reaction rate for PNT-F is faster than PNT-N. This observation might be an important factor when high deposition rates are required.

V. Poly(fluorinated benzocyclobutene)

The Parylene family has very attractive properties for dielectric use, as was mentioned above. When other types of monomers as possible precursors to be used in this approach were considered, the isomeric o-xylylene structure (**Scheme 5**) came to mind as a candidate. This pathway involves the thermolysis of benzocyclobutene derivatives to generate a reactive o-dieneoid intermediate which could then undergo a Diels-Alder reaction of two quinodimethane molecules, one as a diene and one as a dienophile, to yield an intermediate spirodimer. The spirodimer would then fragment to give a benzylic diradical species which may undergo intramolecular coupling to give dibenzocycloocta-1,5-diene or oligomerize to provide poly(o-xylylene). If two of these benzocyclobutene units were joined together they should undergo polymerization upon heating to yield an insoluble, crosslinked system.

In solution, benzocyclobutenes undergo thermally induced ring-opening to produce o-quinodimethane (o-xylylene) inter- mediates which then undergo cycloaddition or dimerization reactions (27). These transformations have been used in solution or melt polymerization reactions where the monomers contain one or more benzocyclobutene groups per molecule (28). Tan and co-worker have reported a thermosetting system based on the Diels-Alder cycloaddition between the terminal benzocyclobut- ene units and alkyne groups (29).

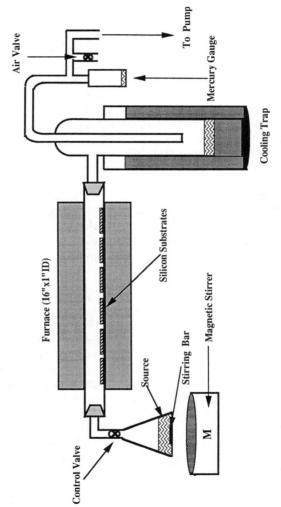

Figure 6. The current CVD system for deposition of PNT-N and PNT-F films.

Polymers prepared from some multifunctional benzocyclobutene monomers exhibit a combination of low dielectric constant and low dissipation factor, slight moisture sensitivity, and good thermal stability, leading to the use of these materials in microelectronic applications (30,31).

Under appropriate thermal conditions, the strained four-membered ring of benzocyclobutene undergoes electrocyclic ring-opening. The temperature at which such a concerted process occurs depends primarily on the substituents of the alicyclic, rather than the aromatic portion of the molecule.

Scheme 5

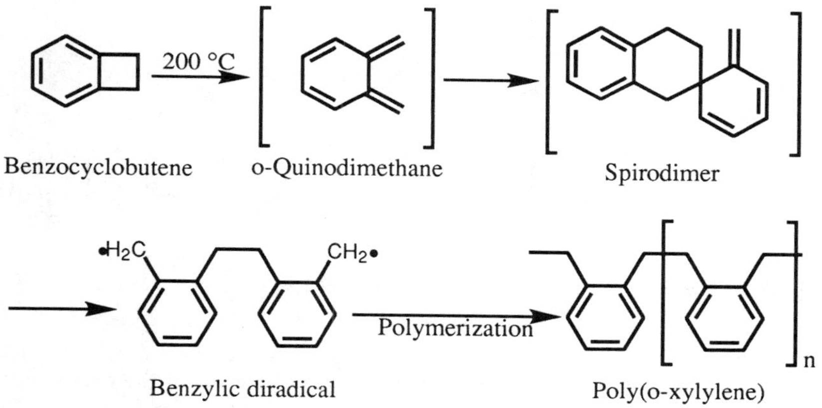

Benzocyclobutene   o-Quinodimethane   Spirodimer

Benzylic diradical   Poly(o-xylylene)

To date, the systems reported have used R groups (**Scheme 6**) which are oligomeric and, therefore, have little or no volatility. If, however, R were to be made small enough that the mass of the monomer was close to that of paracyclophane, it should be possible to vapor deposit polymers derived from such monomers.

Scheme 6

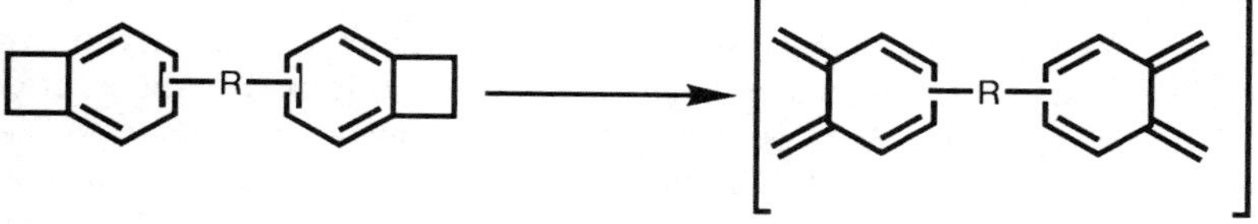

Using a fluorinated benzocyclobutene monomer should provide at least one advantage over the already promising properties of fluorinated poly(p-xylylene). All the good properties such as low dielectric constant and low affinity for water should remain but the thermal stability should be enhanced because of the crosslinking which would accompany the generation of these films. It is also possible that the coefficient of thermal expansion will also be reduced. We were gratified to find that poly(fluorinated benzocyclobutene) was able to be synthesized by polymerization of activated species that are generated by UV irradiation of 7,7',8,8'-octafluoro-4,4'-bis (1,2-dihydrobenzocyclobutene) in the vapor state or in solution (**Scheme 7**). The polymer is obtained as a film and is not soluble in common laboratory solvents. The synthesis of the monomer involves a multi-step synthetic sequence which proceeds in high yield and is described in a current published paper (32).

XPS analysis indicates that the film contains fluorine and we are currently trying to prepare enough of this material to characterize its structure by solid state nuclear magnetic resonance.

Scheme 7

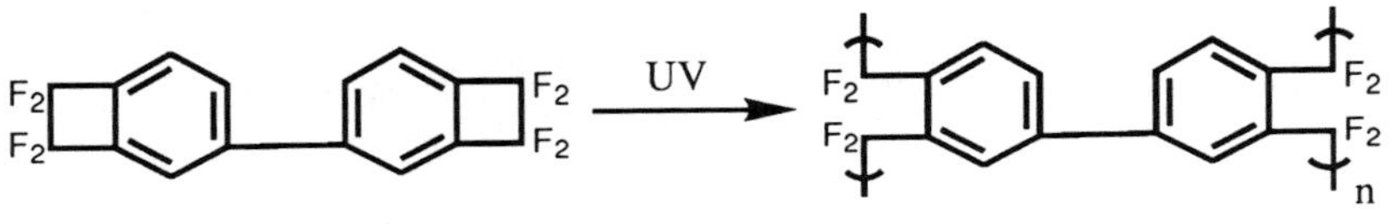

bis-Benzocyclobutene-F_8 Poly(fluorinated benzocyclobutene)

Our original goal was to try to activate 7,7',8,8'-octafluoro-4,4'-bis (1,2-dihydrobenzocyclobutene) thermally to undergo ring-opening of the cyclobutene ring and subsequently polymerize it to form a crosslinked film on the substrate. The cure chemistry of these systems is primarily based upon the fact that under appropriate thermal conditions, the strained four-membered ring of benzocyclobutene undergoes an electrocyclic ring opening. The temperature at which such a concerted process occurs depends principally on the substituents at the alicyclic, rather than the aromatic position. It is predicted that an electron-donating substituent at C_7 and/or C_8 will favor ring-opening, but electron withdrawing groups at those positions will make the ring-opening energetically more demanding (28a). 7,7',8,8'-Octafluoro-4,4'-bis(1,2-dihydrobenzocyclobutene) has four fluorine atoms on each ring and we expected the reaction temperature to be higher than for an unfluorinated system but that we would gain enhanced thermal stability and low dielectric constant after it polymerized. However, the electronic effect is so overwhelming that this compound did not polymerize or undergo Diels-Alder reactions with added dienes or dienophiles at the temperatures or conditions which were useful for benzocyclobutene. However 7,7',8,8'-octafluoro-4,4'-bis(1,2-dihydrobenzocyclobutene) did polymerize when it was irradiated in solution or the vapor state with a UV source. The film collected from the UV polymerization is not soluble in common laboratory solvents. The film is very thin and we were not able to grow thicker films. This result might be caused by absorption of the incident radiation by the film formed on the wall of the quartz reactor, thereby blocking the incoming UV light and preventing the activation of monomer and continuous polymerization. A different reactor geometry and the use of intense laser sources may overcome this difficulty.

Conclusions:

The main objective of our research is to find new precursors and new techniques to vapor deposit thin films of organic polymer with very low dielectric constants for microelectronic interconnection applications. During the dielectric constant measurements, we found that the measured values were very dependent on the quality of the deposited films. Boron doped silicon wafers (Si(100)) coated with poly(naphthalene) or poly(fluorinated naphthalene) to a thickness of 0.5 m were covered with a contact mask containing numerous circular holes 0.8 mm and 1.6 mm in diameter. The wafers were placed in a Airco Temescal CV-8 vacuum chamber and aluminum was deposited by electron-beam evaporation at 10^{-6} Torr. In this manner, circular capacitors with two different diameters were fabricated. Capacitance measurements were carried out with the aid of a Hewlett-Packard, model 4280 A 1-MHz C meter/C-V plotter. Contact was made to the back side of the wafer and via a surface probe to one of the aluminum contacts. The capacitance was measured using a 1-MHz ac signal. The capacitance of 30 individual capacitors were measured and the results averaged. From this capacitance data, the dielectric constants were calculated and averaged using the equation:

$$\varepsilon = dC/A\varepsilon_0$$

in which ε is the dielectric constant, d is the thickness of the film, C is the capacitance, A is the area and ε_0 is the permittivity in a vacuum.

Although the materials we have studied have very low dielectric constant, some of them still fall short of certain requirements, such as thermal stability. Parylenes (PA-N or PA-F) made from different dimers, depending on the type, have dielectric constants ranging from 2.38 to 3.15 and poly(naphthalenes) have dielectric constants lower than 2.5. Teflon AF has a value of 1.9. But VLSI interconnection and packaging applications also require high thermal stability of the films being used. A further consideration is that the diffusion in, and adhesion of metal t. polymer films depends strongly on the thermal stability of the polymer film. For instance, Cu diffusion into PA-N starts at a temperature of 300-350 °C which corresponds, roughly, to the onset of thermal degradation. Adhesion failure between Cu and PA-N also starts at 300 °C. The thickness of PA-N film begins to shrink at 350 °C while annealing in nitrogen(33). The thermostability of these thin films were measured by annealing the film in a nitrogen flow in a tube furnace. The film thickness changed as a function of annealing temperature. The furnace was first stabilized at the annealing temperature and then samples were directly introduced into the hot zone and held at each temperature for 30 minutes. After annealing, samples were removed and cooled to room temperature in air. The decomposition temperature was taken as the temperature where the normalized thickness was reduced by more than 5%. Film thickness was measured with a profilometer (Alpha-step 200) made by the Tencor instrument company.

In **Table I** an overview of the properties of those organic thin films which are being studied by chemical vapor deposition in our laboratory is presented. We can see that exciting possibilities for new dielectrics have been uncovered. These polymers, PA-N, PA-F, PNT-N, PNT-F and poly(bis-BCB-F$_8$), would have been impossible to synthesize or process into thin films by conventional methods because of their limited solubilities.

Table I. An Overview of the Properties of Organic Thin Films

Film	PA-N	PA-F	Teflon AF	PNT-N	PNT-F
Film Deposition	vapor	vapor	vapor	vapor	vapor
Source Material	dimer (solid)	precursor (liquid)	polymer (solid)	precursor (liquid)	precursor (solid)
Toxicity	yes	hazard	hazard	NA	NA
Dielectric Constant	2.60 ± 0.1	2.2 - 2.3	1.93	2.4 ± 0.1	2.3 ± 0.1 (best)
Electric Breakdown	2×10^7 V/M	5×10^7 V/M	5×10^7 V/M	3×10^7 V/M	5×10^7 V/M (best)
Dissociation T(°C)	430 °C in N$_2$	530 °C in N$_2$	360 °C in air	570 °C in N$_2$	NA
Thickness vs. Annealing Temp.	no change to 350 °C in N$_2$	no change to 500 °C in N$_2$	NA	no change to 530 °C in N$_2$	NA
Structure As Deposited	crystalline	crystalline / amorphous	amorphous	crystalline	crystalline
Cracks	yes	no	yes	no	no
Film Thickness	0.1-10 μM	0.1-1 μM	0.1-10 μM	0.5-4 μM	0.5-2 μM
Adhesion to Si	poor	good	poor	good	NA
Adhesion to Al	OK	NA	poor	poor	NA

Additionally, this synthetic approach provides a new route for the synthesis of thin films without the use of toxic solvents. However, there remains the challenge of choosing or synthesizing potential source materials for this methodology. The right precursors should have certain characteristics, such as compatible chemistry, ease of evaporation or sublimation in the vacuum system, no volatile fragments released in the process, and, certainly, the final polymeric film should have the desired thermal stability and dielectric properties. Future work will will continue to expand on this approach including the development of new precursors as well as studying the crystallinity of these materials and its influence on the morphology and utility of these films as dielctric materials.

Acknowledgement:

We thank Prof. Tour for providing details of his synthetic approach prior to publication. Financial support of this effort has been provided, in part, by the IBM Corporation and is gratefully acknowledged.

References

1. Sroog, C. E. *J. Polym. Sci. Macromol. Rev.* **1976**, *11*, 161.
2. Samuelson, G. *Preprints Org. Coat. Plastics Chem.* **1980**, *43*, 446.
3. *Photochemical Vapor Deposition*, Eden J. G., Ed.; Chemical Analysis; John Wiley & Sons, NY 1992, Vol. 122; pp 5-8.
4. Szwarc, M. *J. Chem. Phys.* **1957**, *16*, 128.
5. (a) Gorham W. F. *J. Polym. Sci. A-1* **1966**, *4*, 3076. (b) Gorham, W. F. (to Uion Carbide Corp.), **U.S. Pat.** 3,342,754 (1967).
6. Yeh, Y. L.; Gorham W. F. *J. Org. Chem.* **1969**, *34*, 2366. (b) Gorham, W. F. (to Union Carbide Corp.), **U.S. Pat.** 3,221,068 (1965).
7. Beach,W. F.; Lee, C.; Basset, D. R.; Austin, T. M.; Olson, R., *Encycl. Polym. Sci. Eng.*, 2nd Ed., John Wiley & Sons: NY, **1989**; Vol 17, 990.
8. Joesten, B. L. *J. Appl. Polym. Sci.* **1974**, *18*, 439.
9. Hertler, W. R. *J. Org. Chem.* **1963**, *28*, 2877.
10. Fuqua, S. A.; Parkhurst, R. M.; Silverstein, R. M. *Tetrahedron* **1964**, *20*, 1625.
11. Chow, S. W.; Pilato, L. A.; Wheelwright, W. L. *J. Org. Chem.* **1970**, *35*, 20.
12. Chow, S. W.; Loeb, W. E.; White, C. E. *J. Appl. Polym. Sci.* **1969**, 2325.
13. Dolbier, W. R., Jr.; Asghar, M. A.; Pan, H.-Q. (to Union Carbide Corp.), **U.S. Pat.** 5,210,341(1993).
14 You, L.; Yang, G.-R.; Lang, C.-I.; Wu, P.K.; Lu, T.-M.; Moore, J. A.; McDonald, J. F. P. *J. Vac. Sci. Technol. A* **1993**, *11* (6), 3047.
15. You, L.; Yang, G.-R.; Lu, T.-M.; Moore, J. A.; McDonald, J. F. P. **U.S. Pat.** 5,268,202, (1993).
16. Nason, T. C.; Moore J. A.; Lu, T.-M. *Appl. Phys. Lett.* **1992**, *60*, 1.
17. deWilde, W.; deMey, G. *Vacuum* **1973**, *24*, 307.
18. A brochure detailing the technical properties of Teflon AF can be obtained from the DuPont Corp.,Wilmington, DE.
19. Resnick, P. R. *Polym. Prepr.* **1990**, *31*, 312.
20. (a) Lowry, J. H.; Mendlowitz, J. S.; Subramanian, N. S. *Optical Engineering* **1992**, *31*, 1982. (b) Lowry, J. H.; Mendlowitz, J. S.; Subramanian, N. S. *SPIE* **1990**, *1330*, 142.
21. Blanchet, G. B. *Appl. Phys. Lett.* **1993**, *62*, 478.
22. Nason, T. C.; Moore J. A.; Lu, T.-M. *Appl. Phys. Lett.* **1992**, *60*, 1866.
23. Bergman, R. G. *Acc. Chem. Res.* **1973**, *6*, 2531.

24. Lockhart, T. P.; Cornita, P. B.; Bergman, R. G. *J. Am. Chem. Soc.* **1981**, *103*, 4082.

25. John, J. A.; Tour, J. M. *J. Am. Chem. Soc.* **1994**, *116*, 5011.

26. Chow, S.-W. (to Union Carbide Corp), **U.S. Pat.** 3,268,599 (1966)

27. (a) Boekelheide, V.; Ewing, G. *Tetrahedron Lett* **1978**, *44*, 4245. (b) Schiess, P.; Heitzmann, M.; Rutschmann, S.; Staheli, R. *Tetrahedron Lett.* **1978**, *46*, 4569. (c) Oppolzer, W. *Synthesis* **1978**, 973.

28. (a) Kirchhoff, R. A.; Carriere, C. J.; Bruza, K. J.; Rondan, N. G.; Sammler, R. L. *J. Macromol. Sci.-Chem.* **1991**, *A 28(11&12)*, 1079.
(b) Hahn, S. F.; Martin, S. J.; Mckelvy, M. L. *Macromolecules* **1992**, *25*, 1539.
(c) Kirchhoff, R. A.; Bruza, K. J. *CHEMTECH* **1993**, *23*, 22.

29. (a) Tan T.S.; Arnold F. E. *J. Polym. Sci., Polym. Chem. Ed.* **1988**, *26*, 1819. (b) Walker, K. A.; Markoski, L. J.; Moore, J. S. Macromolecules **1993**, *26*, 3713.

30. Burdeaux, D.; Townsend, P.; Carr, J. J.; Garrou, P. E. *J. Electron. Mat.* **1990**, *19*, 1357.

31. Chinoy, P. B.; Tajadod, J. *IEEE Trans. CHMT* **1993**, *16*, 714.

32. Moore, J. A.; Lang, C.-I.; Lu, T.-M.; Young, G.-R. *Polym. Mat. Sci. Eng.* **1995**, *72*, 437.

33. Yang, G.-R.; Dabral, S.; You, L.; McDonald, J. F.; Lu T.-M. *J. of Electronic Materials,* **1991**, *20*, 571

RECEIVED September 1, 1995

Chapter 31

Plasma Polymerization in Direct Current Glow: Characterization of Plasma-Polymerized Films of Benzene and Fluorinated Derivatives

Toshihiro Suwa, Mitsutoshi Jikei, Masa-aki Kakimoto[1], and Yoshio Imai

Department of Organic and Polymeric Materials, Tokyo Institute of Technology, Meguro-ku, Tokyo 152, Japan

The plasma polymerization by direct current (D.C.) glow discharge method was performed with benzene, fluorobenzene, 1,2,3-trifluorobenezene, and perfluorobenezene. The plasma polymers were characterized by FT-IR and Raman spectroscopies, ESCA, ellipsometry, and contact angle measurements. The applied voltage was significant factor for the chemical structure of the products. The formation of CF_2 or CF_3 owing to molecular rearrangements was hardly observed by this method.

Plasma polymerization is expected to provide unique films since the reaction mechanism is quite different from conventional polymerization. Because no special functions in the starting materials are required, even methane undergoes polymerization under certain conditions and forms highly cross-linked, pinhole-free films.

The plasma reaction has been extensively applied for the modification of surface properties. The plasma polymerized films prepared using fluorine-contained monomer usually lowers the surface energy of the materials (1). These monomers were also utilized frequently for evaluating the mechanism of plasma polymerization by the use of electron spectroscopy for chemical analysis (ESCA). A strong carbon-fluorine chemical bonding causes considerable chemical shifts in the C_{1s} core-level spectra. The degree of this effect depends upon the number of fluorine atoms which bonded to carbon, and hence it would be expected to facilitate quantification of various carbon components such as C-H, CF, CF_2, and CF_3. The plasma polymerization of a series of fluorinated compounds in inductively coupled RF plasma have been widely investigated by D. T. Clark et al. (2,3). They reported that the polymerization of highly fluorinated compounds generally proceeded accompanying eliminations, fragmentations, and rearrangements of initial monomers. Thus, perfluorobenzene plasma, for example, generated CF_2, CF_3, etc. components, which did not exist in the starting monomers.

Plasma for the glow discharge is usually generated by low frequency (50 or 60 Hz), radio frequency (R.F.; 13.56 Hz), or microwave (M.W.; 2.45 GHz). We have recently shown the plasma polymerization of naphthalene and perfluoronaphthalene by a direct current (D.C.) glow discharge (Suwa, T. *Jpn. J. Apply. Phys.* in press.). The films prepared by this method, as indicated there, had

[1]Corresponding author

0097–6156/95/0614–0471$12.00/0

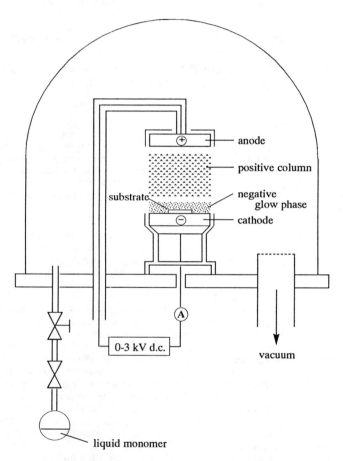

Figure 1 Schematic diagram of the plasma polymerization reactor.

such properties as high film growth and low rearrangement, compared with conventional plasma polymerization. The D.C. glow polymerization have developed coupled with the application as replica films in electron microscopy field (4,5). However, only few attempts have so far been made to explore the chemical properties of the film prepared by this method. In our previous work on naphthalene derivatives, some unique properties, which were unusual in case of normal plasma reaction, were observed.

In this paper, therefore, we confirm that the peculiar results of previous work came from either the characteristics of D.C. glow discharge or the nature of selected monomers. For the purpose of comparison and facility in Elemental Analysis for Chemical Analysis (ESCA), we are concerned with benzene and its fluorinated derivatives, such as fluorobenzene, 1,2,3-trifluorobenezene, and perfluorobenezene.

Experimental

Materials. Fluorinated monomers for this study were purchased from Tokyo Kasei Co. and used without further purification. Since commercial benzene contained thiophen, it was purified by distillation after the treatment with concentrated H_2SO_4. According to the measurement, glass slides, silicon wafers, and silver-coated glass slides were used appropriately as the substrates.

Plasma Deposition. A schematic diagram of the reaction chamber for D.C. glow discharge was shown in Figure 1. The bell-jar-type reactor consisted of two parallel disk electrodes (5 cm in diameter) which were placed vertically to base plate, a high voltage D.C. source, monomer reservoir, a pressure gauge, and a vacuum system. The substrates were placed on the cathode, which was the lower side electrode of the two.

For each run, the experimental procedure employed was as follows. First the reaction chamber was pumped down to less than 10^{-2} Pa. Then the reactor was filled with the monomer gas to a pressure of approximately 10 Pa by the vaporization of liquid monomer. A D.C. voltage (0.4-1.5 kV) was subsequently applied for 5-120 seconds and a plasma polymerized film was formed on the substrate, a surface of which was located in a negative glow phase. Finally the system was evacuated again (< 10^{-2} Pa) for 1 minute and vented to atmospheric pressure.

Analysis. The Fourier transform infrared (FT-IR) spectra were measured by the reflection absorption mode (RAS) with a JEOL JIR-MICRO 6000 equipped with a nitrogen-cooled mercury-cadmium-telluride (MCT) detector. For these measurements, the films were deposited onto 100-nm-thick Ag coated glass slides. Thickness of deposited films were determined by an ellipsometer (Nippon Infrared Instrumental Co., EL-101). The films were prepared on silicon wafer. Electron spectroscopy for chemical analysis (ESCA) was carried out with a ULVAC-PHI-5500MT system. The spectra were acquired using monochromated Al $K\alpha$ (1486.7 eV) radiation at 14 kV and 200 W. The component analysis of the elements was accomplished by data system supplied by ULVAC-PHI, assuming Gaussian peaks with 10% Lorentzian tails. All spectra were cariblated by adjusting the decomposed C_{1s} curve of C-H and/or C-C at energy of 284.6 eV. The sputtering of film surface was done with Ar^+ ions at 2.5 kV and 25 mA. In order to study physical properties of the surface, the contact angles of water and formamide were measured with a contact angle meter (Kyowa Interface Science Co., Model CA-A). The measurements were performed by the sessile drop method. The resulting values were the means of at least six measurements with respective errors of ±1°. A carbon-like film prepared from benzene plasma was preliminarily analyzed by using Raman Spectroscopy. The spectrum was recorded on a JASCO NR-1800 employing an argon ion laser (λ=514.5 nm).

Results and Discussion

IR Spectra. A D.C. glow method has the advantage of setting the different applied voltages easily. It is considered that higher voltage gives more energy to reactive species in the glow phase such as electrons, ions, and radicals and affects the structure of resulting deposits. The plasma polymerization, therefore, was performed by using various applied voltages. Figure 2a-d show the IR spectra of plasma polymerized films of benzene (PPB), perfluorobenzene (PPPB), fluorobenzene (PPFB), and trifluorobenzene (PPTB), respectively. On the whole, spectra varied with changes in the applied voltage during discharge. In case of benzene plasma (Figure 2a), two regions of C-H stretching bands were observed around 3000 cm^{-1}. A small peak at 3000-3100 cm^{-1} is due to the aromatic C-H band, and the latter peak to the aliphatic one. Needless to say, a starting material-benzene has only aromatic C-H bonding. So this result suggests that the aromatic ring of benzene was easily destroyed to a considerable extent even at relatively low voltages. A series of spectra changed gradually with a rise in the applied voltage. On the other hand, new absorption bands appeared in the region of 1600-1800 and 1000-1500 cm^{-1}. The former region is probably attributed to C=O stretching. The other one, which was composed of a sequence of absorption, became broader with the increase of the applied voltage. Although definite assignment of them was difficult to make from these spectra, it was presumed that C-H bending vibration was involved around 1400 cm^{-1}. The formation of carbonyl group means insertion of oxygen in the chemical structure. This phenomenon was also observed in our previous work about naphthalene and we revealed the oxygen existed only in the outermost portion of the films. That was considered to result from the post-reaction of residual radicals generated by a plasma with atmospheric oxygen or water vapor. Strange point to note is that O-H stretching, which appeared at 3300 cm^{-1} for naphthalene plasma polymer, was not able to observed. It might come from the difference of plasma polymerization between these two kinds of aromatic hydrocarbon. This would be supported by Raman spectrum later.

The IR spectra of PPPF were depicted in Figure 2b. As the applied voltage rose, the C-F peak at 1338 cm^{-1} decreased gradually, while new peak centered on 1220 cm^{-1} began to appear in the wide region. Considering the decrease of CF=CF at 1546 cm^{-1}, this spectral variation presumably owed to change of aromatic C-F group into aliphatic one. The C-F groups in various environments made their spectra very vague. Another broad peak in the region of 1600-1800 cm^{-1}. was assigned to C=O stretching. The carbonyl carbon in higher wavenumber region (around 1800 cm^{-1}) would bond to fluorine, whereas fluorine was considered to be eliminated from carbon atoms at higher voltages.

Figure 2c and 2d are IR spectra of PPFB and PPTB, respectively. These spectra have similar tendency with benzene and perfluorobenzene plasma polymer. Both CF=CF stretching of fluorobenzene (1489, 1608 cm^{-1}) and of trifluorobenzene (1513, 1623 cm^{-1}) plasma polymer became small as the applied voltage rose. In addition, a small C-H stretching were observed at lower voltage of each spectra. The existence of carbonyl group was also common phenomenon through four kinds of plasma polymer.

ESCA Spectra. ESCA analysis was performed in order to evaluate the composition of the films. Our interest was particularly the comparison of film surface and the bulk region.

First of all, therefore, a depth profile analysis was carried out for all monomers. Figure 3 shows a depth profiles of PPPF films (approximately 24 nm) prepared at 1.5 kV. The calculated sputtering rate was roughly 0.1 nm/s. For the ESCA analysis, silicon wafer was used as substrate. The profile was composed of carbon, fluorine, oxygen, and silicon. The profile exhibits that oxygen exists at the surface of as-prepared film and is scarcely found within the film. We can recognized that the film is composed of only carbon and fluorine in the bulk, and that oxygen, the existence of which was suggested by IR spectra as well, remained in the neighborhood of the surface. The profiles also indicates that atomic ratio of F/C is much higher at the surface, compared wit the bulk region. A high fluorine concentration in the outermost region is presumably a result of post-reaction. To make this point clear, high resolution measurements were done before and after the argon ion sputtering. The C_{1s} core-level spectrum measured before Ar^+ sputtering is revealed in Figure 4a. There observed a variety of components in the spectrum. They are composed of C-C, C-CF, CF, CF-CF, CF_2 (CFO), CF_3 and a π-π^* shake-up. Since the CFO component, which was detected by IR analysis, was predicted to appear at similar region with CF_2, further separation of these components was not attempted. Argon ion sputtering changed the spectrum remarkably and the peaks in higher binding energy region became small. The quantitative details are summarized in Table I. Similarly to the previous study about plasma polymerized perfluoronaphthalene in D.C. glow, the contribution of CF_2 and CF_3 components to the spectra was very small even in the unsputtered sample. This means the rearrangement of bondings, which has been regarded as usual for the plasma polymerization of fluorocarbon by R.F. or microwave etc. (2,3,6), hardly take place in D.C. glow method.

Table I. The Quantitative Details in C_{1s} Spectrum of PPPB Measured before Ar^+ Sputtering

C_{1s} Components	Binding Energy (eV)	Relative Concentration(%)
C-C	284.6	33.9
C-CF	286.8	36.9
CF	287.6	7.7
CF-CF	289.3	18.7
CF_2 (CFO)	291.8	1.7
CF_3	293.1	0.7
π-π^*	294.7	0.4

We also carried out high resolution measurement with the sample prepared at lower voltage (0.4 kV). No particular difference was observed between the two (spectra not shown). We may therefore, reasonably say, that the diminution of the CF_2 and CF_3 peaks in this spectrum is characteristics of the D.C. glow polymerization of fluorocarbons. Let us now return to an interpretation of high concentration of fluorine at the surface region. The critical difference in spectra measured before and after the sputtering is the disappearance of most of CF and CF-CF components. These components reflect nature of starting monomer, C_6F_6. The chemical structure of the surface is probably close to that of monomer. It is thought that the outermost region of the film was mainly formed by a post-reaction of the residual radicals with monomer gas that remained in the chamber. Interestingly the depth profile displayed that the proportion of fluorine slightly increased again at the film/substrate interface. This could be caused by the difference in plasma condition between initial discharge and the subsequent steady state. Around the interface, trace amount of oxygen was detected as well. Since this peak increased with the appearance of silicon peak, it seems reasonable to suppose that oxygen came from silicon oxide, which had been formed on the surface of the substrate.

ESCA analysis of fluorobenzene and trifluorobenzene plasma polymer was employed in a similar manner as above. The behavior of their depth profiles were, as

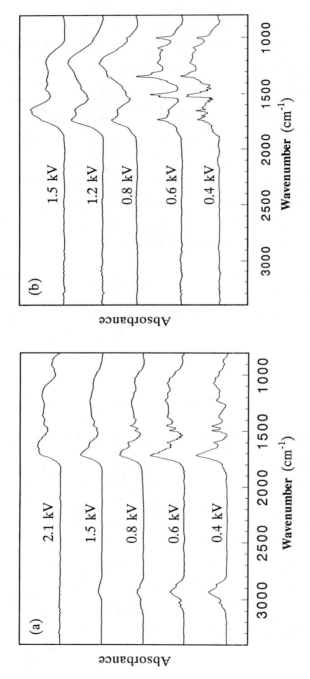

Figure 2 IR spectra of plasma polymerized films prepared at different voltages.
(a): PPB; (b): PPPB; (c): PPFB; (d): PPTB.

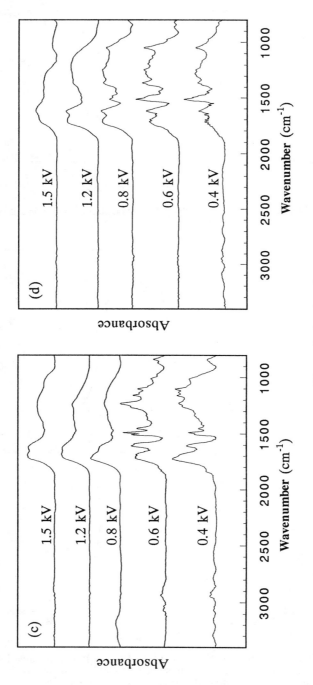

Figure 2. *Continued*

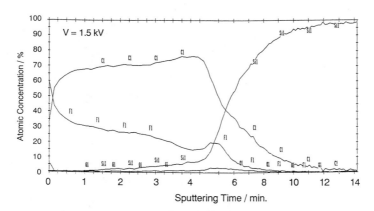

Figure 3 ESCA depth profile of PPPF film deposited on a silicon substrate.

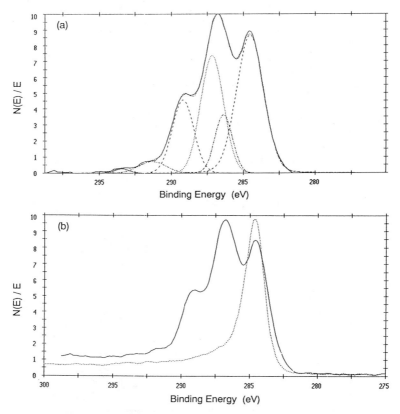

Figure 4   The C$_{1s}$ core-level spectra of PPPF.   (a): the deconvoluted C$_{1s}$ spectrum obtained before the Ar$^+$ exposure; (b): the comparison of the spectra measured before (solid line) and after (dotted line) Ar$^+$ sputtering.

expected, not so different from that of perfluorobenzene except for the atomic concentration of fluorine. These values varied in proportion to the number of fluorine atoms which monomers contain. Figure 5 and 6 show the C_{1s} core-level spectra of the PPFB and PPTB, respectively. They were measured before and after argon bombardment. Compared with the spectra of PPPB, the intensity in higher binding energy region was small even before the bombardment. As can be seen from each spectra, they further decreased in size after the treatment. These results also support that the formation of CF_2 and CF_3 components hardly occur by D.C. glow. Atomic concentration of fluorine in the films is presented in Table II.

Table II. Concentration of Fluorine in the Plasma Polymerized Films Measured before and after Sputtering

Plasma Polymer [a]	Concentration of Fluorine (%)	
	Before	After
PPFB	5	2
PPTB	22	14
PPPB	60	20-30

[a]All films were prepared at 1.1 kV.

Finally PPB was examined. We could not find the difference in the spectra measured before and after the sputtering (Figure 7). Since IR spectra revealed that elimination of the hydrogen from the products proceeded to a considerable extent at higher voltage, It is considered that the film is almost compose of carbon.

Surface Properties of the Plasma Polymerized Films. Contact angle measurement is convenient technique to evaluate the surface properties of films. We measured contact angles of water (θ_W) and formamide (θ_F) on each films which were prepared at 1.1 kV. From the obtained contact angles the surface free energy of each plasma polymerized film was estimated according to Kaelble method (7,8). All the results were presented in Table III.

Table III. Surface Properties of the Plasma Polymerized Films Prepared at 1.1 kV

Plasma Polymer	θ_W (degree)	θ_F (degree)	γ_s^d (mJ/m^2)	γ_s^p (mJ/m^2)	Total Surface Energy (mJ/m^2)
PPB	67.0	42.0	33.7	10.9	44.6
PPFB	71.5	47.3	32.8	8.8	41.6
PPTB	78.7	54.0	32.9	2.4	35.4
PPPB	95.7	72.9	26.5	1.5	28.0

As shown in the table, both contact angles, as expected, increase with increasing the number of fluorine atoms which monomer possess. The surfaces of the films reflect the structure of starting monomers, that is, the amount of surface fluorination are proportion to the F/C ratio of monomer and surface energy is fairly connected with the degree of fluorination. Surface energy are composed of a polar (γ_s^p) and a dispersive component (γ_s^d). They are presented separately in the table. Because of the presence of polar groups on the film surface, γ_s^p is expected to depend on how many polar groups have been incorporated. The results suggest more polar

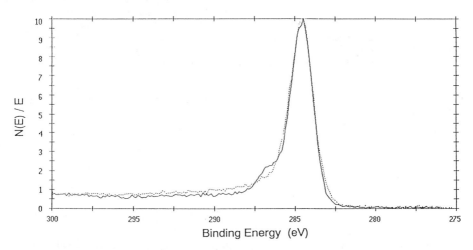

Figure 5. The C$_{1s}$ core-level spectra of PPFB film measured before (solid line) and after (dotted line) Ar$^+$ sputtering.

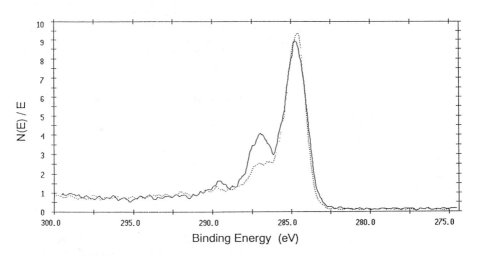

Figure 6. The C$_{1s}$ core-level spectra of PPTB film measured before (solid line) and after (dotted line) Ar$^+$ sputtering.

groups, like C=O, were formed on the benzene polymer than on the fluorinated polymers. This is probably related to the difference of bonding strength between C-H and C-F. The C-H bondings were broken more easily and subsequently formed more polar groups. Contact angle measurements were also employed as a function of applied voltage. However, no particular tendency was observed from them (data not shown).

Film Growth. The deposition rate was very sensitive to the discharge factors such as an applied voltage and a gas pressure. The data we present here were prepared at two kinds of general conditions. As we had noted the effect of the applied voltage, 0.4 kV and 1.5 kV were selected at first. Then target pressures were adjusted at 27 Pa (0.2 Torr) and 4.0 Pa (0.03 Torr), respectively, in order to maintain a stable discharge. To put it more concretely, when the applied voltage was set at 0.4 kV the discharge did not occur at 4.0 Pa. On the other hand, in case of 1.5 kV, the discharge was too vigorous at 27 Pa. Thus these two conditions have different pressure and cannot be compared directly. Figure 8a shows film growth against discharge time at relatively low voltage (0.4 kV). As the diagram indicates, all monomers polymerized approximately at the same rates (0.1-0.3 nm/s). It is interesting to note that these extrapolated lines deviate far from the origin. It admits of two interpretations. One explanation for it is that film growth proceeded more rapidly during initial state of discharge. At this period not so much fragmentation of monomer would take place as subsequent stable discharge period and polymerization reaction would proceed easily. But very soon the reaction would go into the equilibrium state between polymerization and ablation, and it would make the rate of film growth slow down. The other reason is the large amount of absorption of monomer gases on the substrates before the discharge. It is thought that these absorbed molecules were fixed at the substrates by plasma-induced polymerization. ESCA depth profile analysis, as we have seen, indicates that the fluorine concentration around film/substrate interface was different from that of bulk region. By ESCA data we cannot conclude, however, which reason is responsible for the results.

When applied voltage was set at 1.5 kV, specific film growth (2.4 nm/s) was observed in case of perfluorobenzene (Figure 8b). The other three monomers were polymerized at the almost same rates (0.2-0.7 nm/s). These values are comparable to that of common plasma polymerizations induced by radio frequency and microwave. In addition, the extrapolated lines of these data deviate from the origin as well as the that of polymerization at low voltage. It suggests that the large deviation from the origin is probably not due to absorption of monomer molecules on the substrates but due to the rapid polymerization containing little fragmentation and ablation, because the pressure dependence of the absorption was not observed. Although film growth of PPPB was reproducible, we can not so far explain that clearly. The refractive index of films, which was also obtained from ellipsometric measurements, may give a clue to explain the specificity of perfluorobenezene plasma polymer. The refractive indexes of the plasma polymerized films from benzene, fluorobenzene, and trifluorobenezene were approximately 2.0. On the other hand, the index decreased to 1.6 for PPPB. We can, therefore, imagine as follows. The C-F bonding strength is stronger than C-H and large amount of fluorine remained in the film for PPPB. This would prevent the cross-linking of carbon atoms and result in the formation of relatively low density film. It would be closely linked to the rapid film growth.

Raman Spectrum. Judging from the results which we have hitherto mentioned, the plasma polymerized films prepared at higher voltage, except for the oxidized surface, were composed of only carbon. Raman spectroscopy has been frequently used for the characterization of carbon films. The spectrum of PPB prepared at 2.1 kV is shown in Figure 9. We can observe a very broad peak centered at 1550 cm^{-1}. The peak around

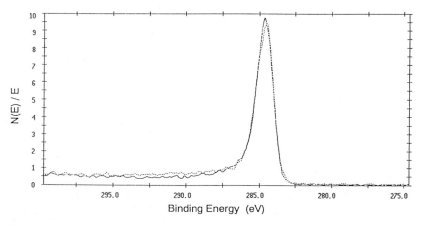

Figure 7. The C_{1s} core-level spectra of PPB film measured before (solid line) and after (dotted line) Ar^+ sputtering.

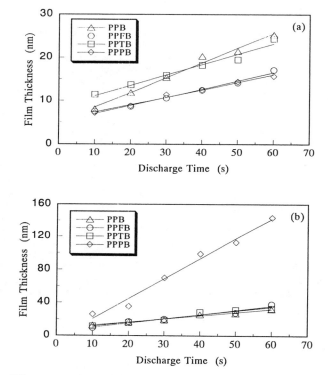

Figure 8. Film growth of each polymer against discharge time performed at 0.4 kV (a) and 1.5 kV(b).

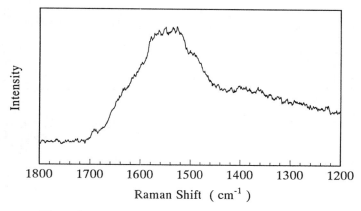

Figure 9. Raman spectrum of PPB prepared at 2.1 kV.

this regain is common for *i*-carbon films. Although this peak can be associated with a sp^2 bonded carbon in an amorphous material (9-16), an unambiguous assignment has not been determined. In contrast to the plasma polymerized naphthalene which had peaks at around 1300-1400 cm^{-1}, it is interesting that PPB has a peak in other region. Details on these difference of carbon films will be presented in future publications.

Conclusion

D.C. glow discharge was used to prepare plasma polymerized films. The IR spectra showed that the chemical structure of the products were affected by the applied voltage during discharge and that plasma polymerization of benzene at higher voltage produced hydrogen-free carbon film. ESCA analysis revealed that molecular rearrangements forming CF$_2$ or CF$_3$ fluorinated monomers, which has been considered as familiar reaction in case of conventional plasma polymerization, did not occur by this method. The surfaces of films were more fluorinated than the bulk regions and the surface energy depended on starting monomers.

Acknowledgments

This work has been performed under the support of "Special Coordination Funds for Promoting Science and Technology" from the Science and Technology Agency of Japan.

Literature Cited

1.   Inagaki, N.; Nakanishi, T.; Katsuura, K. *Polym. Bull.*, **1983,** *9*, 502.
2.   Clark, D. T.; Shuttleworth, D. *J. Polym. Sci., Polym. Chem.*, **1979**, *17*, 1317.
3.   Clark, D. T.; Shuttleworth, D. *J. Polym. Sci., Polym. Chem.*, **1980**, *18*, 27.
4.   Tanaka, A.; Sekiguchi, Y.; Kuroda, S. *J. Electron Microscopy*, **1978**, *27*, 378.
5.   Tanaka, A.; Yamaguchi, M.; Iwasaki, T.; Iriyama, K. *Chem. Lett.*, **1989**, 1219.
6.   Munro, H. S.; Till, C. *J. Polym. Sci., Polym. Chem.*, **1984**, *22*, 3933.
7.   Kaelble, D. H. *Proc. 23rd Int. Cong. of Pure and Appl. Chem.8*; Butterworths: London, 1971; pp.265-302.

8.   Kaelble, D. H.; Dynes, P. J.; H. Cirlin, E. *J. Adhession*, **1974**, *6*, 23.
9.   Dillon, R. O.; Woollam, J. A.; Katkanant, V. *Phys. Rev.*, **1984**, *B29*, 3482.
10.   Couderc P.; Catherine, Y. *Thin Solid Films*, **1987**, *146*, 93.
11.   Sato, T.; Furuno, S.; Iguchi, S.; Hanabusa, M. *Jpn. J. Appl. Phys.*, **1987**, *26*, L1487.
12.   Ramsteiner, M.; Wagner, J. *J. Appl. Phys. Lett.*, **1987**, *51*, 1355.
13.   Yoshikawa, M.; Katagiri, G.; Ishida, H.; Ishitani, A. *Solid State Communications*, **1988**, *66*, 1177.
14.   Houg, P. V. *Diamond Related Mater.*, **1991**, *1*, 33.
15.   Dowling, D. P.; Ahern, M. J.; Kelly, T. C.; Meenan, B. J.; Brown, N. M. D.; O'Connor, G. M.; Glynn, T. J. *Surface and Coating Technology*, **1992**, *53*, 177.
16.   Godbole, V. P.; Narayama, J. *J. Mater. Res.*, **1992**, *7*, 2785.

RECEIVED July 17, 1995

Chapter 32

Syntheses and Properties of Allylated Poly(2,6-dimethyl-1,4-phenylene ether)

Yoshiyuki Ishii, Hiroji Oda, Takeshi Arai, and Teruo Katayose

Designed Products Laboratory, Asahi Chemical Industry Company, Ltd., Kawasaki, Kanagawa 210, Japan

A novel thermosetting polyphenylene ether (PPE) having crosslinkable allyl groups was successfully synthesized from PPE and allyl halides by the modification of the polymer in tetrahydrofuran solution at low temperature under atmospheric pressure. Cured allylated PEE has a low dielectric constant of 2.50, a low dissipation factor of 0.001, high glass transition temperature of 250°C, excellent adhesion to copper foil, and low water absorption. An application of allylated PPE to copper clad laminates reinforced with E-glass cloth was also studied. The resulting allylated PPE laminate has some advantages such as low dielectric constant, high glass transition temperature, low water absorption, light weight, powder-free prepreg, etc. compared with conventional commercial products, such as epoxy resin and polyimide resin.

Low dielectric materials are much interest for applications such as high speed and high frequency multilayer printed circuit boards. Polymeric low dielectric materials have been studied extensively in recent years due to the ease of processability compared to ceramic materials. The practical properties of dielectric material are evaluated by measuring relative signal propagation speed and signal transmission loss.

The relative signal propagation speed for a dielectric material can be written (1):

$$v = k \cdot c / \sqrt{\varepsilon}$$

where v is the relative signal propagation speed, c is the speed of light in vacuum, k is a constant, and ε is the dielectric constant of the material.

For the high frequency region, signal transmission loss can be written (1):

$$\alpha \propto \sqrt{\varepsilon} \cdot \tan \delta \cdot f$$

where α is a signal transmission loss, ε is the dielectric constant, $\tan \delta$ is the dissipation factor, and f is the frequency. From these equations, it is

0097–6156/95/0614–0485$12.00/0

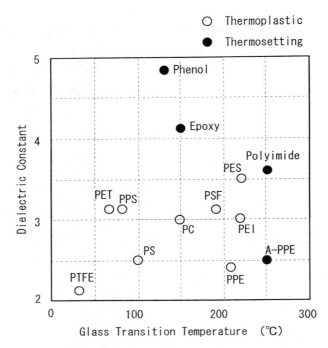

Figure 1. Dielectric Constant and Tg of various resins.

Table 1. Properties of poly(2,6-dimethylphenylene ether)

Item	
Dielectric Constant (at 1MHz)	2.45
Dssipation Factor (at 1MHz)	0.0007
Glass Transition Temperature (°C)	210
Solvent Resistance	
(trichloroethylene)	Soluble
(acid)	Insoluble
(alkali)	Insoluble
Water Absorption (%)	< 0.05

clear that a low dielectric material with a low dissipation factor is ideal for applications in high speed computers and low loss devices.

These dielectric materials also require heat resistance because of exposure to solder at temperatures higher than 260°C. This means that a high glass transition temperature (Tg) is also required. Figure 1 shows the relationship between dielectric constant and glass transition temperature of typical commercial polymers. PTFE is a well-known dielectric material. However PTFE, because of a low Tg of 25°C, is not suitable. On the other hand, a conventional polyimide has a superior glass transition temperature. However the polyimide, with high dielectric constant and dissipation factor, is not suitable for high speed computers and high frequency devices.

Poly(2,6-dimethyl-1,4-phenylene ether) (PPE) has a low dielectric constant, a low dissipation factor, and high glass transition temperature. PPE has other excellent properties, such as low water absorption and solvent-resistance against acid and alkali, as a dielectric material. Table 1 represents some general properties of PPE(2).

One of the disadvantages of PPE is solubility in common organic solvents. PPE has weak solvent resistant to halogenated hydrocarbons and aromatic hydrocarbons. Another disadvantage is a glass transition temperature which is a little low for solder resistance. A route to improve these drawbacks is modification of thermoplastic PPE into crosslinkable PPE, i.e. to introduce crosslinkable groups into the polymer backbone. The following properties are required for a crosslinkable group; non polar in nature, suitable crosslinking temperature, no volatile component during crosslinking reaction, and retention of the inherent advantages of PPE.

There are two types of modification to yield thermosetting PPE: polymerization of a monomer having a crosslinkable group, and introduction of a crosslinkable group onto the PPE backbone via a polymer reaction. Hay studied the polymerization of 2-allyl-6-methyl phenol by oxidative coupling (3). Price et al. also studied the polymerization of allyl functionalized 4-bromophenol (4). However, these polymerizations gave only low molecular weight oils as materials. Moreover, the polymer prepared from phenols with allyl substitutes is unstable in air, becoming insoluble in 2 or 3 weeks. On the other hand, Tsuchida et al. reported the oxidative polymerization of 2,6-bis(3-methyl-2-butenyl) phenol. The degree of polymerization was very low and some of the double bonds were destroyed during polymerization (5). This failure seems to be caused by side reaction of the 3-methyl-2-butenyl group. A syntheses of PPE containing pendant vinyl groups was also studied by Percec et al (6). The synthetic routes involved the chloromethylation of PPE followed by the transformation of the chloromethyl groups into their phosphonium salts and then into vinyl groups by phase transfer catalyzed Wittig reaction. This synthetic route is rather complicated and is hardly to be operated commercially.

Taking into account the desired properties described above, an allyl group ($CH_2 = CH\text{-}CH_2\text{-}$) was chosen as a crosslinkable group. A new type of allylated PPE has been developed for the first time by polymer modification via lithiated PPE in a one step reaction by using n-butyllithium. In this paper, the reaction mechanism for allylated PPE synthesis is described. Unique properties of allylated PPE, such as crosslinking, thermal electric, and other properties, are also described. Furthermore, the application to copper clad laminates is reported.

Experimental

Materials. Poly(2,6-dimethyl-1,4-phenylene ether) (PPE) with a viscosity number of 0.54 as measured in a 0.5 g/dL chloroform solution at 30°C was obtained from Asahi Chemical Industry (M_n =25000; M_w =58000). Butyllithium (1.6M) in hexane, allyl halides, and tetrahydrofuran were used as received.

Allylation of PPE. A solution of PPE (5% by weight) in tetrahydrofuran was reacted under nitrogen with butyllithium. Butyllithium in hexane was added slowly with vigorous stirring. At various time intervals, lithiated PPE was reacted with an equivalent of allyl halide for 30 minutes. The polymer was recovered by precipitation into excess methanol and was dried in vacuum at 80°C for 14 hours.

Analytical Methods. The polymer was then examined by ^{1}H-NMR (AC 200P, Bruker) and FT-IR (JIR-100, JOEL). In the ^{1}H-NMR spectrum (Fig. 2), the allyl group attached onto PPE backbone was confirmed. This result was confirmed by FT-IR; a new absorption peak occurred at 913 cm^{-1}. This peak is the characteristic vinyl deformation band of the allyl group. Gel Permeation Chromatography measurements were performed on a HLC-8020 (Tosoh) with UV detector. Monodisperse polystyrene standards were used for calibration.

The crosslinking reaction of allylated PPE was confirmed by FT-IR; vinyl groups of the allyl group were crosslinked in an addition polymerization. Mechanical properties and thermal properties of allylated film (from chloroform solution) were measured by using tensile tester (TCM-500, Minebea) and thermal mechanical analyzer (TMA-10, Seiko Instrument), respectively. The dynamic mechanic mechanical properties were measured over a frequency range of 1 - 100 Hz and from -150°C to 300°C (Rheovibron® DDV-25FP, Orientec). Samples were prepared by compression of several sheets of allylated PPE film at 200°C for 60 min. Electrical properties at 1 MHz were measured by using impedance analyzer (4192F LF, YHP) with a dielectric test fixture.

Preparation of laminate. The allylated PPE glass-based copper clad laminate was prepared as follows. E-glass cloth was impregnated with allylated PPE based varnish to obtain prepreg for lamination. Six prepreg sheets and a set of copper foil were stacked and then pressed under 20 kg/cm^2 pressure at 200°C for 30 minutes. No postcure was carried out.

Evaluation of laminate. The general properties of the laminate were measured according to JIS standard C-6481.

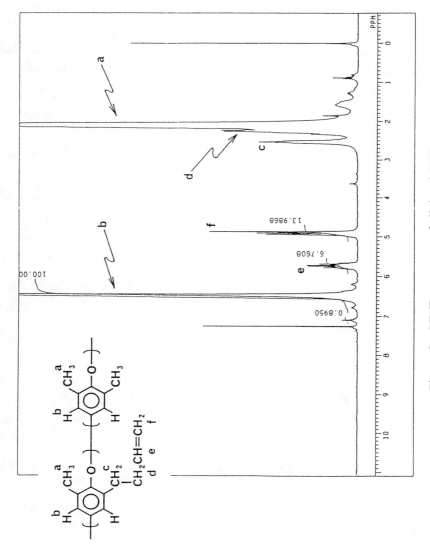

Figure 2. NMR spectrum of allylated PPE.

Table 2. Mole fractions of the moiety substituted onto PPE backbone[a].

| condition | | | substitution unit | | |
| allyl halide | lithiation | | allyl group | | halogen |
	temp(°C)	time (min)	methyl (mol %)	phenyl (mol %)	phenyl (mol %)
I [b]	40	5	2.5	-	13.2
	40	20	5.9	-	9.6
	40	60	10.6	-	4.8
	5	5	4.7	-	15.4
	5	20	3.1	-	9.6
	5	60	6.4	-	14.1
Br	40	5	9.2	3.9	9.3
	40	20	12.3	2.8	7.8
	40	60	15.6	2.9	2.7
	5	5	5.3	4.6	9.0
	5	20	7.8	5.3	9.5
	5	60	7.0	7.1	8.1
Cl	40	5	10.5	2.0	-
	40	20	14.3	1.5	-
	40	60	20.3	0.5	-
	5	5	6.9	0.9	-
	5	20	7.5	0.9	-
	5	60	9.0	0.8	-

a) Mole fractions were determined by ^{1}H-NMR.

b) Addition of allyl iodide caused gelation of the polymer immidiately under all conditions. These results were obtained by extracting with chloroform.

Results and Discussion

Syntheses of allylated PPE. PPE has two sites which can react with butyllithium. Introduction of lithium to PPE occurred at both the ring and the benzylic methyl positions. The former reaction is preferred initially, but isomerization is the benzylic methyl group occurs. Both phenyl lithiated PPE and benzyl lithiated PPE have the possibility to react with the allyl halide. Table 2 represents the mole fractions of the moiety substituted onto PPE backbone. The mole fractions were determined by ^{1}H-NMR. Allylation reactions were carried out at the lithiation temperatures for 30 minutes. From the preliminary allylation results, the reaction was very rapid and produced the corresponding reaction products to the lithiated PPE.

Addition of allyl iodide caused gelation of the polymer immediately under all conditions. Table 2 represents the analysis of the products which were extracted by $CHCl_3$. High degree of substitution for iodide linked directly on phenyl ring of PPE was found when the lithiation was carried out at low temperature (5°C) or at high temperature (40°C) for short (5 minutes).

When allyl bromide was used as allylation reagent, three different structural units formed. Figure 3 represents the relationship between lithiation time, temperatures, and the mole fraction of allylation. Allylation occurred at both the 3,5-position of the phenyl ring and the methyl group. Furthermore, bromination of PPE at the 3,5 position of the phenyl ring occurred. Methyl-substituted unit was formed mainly at high temperature and brominated PPE unit decreased with time. On the contrary, phenyl-substituted units via 3,5-lithiated PPE formed mainly at low temperature. There was no dependence of mole fraction of brominated PPE on time at low temperature. Furthermore, bromination occurred only at phenyl position.

In case of allyl chloride, allylation occurred dominantly at 2,6-position; the 3-allyl-2,6-dimethyl ether unit was found in trace amounts under all conditions. Moreover, chlorine attached onto PPE backbone was not detected by ^{1}H-NMR. Thus, the reaction of lithiated PPE with a series of allyl halides occurs by two competitive reactions. One is the allylation reaction and the other is metal-halogen exchange reaction.

The reaction mechanism of lithiated PPE and a series of allyl halides is explained in large part by metal-halogen exchange. General findings of this reaction are as follows (8): 1) The more organoalkali metal compound, RM, is present at equilibrium, the more readily the R group can support a negative charge, i.e. the more acidic is RH. pKa for hydrocarbons used in this work are Ph-$\underline{H}$ (pKa=39)>Ph-C$\underline{H_3}$ (pKa=37.5)>H_2C=CH-C$\underline{H_3}$ (pKa=36). 2) The rate of reaction is I>Br>Cl>F. For Cl and F the metal-halide exchange is sufficiently slow to allow other reactions to occur. 3) In competition with the rapid metal-halide exchange are metalation and alkylation.

The product for lithiation at low temperature or at high temperature for short times is mainly 3,5-lithiated PPE. This unit is allowed to react with allyl bromide or allyl iodide to form 3,5-halogenated PPE and 3,5-allyl-

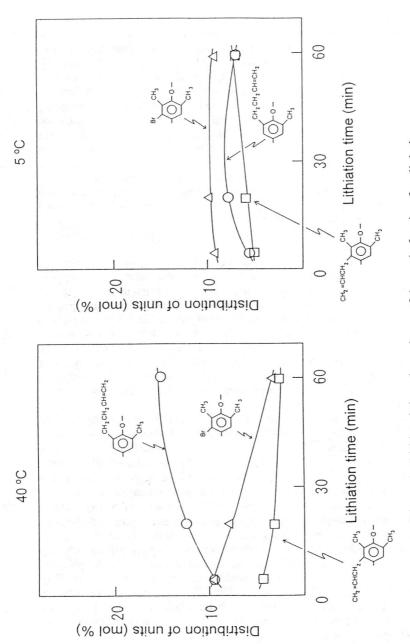

Figure 3. Lithiated time dependence of the mole fraction for allylation at 40 °C and 5 °C.

2,6-dimethyl phenylene ether unit. However, the 3,5-iodinated PPE is activated towards the formation of cross-linked gels by a Wurtz-type coupling. Allyl chloride, on the other hand, does not undergo exchange of chlorine with 3,5-lithiated PPE.

The main product of lithiation at higher temperature for longer periods is PPE lithiated at the side chain, indicating that isomerization of 3,5-lithiated PPE occurred (7). This species reacts with allyl halides to form, exclusively, PPE allylated at the side chain. No halogenation (metal-halogen exchange) was encountered. This is due to the small difference in pKa of Ph-CH$_3$ and allyl group. Scheme 1 represets the estimated reaction mechanism for allylation. Bromine in phenyl positions affects the thermal degradation of the polymer. From thermal gravimetric analysis (TGA), 10% weight loss temperature of allylated PPE containing 4% of bromine is 437°C although that of allylated PPE free from bromine is 456°C. This different is attributed to the dissociation energy of the weak Ph-Br bond (72 kcal/mol) compared to Ph-H bond (112 kcal/mol). In order to obtain excellent thermal stability of the polymers, lithiation conditions were kept constant at 40°C for 60 minutes in this work.

The efficiency of lithiation by nBuLi under the conditions investigated by examined by varying the butyllithium PPE molar ratio. The relationship between the degree of substitution by the allyl group shows good linearity at less than 25% of the degree of substitution. The result indicates that allylation proceeds almost quatitatively and that the efficiency of lithiation is more than 80%. See Figure 4.

Figure 5 represents the molecular distribution of (a) PPE, (b) recovered PPE which was obtained from lithiated PPE by quenching with methanol, and (c) allylated PPE. Allyl bromide was used and lithiation conditions were for 60 minutes at 40°C in this study. The molecular weight distribution profile of the samples were examined by GPC. These samples were essentially the same, except for a slight increase of the molecular weight with the progress of the reaction. This result suggests that the mechanism of increase in molecular weight occurs by two reactions, i.e. metalation and allylation.

The characteristics of the allylation reactions, such as one step reaction, reaction time less than 2 hours, reaction temperature less than 40°C, atmospheric pressure, and efficiency of n-butyllithium of more than 80%, suggest that it may be possible for this process to be operated commercially.

Properties of Allylated PPE. A series of properties were investigated using the polymer allylated at the methyl group. Figure 6 represents the crosslinking reaction of neat allylated PPE and that of allylated PPE in the presence of a peroxide. The crosslinking reaction was examined by the decrease of absorption of the vinyl group in the allyl group and the polymerization between intra- or intermolecules. This study used 3% by weight of 2,5-dimethyl-2,5-di(t-butylperoxy)-3-hexene as a peroxide. Neat resin was thermally cured above the temperature of 250°C for 30 minutes. On the other hand, crosslinking reaction proceeds easily using a peroxide at 180°C for 30 minutes. It should be noted that the curing

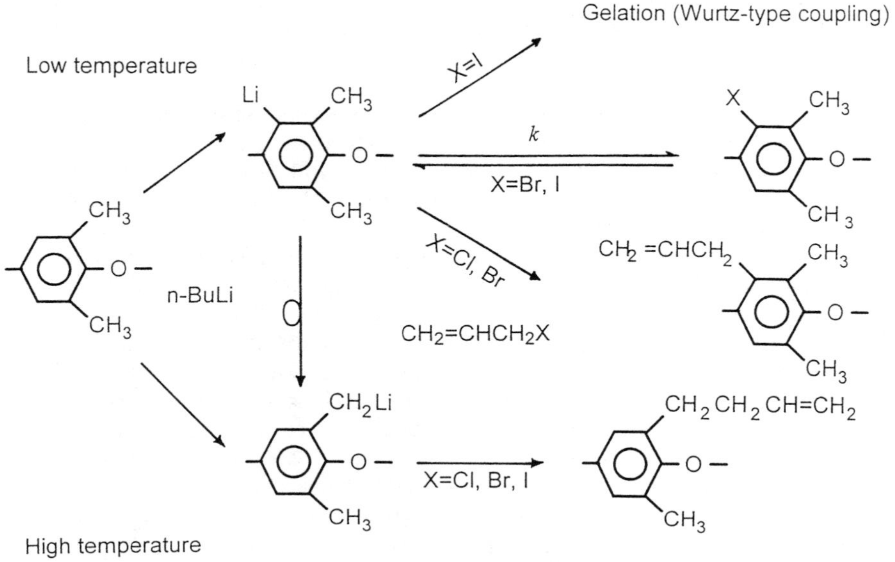

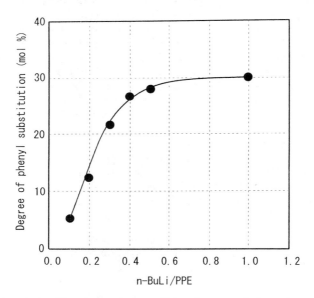

Scheme 1. Proposed mehcanism for allylation of PPE.

Figure 4. The dependence of the degree of phenyl
           substitution by allyl group from allyl bromide
           on n-BuLi/PPE ratio.

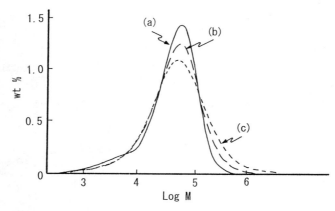

Figure 5.   Molecular weight distributuion of PPE(a), recovered PPE which was obtained from lithiated PPE by quenching with methanol(b), and allylated PPE(c).

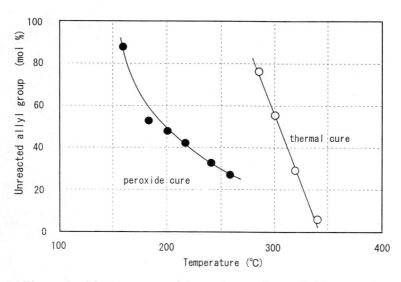

Figure 6.   The temperature dependence of crosslinking reaction.

Table 3. Properties of PPE and cured allylated PPE.

Property	Condition	Unit	PPE	A-PPE (cured)
Electric				
Dielectric constant	1 MHz		2.45	2.5
Dissipation factor	1 MHz		0.0007	0.001
Thermal				
Glass transition temp (DMA)	10 °C/min	°C	210	250
Solvent Resistance				
Trichloroethylene			Soluble	Insoluble
Toluene			Soluble	Insoluble
Acid			Insoluble	Insoluble
Alkali			Insoluble	Insoluble
Physical	23°C, 24hr	%		
Water absorption	23°C	kg/cm	<0.05	<0.05
Peel strength	23°C		N/A	1.7
Specific gravity			1.06	1.06

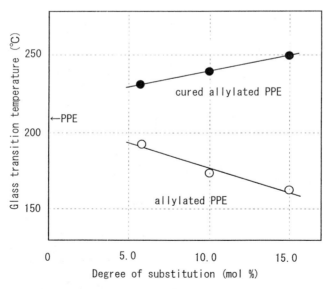

Figure 7.   Glass transition behavior of neat allylated
            PPE and cured allylated PPE in the
            presence of a peroxide.

temperature in the presence of a peroxide is the same as that of commercial epoxy laminates. Cured allylated PPE shows superior solvent resistance against common organic solvents which are good solvents for PPE. Table 3 lists the solubility of PPE and cured allylated PPE with a perioxide. PPE has good solubility against halogenated hydrocarbons and aromatic hydrocarbons. On the contrary, cured allylated PPE is insoluble due to the crosslinking effect.

We further studied the glass transition behavior and mechanical properties of these systems. Figure 7 represents glass transition behavior of allylated PPE and cured allylated PPE. The introduction of the allyl group lowered the glass transition temperature of the polymer. An approximately linear relationship exists between the glass transition temperature and the degree of substitution of allyl group. The Tg of bulk PPE (at 210°C) is presumably associated with main chain orientational relaxation (9). Introduction of allyl group is supposed to make this relaxation easy due to the increase of environmental free volume near the PPE backbone. On the other hand, introduction of cured allylated PPE showed the opposite result (filled circle) by raising the glass transition temperature itself. This result is due to the restriction of the main chain by crosslinking. Fig. 8 and Fig. 9 represent tensile modulus and tensile strength of allylated PPE (open circles) and cured allylated PPE (filled circles) in the presence of a peroxide. There is no difference in tensile modulus. On the contrary, tensile strength is highly influenced by the degree of substitution, i.e. the introduction of the allyl group raises the extensibility of the polymer. Tensile strength can be also raised if the polymer is cured in the presence of a peroxide. The tensile strength of cured allylated PPE, shown in Fig. 9, is assumed to be associated with the entanglement effect due to the introduction and crosslinking of the allyl group. This entangle effect may result in a film forming property of the polymer. Transparent film can be obtained easily from allylated PPE by means of solvent cast method. It should be noted that PPE has poor film forming ability and only opaque and cracked film is obtained by the solvent cast method.

The viscoelastic properties were measured by a dynamic mechanical analyzer using the allylated PPE with a degree of substitution of 15 mole %. Figure 10 and Fig. 11 represent the viscoelastic behavior of allylated PPE and cured allylated PPE at 20 Hz. Cured allylated PPE sample was prepared by compression at 200°C for 60 minutes with 3 wt % of a peroxide. In this study, 2,5-dimethyl-2,5-di(t-butylperoxy)-3-hexene was used as a peroxide. The α relaxation for allylated PPE occurred at 210°C which is similar to that of PPE above 200°C. However, a new shoulder of tan δ appeared at 260°C. The storage modulus was also raised gradually above 270°C. This phenomena is attributed to the crosslinking reaction of allylated PPE during the measurement. Allylated PPE has no β relaxation assigned to hindered tortional oscillations of main chain phenylene units. This result is probably due to the unsymmetrical introduction of the allyl group and is in agreement with Cayrol's work that a series of 2-methyl-6-alkyl-substituted phenylene ether has no β relaxation attributed to packing effect (10).

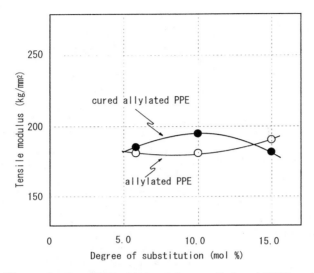

Figure 8. Tensile modulus of neat allylated PPE and
           cured allylated PPE in the presence of a
           peroxide.

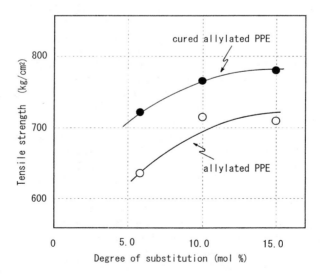

Figure 9. Tensile strength of neat allylated PPE and
           cured allylated PPE in the presence of a
           peroxide.

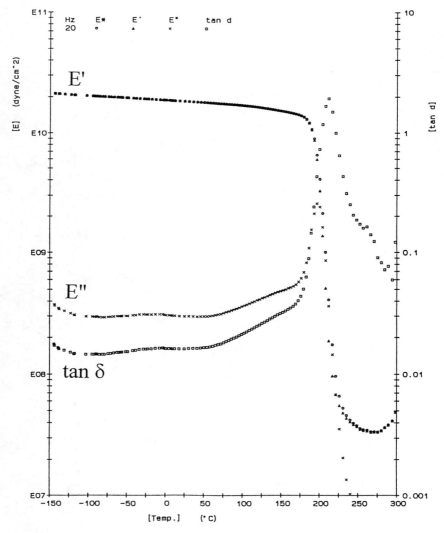

Figure 10. Dynamic mehanical behavior of neat
allylated PPE measured at 2 °C / min
heating rate and 20 Hz frquency.

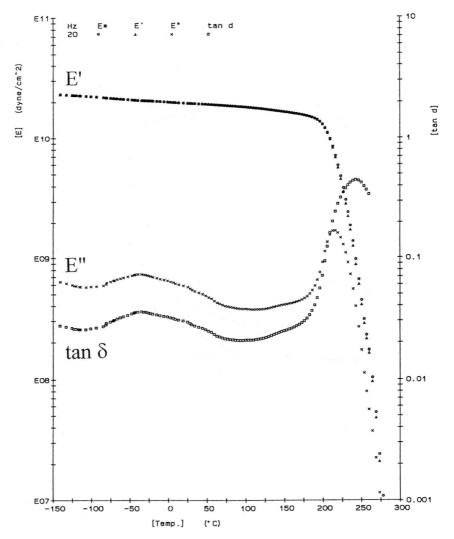

Figure 11.   Dynamic mehanical behavior of cured
allylated PPE measured at 2 °C / min
heating rate and 20 Hz frquency.

The effect of crosslinking was drastic for cured allylated PPE. The intensity of tan δ is lowered and the top of the peak shifted to 250°C. This is due to the crosslinking effect on α relaxation. A new broad relaxation can be found from -100°C to 50°C. This peak is probably either a fine structure from peroxide attached to allylated PPE or the crosslinked moiety of allyl group.

Table 3 lists the electrical properties, thermal properties, chemical properties, and physical properties of PPE and cured allylated PPE. With regard to allylated PPE, the low dielectric constant of 2.5 and low dissipation factor of 0.001 of allylated PPE were obtained without sacrificing the excellent dielectric properties of PPE. It should be noted that these electical properties did not show any meaningful changes after a water absorption test. The resins have these excellent properties because of the absence of polar functional groups. These characteristics have never been matched by conventional thermosetting polymers, such as epoxies, polyimides, and cyanate esters, which cure through reactions of polar functional groups. Light specific gravity of 1.06 is also favorable for the material of mobile communication devices.

Application to laminate. Fabrication of prepreg and copper clad laminate can be done using processes which are common for conventional thermosetting resins. Prepreg is fabricated using a conventional treater. Prior to prepreg fabrication, a peroxide is added. Copper clad lamination for the resin can be done at temperatures ranging from 170 to 200°C for 30 minutes to 2 hours in a similar cycle for epoxy laminates. Although most high Tg resins require a postcure, allylated PPE requires no postcure. Table 4 represents general properties of allylated PPE resin laminate with these of polyimide and FR-4 laminates. Our targeted dielectric constant of 3.0 and Tg of 250°C are achieved in allylated PPE laminates. It should be noted that allylated PPE prepreg has superior storage stability of more than 1 year at 23°C, indicating that allylated PPE shows poor polymerization ability compared with the products of previous work (3,4). Another feature is powder free prepreg due to the film forming property of allylated PPE. Allylated PPE prepreg is quite different from the powder-like prepreg of epoxy resins.

Conclusions

Allylated PPE, a new high performance thermosetting resin, has been developed for printed circuit board applications. Allylation reactions show excellent characteristics such as a one step reaction, short reaction time, low reaction temperature, normal atmospheric pressure, and an efficiency of butyllithium of more than 80%. Cured allylated PPE shows excellent solvent resistance and superior glass transition temperature. Excellent electrical properties are obtained without sacrificing the dielectric properties of PPE. Fabrication of prepreg and copper clad laminate can be done using processes which are common for conventional thermosetting resins. This unique resin has a low dielectric constant, a low dissipation factor, high Tg, excellent adhesion to copper foil, and low water absorption. This new resin should be of great value in high

Table 4. Application of allylated PPE for printed circuit boards.

Property	A-PPE Resin Laminate	Polyimide Resin Laminate	PTFE Laminate
Electric			
Dielectric constant (at 1MHz)	3.0-3.5	4.4-4.7	2.5-2.7
Dissipation factor (at 1MHz)	0.002-0.003	0.015	0.001-0.0015
Thermal			
Glass transition temp (°C)	250	260	25
Chemical			
Flammability	V-0	V-0	V-0
Solvent resistance			
(Trichloroethylene)	Insoluble	Insoluble	Insoluble
(Acid)	Insoluble	Insoluble	Insoluble
(Alkari)	Insoluble	Decomp	Insoluble
Physical			
Coefficient of thermal expansion Z-axis (ppm/ °C)	80-100	40	240
Water absorption (%)	0.1-0.2	0.3	0.05
Peel strength (kg/cm)	1.4-1.6	1.2-1.6	2.2
Specific gravity	1.44-1.64	1.86	2.24

speed/high frequency printed circuit board applications. It should be also useful in applications which require good thermal properties, low water absorption, and good adhesion, to copper foil.

References

1. Takahashi, A.; Nagai, A.; Wajima, M.; Tsukanishi, K.; "Low Dielectric Material for Multilayer Printed Wiring Boards", IEEE transactions of components, hybrid, and manufafuing technology. 1990, 13(4), 1115.
2. Aycock, D.; Abolins, V.; White, D. M.; Encyclopedia of Polymer Science and Engineering; John Wiley & Sons: New York, 1985, Vol 13; 1-30.
3. Hay, A. S,; J. Polym. Sci. **1962**, 58, 581.
4. Kurian, C. J.; Charles, C, Price.; J. Polym. Sci. **1961**, 49, 267.
5. Nishide, H.; Minakata, T.; Tsuchida, E.; Makromol. Chem. **1982**, 183, 1889.
6. Percec, V.; Auman, B.C.; Makromol. Chem. **1984**, 185, 2319.
7. Chalk, A. J.; Hay, A. S.; J. Polym. Sci. A-1. **1969**, 7,691.
8. Wardell, J. L.; Comprehensive Organometallic Chemistry; Pergamon: London, 1982; Vol.1; 43-120.
9. Karasz, F. E.; MacKnight, W. J.; Stoelting, J.; J. Appl. phys. **1970**, 41(11), 4357.
10. Cayrol, B.; Eisenberg, A.; Harrod, J. F.; Rocaniere, P.; Macromolecules. **1972**, 5, 6.

RECEIVED July 7, 1995

Chapter 33

Synthesis and Photochemistry of a 2,6-Dialkoxyanthracene-Containing, Side-Chain-Substituted Liquid-Crystalline Polymer

David Creed[1], Charles E. Hoyle[2], Anselm C. Griffin[1], Ying Liu[1], and Surapol Pankasem[2]

[1]Department of Chemistry and Biochemistry and [2]Department of Polymer Science, University of Southern Mississippi, Hattiesburg, MS 39406

The synthesis and several aspects of the photophysical and photochemical behavior of a methacrylate polymer substituted, *via* a flexible methylene chain, with a 2,6-dialkoxyanthracene chromophore are reported. Polarized light microscopy and DSC indicate the polymer is liquid crystalline. The UV-Vis and fluorescence spectra of the polymer indicate both ground and excited state aggregation of the chromophores as evidenced by perturbations of these spectra relative to those of a simple model compound. Chromophore association in the ground state seems to occur even in a good solvent, dichloromethane. Chromophore aggregation effects are enhanced in films of the polymer, particularly after heating or when poor solvents are added to a solution of the polymer in dichloromethane. Steady state and time-resolved fluorescence experiments indicate significant self-quenching of the excited anthracene chromophore even in highly diluted solutions in dichloromethane and the presence of a long-lived (τ = 30.6ns) intramolecular excimer or excited aggregate and a short-lived (τ = 2.0ns) anthracene singlet state. At higher concentrations an intermolecular excimer or excited aggregate is observed. Fluorescence spectra of films indicate the presence of weakly emissive excimers or excited aggregates. Two types of 4 + 4 photocycloaddition products are most likely formed upon UV or visible light irradiation. Photooxidation products are also formed when irradiation is carried out in the presence of oxygen.

The photochemistry of liquid crystalline (LC) polymers is of fundamental interest because LC states combine some of the characteristics of both highly ordered crystals and disordered but mobile liquids. Crystals are 3-dimensionally ordered but the molecules in a crystal can only undergo relatively small movements about their

0097–6156/95/0614–0504$12.00/0

equilibrium positions in the lattice. In an isotropic liquid, the molecules can move much more freely but there is no long-range order. LC materials combine 1- or 2-dimensional order with some degree of fluidity. LC materials also have a variety of applications or potential applications in which they are inadvertently or deliberately subjected to ultra-violet and visible irradiation with frequently unpredictable effects on their structures and properties. Irradiation may be deliberately used to modify structural properties or in photoimaging applications of a LC polymer. Alternatively, a structural polymer may be degraded by sunlight, or a polymer designed for non-linear optical (NLO) applications may be sensitive to the radiation source used in the NLO-based device. All of these considerations have prompted us to begin a comprehensive study of the influence of LC structure on the photochemistry of polymers.

We have recently reported several aspects of the photochemical and photophysical properties of LC polymers (*1*), including main- (*2*) and side-chain (*3*) substituted poly(aryl cinnamates) and main-chain poly(stilbene-bis-carboxylates) (*4*). In all of these materials, a conjugated chromophore, either aryl cinnamate or stilbene bis-carboxylate, serves as the rigid mesogenic group in the polymer. Among the most interesting effects observed to date is the formation of chromophore aggregates which results in the observation of perturbed UV-Vis absorption spectra of all these materials, wavelength dependent photochemistry (*5*) of the poly (aryl cinnamates), presumably as a consequence of the chromophore heterogeneity that accompanies incomplete or cooperative chromophore aggregation, and hyperchromism upon initial irradiation of fluid LC phases of these materials (*1-3*), that is attributed to disruption of chromophore aggregates as photoproducts begin to form and affect phase behavior. The extent of aggregation is dependent on the LC mesophase. Smectic mesophases, which have 2-dimensional ordering with layering of the mesogenic (chromophoric) groups, show more dramatic spectral perturbations (*3*) than do nematic mesophases, which have only 1-dimensional ordering of the mesogens. We suspect that aggregation driven shifts of absorption spectra to longer wavelengths may contribute to the environmental degradation of many polymers. In this paper we report the synthesis and some preliminary observations of the photophysics and photochemistry of a side-chain substituted LC polymer, **4**, containing a different, photochemically reactive 2,6-dialkoxyanthracene chromophore. Our studies of this type of material were motivated by the well known reversible 4 + 4 photocycloaddition chemistry of anthracene (*6*), the ability of anthracene to fluoresce, which is an advantage in studying the fate of excitation energy absorbed by the polymer, and the many previous reports of the photophysics and photochemistry of anthracenes and anthracene containing polymers.

Experimental

Materials and Characterization. All starting materials were obtained from Aldrich. Methacryloyl chloride was distilled under reduced pressure before use. 2,2'-azobis(isobutyronitrile) was recrystallized from methanol and dried *in vacuo*. Dimethylformamide (DMF) and benzene were dried over CaH_2 and distilled (DMF

Scheme. Synthesis of Polymer, 4.

under reduced pressure). Tetrahydrofuran was dried over $LiAlH_4$ and distilled under argon. All other solvents were spectroscopic grade. UV-Vis, fluorescence, FT-IR, and NMR spectra were obtained using Perkin Elmer Lambda 6, Spex Fluorolog-2, Mattson Polaris, and 300 MHz Bruker spectrometers respectively. Phase transitions were observed with a Reichert Thermovar microscope and Mettler FP5/52 hot stage. Differential scanning calorimetry (DSC) was done using a Perkin Elmer DSC7 at a heating rate of 10°C min^{-1} and gel permeation chromatography (GPC) using a Waters model 410 differential refractometer and model 6000 solvent delivery system. The column used was calibrated with polystyrene standards. Films were spin cast from solutions of the polymer in CH_2Cl_2 on a Headway Research EC101D photoresist spinner. Temperature controlled studies and UV-Vis irradiations were done as previously reported (*1-4*). Polymer films reach the desired temperature within ca. 20 min in the hot stage used in our UV-Vis spectrophotometer. Fluorescence lifetimes were obtained using an Edinburgh Instruments FL 900 CDT fluorescence lifetime spectrometer with a nF 900 nanosecond flashlamp which produces a light pulse of 0.8ns FWHM using nitrogen as a fill gas. Fluorescence quantum yields were obtained upon 265nm excitation using the Spex Fluorolog with 9,10-diphenylanthracene as a fluorescence standard.

2-Decyloxy-6-(6'-hydroxy)hexyloxyanthraquinone, 1, 2,6-Dihydroxyanthraquinone, (10g) and anhydrous K_2CO_3 (20g) were refluxed in DMF (50 ml) under nitrogen for 1.5 hours. 1-Bromodecane (7.5g) was added dropwise over the first 40 minutes. The product was precipitated in aqueous NaOH, washed with NaOH, and recrystallized from glacial acetic acid. The crude product, 2-decyloxy-6-hydroxyanthraquinone, (3g), 6-bromohexanol (2.8g), and anhydrous K_2CO_3 (4.5g) were refluxed in DMF (30ml) for two hours under N_2. The mixture was poured into cold water, filtered, dried, and purified by column chromatography (silica gel/CH_2Cl_2). Yield of yellow solid, **1**, 1g, 27%. ^1H-NMR (CDCl$_3$ vs. TMS) δ 8.25/8.21, 7.70, 7.24/7.20 (6H, aryl), 4.15 (4H, m, -OCH$_2$-), 3.69 (2H, t, HOCH$_2$-), 0.89-1.88 ppm (27H, aliphatic) .

2-Decyloxy-6-(6'-hydroxy)-hexyloxyanthracene, 2. The product (0.83g) from the previous reaction and NaBH$_4$ (0.26g) were refluxed in 2-propanol (15ml). The mixture was poured into water and the precipitate filtered, washed with ether, treated with boiling conc. HCl (20ml), filtered again and washed thoroughly with water and ether and dried *in vacuo*. The yellow solid product (0.6g), an anthrone intermediate not separately characterized, was subjected to the same treatment with NaBH$_4$. The crude product from this second reduction was recrystallized twice from CHCl$_3$, washed with ether and dried (vacuum oven). Yield of yellow powder, **2**, 0.46g, 59%. UV-Vis (95% ethanol), 261, 292, 306, 320, 336, 362, 381, and 401nm; ^1H-NMR (CDCl$_3$ vs. TMS) δ 8.17, 7.86, 7.16/7.12 (6H, s, s, d, aryl), 4.11 (m, 4H, OCH$_2$), 3.69 (t, 2H, HOCH$_2$), 1.87, 1.54, 1.29, 0.88 ppm (27H, m, aliphatic).

6-(2-(6-Decyloxy)anthroxyl)hexyl Methacrylate, 3. Freshly distilled methacryloyl chloride (0.3ml) in dry THF (10ml) was added dropwise over a period of 30 min. to the product, **1**, (0.46g) from the previous step and pyridine (2ml) in THF (30ml). The mixture was stirred for 4 hours, poured into water and extracted with ether. The extracts were washed (dilute HCl, K_2CO_3, water) and dried (Na_2SO_4). After solvent removal, the yellow product was purified by column chromatography (silica gel/CH_2Cl_2). Yield of **3**, 0.23g, 43%. FT-IR (KBr) 3057, 2921, 2852, 1720, 1628, 1475, 1198, 888cm^{-1}. ^{1}H-NMR (CDCL$_3$ vs. TMS) δ 8.16, 7.84/7.79, 7.15 (6H, s, d, m, aryl), 6.10 (1H, s, =CH), 5.60 (1H, s, =CH), 4.20-4.05 (6H, m, -OCH$_2$-), 1.90-0.88 ppm (27H, m, =CCH$_3$CO-, aliphatic Hs)

Poly (6-(2-(6-decyloxy)anthroxyl)hexyl methacrylate), 4, The monomer, **3**, (0.6g) and AIBN (3.8mg, 2 mole %) were dissolved in dry benzene (ca. 9ml) and purged with N_2. The mixture was kept at 60°C for 24 hours and the product precipitated from methanol four times and dried *in vacuo*. Yield of yellow solid, **3**, 0.4g, 67%. DSC g80°CLC167°C. FT-IR (KBr) 3058, 2922, 2853, 1727, 1620, 1464, 1201, 886cm^{-1}. ^{1}H-NMR (CDCl$_3$ vs. TMS) δ 7.90 (2H, m, ArH), 7.61 (2H, s, ArH), 7.03 (4H, m, ArH), 3.88 (6H, m, -OCH$_2$-), 1.90-0.90 ppm (32H, m, aliphatic Hs). Found, C 78.72%; H 8.80%. Calculated, C 78.76%; H 8.88% The molecular weight of the polymer is ca. 62,000 based on GPC measurements with polystyrene standards.

2,6-Didecyloxyanthracene. This model compound was synthesized as described in the literature (7). Yellow crystals from CHCl$_3$, m.p. 140-141°C [lit. 141-143°C (7)], UV-Vis in CH_2Cl_2 (log ε) 264 (5.435), 295 (3.446), 309 (3.726), 324 (3.869), 340 (3.656), 364 (3.470), 384 (3.804), 405nm (3.878).

Results and Discussion

Polymer Synthesis and Characterization. The 2,6-disubstitution pattern for the target polymer was chosen to increase the probability that the polymer would exhibit LC behavior. The polymer was synthesized from 2,6-dihydroxyanthraquinone (Scheme) by first monoalkylating with 1-bromodecane then alkylating the other hydroxy group with 6-bromohexanol. The resultant dialkylanthraquinone, **1**, was then reduced in two steps (an intermediate anthrone was not characterized) using sodium borohydride to give the dialkoxyanthracene, **2**, with a 6'-hydroxy substituted alkoxy substituent. Esterification with methacryloyl chloride to give monomer, **3**, was followed by AIBN initiated polymerization to give polymer, **4**. The first step of the polymer synthesis, monoalkylation of 2,6-dihydroxyanthraquinone, afforded symmetrically dialkylated anthraquinones as by-products. These impurities persisted in the subsequent reactions but were finally removed in the polymer purification step, presumably because they remain in solution when the polymer is precipitated out of benzene with methanol. Because of the difficulty in removing symmetrically substituted impurities from the various small molecule intermediates, we only report elemental analysis for the polymer and not for the intermediates. However, satisfactory UV-Vis, FT-IR, and NMR data were obtained for all the intermediates.

The number average molecular weight of the polymer is 62,000 by GPC relative to polystyrene standards. The pure polymer when heated above 167°C is a yellow, viscous, non-birefringent liquid, presumably the isotropic melt. Upon cooling below about 158°C (T_i) a highly viscous birefringent phase is obtained that can be sheared. Hence, the polymer is liquid crystalline. However, the texture observed using the light microscope is not well enough defined, even after annealing for several hours just below T_i, to enable us to assign the phase as either nematic or smectic. The DSC heating curve shows a T_g at about 80°C and a clearing (isotropization) transition at 167°C. On cooling a sharp exotherm is seen at 158°C with a shoulder on the low temperature side indicating the possibility of a second (smectic?) mesophase.

UV-Vis Spectra. The UV-Vis spectrum of the polymer, **4**, in CH_2Cl_2 is very similar to that of the model compound, 2,6-didecyloxyanthracene (Fig. 1). Both compounds have a strong ^1L_b band at around 260nm and weaker, structured ^1L_a absorption out to about 425nm. The principal difference is the strong shoulder at 257nm in the polymer spectrum, absent for the model compound. We believe this is due to aggregation of the anthracene chromophores in the polymer that is occurring even in the 'good' solvent, CH_2Cl_2. Other explanations, for example a contribution to the absorption from the polymer chain, seem unlikely since this band must be comparable, in molar absorptivity, to that of the strongly allowed anthracene transition. That this aggregation must be *intramolecular* is demonstrated by the absence of a dilution effect on the spectrum shown in Figure 1. An 'as cast' film of the polymer shows a greatly perturbed UV-Vis spectrum (Fig. 2). The spectrum below 300nm is quite different from that of the polymer in CH_2Cl_2 and there are subtle changes to the bands above 300nm. Whereas the long wavelength absorption of the model compound and the polymer in dichloromethane terminates at ca. 425-430 nm (Fig. 1), the absorption of the polymer film can be seen to extend out to ca. 440 nm, although accurate measurement is hampered by baseline changes caused, presumably, by scattering off the film. Further quite dramatic changes are seen when the polymer film is heated up to 155°C (in six steps over a period of ca. 2 hours). Unfortunately, we are unable to reach T_i and thereby run UV-Vis spectra in the isotropic phase with our present heating equipment. On the second heating and cooling cycle (and on subsequent cycles - not shown), the spectra become reproducible (Fig. 3) and quite different from the solution and 'as cast' spectra. Cooling the polymer film never leads to the same spectrum that is observed for the newly cast film (Fig. 2). Heating leads to a loss of the new (not seen in CH_2Cl_2) maximum (235nm) and an increase in the 'isolated chromophore' band (260nm). We believe this change reflects the dissociation of weak complexes upon heating. We can generate a spectrum similar to that observed on the second cooling cycle by adding an equal volume of the poor solvent, methanol, to a solution of the polymer in CH_2Cl_2 (Fig. 4). We have previously reported (*1-4*) quite dramatic changes in the UV-Vis spectra of aryl cinnamate and stilbene containing polymers that occur in fluid and glassy LC phases of thin films of these polymers and in poor solvents. We believe all these changes are due to chromophore aggregation that is enhanced in the

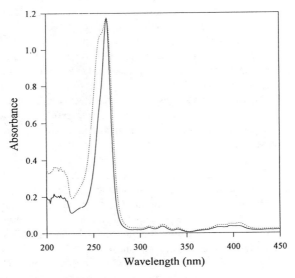

Figure 1. UV-Vis absorption spectra of polymer (dotted line) and model compound (solid line) in dichloromethane at room temperature.

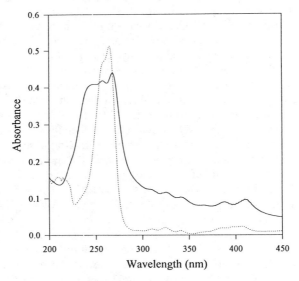

Figure 2. Normalized UV-Vis absorption spectra of polymer in dichloromethane (dotted line) and as an 'as cast' film (solid line) at room temperature.

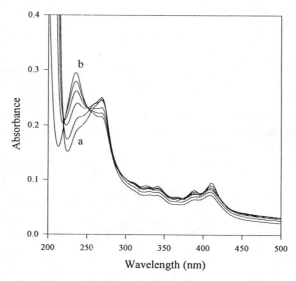

Figure 3. UV-Vis absorption spectra of a thin polymer film during the second cooling from 155 (a) through 140, 122, 100, 81, and 51°C (b).

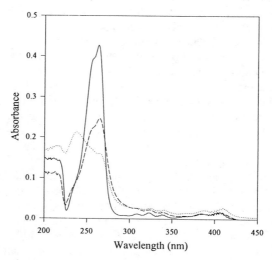

Figure 4. UV-Vis absorption spectra of polymer in good and poor solvents at room temperature. CH₂Cl₂ (solid line), CH₂Cl₂ and CH₃OH, 2:1 (dashed line), and CH₂Cl₂ and CH₃OH, 1:1 (dotted line). The polymer concentration is identical in each solvent.

LC phases and when poor solvents are added to solutions of the polymers in good solvents. Others (8-11) have also reported perturbations of the UV-Vis spectra of films of LC polymers.

Fluorescence. The model compound, 2,6-didecyloxyanthracene, has a single fluorescence band at 441nm in dilute solution in dichloromethane (Fig. 5). This emission shows single exponential decay with a lifetime of 23.7ns under nitrogen. A previous report (7) indicates very little vibrational structure in the fluorescence of this compound in the hydrocarbon solvent, methyl cyclohexane. The more polar solvent, dichloromethane, would tend to reduce vibrational structure relative to a hydrocarbon solvent. It was necessary for us to use dichloromethane in order to be able to compare fluorescence spectra of the model compound and polymer in the same solvent. A dilute solution (4 x 10^{-7} g ml^{-1}) of the polymer, **4**, in dichloromethane shows very similar unstructured fluorescence with a peak maximum at somewhat longer wavelength (447nm) but significantly broader with a weak 'tail' extending out to almost 700nm (Fig. 5). In contrast, the fluorescence of the model compound drops to near zero at about 600nm. Examination of the normalized spectra of the model compound and the polymer (Fig. 5) suggests that the latter has a broad emission with λ_{max} at 525-575nm in addition to the model like fluorescence at ca. 450nm. Further dilution of the polymer solution results in no further change in the shape of the fluorescence spectrum. The fluorescence quantum yields of the model compound and polymer are 0.77 and 0.15 respectively, suggesting the excited anthracene chromophore is heavily quenched even under conditions (high dilution) where intermolecular interactions do not occur during the lifetime of the singlet states. At higher concentrations (1.5 x 10^{-5} g ml^{-1}) the polymer shows a distinct shoulder at about 550nm (Fig. 5). It seems very likely that both the weak red-shifted emission in very dilute solution and the distinct shoulder at 550nm in more concentrated solution (Fig. 5) are due to intramolecular and intermolecular excimers or excited aggregates respectively. These results are supported by fluorescence lifetime experiments which indicate that two species with lifetimes of 2.0 and 30.6 ns contribute to the polymer fluorescence at 450nm in dilute solution. The wavelength dependence of these species suggest the shorter lived emission is due to the excited anthracene chromophore and the longer lived to the excimer or excited aggregate.

　　　Addition of 'poor solvents' to the more concentrated solutions of the polymer in dichloromethane leads, in the case of hexane, to initial enhancement of the 550nm emission followed by the loss of this emission at higher hexane concentrations , or, in the case of dioxane, to suppression of the 550nm emission at all proportions of dioxane. Since increasing proportions of poor solvents must ultimately favor ground state aggregation by both chain coiling and intermolecular association, the 550nm emission may be due to a true excimer or so-called 'pre-formed excimer' rather than to an excited aggregate. Interestingly, only broadened red-shifted emission but no distinct 550nm 'shoulder' is seen in pure polymer films (Fig. 6) and concentrated (25% by weight) dispersions of the model compound in polystyrene, conditions under which ground state aggregation is presumably favored. Heating the polymer film above T_i and then cooling down to room temperature results in further broadening of the fluorescence spectrum (Fig. 6). The absence of well defined excimer or excited aggregate fluorescence in the steady state emission spectra of films perhaps reflects rapid photochemical reaction, self-quenching or other radiationless decay of the excited singlet state in the (condensed) film. Two emissions, one (unstructured) from an excimer, the other (structured) from an excited aggregate have been reported (12) from anthracene. It does not seem improbable that two or more excimers or excited aggregates might ultimately be observable in the present, more complex, case. We are currently attempting to try to resolve this more complex behavior using a combination of steady state and time-resolved fluorimetry. Preliminary lifetime experiments indicate that 'as cast' films of the polymer have at least three emitting species whose proportions and lifetimes are dependent on the thermal history of the film. However, definitive assignments of the origin(s) of emissions in complex cases such as this may not be possible.

Photochemistry. Irradiation of the polymer, **4**, under N_2 in solution or in the film results in loss of the characteristic absorption of the anthracene chromophore (Fig. 7). Films become partially insoluble. Loss of absorption above 350nm is much greater than the loss between 280 and 350nm (Fig. 8). This latter observation and the appearance of a weak fluorescence band at about 350nm in samples irradiated under N_2 in dichloromethane leads us to believe that <u>two</u> 4 + 4 cycloadducts are formed; one from the usual 9,9' to 10,10' coupling of two anthracenes, the other from a less common 1,9' to 4,10' coupling leading to dimer, **5**. This latter reaction has been previously reported (13) from irradiation of the model compound, 2,6-didecyloxyanthracene. The 4 + 4 photocycloadduct, (**5** is the general structure), has a 6,7-dialkyl-2-alkoxynaphthalene chromophore in addition to the 3,4-dialkylalkoxybenzene chromophore that is the only contributor to the weak absorption of the 9,9' to 10,10' dimer. 2-Methoxynaphthalene, a model for the naphthalene chromophore in **5** has absorption and fluorescence at the wavelengths where these are observed for the irradiated polymer. In addition to the weak emission assigned to the naphthalene chromophore in **5**, we observe that irradiation of the polymer in the more concentrated solution results in loss of the distinct excimer emission band at 550 nm. Presumably photocycloaddition occurs until the residual, unreacted anthracenes have no neighbors with which to interact *via* excimer and dimer formation. Naphthalene like emission is not seen in irradiated films most probably because of efficient energy transfer from the alkoxynaphthalene chromophore of **5** to residual anthracene chromophores. There is an excellent overlap between the naphthalene emission in the model compound, 2-methoxynaphthalene and the absorption spectrum of the model compound, 2,6-didecyloxyanthracene, ideal conditions for dipole-dipole ('Forster type') singlet-singlet energy transfer. Irradiations carried out in the presence of oxygen in both solution and thin films lead to more complex UV-Vis spectral changes. Yellowing is much more noticeable and is attributed to the well known tendency of anthracenes to undergo photooxidation across the 9,10 positions leading initially to endoperoxides but, ultimately, to more highly colored 9-anthrone and 9,10-

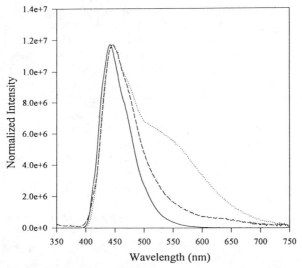

Figure 5. Normalized corrected fluorescence spectra of model compound (1.8 x 10^{-6} g ml^{-1}, solid line), dilute polymer (dashed line, 4.0 x 10^{-7} g ml^{-1}), and more concentrated polymer (dotted line, 1.5 x 10^{-5} g ml^{-1}) excited at 265nm in dichloromethane at room temperature.

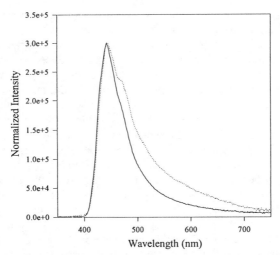

Figure 6. Normalized corrected fluorescence spectra of polymer film, at room temperature. 'As cast' film (solid line), and after annealing above T_i (dotted line).

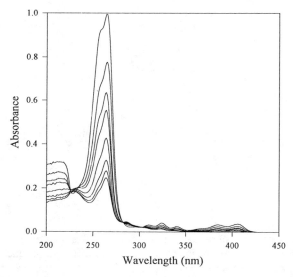

Figure 7. UV-Vis absorption spectra (200-450nm) of a dilute solution of polymer after irradiation (366nm) in dichloromethane under nitrogen at room temperature. From top to bottom: 0, 0.5. 1.0, 1.5, 2.0, 3.0, 5.0, and 8.0min.

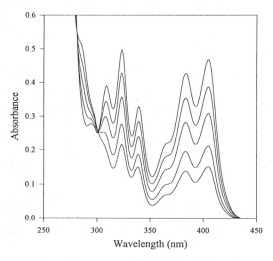

Figure 8. UV-Vis absorption spectra (250-450nm) of a dilute solution of polymer after irradiation (366nm) in dichloromethane under nitrogen at room temperature. From top to bottom: 20s, 2, 4, 8, and 40min.

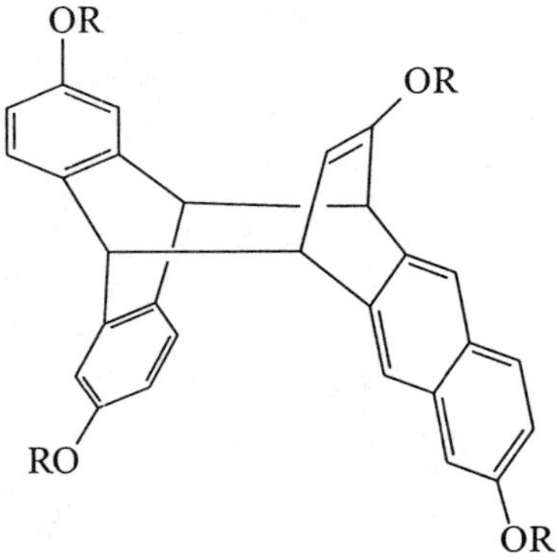

5

anthraquinone secondary photooxidation products. The quantitative relationshipsbetween aggregate formation, ratios of different product types, and other mechanistic and photophysical aspects of the photochemistry of this polymer remain under active investigation.

Acknowledgments

This work was supported by the NSF EPSCoR program, the State of Mississippi, and the University of Southern Mississippi.

Literature Cited

1. Creed, D.; Cozad, R. A.; Griffin, A. C.; Hoyle, C. E.; Jin. L.; Subramanian, P.; Varma, S. S.; Venkataram, K. In *Polymeric Materials for Microelectronics Applications: Science and Technology* (eds. Ito, H.; Tagawa, S.; Horie, K.), ACS Symposium Series, **1994**, *579*, 13-26.
2. Creed, D.; Griffin, A.C.; Gross, J.R.D.; Hoyle, C.E.; Venkataram, K. *Mol. Cryst. Liq. Cryst.* **1988**, *155*, 57-71.
3. Singh, S.; Creed, D.; Hoyle, C.E.; *Proc. SPIE - Int. Soc. Opt. Eng.,* **1993**, *1774*, 2-11.
4. Creed, D.; Cozad, R.A.; Hoyle, C.E.; Morris, J.C.; Jackson Jr., W.J. *Proc. SPIE - Int. Soc. Opt. Eng.,* **1993**, *1774*, 69-73.
5. Creed, D.; Griffin, A.C.; Hoyle, C.E.; Venkataram, K. *J. Amer. Chem. Soc.,* **1990**, *112*, 4049-4050.

6. Bouas-Laurent, H.; Castellan, A.; Desvergne, J.-P. *Pure Appl. Chem.*, **1980**, *52*, 2633.

7. Brotin, T.; Desvergne, J.-P.; Fages, F.; Utermohlen, R.; Bonneau, R.; Bouas-Laurent, H. *Photochem. Photobiol.*, **1992**, *55*, 349-358.

8. Keller, P. *Chem. Mater.*, **1990**, *2*, 3-4.

9. Noonan, J. M.; Caccamo, A. F. In *Liquid Crystalline Polymers*; (eds,, Weiss, R. A.; Ober, C.K.), ACS Symposium Series, **1990**, *435*, 144-157.

10. Whitcombe, M. J.; Gilbert, A.; Mitchell, G. R.; *J. Polym. Sci., (A), Polym. Chem.*, **1992**, *30*, 1681-1691.

11. Gangadhara, Kishore, K.; *Macromolecules*, **1993**, *26*, 2995-3003.

12. Chandross, E. A.; Ferguson, J.; McRae, E. G.; *J. Chem. Phys.*, **1966**, *45*, 3546-3553.

13. Fages, F.; Desvergne, J.-P.; Frisch, I.; Bouas-Laurent, H. *J. Chem. Soc. (Chem. Commun.)*, **1988**, 1413-1414.

RECEIVED July 7, 1995

Chapter 34

Hybrid Polyimide–Polyphenylenes by the Diels–Alder Polymerization Between Biscyclopentadienones and Ethynyl-Terminated Imides

Uday Kumar and Thomas X. Neenan

AT&T Bell Laboratories, 600 Mountain Avenue, Murray Hill, NJ 07974

The Diels-Alder reaction between biscyclopentadienones and pre-imidized ethynyl terminated diimides yields soluble, thermally stable polymers. The ethynyl terminated monomers were prepared by the reaction of commercially available 3-ethynylaniline with a series of dianhydrides. Polymerization of these monomers with substituted cyclopentadienones yielded soluble polymers with M_n and M_n/M_w in the range 8000-28000 and 1.7-4.4 respectively. The molecular weights of the polymers were strongly dependent on the reactant concentrations. The polymers have dielectric constants in the range 2.8-2.9. The materials retain many of the useful properties of polyimides, but require no curing step, suggesting their use as specialized organic dielectrics.

The demand of the microelectronic industry for polymeric materials as organic dielectrics has fueled much novel work in polyimides (*1,2*). Polyimides have an outstanding combination of properties, including thermal and chemical stability, excellent processability and good adhesion to a variety of surfaces. Polyimides suffer from several disadvantages including the evolution of water upon thermal curing of the precursor polyamic acids at high temperature, and the tendency for the cured polyimides to absorb moisture when exposed to conditions of high humidity. Substantial progress has been made in addressing these problems, particularly by the introduction of fluorinated substituents into the polyimide structure (*3-5*). However efforts have continued to design new materials which combine low dielectric constants with mechanical and processing characteristics comparable to present commercial materials e.g Kapton.

In our search for new materials as photodefinable organic dielectrics, we recently revisited the work of Stille and Harris on the preparation of polymers by the Diel-Alder polymerization of m- and p-diethynylbenzene with biscyclopentadienones. The original work (*6-12*) centered on the copolymerization of m- or p-

0097–6156/95/0614–0518$12.00/0

diethynylbenzene (**1** and **2**) with 3,3'-(oxy-*p*-phenylene)bis(2,4,5-triphenyl-cyclopentadienone) (**3**) (Scheme I), and yielded thermally stable, aromatic polymers (**4**) with high glass transition (Tg) temperatures. Of particular note was the unusually high solubility of these highly aromatic materials in common organic solvents such as toluene and chloroform. We extended the Stille-Harris reaction to the preparation of a broad class of new polymers through the use of a variety of diethynyl aromatic dienophiles (*13*). Specifically we showed that a range of both terminal and internal diethynyl aromatics could be used, but that terminal acetylenes yielded higher molecular weights. The polymerization reaction tolerated a variety of substituents on the aromatic ring of the dienophile, including fluorine, silicon and sulfur.

Here we extend the synthetic approach to the preparation of copolymers from the Diels-Alder polymerization of biscyclopentadienones and ethynyl terminated, pre-imidized monomers. The latter materials are readily prepared in a two step, one pot synthesis from commercially available starting materials. We discuss the properties of the materials prepared, and show that this approach offers a new way to prepare soluble, thermally stable polymers. Our intent was to prepare materials which combined the advantages of polyphenylenes (low dielectric constant, hydrolytic stability) with those of polyimides (good mechanical and adhesive properties). In contrast to most conventional polyimides, our new approach yields materials in a pre-imidized form, which have several technological advantages. Poly(amic acid) solutions are notorious for molecular weight and polydispersity instabilities, imididization requires heating poly(amic acids) to temperatures of ~ 330°C, and the acids corrode copper and other metals.

Experimental

Synthesis of monomers 7-11. A typical procedure for the preparation of the ethynyl terminated imide **8** is given below. 2.57 g (22.0 mmol) of ethynylaniline (**12**) was added to a suspension of 10.4 mmoles of 3,3',4,4'-diphenylsulfonetetracarboxylic dianhydride (**14**) in 80 mL of anhydrous THF under argon. The resulting mixture was stirred at 50°C for three hours. Upon cooling the reaction mixture, 20 mL acetic anhydride and 15 mmoles of anhydrous sodium acetate was added, and the resulting mixture stirred at 70°C for 5 h. Compound **8** slowly precipitated as an off-white solid. The reaction mixture was cooled, poured into ice cold water, filtered, washed with water, THF and methylene chloride and dried on a vacuum pump. Compounds **7, 9-11** were prepared similarly. In the case of monomer **9**, the reaction mixture after being poured into water was extracted with chloroform, the organic extract washed with brine and dried with anhydrous $MgSO_4$ and the solvent removed. Full experimental and characterization data will be reported elsewhere.

Polymerization procedure. A representative procedure for the polymer preparations is as follows: A 25 mL Schlenk tube was charged with a suspension of **5** (0.86 mmol), **7** (0.86 mmol) and 5 gm of N,N-dimethylacetamide (DMAC), and the mixture was degassed by three freeze, pump, thaw cycles. The tube was sealed and placed in a oil bath at 200°C for 36 h. The solids dissolved within an hour and the deep magenta solution slowly became very viscous. At the end of the reaction period, the tube was cooled, about 60-100 mg of phenylacetylene added to the (light)

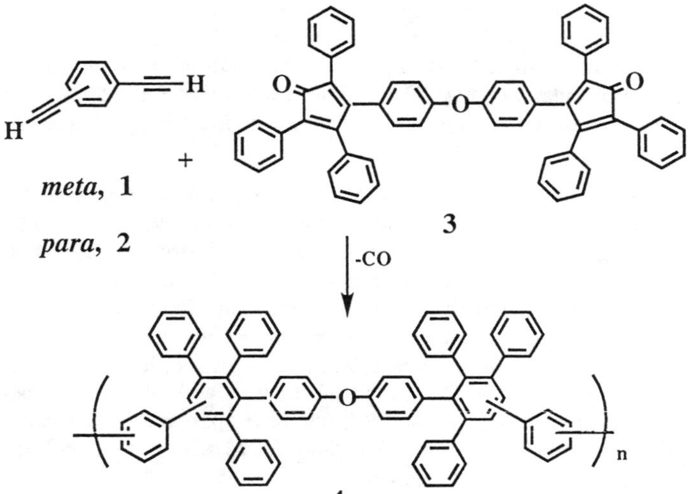

Scheme I. Stille-Harris polymerization of diethynylbenzenes with biscyclopentadienones.

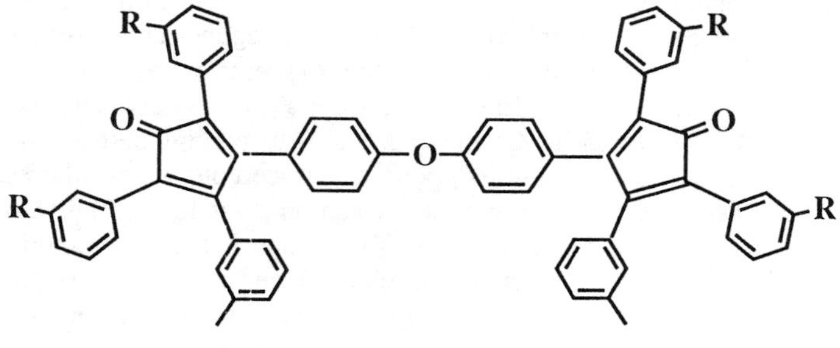

5, R = H   6, R = CH₃

Fig. 1.   Structures of biscyclopentadienones 5 and 6.

magenta colored polymer solution, the tube was resealed, and heated at 200°C for another 3 h. The reaction mixture was diluted with THF and precipitated into hexane/acetone (3:1). Polymer **20a** was purified by dissolving in THF, reprecipitating into hexane/acetone, filtering and drying in a vacuum pump at 70°C for 24 h. The yields of polymers along with other polymer properties are listed in Table 1.

Polymer Characterization.

Results and Discussion

Monomers. Two biscyclopentadienones (**5-6**, Fig. 1) were prepared as described previously (*13*). A series of five ethynyl-terminated imide monomers (**7-11**, Scheme II) were prepared via a two step procedure, involving first the reaction of the amino group of ethynylaniline on the bisanhydride rings of **13-17**, followed by base catalysed ring closure of the intermediate bis-amic acids (*14*). With the exception of the hexafluoro-isopropylidene bridged monomer **9**, the ethynyl terminated monomers were insoluble in all common organic solvents, making characterization by NMR impossible. However the IR C-H stretch at 3250-3270 cm^{-1} confirmed the presence of a carbon-carbon triple bond, and each monomer showed a carbonyl imide stretch at 1715-1725 cm^{-1}. The absence of a carbon-carbon triple bond absorption in the IR ~ 2100 cm^{-1} is due to the presence of symmetry in the molecules. Litt has recently reported similar behavior in di-p-ethynylbenzoyl esters (*15*). The ^{1}H and ^{13}C NMR spectrum of hexafluoroisopropylidene-3,3′,4,4′-bis(phthalimide-N-3-ethynyl-benzene) **9** is consistent with the structure and shows acetylenic proton and carbon resonances at 3.13 and ~ 80 ppm respectively. In light of the extensive literature available on ethynyl terminated polyimide oligomers as melt processable thermosets (*16*), we briefly examined the thermal behavior of monomers **7-11** by differential scanning calorimetry. The bis acetylenes **7-8** and **10-11** do not show a melting endotherm, but exhibit a strong exotherm in the DSC at 238, 250, 309 and 241°C respectively, due to reaction of the ethynyl groups. Litt (*15*) has recently reported that di-p-ethynylbenzoyl esters behave similarly, undergoing exothermic cross-linking without melting in the solid state at 200 - 250°C. Diacetylene **9** does show a distinct melting endotherm at 212°C followed by a broad exotherm at 221°C corresponding to reaction of the acetylenes.

Polymerization Reactions. We chose to carry out the polymerizations between acetylenes **7-11** and the biscyclopentadienones **5-6** to form the polymers **18-27** in Schlenk tubes so that we could visually monitor the reaction (*17*). In a typical polymerization, a 1:1 molar mixture of **5** and **7** was prepared in n,n-dimethylacetamide (DMAc) in a sealed tube. The mixture was degassed by a series of freeze pump thaw cycles, sealed, and immersed in a thermoregulated bath held at 200°C. (The diacetylenes, with the exception of **9** are initially insoluble in dimethylacetamide, but dissolve as the reaction proceeds). The polymerization reaction was monitored by the disappearance of the intense purple color of **5**, appearance of a light brown coloration and an increase in viscosity of the reaction mixture. The polymerization reaction was cooled, 50 mg of phenylacetyene was added, the tube resealed and heated for a further period to terminate the polymerization reaction. A color change from magenta to yellow (generally within

Table 1. Molecular weight dependence on monomer concentration

Polymer	mmol	mmol	Solvent grams	Yield %	M_w ($\times 10^3$)	M_n ($\times 10^3$)	D
	Monomer **8**	Monomer **5**					
18a	0.86	0.86	5.0	95	38	16	2.5
18b	1.24	1.24	4.0	92	18	11	1.7
	Monomer **8**	Monomer **6**					
19	1.30	1.30	4.3	98	28	16	1.7
	Monomer **7**	Monomer **5**					
20a	0.86	0.86	5.0	96	74	25	3.0
20b	1.42	1.42	4.5	51	19	8	2.4
	Monomer **7**	Monomer **6**					
21	1.30	1.30	4.3	99	37	17	2.2
	Monomer **9**	Monomer **5**					
22a	0.86	0.86	5.0	96	42	22	1.9
22b	1.24	1.24	4.0	94	35	20	1.7
	Monomer **9**	Monomer **6**					
23	1.24	1.24	4.0	88	50	28	1.8
24a	0.86	0.86	5.0	100	22	10	2.2
24b	1.40	1.40	4.5	87	33	10	3.3
	Monomer **10**	Monomer **6**					
25	1.42	1.42	5.0	94	86	20	4.4
	Monomer **11**	Monomer **5**					
26a	0.78	0.78	4.5	99	19	10	1.8
26b	1.42	1.42	4.5	96	24	11	2.2
	Monomer **11**	Monomer **6**					
27	1.42	1.42	4.5	98	41	21	2.0

Solvent was dimethyl acetamide; M_w = Weight average molecular weight; M_n = Number average molecular weight; D = Polydispersity.

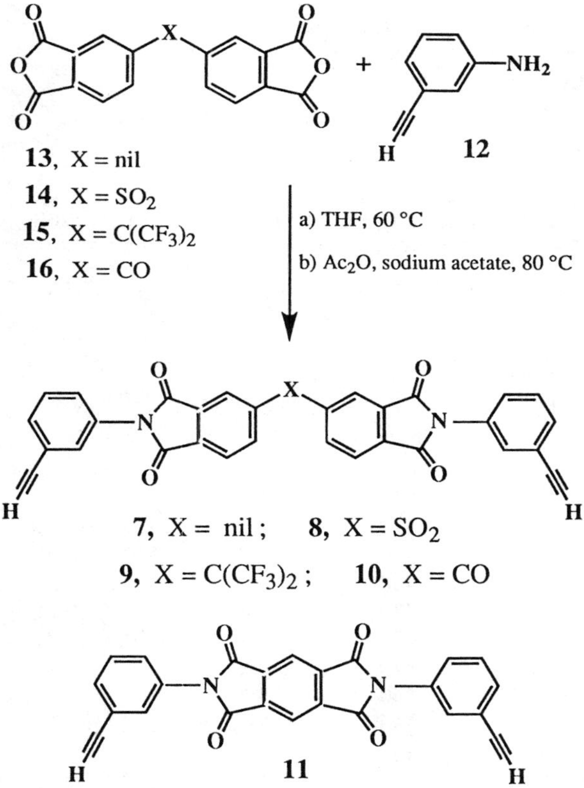

13, X = nil
14, X = SO$_2$
15, X = C(CF$_3$)$_2$
16, X = CO

12

a) THF, 60 °C

b) Ac$_2$O, sodium acetate, 80 °C

7, X = nil ;   8, X = SO$_2$

9, X = C(CF$_3$)$_2$;   10, X = CO

11

Scheme II.   Preparation of ethynyl-terminated imide monomers.

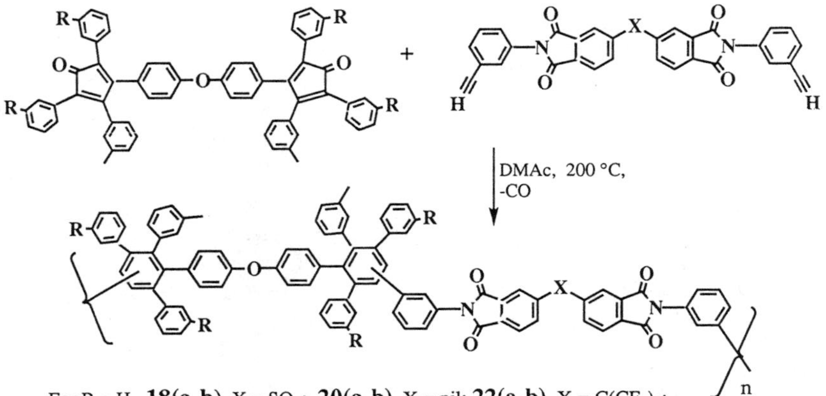

For R = H, **18(a-b)**, X = SO$_2$; **20(a-b)**, X = nil; **22(a-b)**, X = C(CF$_3$)$_2$;
24(a-b), X = CO;. For R = CH$_3$, **19**, X = SO$_2$; **21**, X = nil; **23**, X = C(CF$_3$)$_2$; **25**, X = CO.
Polymers **26a-b** (R = H) and **27** (R = CH$_3$) have monomer **11** as the imide component.

Scheme III.   Preparation of polymers **18-27**.

Table 2. Physical properties of polyimides 18-27

Polymer	Tg(°C)	[a]TGA(°C) (in air)	[a]TGA(°C) (in argon)	[b]UV l$_{max}$ (a)	[c]UV(AU) (cut off)
18a	270	440	440		
18b	260	420	420	252 (113.0)	344
19	245	420	420	252 (104.5)	334
20a	275	450	470		
20b	250	470	470	252 (119.6)	370
21	245	450	450	256 (113.4)	366
22a	265	470	470		
22b	265	440	450		
23	245	330	380	248 (103.0)	320
24a	245	430	440		
24b	245	400	400	252 (120.9)	336
25	240	490	460	254 (116.2)	348
26a	280	470	420		
26b	280	450	430	252 (107.6)	344
27	260	380	450	254 (112.7)	

a. Onset temperature for weight change in thermogravimetric analysis (TGA) for polymers.
b. Absorbance measured in THF solution. The value in parenthesis is the absorptivity.
c. The absorbance value for an absorptivity of 1.0 measured in THF solution.

fifteen minutes) served as an indication of reaction termination. Dilution of the solution with THF, followed by precipitation into hexanes/acetone gave polymer **20a** as a fibrous, off-white material. The preparation of the other polymers followed the same procedure. See Scheme III.

Characterization of the polymers. The polymers **18-27** are all white/tan fibrous materials which are readily soluble in a variety of solvents including THF, p-dioxane, cyclohexylbenzene, N-methylpyrrolidinone, toluene, xylene, chloroform, chlorobenzene or N,N-dimethylformamide. Solutions of up to 15 weight % can be easily prepared, and the solutions are indefinitely stable to precipitation. The solubility is limited by solution viscosity and is not inherent to the polymer. Table 1 shows the strong dependence of the molecular weights of **18-27** on monomer concentration. Typical of condensation polymerizations, an increase in monomer concentration results in an increase in molecular weight and polydispersity (PD) of the polymers. In general the molecular weight of the imide copolymers were lower than those prepared in our earlier study from biscyclopentadienones and diethynylaromatics (*13*). In several instances (**18b,19**) the polymers precipitated from solution during the polymerization, resulting in lower molecular weights.

NMR spectra of polymers. Analysis of ^{1}H and ^{13}C NMR spectra of the poly(phenylene-imide)s was relatively straightforward, with the ratio of aliphatic to aromatic protons generally agreeing well with the structures of the constituent monomer units. In all instances, the resonance for the carbonyl carbon of the cyclopentadienone monomers at ~ 200 ppm disappeared, to be replaced by an increase in the complexity of the aromatic region due to the formation of the new (Diels-Alder produced) phenyl rings. ^{1}H NMR analysis of the polymers derived from **9** showed the loss of the acetylene proton at 3.13 ppm.

The higher molecular (M_n ~ > 40,000) polymers formed thick (1-7 µm) coherent films on a variety of substrates (silicon, quartz, aluminum oxide). Table 2 summarizes the UV data for polymers **18-27**; most polymers show a maximum in absorbance in the UV spectrum at ~ 250 nm. Of most interest to us was the low absorbances of the materials at 364 nm (typically ~ 0.015-0.025 absorbance units/µm) suggesting that photodefinition of the polymers may be possible.

Initial measurements of the dielectric properties of the polymers as 7 µm thick films on aluminum confirmed that the polymers have dielectric constant values in the range of 2.6-2.9, comparing favorably with typical values of 3.0-3.5 for polyimides.

Thermal Properties. The thermal properties of the polymers are summarized in Table 2. All polymers showed a weak glass transition (Tg) temperature in the range of 225-270°C, which is in the range typical of soluble polyimides (*18*). The temperatures reported in columns 3 and 4 of Table 2 are the onset temperatures for weight loss (except for **23** and **27**) measured by thermogravimetric analysis. Poly(phenylene-imide) **23** shows a weight gain of 2% in argon and air and polymer **27** a 1% increase in weight in air due to oxidation of the methyl groups. On further heating **23** and **27** start losing weight at around 450°C and followed a pattern similar to other poly(phenylene-imide)s.

Conclusion. The Diels-Alder polymerization of biscyclopentadienones and ethynyl terminated bis-imides yields soluble polymers of moderate molecular weight. The glass transition temperature of the polymers are in the range of 250-275°C, and the materials have low absorbance in the UV at 364 nm. Since the monomers contained preimidized groups, these poly(phenylene-imides) need no curing step. These polymers may find use as novel coatings or as organic dielectrics. We are continuing to explore these possibilities.

Acknowledgements. We thank Timothy M. Miller for helpful discussions and Wai Tai for the GPC measurements.

Literature Cited

1. Satou, H.; Suzuki, H.; Makino, D. In *Polyimides* Wilson, D.; Stenzenberger, H. D.; Hergenrother, P. M. Eds.; Chapman and Hall: NY, 1990, Chapter 8.
2. Stone, D, S.; Martynenko, Z. *Polymers in Electronics, Fundamentals and Applications*; Elsevier: New York, 1989.
3. Hougham, G.; Tesoro, G.; Shaw, J. *Macromolecules* **1994**, *27*, 3642.
4. Goff, D, L.; Yuan, E. L.; Long, H.; Neuhaus, H. J. Organic Dielectric Materials with Reduced Moisture Absorption and Improved Electrical Properties. In *Polymeric Materials for Electronics Packaging and Interconnection*; Lupinski, J. H.; Moore, R. S. Eds.; American Chemical Society: Washington, DC. 1989; pp 93-100 .
5. St Clair, T. L. In Polyimides Wilson, D.; Stenzenberger, H. d.; Hergenrother, P. M. Eds.; Chapman and Hall: NY, 1990, Chapter 3.
6. Stille, J. K. Noren, G. K. *Macromolecules* **1972**, *5*, 49.
7. Stille, J. K.; Gilliams, Y. *Macromolecules* **1971**, *4*, 515.
8. Stille, J. K.; Noren, G. K. *J. Polym.Sci., Part B* **1969**, *7*, 525.
9. Stille, J. K. *J. Macromol. Sci. Chem.* **1969**, *3*, 1043.
10. Stille, J. K.; Rakutis, R. O.; Mukamal, H.; Harris, F. W. *Macromolecules* **1968**, *1*, 431.
11. Mukamal, H.; Harris, F. W.; Stille, J. K. *J. Polym. Sci., Part A-1* **1967**, *5*, 2721.
12. Stille, J. K.; Harris, F. W.; Rakutis, R. O.; Mukamal, H. *J. Polym. Sci., Part B* **1966**, *4*, 791.
13. Kumar, U.; Neenan, T. X. *Macromolecules* **1995**, *28*, 124.
14. (a) Unroe, M. R. ; Reihhardt, B. A. *J. Poly. Sci. Part. A, Polymer Chem.* **1990**, *28*, 2207.
15. Melissaris, A. P; Litt, M. H. *Macromolecules* **1994**, *27*, 2675.
16. Hergenrother, P. M. In *Polyimides* Wilson, D.; Stenzenberger, H. D.; Hergenrother, P. M. Eds.; Chapman and Hall: NY, 1990, Chapter 6.
17. The earlier polymerization procedure reported by Stille and Harris involved the Diels-Alder condensation between biscyclopentadienones and diacetylenes in a Parr reactor at ~ 200°C and at pressures of ~ 200 psi.
18. Bell, V. L.; Stump, B. L.; Gager, H. *J. Polym. Sci. Chem. Ed.* **1976**, *14*, 2275.

RECEIVED July 7, 1995

Chapter 35

Polysiloxane Thermoplastic Polyurethane Modified Epoxy Resins for Electronic Application

Tsung-Han Ho[1] and Chun-Shan Wang[2]

[1]Department of Chemical Engineering, National Kaohsiung Institute of Technology, Kaohsiung, Taiwan 807, Republic of China
[2]Department of Chemical Engineering, National Cheng Kung University, Tainan, Taiwan 701, Republic of China

A stable dispersion of polysiloxane thermoplastic polyurethane (TPU) particles in an epoxy resin matrix was achieved via the epoxy ring opening with isocyanate groups of urethane prepolymer to form an oxazolidone. The effect of structure and molecular weight (MW) of polysiloxane TPU in reducing the stress of electronic encapsulant were investigated. The mechanical and dynamic viscoelastic properties of polysiloxane TPU modified epoxy networks were also studied. A "sea-island" structure was observed via SEM. The devices encapsulated by the TPU-modified EMC were also evaluated by the thermal shock cycling test. The dispersed polysiloxane TPU rubbers effectively reduce the stress of cured epoxy resins by reducing flexural modulus and the coefficient of thermal expansion (CTE), while the glass transition temperature (Tg) is increased because of rigid oxazolidone structure formation.

Epoxy molding compounds (EMCs) have been widely used as encapsulation material for semiconductor devices. Because of its excellent heat, moisture, solvent, and chemical resistance, superior electrical and mechanical properties, and good adhesion to many substrates, o-cresol-formaldehyde novolac epoxy (CNE) is the resin typically employed to encapsulate microelectronics devices. Upon cure, this multifunctional epoxy resin provides a densely cross-linked protective layer; however, it is relatively brittle.

The trend of electronics equipment is being miniaturized and becoming thinner, at the same time the scale of integration of large scale integrated circuits (LSICs) is continuing upward, forcing the design of large chips, finer patterns, and high pin counts that are more susceptible to internal stress failure. The prevailing surface mount technology (SMT) also generates thermal stress to devices. Internal stress causes package cracking, passivation layer cracking, aluminum pattern deformation,

0097–6156/95/0614–0527$12.00/0

etc (*1-3*). Therefore, the development of a low-stress EMC is required for high-reliability semiconductor devices. The sources of internal stress resulted from the use of plastic encapsulants are considered to be shrinkage of the plastic upon curing process and thermal mismatch between the resin and the device. In the case of EMC encapsulation, the first source of shrinkage has been relatively minor, whereas the second one, which is caused by the difference in thermal expansion coefficients between resin and silicon chip, is the dominant factor. The internal stress on IC devices encapsulated by EMCs as produced by the difference between thermal expansion coefficients can be approximated by the product of flexural modulus and the thermal expansion coefficient (*4*). Therefore, it is necessary to reduce the thermal expansion coefficient and the flexural modulus of EMCs to reduce internal stress.

Reductions of internal stress by lowering of either the thermal expansion coefficient or the flexural modulus of the encapsulant have been reported (*5-11*). Increasing the amount of silica filler used in an encapsulant effectively lowers the thermal expansion coefficient; however, this approach not only increases the elastic modulus, but also increases the viscosity of the resin composition, resulting in poor moldability. Lowering of the flexural modulus by modification with a rubber in a "sea-island" two-phase structure is considered to be more desirable than a one-phase structure in view of the thermal property needs (*12*). Traditional modifiers, which can reduce the elastic modulus of the cured epoxy resins, include reactive liquid rubber such as carboxy terminated butadiene-acrylonitrile copolymer (CTBN) have been reported (*13,14*). When CTBN modifiers are incorporated into resin compositions, a two-phase morphology consisting of relatively small rubber particles dispersed in a resin matrix is generated that toughens epoxy resins. However, phase separation depends upon the formulation, processing, and curing conditions. Incomplete phase separation can result in a significant lowering of the glass-transition temperature (Tg) (*15*). Moreover, their presence in epoxy molding compounds raises the thermal expansion coefficient of the resultant EMC. The relatively high Tg of butadiene-acrylonitrile copolymer also limits their low-temperature applications. Polysiloxanes are known for their excellent thermal and thermooxidative stabilities, moisture resistance, good electric properties, low stress, and lower Tg (-123 °C) values than conventional elastomers. Low-stress EMCs modified by vinyl or hydride terminated polydimethyl siloxanes reported by the authors (*16,17*) have effectively reduced the stress of cured epoxy resins by reducing flexural modulus and the coefficient of thermal expansion, while the Tgs were hardly depressed.

In this work, a stable dispersion of polysiloxane thermoplastic polyurethane (TPU) particles in an epoxy resin matrix was achieved via the epoxy ring opening with isocyanate groups of urethane prepolymer to form an oxazolidone. The effect of structure and molecular weight of polysiloxane TPU in reducing the stress of electronic encapsulant were investigated. The mechanical and dynamic viscoelastic properties of polysiloxane TPU modified epoxy networks were also studied. The devices encapsulated by the TPU-modified EMC were also evaluated by the thermal shock cycling test.

Experimental

Materials. All reagents and solvents were extra pure reagent grade or were purified by standard methods before use. The control epoxy resin was o-cresol-formaldehyde novolac epoxy resin (CNE) [Quatrex3330, Dow Chemical Co., epoxy equivalent weight (EEW) 192]. A phenol-formaldehyde novolac resin was used as curing agent with an average hydroxyl functionality of 6 and a hydroxyl equivalent weight of about 104 (Schenectady Chemical, HRJ-2210). Methylene bis(4-phenylisocyanate) (MDI) [Multrathane, Mobay Chemical] was distilled under reduced pressure (170°C at 0.05 mmHg). Polytetramethylene ether glycol (PTMG) [Polymeg, Quaker Oats] was degassed under vacuum at 65°C and 2 mm Hg for 3 h to remove any absorbed water, then stored over type 4 Å molecular sieves. Polypropylene ether glycol (PPG) and polyethylene ether glycol (PEG) [Naclai Tesque Inc. Kyoto Japan] were degassed in the same manner as the PTMG. Polydimethylsiloxane-α,ω-diol (PDMS) with hydroxy number 62 (XF-6001) and 112 (X-22-160AS), giving a calculated MW of 1810 and 1002 respectively, were purchased from Shin-Etsu Chemical Co., Ltd., Japan. The polydiols used in the manufacture of polyurethanes, whose structure, MW, and designations are listed in Table I. The structural formulas of CNE and MDI are also shown in the following. Stannous octoate was used as a catalyst in the syntheses of isocyanate terminated PDMS diols. 2-Phenylimidazole was used as a catalyst in epoxide-isocyanate reaction to form oxazolidone from the epoxide-isocyanate reaction. Ph$_3$P was the triphenylphosphine that was used as a curing accelerator.

Table I. Polydiols Used in the manufacture of Polyurethanes

Common name	Structure	Molecular weight	Sample Designation
polyethylene ether glycol (PEG)	$HO-(-CH_2CH_2O-)_{\overline{n}}H$	2000	PEG
polypropylene ether glycol (PPG)	$HO-(-CH_2CH_2O-)_{\overline{n}}H$ with CH_3	2000	PPG
poly(tetramethylene ether) glycol (PTMG)	$HO-(CH_2CH_2CH_2CH_2O)_{\overline{n}}H$	2000	PTMG
polydimethyl siloxane-α,ω-diol (PDMS)	$HO-R-\underset{CH_3}{\overset{CH_3}{Si}}-O-[\underset{CH_3}{\overset{CH_3}{Si}}-O]_n-\underset{CH_3}{\overset{CH_3}{Si}}-R-OH$ $R=(CH_2)_3$	1810 / 1002	S1810 / S1002

General Procedure for the Synthesis of Urethane Prepolymer.

Urethane Prepolymer. To a flame-dried 500 mL four-neck round-bottom flask, equipped with a water cooled condenser with a capped CaCl₂ drying tube, a thermometer, N₂ inlet, a 150 mL addition funnel, and mechanical stirrer was charged 1.0 mole of MDI and heated to 85°C. To the MDI was added 0.5 mole of polyol (PEG, PPG, or PTMG) dropwise (i.e., NCO : OH = 2: 1) under a nitrogen atmosphere. The mixture was stirred and maintained at 85°C until the absorption peak of OH group in infrared (IR) spectra had disappeared.

PDMS-based Urethane Prepolymer. The reaction flask was equipped the same as mentioned above. The PDMS-based urethane prepolymer was synthesized by adding 0.5 mole of PDMS with 0.15 wt% stannous octoate dropwise to 1.0 mole of MDI over a period of 1 h while maintaining the reaction temperature at 65°C under a nitrogen atmosphere. Completion of the reaction was confirmed by IR for the disappearance of the absorption peak of OH group.

General Procedure for the Preparation of TPU-modified CNE. To a 1 L four-neck round-bottom flask, equipped with a CaCl₂ drying tube, heating mantle, N₂ inlet, stirrer, thermocouple, and temperature controller was added 400 g of *o*-cresol-formaldehyde novolac epoxy resin (CNE). The epoxy resin was heated to 150°C and then vigorously stirred and dehydrated under vacuum (< 10 mm Hg) until the water content was less than 0.01% (measured by Karl Fischer). The reaction temperature was then raised to 160°C and 350 ppm (based on CNE) of 2-phenylimidazole was added. To the stirring CNE was added 57.2 g of urethane or PDMS-based urethane prepolymer. The reaction temperature was held at 160°C for 2 h. Completion of the epoxide-isocyanate reaction was confirmed by infrared spectroscopy for the disappearance of -NCO absorption. The resulted TPU- or Polysiloxane-TPU-modified epoxy resin contained ca. 12.5 wt % dispersed rubber and had an EEW of ca. 230.

Curing Procedure of Epoxy Resins. Various rubber-modified epoxy resins were mixed with a stoichiometric amount of curing agent and Ph₃P in a mill at moderate temperature to give a thermosettable epoxy resin powder. The resin powder was cured in a mold at temperature of 150°C and 50 kg/cm² for a period of 1 h and then postcured at 180°C for 2 h and 210°C for 3 h to obtain a cured specimen.

Measurement and Testing. Infrared spectra were recorded with a Perkin-Elmer 16PC FTIR spectrophotometer operated with a dry air purge. Signals of four scans at a resolution of 4 cm⁻¹ were averaged before Fourier transformation. EEW of epoxy resins were determined by the HClO₄/potentiometeric titration method. Dynamic viscoelastic properties were performed on a Rheometrics RDA-II rheometer between -150 and 250°C, with a heating rate of 5 or 2°C/step at a frequency of 1 Hz. The rectangular torsion mode was chosen and the dimensions of the specimen were 51 (L) × 12.7 (W) × 0.76 (T) mm³. The storage modulus G' and tan δ were determined. The JEOL JSM-6400 scanning electron microscope was employed to examine the morphology of cured rubber-modified samples fractured cryogenically in liquid nitrogen. The fracture surfaces were vacuum coated with gold. Flexural properties of cured resins were obtained at a Shimadzu AGS-500 universal testing machine. Flexural strength and modulus were obtained at a crosshead speed of 2 mm/min according to ASTM D790-86. A three-point loading system was chosen and rectangular bar specimens, 80 (L) $\times$ 10 (W) $\times$ 4 (T) mm^3, were molded directly by a transfer molding process. The coefficient of thermal expansion (CTE) was measured with a DuPont 943 thermal mechanical analyzer (TMA) in accordance with ASTM E831-86. A specimen 4-mm in length was used at a heating rate of 5°C/min. Normally, the thermal expansion increases with the increase in temperature and the CTEs were calculated from the slope. An abrupt change in slope of the expansion curve indicates a transition of the material from one state to another. A thermal shock cycling test was carried out by the following procedure (*18*): the device used was a 14-pin LM 324 quad operational amplifier with a single passivation layer. The device was encapsulated with an encapsulation formulation by a transfer molding process and subjected to a thermal cycling test. A cycle consisted of -65°C $\times$ 15 min and 150°C $\times$ 15 min. The devices were inspected by an optical microscope for cracks after 250, 500, 750, 1000, 1500, 2000, 2500, 3000, 3500, 4000, and 4500 cycles. Any crack observed in the encapsulated device was counted as the failure of that device. The percentage of devices that failed (cracked), as a function of cycles, is plotted.

Results and Discussion

The incorporation of TPU or polysiloxane TPU as a soft-segment into an epoxy resin matrix can be achieved in a two-step reaction. The first step involves the synthesis of urethane prepolymer or PDMS-based urethane prepolymer are shown in Figure 1. The equation for the first step indicates the formation of an isocyanate terminated soft segment which is then grafted onto epoxy resin in the second step via the epoxy ring opening in the presence of a catalyst (2-phenylimidazole) to form an oxazolidone to give a stable dispersion of TPU particles in an epoxy resin matrix as shown in Figure 2. These resulting TPU modified CNEs are multifunctional epoxy resins which show similar reactivity to other epoxy resins and can be cured with various curing agents.

Infrared Spectroscopy. Typical FTIR survey spectra of PDMS diol (X22-160AS), PDMS-based urethane prepolymer, and polysiloxane TPU modified CNE are shown in Figure 3. The completion of isocyanate terminated prepolymer formation was confirmed by the disappearance of the hydroxy group absorption peak at 3500 cm^{-1} and the appearance of the absorption peak of urethane at 3316 cm^{-1} (N-H) and 1735 cm^{-1} (C=O). The resulting isocyanate terminated urethane prepolymer was then grafted onto epoxy resin via the epoxy ring opening in the presence of a catalyst (2-phenylimidazole) at 160°C to form an oxazolidone. The completion of this reaction was confirmed by the disappearance of the -NCO group absorption peak at 2250-2270 cm^{-1}. The trimerization of -NCO to form an isocyanurate, a side reaction which can take place simultaneously to the oxazolidone formation, is known to be favored at a low reaction temperature. However, at a reaction temperature of 160°C and with 2-

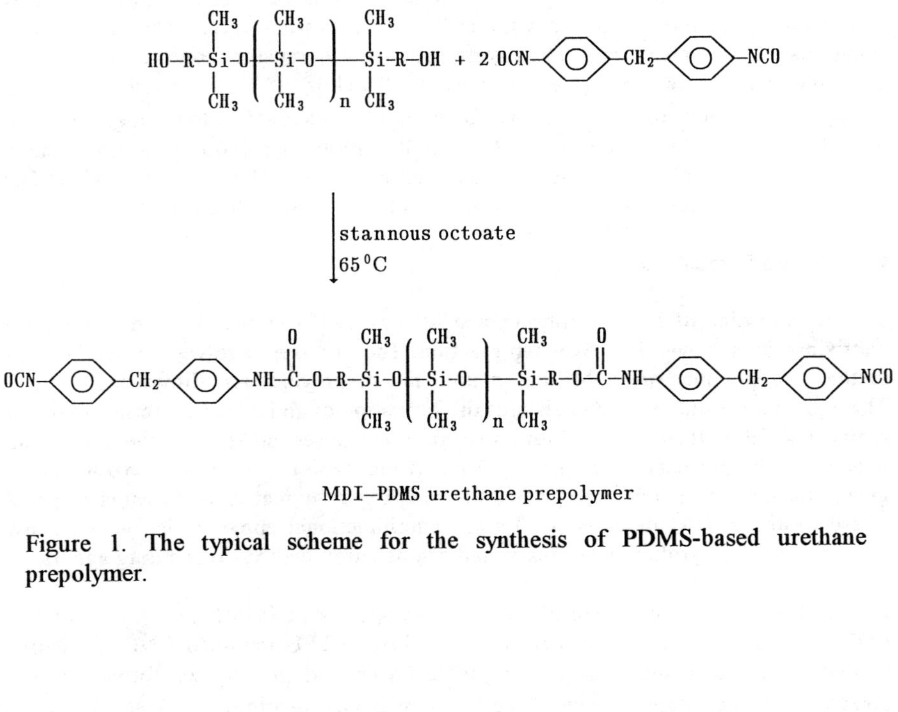

MDI—PDMS urethane prepolymer

Figure 1. The typical scheme for the synthesis of PDMS-based urethane prepolymer.

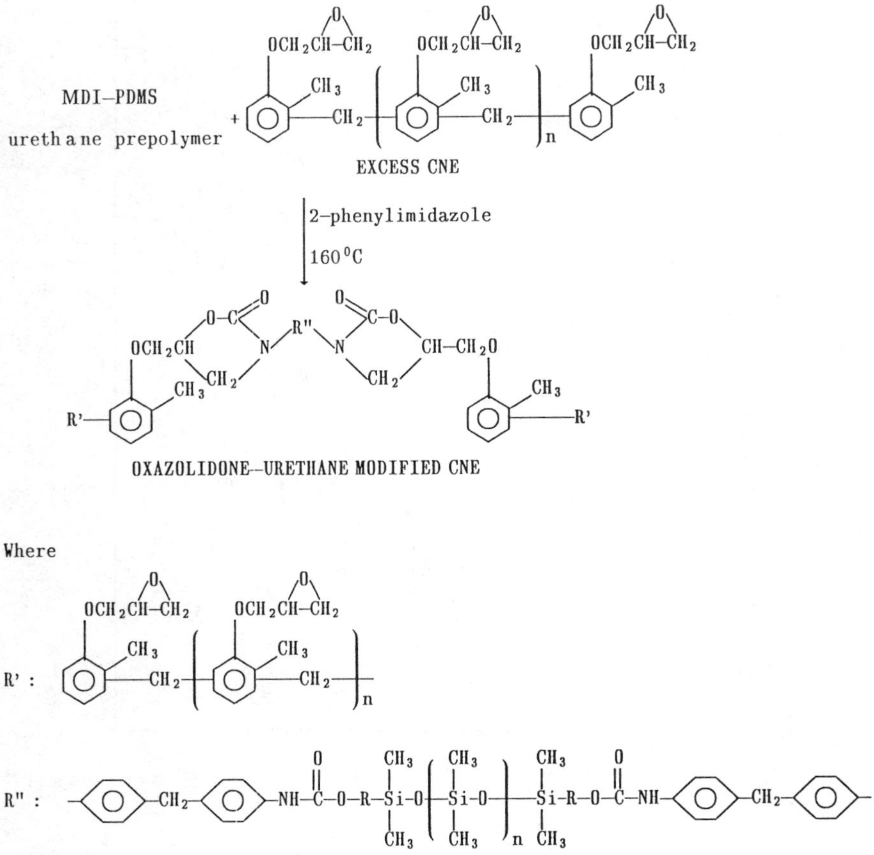

Figure 2. The typical scheme for the preparation of polysiloxane TPU modified CNE via epoxide-isocyanate reaction.

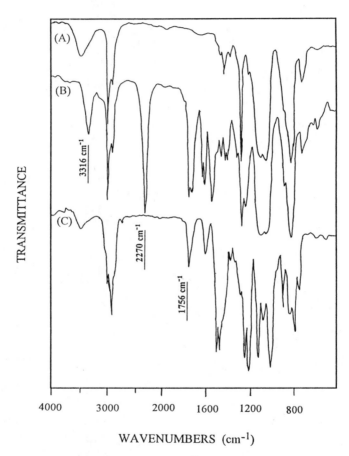

Figure 3. FTIR spectra of (A) polydimethylsiloxane-α,ω-diol (X-22-160AS);
(B) PDMS-based urethane prepolymer; (C) polysiloxane TPU modified CNE.

phenylimidazole catalyst, the oxazolidone formation is predominant. A typical IR spectra of polysiloxane TPU modified CNE are given in Figure 3, which shows the IR characteristic absorption peak of oxazolidone at 1756 cm^{-1} and without any absorption peak of isocyanurate at 1700-1710 cm^{-1}.

Dynamic Viscoelastic Properties. The dynamic viscoelastic spectra are shown in Figure 4 and 5. The sample designations correspond to those in Table I. Dynamic viscoelastic analysis can give information on the microstructure of cured rubber modified epoxy resins. Figure 4 shows the storage modulus G' and tanδ curves for the resins modified with various TPU particles and Figure 5 shows similarly for cured polysiloxane TPU modified CNE. Storage moduli decreased with modification for all samples. The tanδ curves for the control epoxy network exhibit two major relaxations observed in most epoxy polymers (*19*): a high-temperature or α transition corresponds to major Tg of the cured epoxy resins above which significant chain motion takes place; the low temperature or β transition is attributed predominantly to the motion of the CH$_2$-CH(OH)-CH$_2$-O (hydroxyether) group of the epoxy. Figure 4 shows that the α-relaxation peak in the tan δ curve became broader with PEG or PPG modification, but the peak position and the magnitude of the α-relaxation peak hardly changed with respect to the control resin. However, the peak position of the α-relaxation peak in the tan δ curve for the resin modified with PTMG shifted toward higher temperature. This result can be attributed to the solubility between the epoxy resin and the TPU particles. The solubility of TPU particles in the epoxy resin decrease with increasing carbon number in the polyol and this lower solubility has resulted in a complete phase separation. The β-relaxation peak in the tan δ curves at ca. -100 to -50°C was observed without additional peak for TPU. These results can be explained in that Tg's of the pure TPUs are around the β-relaxation peak of cured epoxy resin. Figure 5 shows that the peak position of the α-relaxation peak in the tan δ curves for the resins modified with polysiloxane TPU shifted toward higher temperature. This result can be attributed not only to a complete phase separation of polysiloxane TPU particles but also to the formation of oxazolidone structure via the epoxy ring opening with isocyanate groups of urethane. In addition to α and β peaks, these curves all show an additional markedly small peak from -130 to -105°C with a center near -120°C corresponding to the Tg of the polysiloxane phase. This small peak further supports the multi-phase separation. The dynamic viscoelastic properties of the cured rubber-modified epoxy resin systems including the major Tg and the rubber Tg are shown in Table II.

Morphology. SEM photomicrographs of cold snap surfaces for the control and five TPU modified resins are given in Figure 6. A sea-island structure is observed in all rubber modified resins. No matter it is TPU or polysiloxane TPU modification, the spherical shape of rubber particles are quite the same. The sizes of rubber particle are 0.2~1 μm and slightly proportional to the MW and the structure of the TPU modifiers. The polysiloxane TPU particles have slightly larger size than the polyol TPU particles. These can be attributed to the compatibility of TPU and CNE.

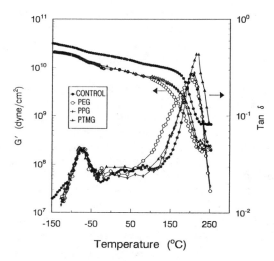

Figure 4. Dynamic viscoelastic analysis for the control and various polyol TPU modified epoxy resins (no filler). The curing agent is phenol-formaldehyde novolac (HRJ-2210).

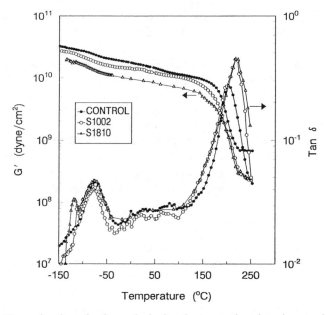

Figure 5. Dynamic viscoelastic analysis for the control and various polysiloxane TPU modified epoxy resins (no filler). The curing agent is phenol-formaldehyde novolac (HRJ-2210).

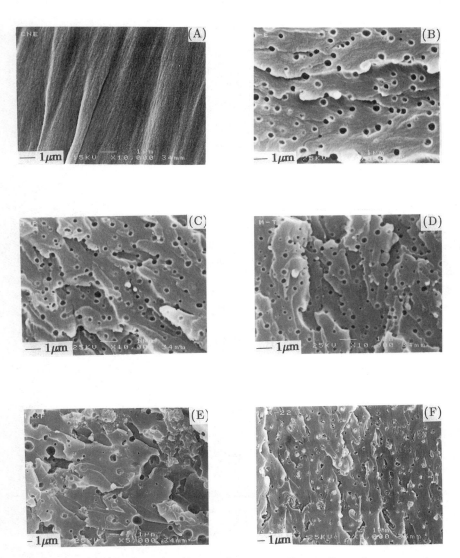

Figure 6. Morphologies of cold-snap fracture surfaces of cured epoxy resins modified with various TPU particles (12.5 wt %). (A) unmodified; (B) PEG; (C)PPG; (D) PTMG; (E) S1810; (F) S1002.

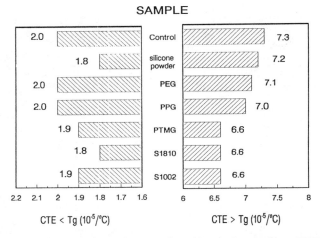

Figure 7. Coefficient of thermal expansion (CTE) for various TPU modified epoxy resin encapsulants.

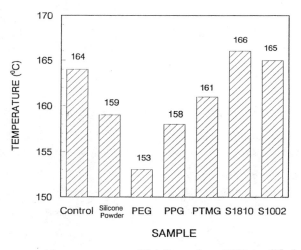

Figure 8. Glass-transition temperature (Tg) for various TPU modified epoxy resin encapsulants.

Table II. Dynamic Viscoelastic Properties of Cured Neat Rubber-Modified
Epoxy Resin System

Sample Designation	Tg^a,°C Matrix	Shear Modulus, 20°C, $\times 10^9$dyne/cm²	Tg^b,°C Rubber
Control[c]	204	16.0	--
PEG	204	9.1	-67
PPG	206	9.1	-68
PTMG	217	9.0	-69
S1810	220	9.0	-119
S1002	218	13.5	-113

[a]Peak of tan δ at higher temperature.

[b]Peak of tan δ at higher temperature.

[c]Control: unmodified CNE.

Encapsulation Formulation. A control resin and CNE modified with various TPU rubbers or silicone powder were formulated into seven electronic encapsulating formulations. The formulations were each cured at 175 °C for 4 h. The detail formulations are given in previous paper (*4*). The thermal mechanical properties of the cured encapsulating formulations were determined by the following tests.

Thermal Mechanical Properties. Figure 7 shows the thermal mechanical properties. The CTE in the glassy state below the Tg was taken from 60 to 100 °C and the CTE above the Tg was taken from 200 to 240 °C. For the CTE below Tg, all rubber modified encapsulants have approximately equal to or slightly lower CTE than the unmodified resin and this will result in a small difference in CTEs between encapsulant and silicon chip.

Tgs were determined from tangents of the CTE as a function of temperature at 100 and 200°C. The results are shown in Figure 8. The Tgs are slightly lower for polyol TPU modification, while Tgs are slightly higher for polysiloxane TPU modification compared to the unmodified encapsulant. This observation is consistent with the viscoelastic investigation.

Flexural Test Properties. Figure 9 shows the result of the flexural test. The flexural moduli of the cured encapsulants were reduced markedly no matter whether it is modified with polyol TPU or polysiloxane TPU. The flexural strengths of polysiloxane TPU modified CNE were approximately equal to or slightly larger than that of the control resin, however, the flexural strength was significant reduced with the silicone powder modification. This result indicates that physical blending, although improves the toughness of cured epoxy resin, lowers it's strength.

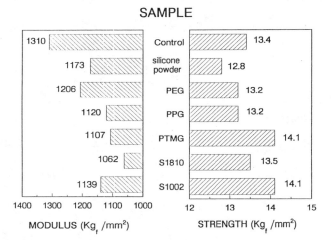

Figure 9. Flexural test properties for various TPU modified epoxy resin encapsulants.

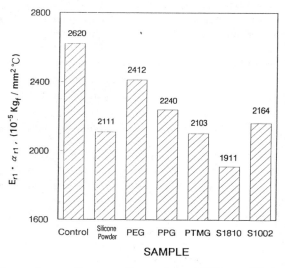

Figure 10. Comparison of stress for various TPU modified epoxy resin encapsulants.

Internal Stress. The internal stress of IC devices encapsulated by EMCs is closely related to the product of flexural modulus (Er1) and thermal expansion coefficient (αr1) below the Tg of the cured EMCs (*16*). Figure 10 shows the internal stress of the control and various rubber-modified encapsulants. The polysiloxane TPU modified EMC (S1810) and TPU modified EMC (PTMG) which not only has a lower stress but also has a higher Tg than that of the control, were chosen to encapsulate the semiconductor devices. The encapsulated devices were subjected to a thermal shock cycling test in comparison with the control resin.

Thermal Shock Cycling Test. The thermal shock cycling test involves cycling the encapsulated devices at -65 and 150 °C and observing the crack (failure) of encapsulated devices at various intervals (after 250, 500, 750, 1000, 1500, 2000, 2500, 3000, 3500, 4000, and 4500 cycles). Any crack that occurred in a device is counted as failure for that device. The percentage of devices that failed vs. test cycles is given in Figure 11 .The result indicates that for 50 % of the devices, failure happened after 3500 cycles for the S1810 modified EMC and after 2500 cycles for the PTMG modified CNE compared to after 750 cycles for the control resin. S1810 modified EMC has much better thermal shock resistance.

Conclusion

A process was developed to incorporate stable dispersed polysiloxane TPU particles in an epoxy resin matrix that not only greatly reduces the stress of cured EMCs but also increases Tg of the resultant EMC. Reduction in internal stress of the encapsulant was greatly affected by the structure and molecular weight of the TPU. Electronic devices encapsulated with the polysiloxane TPU modified EMCs have exhibited excellent resistance to the thermal shock cycling test and have resulted in an extended device used life.

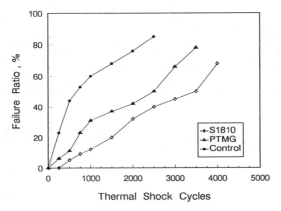

Figure 11. Thermal shock cycling test results. Temperature cycle: -165 °C × 15 min and 150 °C × 15 min.

Acknowledgment

Financial support of this work by the National Science Council of the Republic of China is gratefully appreciated (NSC 83-0405-E006-046).

Literature Cited

1. Kuwata, K.; Iko, K.; Tabata, H. *IEEE Trans. Comp. Hybrids Manuf. Technol.*, **1985**, *CHMT-8*(4), 486.

2. Thomas, R. E. *IEEE Trans. Comp. Hybrids Manuf. Technol.*, **1985**, *CHMT-8*(4), 427.

3. Suhl, D. *IEEE Trans. Comp. Hybrids Manuf. Technol.*, **1990**, *CHMT-13*(4), 940.

4. Ho, T. H.; Wang, C. S. *J. Appl. Polym. Sci.*, **1993**, *50*, 477.

5. Nakamura, Y.; Yamaguchi, M.; Okubo, M.; Matsumoto, T. *J. Appl. Polym. Sci.*, **1992**, *45*, 1281.

6. Hoffman, D. K.; Arends, C. B. U. S. Patent **1987**, 4,708,996.

7. Tyagi, D.; Yilgor, I.; McGrath, J. E.; Wilkes, G. L. *Polymer*, **1984**, *25*, 1807.

8. Nakamura, Y.; Uenishi, S.; Kunishi, T.; Miki, K.; Tabata, H.; Kuwada, K.; Suzuki, H.; Matsumoto, T. *IEEE Trans. Comp. Hybrids Manuf. Technol.*, **1987**, *CHMT-12*(4), 502.

9. Verchere, D.; Pascault, J. P.; Sautereau, H.; Moschiar, S. M.; Riccardi, C. C.; Williams, R. J. J. *J. Appl. Polym. Sci.*, **1991**, *42*, 701.

10. Verchere, D., Sautereau, H.; Pascault, J. P. *J. Appl. Polym. Sci.*, **1990**, 41, 467.

11. Hourston, D. J.; Lane, J. M. *Polymer*, **1992**, *33*(7), 1379.

12. Nakamura, Y.; Uenishi, S.; Kunishi, T.; Miki, K.; Tabata, H.; Kuwada, K.; Suzuki, H.; Matsumoto, T. *IEEE 37th ECC Proceeding*, **1987**, 187.

13. Sultan, J. N.; McGarry, F. J. *J. Appl. Polym. Sci.*, **1973**, *13*(1), 29.

14. Pearson, R. A.; Yee, A. F. *J. Mater. Sci.*, **1986**, *21*, 2475.

15. Kim, D. H.; Kin, S. C. *J. Appl. Polym. Sci.*, **1991**, *31*(5), 289.

16. Ho, T. H.; Wang, C. S. *J. Appl. Polym. Sci.*, **1994**, *51*, 2047.

17. Ho, T. H.; Wang, C. S. *J. Appl. Polym. Sci.*, **1994**, *54*, 13.

18. Kojima, Y.; Ohta, T.; Matsushita, M.; Takahara, M.; Kurauchi, T. *J. Appl. Polym. Sci.*, **1990**, *41*, 2199.

19. Yorkgitis, E. M.; Eiss, Jr. N. S.; Tran, C.; Willkes, G. L.; McGrath, J. E. In *Epoxy Resin Chemistry II*; Dusek, K., Ed.; Advances in Polymer Science 72; Springer-Verlag: New York, NY, **1985**, pp 79-109.

RECEIVED July 7, 1995

INDEXES

Author Index

Allen, Robert D., 255
Arai, Takeshi, 485
Asakawa, Koji, 239
Ban, Hiroshi, 69
Bassett, David, 228
Beach, James V., 355
Breyta, Greg, 21
Buckley, Leonard J., 369
Carter, K., 425
Chandross, E. A., 194
Chin, E., 84,207
Creed, David, 504
Darling, Graham D., 149
Denison, Mark D., 110
DiPietro, Richard A., 255,425
Falcigno, P., 35
Fedynyshyn, Theodore H., 110
Forte, Anthony R., 271
Gabor, Allen H., 281
Galvin, Mary E., 166
Gargiulo, N., 137
Gerena, Linda, 166
Griffin, Anselm C., 504
Griffith, James R., 369
Hanson, J. E., 137
Harley, R., 84
Hasegawa, Masatoshi, 379,395
Hawker, C., 425
Hedrick, J. L., 425
Heffner, Sharon A., 166,180,207
Ho, Tsung-Han, 527
Hofer, Donald C., 21,255
Hofmann, M., 299
Horn, Mark W., 271
Houlihan, F. M., 4,84,207
Hoyle, Charles E., 504
Hsiao, Yu-Ling, 355
Hu, Henry S.-W., 369
Imai, Yoshio, 471
Ishida, Mina, 395
Ishii, Junichi, 395
Ishii, Yoshiyuki, 485

Ito, Hiroshi, 21
Jensen, K. H., 137
Jerome, R., 425
Jiang, Ying, 228
Jikei, Mitsutoshi, 471
Kakimoto, Masa-aki, 471
Kang, Doris, 110
Katayose, Teruo, 485
Kirner, H. J., 35
Knurek, C., 137
Kometani, J. M., 84,137,194,207
Kumar, Uday, 518
Kunz, Roderick R., 255,271
Lang, Chi-I, 449
Liu, Ying, 504
Loy, Douglas A., 355
Lu, T.-M., 449
Mack, C. A., 56
Maher, John, 228
Matano, Takahumi, 395
Matsuo, Takahiro, 318
Meetsma, A., 333
Mertesdorf, C., 35
Mirau, Peter A., 166,180
Miwa, Takao, 395
Mixon, David A., 137,180
Mochizuki, Amane, 413
Moore, J. A., 449
Motta, D., 137
Münzel, N., 4,35,299
Naito, Takuya, 239
Nakamura, Jiro, 69
Nakase, Makoto, 239
Nalamasu, O., 4,84,194,207
Nathal, B., 35
Neenan, Thomas X., 194,518
Nogi, Norihiko, 318
Novembre, Anthony E., 4,137,180
Ober, Christopher K., 281
Oda, Hiroji, 485
Okabe, Yoshiaki, 395
Palmateer, Susan C., 271

Pankasem, Surapol, 504
Pingor, D. A., 137
Puyenbroek, R., 333
Reichmanis, Elsa, 4,166
Rothschild, Mordechai, 271
Rousseeuw, B. A. C., 333
Russell, T. P., 425
Sakamizu, Toshio, 124
Sanchez, M., 425
Schacht, H. T., 35
Schaedeli, U., 299
Schulz, R., 35
Shida, Naomi, 239
Shindo, Yoichi, 379,395
Shirai, Masamitsu, 318
Shiraishi, Hiroshi, 124
Sinta, Roger, 110
Slater, S. G., 4,35
Snow, Arthur W., 369
Sooriyakumaran, R., 21
Sugimura, Tokuko, 379,395
Suwa, Toshihiro, 471
Tada, Tsukasa, 239

Takahashi, Akio, 395
Takeichi, Tsutomu, 439
Tanaka, Akinobu, 69
Tanikawa, Masaaki, 439
Tarascon, R., 4
Thackeray, James W., 110
Timko, A. G., 4
Tinguely, E., 299
Tsunooka, Masahiro, 318
Ueda, Mitsuru, 413
Ueno, Takumi, 124
Ushirogouchi, Tohru, 239
van de Grampel, J. C., 333
van der Drift, E. W. J. M., 333
Vekselman, Alexander M., 149
Wallraff, Gregory M., 255
Wan, I- Y., 255
Wang, Chun-Shan, 527
Waymouth, Robert M., 355
Yang, G.-R., 449
Zettler, A., 35
Zhang, Chunhao, 149

Affiliation Index

AT&T Bell Laboratories,
4,84,137,166,180,194,207,518
Asahi Chemical Industry Company,
Ltd., 485
Ciba-Geigy Inc., 299
Cornell University, 281
Delft University of Technology, 333
FINLE Technologies, 56
Geo-Center, Inc., 369
Hitachi Ltd., 124,395
IBM Almaden Research Center, 21,255,425
Massachusetts Institute of Technology,
255,271
Matsushita Electric Industrial Company,
Ltd., 318
McGill University, 149
Mount Holyoke College, 166
NTT LSI Laboratories, 69
National Cheng Kung University, 527
National Kaohsiung Institute of
Technology, 527

Naval Research Laboratory, 369
OCG Microelectric Materials AG, 4,35,299
OCG Microelectronic Materials Inc., 4,35
Rensselaer Polytechnic Institute, 449
Sandia National Laboratories, 355
Seton Hall University, 137
Shipley Company LLC, 110
Stanford University, 355
Toho University, 379,395
Tokyo Institute of Technology, 471
Toshiba Research and Development
Center, 239
Toyohashi University of Technology, 439
Union Carbide Chemical & Plastics, Co.,
Inc., 228
Universite de l'Etat a Liege, 425
University of Groningen, 333
University of Osaka Prefecture, 318
University of Southern Mississippi, 504
Yamagata University, 413

Subject Index

A

Absolute chain length, calculation, 95t,100,101f

Absorption band shift method, single-layer resist for ArF excimer laser exposure containing aromatic compounds, 239–254

Acetal(s)
proton-assisted decomposition mechanism, 40–41
role in thermal decomposition, 38f,39

Acetal-blocked poly(vinylphenols), structure–property relationship, 35–53

Acetoxystyrene, hydrogen bonding in sulfone- and N-methylmaleimide-containing resist polymers, 166–179

Acid, photogeneration, 208–209

Acid-catalyzed imaging mechanisms, examples, 21–22

Acid-catalyzed rearrangement reactions, use for resist system design, 229

Acid diffusion
concentration dependence of diffusivity, 60–61
effect on resist performance, 76–82
equation, 60–61
evaluation
electrochemical method, 70–71,74–76
ion conductivity method, 70,76,78f
mask contact replication method, 70–73f
importance in chemical amplification, 2–3,69
initial acid distribution, 60
lithographic effects in chemically amplified resists, 56–66
modeling, 63–66
reaction–diffusion system, 62
role in lithography, 121,122f
temperature dependence of diffusivity, 60

Acid generators, water-soluble onium salts, 124–135

Acid loss
mechanisms, 61–62
types, 61

Acid surface depletion effect on phenolic polymers
acid diffusion effect on lithography, 121,122f
Arrhenius behavior for acid loss, 117,119–121
experimental description, 111,113
photoacid generating efficiency, 112f–115,118t
polymer matrix effects on diffusion, 120f,121
sources, 111–112
thermal processing effect on acid loss, 114f–118t

Acrylic-based 193-nm single-layer resist, properties, 247–252

Acrylic polymers, design, 255–269

Adhesive and photosensitive polymer, application of polyisoimide as polyimide precursor, 413–424

Alkyl-substituted sulfonium salts, acid generators for chemically amplified positive resists, 124–135

Allylated poly(2,6-dimethyl-1,4-phenylene ether)
applications, 501–503
dielectric constant–glass transition temperature relationship, 486f,487
dynamic mechanical behavior, 497,499–500
experimental description, 487–488
glass transition behavior, 496–497
^{1}H-NMR spectra, 488,489f
properties, 493,496–503
synthesis, 490–495
temperature dependence of cross-linking reaction, 493,495f
tensile modulus, 497–498
tensile strength, 497–498

Amine(s), functional development, 159,161,162f

Amine-free environment, approach, 2

Amine photogenerator, application of polyisoimide to photosensitive polyimide system, 421–424

Annealing concept for environmental
  stabilization of chemically amplified
  resists
  approach, 23–24
  experimental description, 22–23
  glass transition temperature reduction,
    24–28f
  high-temperature bake, 27–32
  schematic representation, 22
APEX, environmental stabilization, 31
ArF excimer laser exposure containing
  aromatic compounds, single-layer
  resist, 239–254
Aromatic compounds, single-layer resist
  for ArF excimer laser exposure, 239–254
Aromatic polyimides
  disadvantages, 413
  factors affecting properties, 395
Arrhenius behavior, acid loss, 117,119–121
Arylmethyl sulfone photoacid generators
  advantages and disadvantages, 138
  experimental description, 145–146
  intermolecular abstraction, 140–141
  internal hydrogen donor effect, 139–140
  photochemistry, 138–139
  polar effects, 141–142

B

Base-solubilizing components of
  photoresist resins,
  styrylmethylsulfonamides, 194–205
Benzene derivatives, characterization of
  plasma-polymerized films, 471–483
Benzhydryl ammonium salts, *See*
  Quaternary ammonium salt photobase
  generators
Benzophenone-type polyimides,
  photochemistry, 386f–394
Biphenyl-type polyimides, photophysics,
  381–385,387
Biscyclopentadienones and ethynyl-
  terminated imides, preparation
  of hybrid polyimide–polyphenylenes by
  Diels–Alder polymerization, 518–525
Block copolymer(s), knowledge, 281

Block copolymer–homopolymer resist
  mixtures, use as resist materials, 287,290
Block copolymer resist materials, silicon-
  containing, *See* Silicon-containing
  block copolymer resist materials
tert-Butoxycarbonyl, process issue
  analysis with chemically amplified
  positive resists, 4–18
tert-Butoxycarbonyl-based deep-UV
  resist, chemical amplification, 84,86
tert-Butoxycarbonyl-blocked polymer,
  acidolysis, 58
(*tert*-Butoxycarbonyl)oxy, role in
  chemically amplified positive resists
  system properties, 5
[(*tert*-Butoxycarbonyl)oxy]styrene-based
  polymers, use of 4-[(methanesulfonyl)-
  oxy]styrene for property improvement,
  207–226
(*tert*-Butoxycarbonyl)styrene resins,
  shrinkage, 195
tert-Butyl group, acidolytic removal,
  208–209

C

Camphorsulfonic acid, surface depletion
  effect on phenolic polymers, 110–122
Carboxy-terminated butadiene–acrylonitrile
  copolymer modifiers
  glass transition temperature, 528
  reduction of internal stress, 528
Catalytic action, role in high sensitivity
  in chemically amplified resists, 2
Chain length relationship
  clearing dose, 85
  Hammett σ values, 88,89f
Chemical amplification
  concept, 2
  description, 334–335
  examples, 334–336
Chemical amplification based on photoacid-
  catalyzed deprotection of polymers,
  design of highly sensitive resist
  systems for use in short-wavelength
  lithographic technology, 228

Chemical amplification concept, foundation for design of advanced resist systems for lithographic imaging, 21

Chemical vapor deposition, description, 451

Chemically amplified deep-UV resists, photogenerated acid strength–postexposure delay effect relationship, 84–107

Chemically amplified photoresists, design using dienone–phenol rearrangements reaction, 228–235

Chemically amplified positive resists
  process issue analysis, 4–18
  role of (*tert*-butoxycarbonyl)oxy on properties, 5
  typical design, 27
  water-soluble onium salts as acid generators, 124–135

Chemically amplified resist(s)
  acid diffusion effect, 1,60–61,69–82
  acid loss, 61–62
  advantages, 3
  amine-free environment, 2
  annealing concept for environmental stabilization, 21–32
  applications, 149
  composition, 2
  concept, 56–57
  delay problem, 21–22
  examples using nonionic acid generators, 124
  functional imaging, 149–163
  kinetics, 57–59
  lack of environmental stability, 110
  lithographic effects of acid diffusion, 56–66
  polymer alteration, 151
  polysiloxane applications, 333–353
  relief and functional imaging, 149–151
  resist properties, 1–2
  role of catalytic action in high sensitivity, 2
  sensitivity, 2
  steps in formation of latent images, 69
  use for manufacturing-scale deep-UV lithography, 194

Chemically amplified resist based on phenoxyethyl protection of poly(vinylphenol), advantages, 228

Clearing dose, relationship to chain length, 85,87*f*

Coefficient of thermal expansion, polymer dielectrics for microelectronics, 367–368

Cold wall vapor deposition, description, 452–453

Computer chips, fabrication, 449

Conventional photolithography, improvements, 1

Copolymers with deep-UV imageable blocks, use as resist materials, 286–287,288–289*t*

Cresol–formaldehyde novolac epoxy, properties, 527

Cross-link site for polyimides
  internal acetylene unit, 439–448
  use to improve properties, 439

D

Deep-UV photoresists, structure–property relationship of acetal- and ketal-blocked poly(vinylphenols), 35–53

Deep-UV technology, design of chemically amplified positive resist systems, 5

Delay problem, chemically amplified resists, 21–22

Deposition method, liquid phase, plasma-developable photoresist system based on polysiloxane formation at irradiated surface, 318–331

Design, 193-nm positive resist, 255–269

2,6-Dialkoxyanthracene-containing side chain substituted liquid-crystalline polymers
  characterization, 508–509
  experimental description, 505,507
  fluorescence, 512–514*f*
  photochemistry, 513,515–516
  preparation, 506–508
  previous studies, 505
  synthesis, 508
  UV–visible spectra, 509–512

Diazonaphthoquinone, application of
    polyisoimide to photosensitive
    polyimide, 417,419–422f
Diblock copolymer, phase behavior,
    282–284
Dielectric(s), polymer, microelectronics,
    367–368
Dielectric constants
    calculation, 467
    reduction methods, 367–369
Dielectric loss factor, calculation, 70
Diels–Alder polymerization between
    biscyclopentadienones and
    ethynyl-terminated imides, hybrid
    polyimide–polyphenylenes, 518–525
Dienone–phenol rearrangements reaction
    acid-catalyzed 4-alkyl-4-acetoxy-
        cyclohexadienone rearrangement
        reaction, 234–235
    chemistry, 234
    experimental description, 229,231
    film loss, 235
    imaging experiments, 235
    instrumentation, 229
    polymerization of monomer, 233
    proton NMR
        4-hydroxy-4-methylcyclohexadien-1-
            one, 230f,231
        monomer, 230f,231,233
        polymer, 232f–234
    sensitivities, 235
    synthesis and polymerization of
        4-(4-methylcyclohexadien-1-one)
        methacrylate, 229,233
Diffusion
    effect of polymer matrix, 120f,121
    importance in chemical amplification, 2–3
Diffusion coefficient, calculation, 70
Diffusion length, relationship with
    concentration of acids, 80f–82
{[(4,5-Dimethoxy-2-nitrobenzyl)oxy]
    carbonyl}-2,6-dimethylpiperidine,
    application to photosensitive polyimide
    system, 421–424
Dimethylsilyldimethylamine, optimization
    of positive-tone 193-nm silylation
    resist process, 271–279

Direct current glow discharge, preparation
    of plasma-polymerized films, 471–483
Dissolution rate selectivity, role in
    stability, 15
Dry development, top surface imaged
    resists for 193 nm, 274–276f
Dry development techniques employing
    reactive ion etching pattern transfer,
    advantages for sub-0.25-μm
    imaging, 237
Dry etch barriers, top surface imaging
    approach based on light-induced
    formation, 299–315
Dyes, functional development, 161–163f

E

e-beam lithographic technology, design of
    chemically amplified positive resists
    systems, 5
Electrochemical method
    evaluation of acid diffusion, 71,74–76
    experimental procedure, 70
Electron spectroscopy for chemical analysis
    (ESCA) spectroscopy, characterization of
    plasma-polymerized films of benzene
    and fluorinated derivatives,
    474–475,478–480,482f
Electronics application, polysiloxane
    thermoplastic polyurethane modified,
    527–541
Electronics equipment, trend, 527
Environmental stabilization of chemically
    amplified resists, annealing concept,
    21–32
Environmentally friendly polysilane
    photoresists
    branching vs. photosensitivity, 359,362f
    characterization, 358–359
    cross-linking, 359
    effect of tert-butyl group, 361,364
    experimental procedure, 357–358
    future work, 364–365
    oxygen etch behavior, 361
    photobleaching, 359,360f
    side groups, 357,360f,363
    solubility, 361,363f

Environmentally friendly supercritical CO_2 resist, development, 296

Epoxy molding compounds
need for low-stress compounds, 528
properties, 527

Epoxy resins, polysiloxane thermoplastic polyurethane modified, *See* Polysiloxane thermoplastic polyurethane modified epoxy resins for electronic application

Epoxy series, chemically amplified resist applications, 349–351

ESCAP, annealing concept for environmental stabilization, 21–32

Etch resistance, process issue analysis with chemically amplified positive resists, 16–17

Ethynyl-terminated imides and biscyclopentadienones, preparation of hybrid polyimide–polyphenylenes by Diels–Alder polymerization, 518–525

Exposure tool environment, role in stability, 15

F

Fabrication of computer chips
components, 449
research areas, 449

Flexural test, polysiloxane thermoplastic polyurethane modified epoxy resins for electronics application, 539–540

Fluorescence, 2,6-dialkoxyanthracene-containing side chain substituted liquid-crystalline polymers, 512–514*f*

Fluorinated derivatives, characterization of plasma-polymerized films, 471–483

Fluorinated polymers, microelectronic applications, 367

Fluorine-containing epoxies and acrylics, properties, 370

Fluorine content, fluoropolymers with low dielectric constants, 369–377

Fluoropolymers with low dielectric constants
dielectric constants, 375–376
experimental procedure, 370–371,373

Fluoropolymers with low dielectric constants—*Continued*
monomer characterization, 373,374*f*
monomer preparation, 372–373
polymer characterization, 375
polymer preparation, 372–373
previous studies, 369–370
reduction of dielectric constants, 376–377
structure–property relationships, 375–376

Fourier-transform IR spectroscopy, characterization of plasma-polymerized films of benzene and fluorinated derivatives, 474,476–477*f*

Functional imaging with chemically amplified resists
description, 151
experimental procedure, 152–153
exposure-controlled sorption
calcium hydroxide, 156–158*f*
other metal ions, 157,159,160*f*
functional development
with amines, 159,161,162*f*
with dyes, 161–163*f*
with metal ions, 156
instrumentation, 151–152
postexposure bake reactions, 154–155
reactive ion etching, 159
refractive index, 159,160*f*
relief development, 154*f*–156
resist evolution, 153,155

G

Glass transition, relationship to protecting group structure, 46*f*–48

Glass transition temperature, reduction for annealing of resist films, 24–28*f*

Graft copolymers, chemically amplified resist applications, 341–346

Growth of diffusion layer, calculation, 71

H

High-performance chips, development requirements, 451

High-performance plastics, potential applications, 413

High-performance polymers
  advantages, 425
  role as packaging materials, in
    manufacture of microelectronic
    devices and components, 425
High-performance positive-tone resists,
  demand, 194
High-resolution proton NMR, hydrogen
  bonding in sulfone- and N-methyl-
  maleimide-containing resist polymers
  with hydroxystyrene and acetoxystyrene,
  170f,171,173,177
Homopolymer–block copolymer resist
  mixtures, use as resist materials, 287,290
Hot wall vapor deposition, description, 453
Hybrid polyimide–polyphenylenes by
  Diels–Alder polymerization between
  biscyclopentadienones and ethynyl-
  terminated imides
  characterization, 525
  experimental description, 519
  monomers
    characterization, 521
    preparation, 520f,521,523
  NMR spectra, 524t,525
  polymerization procedure, 519,521,522t
  polymerization reactions, 521,524
  previous studies, 518–520
  synthetic procedure, 519
  thermal properties, 525
Hydrogen bonding in sulfone- and
  N-methyl-maleimide-containing resist
  polymers with hydroxystyrene and
  acetoxystyrene
  blend miscibility, 169,171
  effect of sulfone, 175t,177–178
  experimental description, 167–168
  future work, 178–179
  high-resolution proton NMR,
    170f,171,173,177
  importance, 167
  strength and mode vs. concentration,
    173,175–178
  structures, 168–169
  two-dimensional nuclear Overhauser effect
    spectra, 172f–174f,177

Hydrogenated poly(p-vinylphenol) matrix,
  surface depletion effect on phenolic
  polymers, 110–122
Hydrosiloxane–triallyl ether resins,
  fluoropolymers with low dielectric
  constants, 369–377
Hydrosilylation, Si–C bond formation, 370
Hydroxystyrene, hydrogen bonding in
  sulfone- and N-methylmaleimide-
  containing resist polymers, 166–179

I

Imageability, importance, 368
Insulating layer in electric systems,
  requirements, 369
Integrated circuits, effort to design and
  fabricate reduced-size elements, 1
Interconnect dielectric materials,
  requirements, 449–450
Intermolecular charge transfer of
  polyimides
  donor–acceptor alternative sequence in
    aromatic chains, 396t–400f
  dynamic mechanical properties,
    403,409–410f
  effect of casting temperature, 403,407f
  effect of molecular stacking,
    403,409–410f
  experimental description, 395–397
  in-plane molecular orientation
    effect of film thickness,
      403,405–406f,409,411
    effect of thermal treatment,
      399,402–404f
    measurement, 399,401f
  molecular aggregation vs. intermolecular
    charge-transfer fluorescence,
    399,400–401f
Intermolecular donor–acceptor interaction,
  presence in solid-state aromatic
  polyimides, 395
Internal acetylene unit, cross-link site
  for polyimides, 439–448
Internal stress, reduction methods, 528

Intramolecular charge transfer of polyimides
experimental description, 380
material selection, 379–380
photochemistry, 386*f*–394
photophysics, 381–385,387
Ion conductivity method
evaluation of acid diffusion, 76,78*f*
experimental procedure, 70
IR spectroscopy, polysiloxane thermoplastic polyurethane modified epoxy resins for electronic application, 531,534–535
Irradiated surface, plasma-developable photoresist system based on polysiloxane formation, 318–331

J

JASIC, chemically amplified resist applications, 345,347–349

K

Ketal(s), proton-assisted decomposition mechanism, 40–41
Ketal-blocked poly(vinyl phenols), structure–property relationship, 35–53
Kinetics, chemically amplified resists, 57–59

L

Latent image contrast, resist performance–acid diffusion relationship, 77,78*f*
Latent images of chemically amplified resists, steps in formation, 69
Lateral distribution of decomposed dissolution inhibitor, calculation, 76
Light-induced formation of dry etch barriers, top surface imaging approach, 299–315

Liquid-crystalline polymers
applications, 505
2,6-dialkoxyanthracene-containing side chain substituted, *See* 2,6-Dialkoxyanthracene-containing side chain substituted liquid-crystalline polymers
photochemistry, 504–505
properties, 504–505
Liquid-phase deposition method, plasma-developable photoresist system based on polysiloxane formation at irradiated surface, 318–331
Lithographic effects of acid diffusion in chemically amplified resists
acid diffusion, 60–61
acid loss, 61–62
experimental description, 57
kinetics, 57–59
modeling, 63–66
reaction–diffusion system, 62
Lithographic technologies, resist properties, 1–2
Low dielectric constant(s), fluoropolymers, 369–377
Low dielectric constant polymers, vapor depositable, *See* Vapor-depositable low dielectric constant polymers
Low dielectric materials, applications, 485

M

Mask contact replication method
evaluation of acid diffusion, 70–73*f*
experimental procedure, 70
Materials designed for use as resists of the future, requirements, 281
Meta-isomer resins, glass transition temperature, 24–28*f*
Metal(s) at polymer surface, selective formation, 319
Metal ion(s), functional imaging with chemically amplified resists, 149–163
Metal ion developed resist
reactive ion etching, 159
refractive index, 159,160*f*

Metal oxides, selective formation, 319

Methacrylate polymers
design considerations, 256–269
optical transparency, 256,258f

Methanesulfonic acid, surface depletion
effect on phenolic polymers, 110–122

4-[(Methanesulfonyl)oxy]styrene
absorbances, 215
base-catalyzed hydrolytic cleavage, 215
developability, 215,220–221
effect of base-soluble 4-hydroxystyrene
unit, 215,221,222f
effect of photoacid generator,
221,223–226
experimental description, 207,209
lack of deprotection during acidolytic
thermal processing, 215,218f
lithography procedure, 211,214t,216–217t
monomer synthesis, 208–209
polymer synthesis, 210,212–213t
resist IR study procedure, 211
resist NMR study procedure, 211,215
thermal analysis procedure, 210

N-Methylmaleimide-containing resist
polymers with hydroxystyrene and
acetoxystyrene, hydrogen bonding,
166–179

α-Methylstyrene–styrene–styrene
copolymers, preparation of polyimide
nanofoams, 425–438

Microdevices, role of optical
lithography, 299

Microelectronics, polymer dielectrics,
367–368

Microelectronics industry, demand for
polymeric materials, 518

Microlithographic evaluation,
silicon-containing polymers, 333–353

Microlithographic technology, advances,
255–256

Miscibility in novolac–poly(2-methyl-1-
pentenesulfone) resists
effect of glass transition temperature,
182t,183
effect of thermal degradation temperature,
182t,183
experimental description, 181–183

Miscibility in novolac–poly(2-methyl-1-
pentenesulfone) resists—Continued
film quality improvement approaches,
192–193
proton NMR assignment, 183–184
reproducibility in blend formation,
189,190f,192
solid-state cross-polarization magic-
angle spinning spectra, 189,191f,t
strength of intermolecular interaction,
185,187f,189,191f,t
structure–property relationship,
185,188–189
two-dimensional nuclear Overhauser
effect spectra, 185–190

Modeling of lithographic effects of acid
diffusion in chemically amplified resists
dose-to-clear vs. acid diffusivity, 63
parameters, 63–64
resist line width
vs. diffusivity, 64,65f
vs. focal position, 64,66f
sidewall angle vs. diffusivity, 64,65f

Molecular architecture, fluoropolymers
with low dielectric constants, 369–377

Morphology, polysiloxane thermoplastic
polyurethane modified epoxy resins for
electronics application, 535,537f

N

Nanofoams
preparation from styrenic block
copolymers, See Polyimide nanofoams
prepared from styrenic block
copolymers
reduction of dielectric properties of
polyimides, 426

NMR spectroscopy
hybrid polyimide–polyphenylenes by
Diels–Alder polymerization between
biscyclopentadienones and
ethynyl-terminated imides, 524t,525
miscibility in novolac–poly(2-methyl-1-
pentenesulfone) resists, 180–193

NORSOX, chemically amplified resist
applications, 345,347–349

Novolac–poly(2-methyl-1-pentenesulfone) resists, miscibility, 180–193

Novolac–poly(olefinsulfone) blend
advantages, 180
amplification, 181
development, 180
processing problems, 181

Nuclear Overhauser effect spectroscopy, hydrogen bonding in sulfone- and N-methylmaleimide-containing resist polymers with hydroxystyrene and acetoxystyrene, 172f–174f,177

Nucleophilicity and basicity of sulfonate anion, free H^+ effect, 85,87–88

O

Optical lithography using 193-nm radiation, use for manufacture of integrated circuits with sub-0.25-μm features, 271

P

Packaging materials in manufacture of microelectronic devices and components, requirements, 425

Parylene, *See* Poly(p-xylylene)

Parylene F, *See* Poly(tetrafluoro-p-xylylene)

Perfluoroalkyl groups, reduction of dielectric properties of polyimides, perfluoroalkyl groups, 425–426

Phase behavior, polymers, 282–284

Phenolic polymers, effect of acid surface depletion, 110–122

Phenolic resin–oligomeric dimethylsiloxane resists, use as resist materials, 290

Photoacid and photobase generators
arylmethyl sulfone photoacid generators, 138–142
experimental description, 138,145–146
quaternary ammonium salt photobase generators, 143–145f

Photoacid-generating efficiency, 112–115,118

Photoacid generation, reaction, 57

Photoacid generators
role in postexposure delay effect, 85
thermal decomposition mechanism, 85–86

Photoacid strength, role in stability, 11–13

Photocatalyst generators
demand, 137–138
design considerations, 138
history, 137

Photochemically generated catalysts and initiators, *See* Photocatalyst generators

Photochemistry
2,6-dialkoxyanthracene-containing side chain substituted liquid-crystalline polymers, 513,515–516
liquid-crystalline polymers, advantages, 504–505
relationship with charge transfer of polyimides, 386f–394

Photodesulfonylation, description, 138

Photogenerated acid strength–postexposure delay effect relationship with positive-tone chemically amplified deep-UV resists
anomalous postexposure delay effect, 105,106–107f
experimental description, 88,90–95
lithography
absolute chain length calculation, 95t,100,101f
conditions, 91t
Hammett plots of relative chain length, 93,96–99
mechanistic models
anomalous substituent effect, 100,102–103
turnover effect, 98,100
turnover effect evaluation, 98,99f
molecular polarizability–anomalous postexposure delay substituent effect relationship, 102,104
thermal stability of photoacid generator, 93–95t

Photolithography, conventional, improvements, 1

Photophysics, relationship with charge transfer of polyimides, 381–385,387

Photoreaction of photoacid generator
kinetics, 57–59
mechanisms, 57–58
Photoresist(s), environmentally friendly
polysilane, *See* Environmentally
friendly polysilane photoresists
Photoresist resins, versatile base-
solubilizing components, 194–205
Photoresist system based on polysiloxane
formation at irradiated surface,
plasma developable, *See* Plasma-
developable photoresist system based on
polysiloxane formation at irradiated
surface
Photoresist technology, advances, 255–256
Photosensitive polymer, application of
polyisoimide as polyimide precursor,
413–424
Photosensitive silicon-containing graft
copolymers, use as resist materials,
290–291
Plasma-developable photoresist system
based on polysiloxane formation at
irradiated surface
copolymer preparation, 320–322*f*
copolymer structure, 322*f*,323
effect of alkoxysilane structure, 327,329*t*
effect of methyltriethoxysilane,326*f*–329*f*
experimental description, 319–323
irradiation, 326*f*,327
mechanism, 319,322*f*
monomer preparation, 319–320
oxygen plasma etching, 327,330*f*
polymer structure effect, 327,330*f*
Plasma polymerization
advantages, 471
in direct current glow
apparatus, 472*f*,473
ESCA spectra, 474–475,478–480,482*f*
experimental description, 473
film growth, 481,482*f*
Fourier-transform IR spectra,
474,476–477*f*
previous studies, 471,473
Raman spectrum, 481,483
surface properties of films, 479,481
modification of surface properties, 471

Plasma processing of polymers, formation
of low dielectric constant polymers, 368
Poly[amic acid(s)], polymerization, 379
Poly(amic acid) precursor, preparation and
thermal imidization, 428,431
Poly(amic alkyl ester), preparation and
thermal imidization, 428,432
Poly(*tert*-butyl acrylate-*co*-hydroxy-styrene)
resist, environmental stabilization, 31
Poly[(*p-tert*-butylphenyl)silane]
characterization, 364,365*f*
synthesis, 361,364
Poly(di-*tert*-butyl fumarate-*co*-styrene)
resist, functional imaging, 149–163
Poly(2,6-dimethyl-1,4-phenylene ether)
disadvantages, 487
modification to yield thermosetting
polymer, 487
properties, 486*t*,487
Poly(fluorinated benzocyclobutene)
preparation, 463,465–467
properties, 465–466,468
structure, 463,465
Polyimide(s)
advantages as thermally stable
polymers, 439
applications, 413
disadvantages, 518
intermolecular charge transfer, 395–411
microelectronic applications, 367
photophysics, photochemistry, and
intramolecular charge transfer, 379–394
problems, 439
properties, 518
Polyimide(s) having internal acetylene
units
cross-linking of (4,4'-diaminodiphenyl)-
acetylene, 441,443–445
experimental description, 440–441
IR spectra of polymers, 441,442*f*
monomer structures, 440,442*f*
polymer structure, 441,442*f*
previous studies, 440
properties
glass transition temperature, 445–447*f*
stress–strain curves, 445,447*f*
tensile modulus, 445,448*f*

Polyimide nanofoams prepared from styrenic block copolymers
amino termination of styrene and α-methylstyrene oligomers, 428,430
characteristics
aromatic amine functional thermally labile oligomers, 428
polyimide-based block copolymers, 428–429,433
dynamic mechanical spectra, 433,434*f*
experimental description, 426–427
poly(amic acid) precursor preparation and chemical imidization, 428,431
poly(amic alkyl ester) preparation and thermal imidization, 428,432
thermal gravimetric analysis, 427,434*f*
Polyimide–polyphenylenes by Diels–Alder polymerization between biscyclopentadienones and ethynyl-terminated imides, *See* Hybrid polyimide–polyphenylenes by Diels–Alder polymerization between biscyclopentadienones and ethynyl-terminated imides
Polyimide precursor to polymer adhesive and photosensitive polymer, polyisoimide application, 413–424
Polyisoimide
adhesion strength evaluation procedure, 414
application to photosensitive polyimide system
amine photogenerator, 421–424
diazonaphthoquinone, 4l7,419–422
description, 413
evaluation as high-temperature adhesives, adhesion properties, 416–419*f*
experimental description, 414
photosensitivity measurement procedure, 414
polymer synthesis, 416
polymer synthetic procedure, 414
previous studies, 413
property measurement procedure, 415
Polymer(s)
importance in development of resist technology, 166
microelectronics applications, 367

Polymer(s)—*Continued*
phase behavior, 282–284
vapor-depositable low dielectric constant polymers, *See* Vapor-depositable low dielectric constant polymers
Polymer alteration, chemically amplified resists, 151
Polymer(s) as resists in microlithography, design, 333
Polymer dielectrics for microelectronics applications, 367
coefficient of thermal expansion, 367
dielectric constant reduction methods, 367–368
imageability, 367
Polymer matrix, effect on diffusion, 120*f*,121
Polymer resins, requirements for sub-0.2-μm imaging, 237
Polymer type, fluoropolymers with low dielectric constants, 369–377
Polymeric binder in two-component positive deep-UV photoresists, structure–property relationship of acetal- and ketal-blocked poly(vinylphenols), 35–53
Poly(2-methyl-1-pentenesulfone)–novolac resists, miscibility, 180–193
Poly(α-methylstyrene), preparation of polyimide nanofoams, 425–438
Poly(naphthalene) and poly(fluorinated naphthalene)
apparatus for preparation, 462–464*f*
preparation, 462
properties, 463,468
structure, 461–462
Poly(olefinsulfone)–novolac blend, *See* Novolac–poly(olefinsulfone) blend
Polyphenylene–polyimides by Diels–Alder polymerization between biscyclopentadienones and ethynyl-terminated imides, *See* Hybrid polyimide–polyphenylenes by Diels–Alder polymerization between biscyclopentadienones and ethynyl-terminated imides

Polysilane(s)
dehydrogenative coupling on
  phenylsilane, 356
potential applications, 355
synthetic problems, 355–356
use as positive resists, 355
Polysilane photoresists, environmentally
  friendly, *See* Environmentally friendly
  polysilane photoresists
Polysiloxane(s)
advantages and disadvantages, 353
epoxy series, 349–351
experimental description, 336
experimental procedure
  microlithography, 339–341
  synthesis, 336–340
graft copolymers, 341–346
JASIC, 345,347–349
NORSOX, 345,347–349
polysilsesquioxanes, 351–352
properties, 370,528
selective formation, 319
Polysiloxane formation at irradiated
  surface, plasma-developable
  photoresist system, 318–331
Polysiloxane thermoplastic polyurethane
  modified epoxy resins for electronic
  application
dynamic viscoelastic properties,
  535,536*f*,539*t*
encapsulation formulation, 539
experimental description, 528–534
flexural test properties, 539–540
internal stress, 540–541
IR spectroscopy, 531,534–535
morphology, 535,537*f*
previous studies, 528
synthesis
  prepolymer, 531,532*f*
  resin, 531,533*f*
synthetic procedure, 530
thermal mechanical properties, 538*f*,539
thermal stock cycling test, 541
Polysilsesquioxanes, chemically amplified
  resist applications, 351–352
Polystyrene, preparation of polyimide
  nanofoams, 425–438

Poly(tetrafluoro-*p*-xylylene)
chemical vaporization preparation
  method, 456–457,459
optical micrographs, 458–460*f*
precursor preparation method, 456
previous preparation methods, 455–456
properties, 456–457,468
Polyurethane modified epoxy resins for
  electron application, *See* Polysiloxane
  thermoplastic polyurethane modified
  epoxy resins for electronic application
Poly[vinylphenol(s)], structure–property
  relationship, 35–53
Poly(vinylphenol) resist, optimization of
  positive-tone 193-nm silylation resist
  process, 271–279
Poly(*p*-xylylene)
apparatus for preparation, 453–454
preparation, 453,455
properties, 453,458*f*,468
Positive deep-UV photoresists, structure–
  property relationship of ketal and
  ketal-blocked poly(vinylphenols),
  35–53
Positive deep-UV resists
advantages, 35
protecting groups, 37
technical hurdles, 35,37
Positive resist design, 193 nm
building block approach, 257,259
challenge, 256
etch-resistant versions, 261–269
experimental description, 256–258*f*
properties, 255
version 1 resist, 259–261
Positive-tone 193-nm silylation resist
  process, optimization, 271–279
Positive-tone chemically amplified deep-
  UV resists, photogenerated acid
  strength–postexposure delay effect
  relationship, 84–107
Postexposure bake, environmental
  stabilization of chemically amplified
  resists, 21–32
Postexposure bake process step, role in
  sensitivity of chemically amplified
  positive resist systems, 5

Postexposure delay effect
  description, 2,85
  methods to reduce effects, 85
  process issue analysis with chemically
    amplified positive resists, 9–15
Postexposure delay effect–photogenerated
    acid strength relationship with
    positive-tone chemically amplified
    deep-UV resists, lithography, *See*
    Photogenerated acid strength–
    postexposure delay effect relationship
    with positive-tone chemically amplified
    deep-UV resists
Postexposure silylation process,
    studies, 318
Printed wiring board industry, advantages
    of using self-developing photoresist,
    356–357
Process issue analysis with chemically
    amplified positive resists
  etch resistance, 16–17
  experimental description, 5–7
  factors affecting stability
    dissolution rate selectivity, 15
    exposure tool environment, 15
    photoacid strength vs. stability, 11–13
    protecting group, 13
    swing curve, 14f,15
  postexposure delay
    carbonyl C_1s intensity vs. exposure
      and postexposure bake, 9,10f
    carbonyl stretch vs. postexposure
      delay time intervals, 7,10f
    postexposure delay induced surface
      residue, 7,8f
  sensitivity to postexposure bake, 17
  shrinkage, 16–17
  substrate contamination, 14f–16
Properties
  allylated poly(2,6-dimethyl-1,4-phenylene
    ether), 493–503
  intermolecular charge transfer of
    polyimides, 403,409–410f
  polysiloxane thermoplastic polyurethane
    modified epoxy resins for electronics
    application, 535–536,538–541

Protecting group
  role in stability, 13
  secondary reactions, 49–50
Protecting group structure relationship
  glass transition, 46f–48
  thermal decomposition, 38–47
Proton-assisted decomposition, mechanism,
    40–41

Q

Quaternary ammonium salt photobase
    generators
  experimental description, 145–146
  Newman projection of calculated favored
    conformation, 143–144
  photochemistry, 143–144
  photogeneration, 143
  photolysis, 144,145f

R

Raman spectroscopy, characterization of
    plasma-polymerized films of benzene
    and fluorinated derivatives, 481,483
Reaction–diffusion system
  description, 62
  modeling, 63–66
  solution, 62
Reactive ion etching, metal ion developed
    resist, 159
Refractive index, metal ion developed
    resist, 159,160f
Relative chain length, Hammett plots,
    93,96–99
Relative signal propagation speed for
    dielectric material, calculation, 485
Resist materials
  development for sub-0.25-μm imaging,
    237–238
  need for improvement, 180
  parameters affecting application, 334
  silicon-containing block copolymer, *See*
    Silicon-containing block copolymer
    resist materials

Resist performance–acid diffusion relationship
latent image contrast, 77,78f
model of latent image formation, 76–77
resolution–sensitivity relationship, 77,79–82
Resist properties, lithographic technologies, 1–2
Resolution
calculation, 79
relationship to resolution, 77,79–82

S

Selective formation of polysiloxanes, metal oxides, and metals at polymer surface, studies, 319
Self-developing resists, description, 355
Sensitivity
calculation, 79
chemically amplified positive resists, postexposure bake process step, 5
postexposure bake, process issue analysis with chemically amplified positive resists, 17
relationship to resolution, 77,79–82
role of catalytic action for chemically amplified resists, 2
Shrinkage, process issue analysis with chemically amplified positive resists, 16–17
Si–C bond formation, hydrosilation, 370
≡Si–OR bond, stability, 290
Signal transmission loss, calculation, 485,487
Silicon-containing block copolymer resist materials
bilevel resist scheme, 284–285
block copolymer–homopolymer resist mixtures, 287,290
copolymer with deep-UV imageable blocks, 286–289t
e-beam sensitive silicon-containing block copolymers, 291–292
e-beam sensitive silicon-containing graft copolymers, 291
experimental description, 281–282

Silicon-containing block copolymer resist materials—*Continued*
first use, 285–286
for 193-nm lithography
advantages of block over random copolymers, 295
description, 293
environmentally friendly supercritical CO_2 resist development, 296
experimental description, 293–296f
polymerization rate, 295
phase behavior of polymers, 282–284
phenolic resin–oligomeric dimethylsiloxane resists, 290
photosensitive silicon-containing graft copolymer, 290–291
≡Si–OR bond stability, 290
Silicon-containing graft copolymers, first use as resist materials, 286
Siloxane-containing polymers, advantages, 370
Single-layer approaches to top surface imaging, description, 301
Single-layer resist for ArF excimer laser exposure containing aromatic compounds
absorbance of aromatic compounds at 193-nm light using molecular orbital calculation, 241–243f,245t
acid-sensitive polymers, 246t–248f
acrylic-based 193-nm single-layer resist properties, 247–252
experimental description, 240
photoacid generators, 242,247
previous studies, 239–240
transparency enhancement at 193 nm for naphthalene-containing polymer, 249,251f,253
transparent resist materials with aromatic rings, 242,244f–246f
Stability, influencing factors, 9,11–15
Stabilization of chemically amplified resists, environmental, annealing concept, 21–32
5B-Steroid dissolution inhibitors, effect on etch resistance, 261–269
Structure, intermolecular charge transfer of polyimides, 395–409,411

Structure–property relationship of acetal-
and ketal-blocked poly(vinylphenols) as
polymeric binder in two-component
positive deep-UV photoresists
Acetal 1, 51–53
bond cleavage vs. stability, 41,43
cross-linking, 43,46–47
cyclic structure vs. stability, 43–45*f*
experimental description, 36*f*,37,39
glass transition temperature vs. degree
of protection, 46*f*,47–48
lithographic performance, 47–53
protecting group structure vs.
glass transition, 46*f*,47–48
thermal decomposition, 38–47
proton-assisted decomposition mechanism,
40–41
secondary reactions of protecting
groups, 49–50
thermolysis vs. degree of protection,
41,44*f*
weight loss vs. temperature, 41,42*f*
Styrene–α-methylstyrene copolymers,
preparation of polyimide nanofoams,
425–438
Styrenic block copolymers, preparation of
polyimide nanofoams, 425–438
Styrylmethylsulfonamides
experimental description, 195–197,199
monomer preparation, 198*f*–200,202
polymer characterization, 200
polymer preparation, 200–202
thermal analysis, 200,203–205
Sub-0.25-μm imaging, development of
new resist materials and processes,
237–238
Substituted poly[(*p-tert*-butylphenyl)
silanes], synthesis, 357–358
Substituted poly(phenylsilanes),
synthesis, 357
Substrate contamination, process issue
analysis with chemically amplified
positive resists, 14*f*–16
Sulfone-containing resist polymers with
hydroxystyrene and acetoxystyrene,
hydrogen bonding, 166–179

Sulfonium salts, alkyl substituted, 124–135
Surface imaging processes,
requirements, 299
Surface mount technology, thermal stress
generation, 527–528
Swing curve, role in stability, 13
Synthesis
allylated poly(2,6-dimethyl-1,4-phenylene
ether), 490–495
2,6-dialkoxyanthracene-containing side
chain substituted liquid-crystalline
polymers, 508
silicon-containing polymers, 333–353

T

Teflon and Teflon AF
preparation, 459,461
properties, 461,468
structures, 459
Tertiary amine photogenerator,
development, 143–146
Thermal decomposition, relationship to
protecting group structure, 38–47
Thermal expansion coefficient, influencing
factors, 395–411
Thermal processing, effect on acid loss,
114*f*–118*t*
Thermal properties, hybrid
polyimide–polyphenylenes by
Diels–Alder polymerization between
biscyclopentadienones and
ethynyl-terminated imides, 525
Thermal stock cycling test, polysiloxane
thermoplastic polyurethane modified
epoxy resins for electronic
application, 541
Thermal stress generation, surface mount
technology, 527–528
Toluenesulfonic acid, surface depletion
effect on phenolic polymers, 110–122
Top surface imaged resists
for 193 nm
cost, 278
dry development, 274–276*f*
experimental description, 271–273

Top surface imaged resists—*Continued*
for 193 nm—*Continued*
lithography, 275,277–279
resist and silylating agent chemistry,
272*f*–274
silylation mask shape, 274,276*f*
process, 271
Top surface imaging approach based on
light-induced formation of dry etch
barriers
additives, 305
experimental description, 311,314–315
imaging concept, 300*f*,301
lithographic evaluation
monomer selection, 307,308*f*
process evaluation, 306*t*,307,309–310
results, 307,311–314*f*
matrix polymers, 303,305–306
photoactive compounds, 305
reactive monomers
preparation, 302–304
properties, 303,304*t*
Top surface imaging concepts,
categories, 301
Triallyl ether–hydrosiloxane resins,
fluoropolymers with low dielectric
constants, 369–377
Triflic acid, surface depletion effect on
phenolic polymers, 110–122
Trimethylbenzhydrylammonium iodide
photochemistry, 143–144
photolysis, 144,145*f*
Turnover effect
evaluation, 98,99*f*
mechanistic models, 98,100
Two-component positive deep-UV
photoresists, structure–property
relationship of acetal- and ketal-
blocked poly(vinylphenols), 35–53
Two-dimensional NMR spectroscopy,
hydrogen bonding in sulfone- and
N-methyl-maleimide-containing resist
polymers with hydroxystyrene and
acetoxystyrene, 166–179
Two-layer approaches to top surface
imaging, description, 301

U

UV-assisted vapor deposition, description, 453
UV–visible spectra, 2,6-dialkoxyanthracene-
containing side chain substituted liquid-
crystalline polymers, 509–512

V

Vapor-depositable low dielectric constant
polymers
chemical vapor deposition approach, 451
experimental approach, 451
future work, 468
poly(fluorinated benzocyclobutene)
preparation, 463,465–467
properties, 465–466,468
structure, 463,465
poly(naphthalene) and poly(fluorinated
naphthalene)
apparatus for preparation, 462–464*f*
preparation, 462
properties, 463,468
structure, 461–462
poly(tetrafluoro-*p*-xylylene)
chemical vaporization preparation
method, 456–457,459
optical micrographs, 458–460*f*
precursor preparation method, 456
previous preparation methods, 455–456
properties, 456–457,468
poly(*p*-xylylene)
apparatus for preparation, 453–454
preparation, 453,455
properties, 453,458*f*,468
precursor requirements, 452
scanning electron microscopy, 450*f*,451
Teflon and Teflon AF
preparation, 459,461
properties, 461,468
structures, 459
thermal stability, 467
vapor deposition process design
cold wall vapor deposition, 452–453
hot wall vapor deposition, 452
UV-assisted vapor deposition, 453

VLSI device design and manufacture
future, 238
requirements, 237
techniques, 227–238
Volume loss due to acidolysis,
problem, 207

W

Water-soluble onium salts for chemically
amplified positive resists
characteristics, 128f–131
experimental description, 125–127,129
lithographic performance, 134–135
previous studies, 124–125

Water-soluble onium salts for chemically
amplified positive resists—*Continued*
radiation-induced acid formation,
131–134
Wurtz coupling of dichlorosilanes,
synthesis of polysilanes, 355–356

X

X-ray technology, design of chemically
amplified positive resist systems, 5

Z

Zip length, values, 426

Production: Susan Antigone
Indexing: Deborah H. Steiner
Acquisition: Michelle D. Althuis
Cover design: Alan Kahan

Printed and bound by Maple Press, York, P.A

Bestsellers from ACS Books

The ACS Style Guide: A Manual for Authors and Editors
Edited by Janet S. Dodd
264 pp; clothbound ISBN 0–8412–0917–0; paperback ISBN 0–8412–0943–X

Understanding Chemical Patents: A Guide for the Inventor
By John T. Maynard and Howard M. Peters
184 pp; clothbound ISBN 0–8412–1997–4; paperback ISBN 0–8412–1998–2

Chemical Activities (student and teacher editions)
By Christie L. Borgford and Lee R. Summerlin
330 pp; spiralbound ISBN 0–8412–1417–4; teacher ed. ISBN 0–8412–1416–6

Chemical Demonstrations: A Sourcebook for Teachers,
Volumes 1 and 2, Second Edition
Volume 1 by Lee R. Summerlin and James L. Ealy, Jr.;
Vol. 1, 198 pp; spiralbound ISBN 0–8412–1481–6;
Volume 2 by Lee R. Summerlin, Christie L. Borgford, and Julie B. Ealy
Vol. 2, 234 pp; spiralbound ISBN 0–8412–1535–9

Chemistry and Crime: From Sherlock Holmes to Today's Courtroom
Edited by Samuel M. Gerber
135 pp; clothbound ISBN 0–8412–0784–4; paperback ISBN 0–8412–0785–2

Writing the Laboratory Notebook
By Howard M. Kanare
145 pp; clothbound ISBN 0–8412–0906–5; paperback ISBN 0–8412–0933–2

Developing a Chemical Hygiene Plan
By Jay A. Young, Warren K. Kingsley, and George H. Wahl, Jr.
paperback ISBN 0–8412–1876–5

Introduction to Microwave Sample Preparation: Theory and Practice
Edited by H. M. Kingston and Lois B. Jassie
263 pp; clothbound ISBN 0–8412–1450–6

Principles of Environmental Sampling
Edited by Lawrence H. Keith
ACS Professional Reference Book; 458 pp;
clothbound ISBN 0–8412–1173–6; paperback ISBN 0–8412–1437–9

Biotechnology and Materials Science: Chemistry for the Future
Edited by Mary L. Good (Jacqueline K. Barton, Associate Editor)
135 pp; clothbound ISBN 0–8412–1472–7; paperback ISBN 0–8412–1473–5

For further information and a free catalog of ACS books, contact:
American Chemical Society
Product Services Office
1155 16th Street, NW, Washington, DC 20036
Telephone 800–227–5558